FRESHWATER FISH CULTURE

— Volume 2 —

S.K. Sarkar

Department of Zoology
Netaji Nagar College (Day), Kolkata

2014
Daya Publishing House®
A Division of
Astral International Pvt. Ltd.
New Delhi – 110 002

ISBN 9789351240631

Published by   : **Daya Publishing House**®
                   A Division of
                   Astral International Pvt. Ltd.
                   – ISO 9001:2008 Certified Company –
                   4760-61/23, Ansari Road, Darya Ganj
                   New Delhi-110 002
                   Ph. 011-43549197, 23278134
                   E-mail: info@astralint.com
                   Website: www.astralint.com

Laser Typesetting   : **Classic Computer Services**, Delhi - 110 035

Printed at   : **Thomson Press India Limited**

PRINTED IN INDIA

Dedicated to
the Memory of My Mother

BINAPANI

to

Ratna (Sharmila), Kalyani, Swagata (Rimi)

UNIVERSITY OF CALCUTTA

Professor Samir Banerjee
Hiralal Chaudhuri Professor,

DEPARTMENT OF ZOOLOGY
AQUACULTURE RESEARCH UNIT
35, Ballygunge Circular Road
Kolkata-700 019 (India)
Phone: 24614981, (Ext. 306)
Fax : 91-033-2476-4419
E.mail- samirban@vsnl.net
20.11.06

FOREWORD

Dr. S.K. Sarkar has written in detail a book entitled "Freshwater Fish Culture" remarkably felicitous for fishery students. It symbolizes a great deal of effort and my consent goes to the author. Fish culture/aquaculture generally responds attentively to the conditions in the farmers' community. Fish culture provides not only low cost animal protein for the human society but helps foreign exchange earnings as well. Although the basic fundamental aspects offish culture are more or less similar, they also differ in many ways from country to country and even from region to region. Fish culture will differ in important respects depending on the development of farming strategy. Hence it is inevitable to have a freshwater fish culture book that responds commendably to the remarkable features of state-of-the-art culture system.

There is beyond doubt that this book will be more useful for students and farmers to learn fish culture and post-harvest lessons from a book that attempts to reflect the farming strategies for the nation in general and the farmers' community in particular.

This book uniquely affords an opportunity for the students, teachers and fish culturists to make a preference and to make it with the knowledge that the text concerned has the senility of region for culture that will definitely make it appropriate and useful to those who are actively engaged in the subject concerned.

20/11/06

(Prof. Samir Banerjee)

PREFACE

Although fish culture / aquaculture is a touch-and-go business, it is the life and soul of progressive fish culturists. Fish culture encompasses a very extensive water area. It accepts the study of both freshwater and marine fishery resources and their exploitation in full potential, the growth and yield of fish following the adoption of state-of-the-art technologies that have dealt with in different chapters of this book. It covers the relationship between fish cultivation and environmental factors and the manner in which farmers are actively involved in fish culture. It also embraces other kinds of economic activities and employment opportunities such as fishery management for sustained production, harvesting, processing and marketing as well as trade which have grown up as a result of technological development by exploiting aquatic resources. Fish culture activities no longer depend on the relationship between farmer and his aquatic ecosystem but must take into account of the economic interaction between different geographical regions. As a generalization, the success of aquaculture activities entirely depend, to a large extent, on the distribution of commercial fish species, availability of quality ecosystem, standard of living of the rural people, and the various kinds of aquaculture activities around this planet. At present, fish culture has, however, become so important that culture activities are undertaken by fish farmers and fishermen across the world.

In studying the various aquacultural activities for rural development, employment opportunities, and for human nutrition, several aspects nust be covered. First, the methods and means of production must be evaluated, the way in which fish are stocked and reared under any geographical locations such as from high altitudes to plains must be apprehended. Methods of production can transmute not only with the climate and soil but also according to the type of culture adopted, tecmological achievements, and the social characteristics of the farmer who accomplish them. Second, the allotment of aquacultural activities must be evaluated. There are several factors to be taken into attention in the location of aquacultural activities. Allotment may be affected by the techniques adopted which may alter with time as these techniques are gradually improved. Distribution of fish may depends largely on the availability of suitable culture ecosystem, but it may also hinge on comparative cost of production, availability of market for the fish, availability of transportation, economic and political conditions, market fluctuations and government economic policies.

Since population of aquatic species is so diverse and so unevenly distributed over the face of this planet that different regions have different aquacultural activities. Such regional differences lead to the trade between different regions and countries. Thus, fish marketing strategy and fish trade must be studied along with the culture-related activity.

Population explosion is one of the most important issues that must be kept in mind while fish culture and post-harvest technology are considered. Future projections made by the United States indicate that the world population will increase 41 per cent by the end of 2050 to 8.9 billion people. As a corollary, emphasis should be given to culture- and capture-related industries by adopting (i) genetic engineering, (ii) pollution control devices, and (iii) conservation strategies of fish germplasm and fish biodiversity to a large extent.

A text on the subject like Freshwater Fish Culture which is environmentally specific must be considered as great importance. The Second Volume is the continuation of the first one and is a comprehensive and full-fledged text on different major areas of the Freshwater Fish Culture that addresses specifically to the students of fisheries science. A student of fisheries science should have a proper idea of the several issues involved in fish culture.

Each chapter starts with an introductory panel, which summarizes the cotents. It is followed by flowing main text which discusses the main subject concerned. Supplementary informations have been given in panels and tables.

Illustrations play a key role in this book. Diagrams are a special feature of the book and have been authorized to further expand upon themes within the text. At the end of the chapters, some appendixes have been incorporated. Finally, the book ends with a glossary furnishing definitions and brief explanations of important technical terms. Through 26 Chapters (Vol. 2 & 3), however, the book comprehensively covers the aspect of Freshwater Fish Culture programs of the Graduate and Post-Graduate students of fisheries science.

Similar to Volume 1, this Volume has also been presented in a simple style. Clarity has been given top priority throughout. Utmost care has been taken to recite the most intricate ideas in a simple and easy-to-follow style.

Significant advances in fisheries science that are important in fish culture and fisheries continue to be emphasized. The book incorporated many informations motivated by the fact that students of fisheries science (and also fish culturists) desire for a systematic information and a shorter text. It is hoped that the book will be of yeoman's service to fishery students.

A large number of tables and panels have been provided to make the study easy and interesting. While panels have illustrated the various aspects noted in a chapter assembled from different sources such as books, magazines, and websites, tables and figures have been provided to further support the learning process.

PREFACE

Although fish culture / aquaculture is a touch-and-go business, it is the life and soul of progressive fish culturists. Fish culture encompasses a very extensive water area. It accepts the study of both freshwater and marine fishery resources and their exploitation in full potential, the growth and yield of fish following the adoption of state-of-the-art technologies that have dealt with in different chapters of this book. It covers the relationship between fish cultivation and environmental factors and the manner in which farmers are actively involved in fish culture. It also embraces other kinds of economic activities and employment opportunities such as fishery management for sustained production, harvesting, processing and marketing as well as trade which have grown up as a result of technological development by exploiting aquatic resources. Fish culture activities no longer depend on the relationship between farmer and his aquatic ecosystem but must take into account of the economic interaction between different geographical regions. As a generalization, the success of aquaculture activities entirely depend, to a large extent, on the distribution of comnercial fish species, availability of quality ecosystem, standard of living of the rural people, and the various kinds of aquaculture activities around this planet. At present, fish culture has, however, become so important that culture activities are undertaken by fish farmers and fishermen across the world.

In studying the various aquacultural activities for rural development, employment opportunities, and for human nutrition, several aspects nust be covered. First, the methods and means of production must be evaluated, the way in which fish are stocked and reared under any geographical locations such as from high altitudes to plains must be apprehended. Methods of production can transmute not only with the climate and soil but also according to the type of culture adopted, tecmological achievements, and the social characteristics of the farmer who accomplish them. Second, the allotment of aquacultural activities must be evaluated. There are several factors to be taken into attention in the location of aquacultural activities. Allotment may be affected by the techniques adopted which may alter with time as these techniques are gradually improved. Distribution of fish may depends largely on the availability of suitable culture ecosystem, but it may also hinge on comparative cost of production, availability of market for the fish, availability of transportation, economic and political conditions, market fluctuations and government economic policies.

Since population of aquatic species is so diverse and so unevenly distributed over the face of this planet that different regions have different aquacultural activities. Such regional differences lead to the trade between different regions and countries. Thus, fish marketing strategy and fish trade must be studied along with the culture-related activity.

Population explosion is one of the most important issues that must be kept in mind while fish culture and post-harvest technology are considered. Future projections made by the United States indicate that the world population will increase 41 per cent by the end of 2050 to 8.9 billion people. As a corollary, emphasis should be given to culture- and capture-related industries by adopting (i) genetic engineering, (ii) pollution control devices, and (iii) conservation strategies of fish germplasm and fish biodiversity to a large extent.

A text on the subject like Freshwater Fish Culture which is environmentally specific must be considered as great importance. The Second Volume is the continuation of the first one and is a comprehensive and full-fledged text on different major areas of the Freshwater Fish Culture that addresses specifically to the students of fisheries science. A student of fisheries science should have a proper idea of the several issues involved in fish culture.

Each chapter starts with an introductory panel, which summarizes the cotents. It is followed by flowing main text which discusses the main subject concerned. Supplementary informations have been given in panels and tables.

Illustrations play a key role in this book. Diagrams are a special feature of the book and have been authorized to further expand upon themes within the text. At the end of the chapters, some appendixes have been incorporated. Finally, the book ends with a glossary furnishing definitions and brief explanations of important technical terms. Through 26 Chapters (Vol. 2 & 3), however, the book comprehensively covers the aspect of Freshwater Fish Culture programs of the Graduate and Post-Graduate students of fisheries science.

Similar to Volume 1, this Volume has also been presented in a simple style. Clarity has been given top priority throughout. Utmost care has been taken to recite the most intricate ideas in a simple and easy-to-follow style.

Significant advances in fisheries science that are important in fish culture and fisheries continue to be emphasized. The book incorporated many informations motivated by the fact that students of fisheries science (and also fish culturists) desire for a systematic information and a shorter text. It is hoped that the book will be of yeoman's service to fishery students.

A large number of tables and panels have been provided to make the study easy and interesting. While panels have illustrated the various aspects noted in a chapter assembled from different sources such as books, magazines, and websites, tables and figures have been provided to further support the learning process.

This book would have not been possible without the co-operation extended to me time to time by Sharmila (Ratna) and Kalyani. Professor Samir Banerjee of the University of Calcutta has written the foreword of this Volume. In a few condensed lines he has highlighted the uniqueness of this Volume and commended it to teachers, students, progressive fish culturists, government and non-government organizations, planners, and entrepreneurs. The author is inmensely grateful to him. The Central Inland Capture Fisheries Research Institute (CICFRI), Barrackpur, West Bengal, has provided the library facility to the author in bringing out this book. The author is grateful to the CICFRI.

The author looks forward to receive more current informations from readers in the future.

Budge Budge (South 24 Parganas)   **S.K. Sarkar**
West Bengal, Kolkata

CONTENTS

1

General Organizations and Biology of Fishes

Fish culture is no doubt an ancient art and it has been practised in one form or the other since time immemorial. The emergence of fish culture as a profitable venture and nutritional discipline is, however, relatively of recent origin. And within this relatively short period, it has gained a great deal of importance and stature. In fact, most aquafarms the world over, are regarded fish culture in particular and aquaculture in general as the most important aspect of production strategies. Before going to discuss on fish culture in detail, however, it is inevitable to have a gross idea about the general organizations and biology of cultured fish – one of the most versatile concepts of fish culture. The discussion on these subjects is spread out over a series of sections that follows.

1.1 Rationale*

In studying the fish culture and basic bodily function of fishes, it is important to understand that they possess all the main organ systems familiar to that of other vertebrate animals. The important differences between fish and other vertebrates apart from their different evolutionary status lies on the fact that fish generally have no lungs but take up dissolved oxygen in the surrounding water through their gills. Also, fish live in water and the body temperature is similar to that of the surrounding water. All basic principles of animal biology apply to fishes but 'cold-bloodedness' and aquatic mode of life are factors that should always be taken into consideration.

Although both the morphological and anatomical structures of freshwater as well as marine water fishes are extremely variable, it is very difficult to bring them into a common perspective. Freshwater fish have, however, been considered here in a general way so that progressive fish culturists may have a succinct idea about the basic structural forms and biology of fishes when their culture strategies are adopted by farmers. The most important framework of discussion in the present chapter has been made on the basis of several criteria that follows :

1. Morphological features
2. Alimentary canal and associated structures
3. Endoskeleton, muscles and buoyancy

---

* For further study on fish biology, see Bond (1996) and Chonder (1999).

4. Respiratory and circulatory systems

5. Accessory respiratory organs

6. Swimming

7. Kidney and osmoregulation

8. Endocrine and immune systems

9. Reproductive system, reproduction, life history etc.

Ichthyo-biologists call fish as one of the most successful vertebrates. There are more than 20,000 known species of fish across the world, living in such diverse habitats as stagnant ponds, wetlands, ocean depths, sub-zero polar water, flowing rivers, streams, and thermal springs that heat up 48°C.

On the basis of evolutionary relationships, the living vertebrates have been traditionally classified into seven groups such as (1) Jawless fish, (2) Cartilaginous fish, (3) Bony fish, (4) Amphibians, (5) Reptiles, (6) Birds, and (7) Mammals. However, since fishes with a bony skeleton form a large group of vertebrate animals which live in water and enormous varieties of bony fishes are used as food, they will receive the major attention in this chapter. Non-edible fish are also important in academic and research interests which is beyond the scope of the present text. The brief discussion in this chapter will be considered primarily with a few important species of food fish.

The bony fish (Class :Osteichthyes) have largely or partly body skeletons and a gas-filled swim bladder (or in a few species, lungs) for buoyancy. They have various types of bony scales (see later) and are by far the largest and most diverse group of living vertebrates. Nearly one-third live in freshwaters. There are two sub-classes under the class Osteichthyes: (1) Actinopterygii (ray-finned fish) and (2) Sarcopterygii (lobe-finned fish). However, the former includes herring, trout, salmon, eels, perch, carp, cichlids, catfish, snakeheads etc. and are widely distributed across the world. Their paired fins are fan-shaped, and although skeletal elements enter the base of the fin, most support comes from dermal rays that have evolved from rows of bony scales.

Most of the familiar and modern bony fish are teleosts (included under the super order Teleostei of the sub-class Actinopterygii) and the group contains more than 22,000 living species. They vary from tiny gobies less than 1 inch (2.5 cm) to large marlins of 11 feet (3.4 m) or more. The economics of mankind largely depends on them and a number of industries have been established to exploit them. Teleosts present an almost embarrassing – sometimes ridiculous – variety of form; and an astonishing range of adaptation. Some teleosts such as salmon migrate extremes of freshwater and salt water. Enormous varieties of teleost fish are of outstanding practical importance to mankind. Nearly all of the important sport and commercial fish species are teleost.

The study of fish in relation to their origin, evolution and biology is very important to any progressive fish culturists. In general, the term biology refers to the science of living things. It is divided into specialized fields according to the kinds of organism studied:

botany for plants, ichthyology for fishes, zoology for animals, entomology for insects and so on. It is further splitted into levels of study, ranging from molecular biology to whole communities of animals and plants, which form the subject matter of ecology. Fish biology, however, includes morphology, anatomy, various physiological functions such as feeding, digestion, reproduction, development, embryology etc. Hence the study of fish biology is vast, diverse, and complex which is beyond the scope of this book. Much has been written about the biology of many fresh warm water and cold water fish species and a number of bibliographies, books, and review articles are available for the purpose. Interested progressive fish culturists may consult the two books entitled (1) "The Biology of Finfishes and Shellfishes" and (2) "The Biology of Fishes" where the biology and general organizations of a variety of freshwater fish species have been lucidly described. For this reason, a very brief discussion will be made here regarding the biology of some important groups/species of freshwater teleost fishes and many of the broader aspects of their structure and biology have been omitted.

Origin of Freshwater Fish

A growing body of research has suggested that the sea was possibly the original home for fishes and numerous as well as diverse forms found today in rivers, streams, lakes, ponds, and reservoirs have been derived from marine ancestors who were driven to leave the sea through stress of competition. These fishes consumed a variety of natural food organisms and quiet places for successful breeding. Many of these visitors have become permanent residents in the new habitat, and the altered conditions influencing over many generations led to the formation of freshwater races, and as time went on, species have become very distinct from those of their marine counterparts. In course of time, freshwater forms have tended to move apart so as to cover a wide freshwater area, and to become more and more modified, until finally whole families or even larger groups composed exclusively of freshwater fishes as found at the present time.

1.2 The Study of Fish

The study of fishes with respect to their general organizations and biology is of special interest for various reasons. They afford sport for the angler, provide food for millions all-over the world, and have commercial uses as animal food and as raw materials. Many species are kept as pets and have made advanced ameteur ichthyologists out of an army of hobbyists. In the academic field, some scientists concentrate on comperative morphology of fishes, some study their biology and others study their systematic relationships and great evolutionary significance. Many believe that fish behavior is at once directed by their environment and influences their environment.

Our knowledge concerning fishes is incomplete, although laymen, farmers, scientists, naturalists, and others have been recording information for decades. However, thousands of freshwater fishes are known mainly as a few preserved specimens in glass jars with little or no information available for them. For many species that enter culture and capture fisheries across the world, only the basic features of their life histories and production-oriented culture of selected fish species are known. Therefore, there is much yet to be done, much to be explored, analyzed, reported, and shaped into scenarios. The labors that remain for fish culturists/ichthyologists are great and compelling.

1.3 Challenges of The Aquatic Environment

During the long evolution, fishes have adapted to many physical and chemical aspects of the aquatic environment. Water is highly viscous and dense medium, so fishes with a specific gravety close to that of water sinks slowly and must exert little effort to maintain a greater depth. However, rapid movement through water requires significant amount of energy. Water can held at most a few miligram per litre of dissolved oxygen, so fish must get along with little oxygen, develop needs to supply enough oxygen to support active metabolism, or breathe air. The ability of water to dissolve various substances is well known, and these substances provide the ability of the fish to maintain the proper osmo-concentration of its body fluids. Other solutes of significance could be pollutants with harmful effects, significant biological odors, or nutrients involved in primary production by plants. The air-water interface limits the light that can pass through. Pressure increases with depth can limit vertical movement of many species of fish because of limited tolerance.

Fishes have evolved adaptations for lives both active and inactive, for slow and rapid swimming, for gathering of food, for awareness of their environment, and for many successful modes of reproduction.

This chapter will introduce (1) the general morphological features that adapt fishes to live in water, (2) discuss anatomical features that fish culturists will be able to understand and study the relationships among fishes, and (3) informations with regard to reproduction, embryology, and life stages of freshwater fishes.

1.4 Relationship With The Environment and Fish Biology

It should be kept in mind that environmental factors (both abiotic and biotic) react ceaselessly with fish biomass that has the potentiality of sustaining the production of aquatic ecosystems. A variety of environmental factors that have frequently been mentioned elsewhere in this volume, exist in any fish culture ecosystem that definitely encourage the yield of robust and disease-free fish. Environmental factors are, however, commonly linked together during the entire culture period. Interactions also occur among/between different environmental factors. For examples, environmental temperature has a marked influence on the feeding rate and growth characteristic of farmed fish. Similarly, metabolites of nitrogen fertilizers such as ammonia, nitrate, and excretory products of fishes are the most important critical factors that limit fish production. Fish excrete ammonia and carbon dioxide through gills. The amount of excretion is related to oxygen consumption and feeding rate of fish. These products are toxic to fish and their toxicity is dependent on the fluctuation of temperatures. Ammonia is toxic in alkaline water whilst carbon dioxide is toxic in acidic conditions. Carbon dioxide itself tends to acidify water by forming carbonic acid so paradoxically that helps mitigate the ammonia problem. Thus, there is a complex series of reactions among water, carbon dioxide, temperature, and ammonia. These interactions significantly influence fish biology and production because the resultant metabolites have detrimental effects on ecosystem in general and on fish in particular. Generation of toxic metabolites would accumulate in ecosystems resulting in the mass mortality of fish.

Keeping this view in mind, due consideration should be given to establish a favorable relationship between environmental factors and fish yield. But water pollution is of greatest importance. This destructive phenomenon would have to be kept in memory when the biology of fish in culture system is evaluated.

The ability of an ecosystem to yield robust fish is determined by many factors such as pH, temperatures, oxygen, alkalinity, carbon dioxide, nitrogen, phosphorus, calcium, feed and feeding, inorganic and organic compounds etc. that markedly influence the farmer's incentive to produce. Fish production is also influenced by biological factors which has been discussed at length in Chapter 5.

1.5 Morphological Features

Teleost fish have a streamline body. Their body have taken on a vast array of shapes. Water flows over the body easily during swimming, allowing the fish to move through the water with little resistance. Some fish such as *Mastocembellus* and *Anguilla* have a round body which allows them to lie hidden into the mud. However, different morphological features that follow are theoretically important when fishes are grouped/classified but definitely has great significance when culture of commercially important species of fish is considered and become the deciding factor of ecosystem productivity. The various morphological features will now receive our attention.

Scales

scales are bony plates which cover the outer surface of fishes. All bony fishes with scales have so-called bony- ridge scales which are thin and translucent, and their surface is marked with ridges alternating with shallow depressions. They grow on body surfaces and the ridges represent new growth. By counting the groups of ridges it is possible to ascertain the age of the fish.

1. Structure of Scales

Teleost fish are covered with flexible thick or thin scales produced by cells in the skin. The scales overlap each other and form a tough protective coating of the body of fishes. The scales are solid calcareous structures embedded in pockets in the tough dermal layer of the skin. Most of the body (excluding the head region in case of carp) is covered with scales. About 80 per cent of the scale is not visible but can be torn out of the pockets if the skin is scrapped. Ring-shaped structure can be seen in fish scales if they are examined under a simple microscope. Teleost scales are overlapping plates of bone that continues to grow throughout life. They are layered like roof tiles to form a smooth protective surface.

A scale has two parts: visible part and root part. The former has epidermis still attached but the latter type is characterized by concentric circular lines reflecting the growth of the scale. The age of the fish can be estimated by careful examination of the presence of rings on the scales, although one ring does not correspond to one year. When the fish grows rapidly, the circuli are far apart and when the fish grows slowly, the circuli are closed together.

During starvation, the scale may regress in size and show erosion and subsequent growth may not be concentric with the original circuli. Thus, past episodes of starvation are recorded in the scales. Cultured fish in particular often lose scales. Replaced scales are rapidly developed but are devoid of circuli in the rapid replacement growth phase. Scales are, however, useful in fish culture in checking the fish of unknown origin.

1. *Type of Scales:* In teleosts, two types of scales are found such as (i) cycloid scales and (ii) ctenoid scales (Fig. 1.1). In case of ctenoid scales, the hinder edges are provided with rows of pointed outgrowths called *ctenes* or *spines*. Ctenoid scales are found in *Lates, Ophiocephalus, Anabas* etc. Many of our commonest freshwater fishes (such as carp, goldfish, perch etc.) have no pointed outgrowths on the free posterior edge of the scale and are termed as *cycloid scales*. Catfishes have no scales, however.

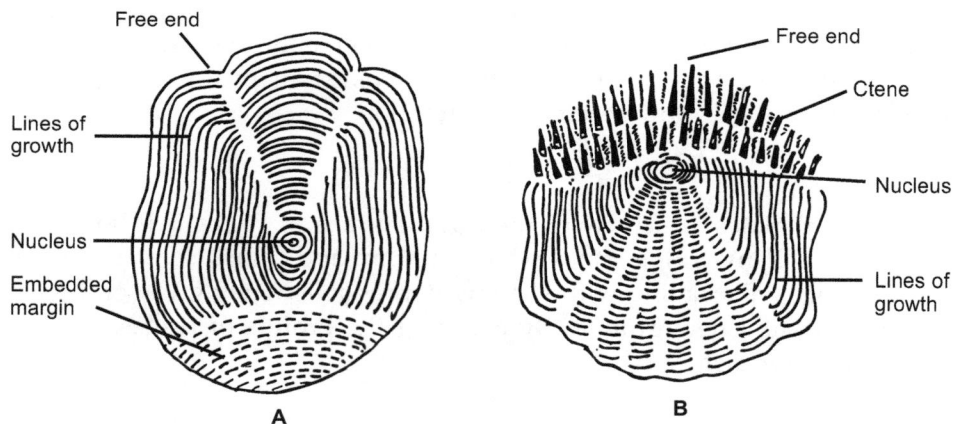

Fig. 1.1 : Scales of fishes : (A) A typical cycloid scale; (B) A typical ctenoid scale

Fins

Fins are flattened appendages which are used in swimming and in fishes, fins are most highly evolved. The fins of fishes are essentially double-flaps of skin with a skeletal support of bony elements in bony fishes. In bony fishes, the fins can be erected and depressed and some can be moved by means of muscles attached to the bony rays.

1. Types of Fins According to their position

Fins are named according to their position on the fish's body (Fig. 1.2). The single or multiple dorsal fins run along the back; the anal fin projects from behind the vent or anus. The caudal or tail fin provides the chief propelling motion in swimming; the paddle-like pectoral fins (each side behind the head) and pelvic fins (beneath the belly) keep the swimming fish in a level posture. The adipose fin, situated at the rear of the back, just forward of the tail fin, is a 'false' fin in that it is made of fatty tissue.

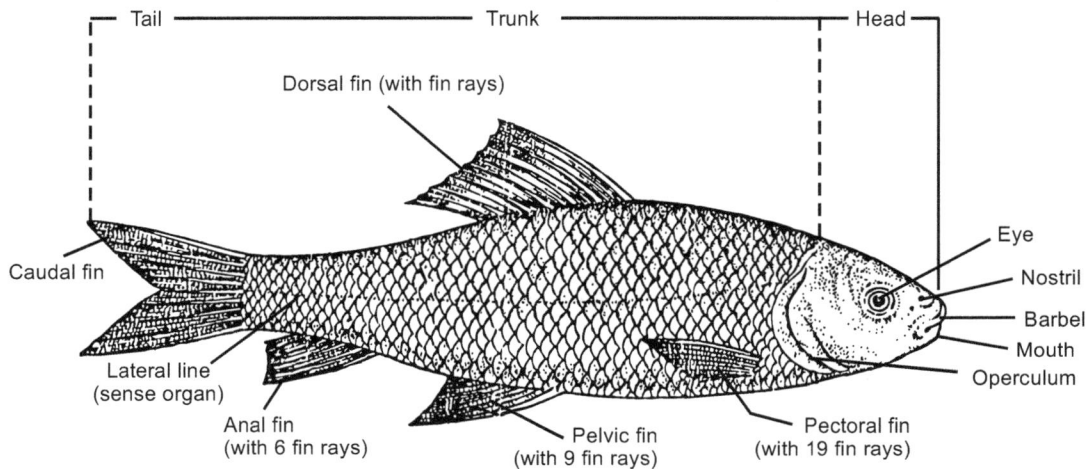

Fig. 1.2 : External features of a typical bony fish (*Labeo rohita*)

2. Structure of Fins

All fins are, however, supported by bony rays which consists of branched columns of little bones bound together by tendons. Muscles at the base of the fin can move the rays to erect the fins. The fins control the direction of movement when the fish is swimming.

Some fins have a protective function, such as the dorsal spines of *Lates* sp., stickleback, *Mystus* spp. etc. and pectoral spines of *Clarias* and *Heteropneustes.*

Smell

Smell is an important sense, enabling fishes to perceive and recognize specific chemicals. However, the nostrils of fish do not open into the back of the mouth and are not used for breathing. They lead into organs of smell which are very sensitive to most teleosts, so that a fish can detect the presence of food in the water from considerable distances. The nostrils are double, so allowing water to pass through the organ of smell. Moreover, fish are very sensitive to a variety of odors in the water which enable them to recognize their habitat.

Sight

Sight is very important in many fish. Eyes of most bony fish have flat cornea, large and round pupil which do not vary in size. There is no eyelids, but the skin – as it crosses the orbit, becomes transparent. Each eye has a thick and spherical lens and it has little power of accomodation because its curvature does not alter. In a few teleostean fish, the lens is drawn in and out by a special structure, called *falciform process,* allowing for clear focussing on objects underwater (Fig. 1.3). Visual recognition of food particles is very important to fish and they are also able to observe activity above the surface of the water. The vision must be excellent because of formation of a large congregation as in schooling, vision seems to be the guiding factor. If a mirror is kept under water, fish will see its image because it tries to pass through it. In most freshwater catfishes, such as *Clarias* and *Heteropneustes,* the eye is very small. Since the two eyes are situated on two sides of the compressed head, vision is monocular, the fish being not capable of focussing both its eyes on the same object at a time. In Gobidae (such as *Boleopthalmus* and *Periopthalmus*), the remarkable eyes are borned on the top of the head and are

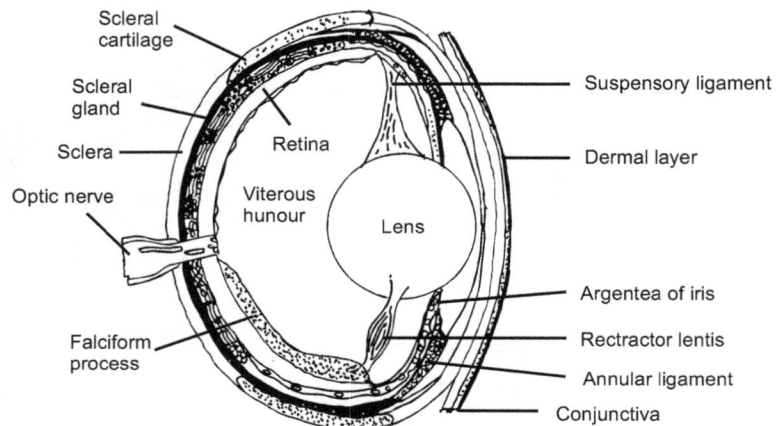

Fig. 1.3 : Section of an eye of teleost

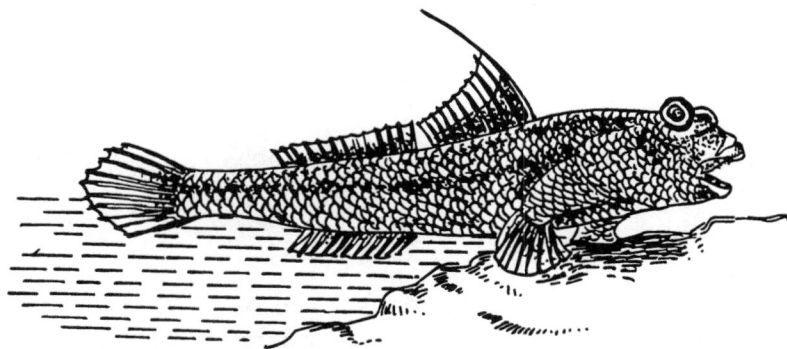

Fig. 1.4 : *Boleopthalamus* sp. (The Mud Skipper). A small goby (fish) of tropical mangrove swamps which is able to move around out of water. Mud-skippers frequently leave the water altogether and lie on the mud, or climb the exposed mangroove roots, using their pectoral fins as levers and the pelvic fin sucker to stay in place. Their eyes are placed high on the head and give good all-round vision. They are very alert to predators, skipping away across the mud to hide in the burrows of crabs or returning to water at the first sign of danger. The length of these semi-aquatic fish varies between 12.0 and 14.0 cm with mottled brown colouration.

projecting and movable (Fig. 1.4). It is believed that when the fish is under water in act of swimming, the upper half on the eye has aerial vision whilst the lower half helps see through water. With the bulging eyes, most of the bony fish can see well and many species are thought to see in colour.

Sense Organs

Fish have the same basic senses as we do : sight, hearing, touch, sense and taste. However, along each side of the body of fish there is a system called as *lateral line system* which is considered as one of the sense organs in fish. The lateral line is a fluid-filled tube running along the fish's flank just below the skin which helps detect movements in water. A disturbance set up in the water will produce vibrations in the fluid of the tube. The lining of the tube contains nerve endings which are highly sensitive to vibrations and which send impulses to the brain. The fish can thus detect the direction and intensity of water movements.

Fish have taste buds in the mouth and over the barbles which are useful for sampling food particles before picked them for swallowing. Fish can detect temperature changes and are very sensitive to odors in the water.

Fish also have a brain connected to the spinal cord which runs along the vertebral column distributing nerves throughout the body. The brain is complex enough to enable the fish to learn simple tasks and fish can be taught to operate demand feeders and will learn to live with a regular management coming to feed at an appropriate time or signal. Fish transferred to new surrounding are disturbed and will explore, taking some time to settle down.

Movement

Movement, or the ability to change position, is one of the fundamental characteristics of living organisms including fish. This capacity is most highly evolved in animal kingdom in response to their need to seek or catch food.

The vertebral column of a fish essentially consists of a number of vertebrae held together by ligaments. The ligaments are loose enough to permit slight sideways movement between each pair of vertebrae so that the spine as a while is somewhat flexible.

The forward movement of the fish through the water is produced by wave-like movements passing down to its body, ending in side-ways lashing movements of the flexible tail. The movements may be a hardly perceptible rippling motion, as in mackerel, or very pronounced as in eels.

The swimming movements are generated by waves of muscular contraction passing from the head to the tail down each side of the body alternately (Fig. 1.5). A sideways and backward thrust of the tail and body against the water develops a sideways and forward movement of the fish in the opposite direction. The final beat of the tail may contribute as much as 40 per cent of the total forward thrust. The sideways movements produced by two successive and similar beating actions are equal and opposite, and reject each other out so that the net effect is to move the fish straight forward. The fish can turn to the right or left by beating its body more strongly to one side than to the other.

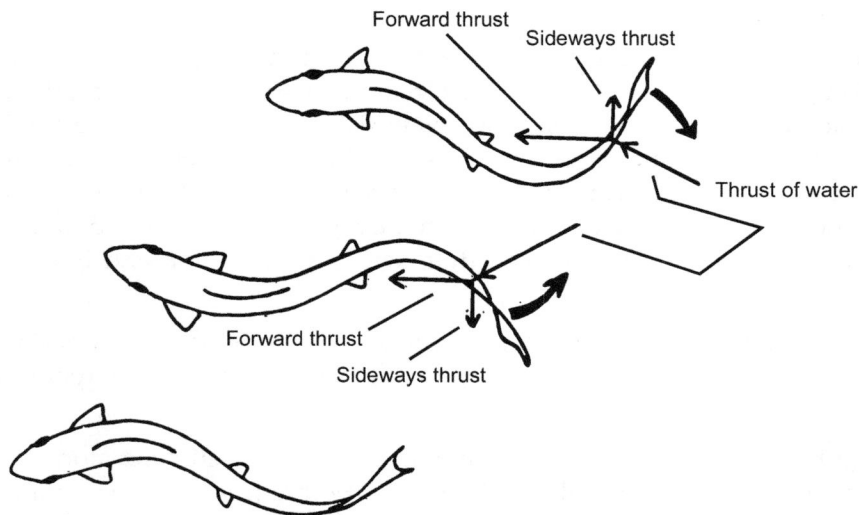

Fig. 1.5 : Diagram to show how swimming movements produce motion; the bold arrows indicate the direction in which the tail is moving

The force required to propel a fish through water is generated by movements of its whole body. In general, the fins control the stability of the fish and the direction of its movements. Rolling and yawning from side to side are controlled by the median fins, which present a large vertical area to the water and resist any tendency to sideways drift. The paired fins act in the same way as the hydroplanes of a submarine, and control the pitch of the fish. The angle at which they are held can be varied by muscular action, resulting the fish to swim downward or upward. The pectoral fins are responsible for pitch control. When the broad surfaces of the paired fins are turned at right angles to the direction of movement, the fin becomes sluggist and terminates its movement.

Migration

Migration is the regular and periodic movement of an animal including fish from the area in which it has been living to another area for feeding, sheltering, and reproduction and then back again. Migration essentially involves a round trip journey always to the same destination. Several species of fish, however, migrate to specific areas to mate, lay eggs, give birth, and raise young.

1. Types of Migration

In the majority of fish species, two major types of migration are generally encountered such as catadromous migration and anadromous migration. Catadromous fishes generally feed and grow in freshwaters but return to the sea to breed. The adult Eel (*Anguilla anguilla*), for example, cross the Atlantic to spawn in deep water to the south of Bermuda, the larvae subsequently making their way in an easterly direction., reaching the coasts of Europe when about two and a half years of age. Here they become transformed into elvers (about 3 years old). Other catadromous forms include *Galaxiids* which are confined to the rivers of the southern extremity of south America, southern Australia and New Zealand for the most part, but a very few species from Australia and New Zealand reverses the habit of its northern relatives and returns to its original home in the sea for reproduction.

In contrast to catadromous, anadromous fishes are those which spend a part of their life in the sea, for such fishes feed and grow in this habitat, ascending the rivers with degree of regularity to spawn. Stargeon (*Acipenser*), Shad (*Alosa*), Salmon, Trout (*Salmo*), and Char (*Salvelinus*) are apt examples of this type of fish. Salmon and trout are closely related species each with roughly the same range in the sea and each entering the rivers to spawn. Some forms of salmon (such as *Sebago* and *Ouananicha*) do not nigrate to the sea and as a corollary, they have become permanent residents in freshwaters. The trout forms freshwater colonies in almost all suitable lakes and rivers which it enters, and many of these permanent residents have become so greatly modified in the course of time that they present an exceptional diversity of forms, size, coloration and the like.

It should be pointed out that both trout and salmon are found in the sea from Iceland and the northern part of Norway southwards to the Bay of Biscay, but whereas the salmon (*Salmon salar*) has succeeded in crossing the ocean and is found on the Atlantic coast of North America, the trout (*Salmo trutta*) which generally does not migrate nearly so far out to sea, is absent from America. Salmon, however, hatch and feed in rivers, then they migrate to the ocean where they grow to maturity. The adult salmon eventually leave the ocean and return to the rivers in which they were born. Here they mate, lay their eggs, and die.

The freshwater colonies of white fish (*Coregonus*) and Char (*Salvelinus*) inhabit the sea and running up the rivers to spawn.

Most freshwater carps feed in different trophic levels and when breeding season approaches during monsoon season, they migrate to shallow areas of rivers reservoirs, and lakes where they congregate into shoals. Females lay eggs in these regions. Here eggs are fertilized by the males and then return to their original habitats.

1.6 Anatomical Features

Anatomy is, in general, the study of the biological structure of all living things. The anatomy of fish (Fig. 1.6) comprises eleven body systems. The first of these encompasses the external or morphological features which has already been described in section 1.5.

The second is the skeletal system which is internal. Next is the muscular systems followed by alimentary system, respiratory system, circulatory system, excretory system, nervous system, endocrine system, reproductive system, and immune system. By simply dissecting a fish by scissors and scalpels, it is possible to study its anatomy. But although dissection was once the central means of anatomical study, it is now becoming increasingly obsolete as dissecting instruments are replaced by computer technology (See Panel 1.1).

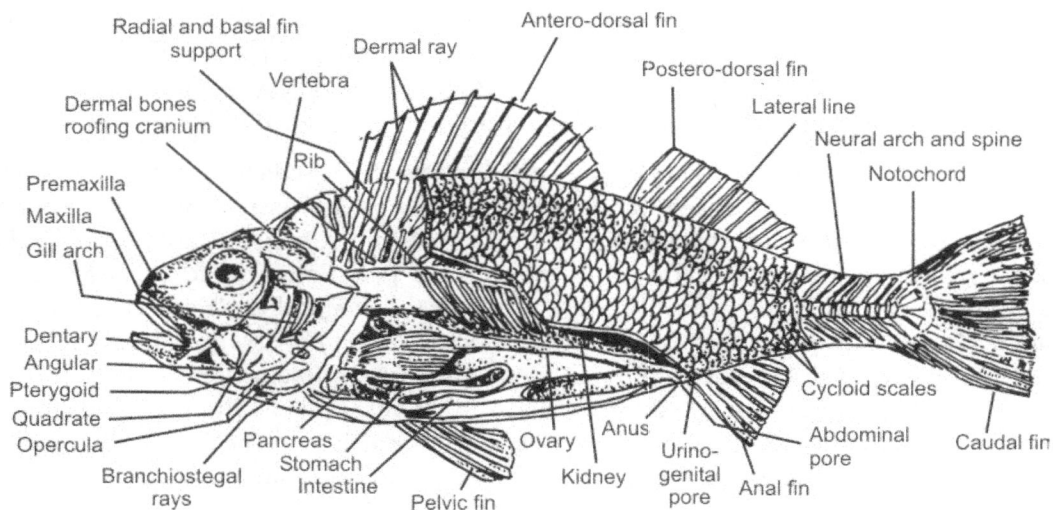

Fig. 1.6 : The general anatomy of a teleostean fish.

In modern anatomy, the structure and function of any organ system are closely related with each other and the study of anatomy continues to be important in biological sciences. Since there is no scope to discuss the modern anatomy at length in the present book, a simple anatomical study of the structure and function of some important systems of fish's body has been lucidly described in several sub-sections that follows.

Endoskeletal System

An endoskeleton is a hard, supportive structure which is found within the soft flesh of an animal (Fig. 1.7). An endoskeleton can be made of silica, calcium carbonate, cartilage, or bone. In fish, however, bone is the primary material of the endoskeleton. In general, fish endoskeleton is divided into an axial and an appendicular skeleton. Included in the axial skeleton are the bones of the head, the backbone, the ribs, and the tail. The appendicular skeleton includes the bones of the girdles which are attached with the backbone.

The skull of bony fish is very complex, comprising numerous bones connected to one another through flexible connections. Their jaw and palate are more movable and their skull bones are thinner. The flexible structure of the skull permits breathing and feeding movements.

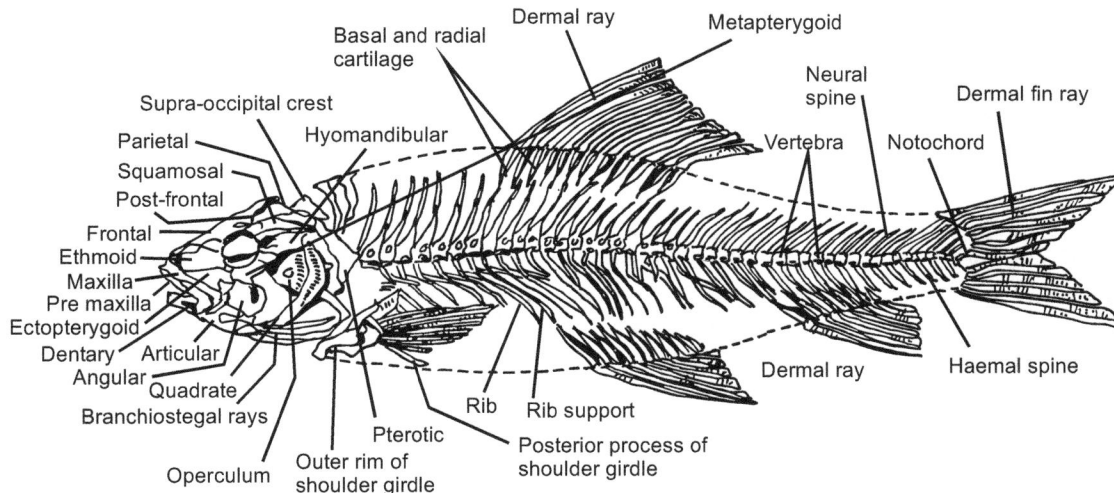

Fig. 1.7 : Endo-skeletal strcuture of a teleostean fish (*Labeo* sp.)

The backbone is made up of numerous vertebrae which have a double cone structure which appears as an 'X' in X- rays. The paired fins are supported on the pectoral girdle which is attached behind the skull and the pelvic girdle which is not directly connected to the skeleton and is a bony bar lying in the belly musculature.

Each vertebra has a dermal spine and a ventral spine which support the median septum to which muscles are attached on either side. In the abdomen region there are ribs supporting wall. Some fish such as carps and cichlids have intra-muscular bones distributed through the main muscles, while others such as catfishes, salmonoids etc. do not have intra-muscular bones and hence the flesh of these fishes is easy to eat.

Muscles

Muscular systems are systems of muscles that are capable of contracting and relaxing, resulting in movement. However, beneath the skin the most prominent and edible part. of the body of the fish comprising about 45 per cent of fresh weight is the swimming muscle.

The muscle is organized into zigzag blocks (or called *myotomes*) arranged on either side of the backbone. The muscle is most important to the consumer.

In each myotome, the muscle fibres run roughly parallel to the backbone between the sheets of connective fibres, called *myocepta,* which act as tendons providing an anchorage for the fibres. When the meat is cooked the myocepta dissolve away, leaving the myotomes to fall apart into the flakes characteristic of fish flesh. The structure of the fish muscle fibres is exactly the same as those of other vertebrates which is familiar in most textbooks.

In contrast to higher vertebrates, fish muscle has a very poor blood supply and has no oxygen-binding pigment – termed as *myoglobin.* This gives fish muscle its characteristic pale white or creamy color. This is termed as *white muscle* and is only used during burst swimming activity. Muscles work anaerobically converting the sugar to lactic acid. The acid generates in the muscle fibres and the cost of getting rid of this is the 'oxygen debt' which the fish has to repay after a burst of exercise. This is what happens when fish struggle during netting or grading, it may take them 24h to get back to normal condition after exhausting exercise and it is usual for fish to die from acidosis due to handling stress if not allowed to recover adequately.

Once contracted, muscle fibres cannot expand unless an additional force comes into play. This force – termed as the *pull of opposing muscle,* helps extend the contracted muscle fibres. For this reason, skeletal muscle is laid down in opposing or antagonistic blocks, with one muscle or groups of muscles pulling against another.

Fish depend on this opposite pull of muscles to propel them through the water. Their muscles are arranged in antagonistic groups called *myotomes,* on either side of the vertebral column. When the myotomes on one side of fish's body contract, the vertebral column bends. Lateral contraction and expansion of the myotomes on each side of the vertebral column bends the body of the fish into a series of waves, which propel it through the water. These waves, beginning at the front end, and travelling forward the tail of the fish, are more numerous in thinner and longer fish.

Buoyancy

Fish are able to stay at a particular depth in the water without expensing much energy, by making their bodies weightless compared with the surrounding water. Most bony fishes, however, achieve buoyancy by using their special structures called *swim bladders*. The swim bladder is an organ containing gas,which is originally developed from the digestive system. swim bladders are either connected to the digestive system and are inflated by gulping air or are filled with gas during the young stages of the fish's life before the swim bladder becomes sealed.

1. Structure

The swim bladder in teleosts is extremely variable in structure, shape, and size because of their adaptation to different mode of life. It is essentially simple and torpedo-shaped and a tough sac-like structure with an overlying capillary network. The swim bladder is composed of 1 or 2 or many chambers. It is situated below the kidneys or between the gonads and above the alimentary canal. The connections of swim bladder with the oesophagus may be retained throughout life – called the *physostomous swim bladder* or may be lost in the adult – called the *physoclistous swim bladder* (Fig. 1.8)

Carps and minnows (Family: Cyprinidae) have anterior and posterior chambers connected by a sphincter. The swim bladder of the feather backs (Family: Notopteridae) is divided into lateral halves, with the two chambers connecting anteriorly. In the herrings (Family: Clupeidae) , the swim bladder has a posterior opening to the exterior near the anus through which gas may be voided.

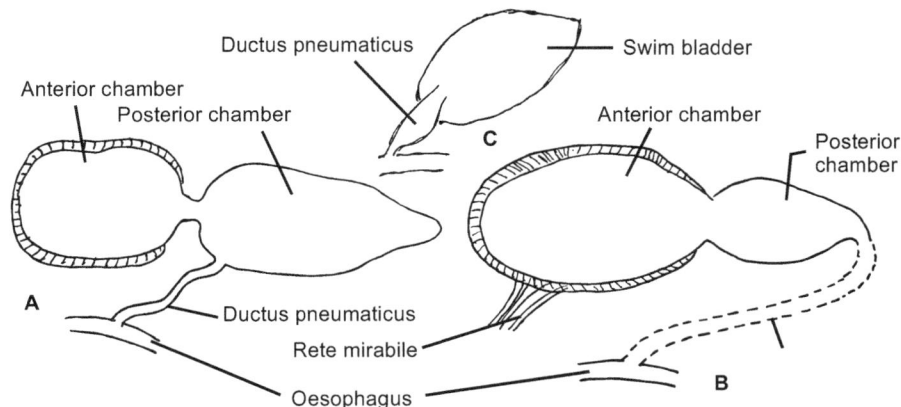

Fig. 1.8 : **Swim bladder of advanced teleostean fish. (A) Physostomous swim bladder; (B) Physocleistous swim bladder. The composition of air in case of physostomous swim bladder is more or less comparable to that of atomospheric air (such as 78 per cent nitrogen, 21 per cent oxygen etc.). But in case of physocleistous swim bladder the composition is variable. The swim bladders of trout and salmon contain nitrogen in high amounts but it does not create any adverse effect on them. Inner wall of the anterior chamber contains red-colored gas gland or so called *red gland* which is provided with rete mirabile – a blood capillary. Gas is entered into the swim bladder through blood capillary. Gas is reabsorbed from the posterior chamber of the swim bladder. The entrance and exit of the gas is controlled by autonomic nervous system. In case of physostomous swim bladder, the atmospheric air is directly entered into the swim bladder via the pharynx and air tube. In certain teleostean fish such as Eel; (C) the ductus pneumaticus has dilated.**

2. *Function*

The swim bladder of fish performs various functions such as hydrostatic, balancing, sound production, sound reception, and respiration.

The concentration and pressure of the air can be controlled by gas diffusing into, or out of, the swim bladder from the blood. As a fish swims downward, the surrounding water increases in pressure and the swim bladder is compressed and deflated. But when the fish swims upward, the surrounding water decreases in pressure and the swim bladder inflates.

Alimentary Canal and Associated Structures

The alimentary canal is the path that food takes through an animal's body, allowing absorption of carbohydrates, fats, and proteins. The canal is also responsible for the mechanical and chemical breakdown of consumed foodstuffs.

The alimentary canal in different fish species is, however, highly adapted to the feeding style and diet of the fish. Herbivorus fish (such as carps) and fish that take in mud and other debris when feeding, have much longer intestine than carnivorus fish species (Fig. 1.9 (a) & (b)) as the diets of the first group take longer time to digest. Carnivorous fish such as catfishes, air-breathing fishes, and snake-heads have well-adapted stomachs, which secrete digestive enzymes and acids to break down proteins in the diet they ingest. Herbivorous and carnivorous fish of the family Cichlidae may not have stomach.

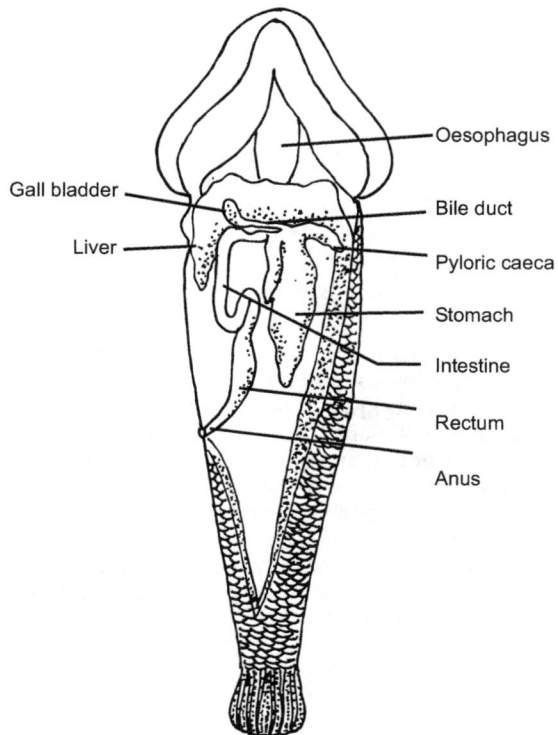

Fig. 1.9 (a) : Digestive system of
Ophiocephalus punctatus.

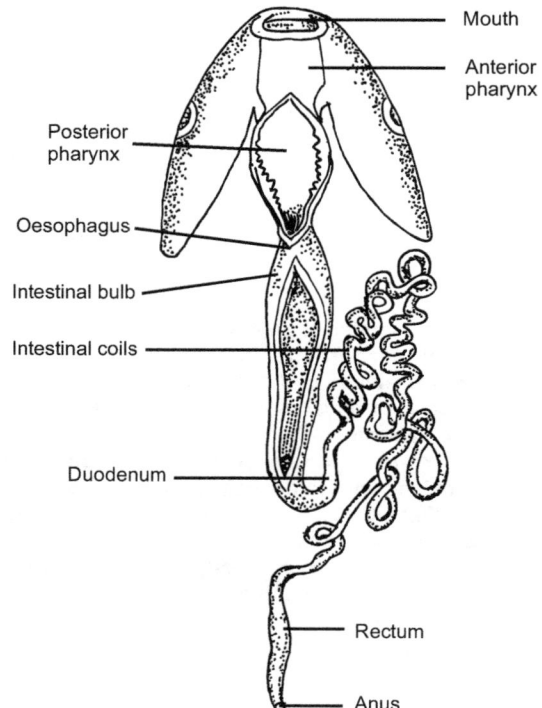

Fig. 1.9 (b) : Digestive system of a typical
bony fish, *Labeo rohita*

1. Teeth : In some carnivorous fishes (such as trout) the mouth is equipped with the large, many small, recurved, and conical teeth – borne on the maxillae, palatines, pre-maxillae, and dentaries which undoubtedly serve to prevent the escape of the animals used as food. The food of the carp consists principally of phyto- and zooplankton populations. Carnivorous fish, on the other hand, consumes arthropods and small fishes.

Teeth are borne on several of the head and face bones. Those in the upper jaw include the pre-maxillary and the maxillary in most of the soft-rayed fishes, but the maxillary does not bear teeth in most of the higher teleosts. Additional teeth are commonly the vomer and palatines. Many species bear teeth on the pterygoids and parasphenoid bones. In the lower jaw, the dentaries are the main toothed bones, but teeth may be present on the basibranchials. In majority of the teleosts, however, teeth are connected to the jaw by collagen where the mineralization is not complete.

2. Pharynx : The pharynx is perforated on each side by four elongated gill-slits, fringed by the tooth-like gill-rackers. Gill-rackers are the inwardly-directed projections of the gill arches and the gill-rackers help in filtering food from the water. Fishes that consume large prey may have gill-rackers that are few in numbers and small, but may carry rough prominences or denticles that aid in holding and swallowing prey. Plankton feeders have an extensive straining shieve formed of long and slender gill-rackers.

The fifth gill-arches of bony fishes are reduced to a single lower element – called the *fifth ceratobranchial* on each side. This bone bears teeth that bite against opposing teeth bone on the upper elements – called the *pharyngobranchials*, of one or all of the four branchial arches. Minnows, Cichlids etc. are apt examples with pharyngeal jaws.

The pharynx is well demarcated into an anterior respiratory part and a posterior masticatory part. While the anterior part is perforated by gill slits that are placed laterally, the posterior part of the pharynx bears pharyngeal teeth which help in crushing the solid foods. These teeth are all alike – generally termed as *homodont*. The root part of the teeth remains embedded in the mucus membrane of the floor of the buccal cavity. Pharyngeal teeth are varied in size and shape, ranging from small conical points to grinding plates.

Although several species of fish have tooth plates or molariform teeth in the mouth and can grind the food with them, pharyngeal teeth appear to be the major apparatus used for mastication among the teleosts. The grass carp, *Ctenopharyngodon idella*, has relatively long, rough-edged pharyngeal teeth that intermesh while tearing the soft plants on which the species feeds.

3. Oesophagus : The pharynx leads into a short and distensible oesophagus which ends in a sphincter – prevents the respiratory water current from entering into the stomach. Short and distensible oesophagus permits to swallow relatively large objects; however, microphagus fishes have less distensible tubes. Oesophageal walls are generally equipped with both longitudinal and circular muscles, which function in swallowing.

4. Stomach : In most fishes, a stomach is present, varying in shape and structure according to the diet of the various species. The stomach of some carnivorous fishes, for instance, is muscular, sac-like structure (snake-heads) or J-shaped bag (trout and salmon) and consists of two parts: cardiac and pyloric. The stomach is, however, absent in most cyprinids such as minnows, pipe fishes, *Labeo, Catla*, grass carp, silver carp etc. In these fishes, the oesophagus empties directly into the intestine.

5. Intestine : From the stomach, a narrow, elongated, and more or less of uniform diameter intestine arises. The length of the intestine in teleost fishes is generally correlated to the amount of indigestible material ingested. Carnivores generally have short intestine (Table 1.1), it is passed at first forward as the duodenum and then passes backwards without convolution. Herbivorous fishes and those that consume detritus and mud have intestines several times the body length and forms a number of coils. Elongation of intestine and its coiling in many teleostean fish are related to its herbivorous food habit.

Omnivorous fishes have intestine of intermediate length. Sturgeons, bowfin (*Amia* sp.), and gars (*Lepisosteus* sp.) possess a spiral intestine, in which the absorptive surface significantly increased by the cork-screw course of a fold of tissue down the length of the organ. The terminal part of the intestine is dilated and forms a thin-walled sac – termed as *anus* – situated in front of the urinogenital opening and in front of the anal fin.

Table 1.1 : Ratio of Interstine Lenght (IL) to Body Length (BL) in Some Freshwater Fishes

Species	IL/BL	Remark
Carassius auratus (Gold fish)	2.3	Phytoplankton and detritus feeder
Ctenopharyngodon idella (Grass carp)	2.5	Herbivorous
Hypophthalmicthys molitrix (Silver carp)	13.0	Phytoplankton feeder
Labeo rohita (Rohu)	9.0	Planktophagus
Catle catle (Catla)	7.6	Planktophagus
Cirrhinus mrigala (Mrigal)	4.5	Planktophagus
Cyprinus carpio (Common carp)	5.7	Omnivorous
Oreochromis moaasmbicus (Tilapia)	3.6	Omnivorous
O. niloticus (Tilapia)	3.8	Omnivorous

The pyloric caeca, digestive glands, and swim-bladder are attached to or associated with the intestine.

(i) Pyloric Caeca : On the intestine of bony fishes, just beyond the pyloric end of the stomach, there may be from one to many blind glandular sacs – called *Pyloric caeca* (Fig. 1.10). The number of pyloric caeca is variable: it is two in case of snake- heads (such as *Ophiocephalus* spp.) and mullets (such as *Mugil* ap.), it is one in case of bichir (*Polypterus* sp.) the yellow perch (*Perca flavescens*) has three, and salmon have 40 pyloric caeca. In others, such as mackerels, the number of these caeca may be 200 or more. In sturgeons, many caeca form a single mass, but only a single duct leads to the intestine. However, caeca of different species vary considerably in size, state of branching, and the connection with the intestine. The functions of pyloric caeca probably involve both digestion and absorption.

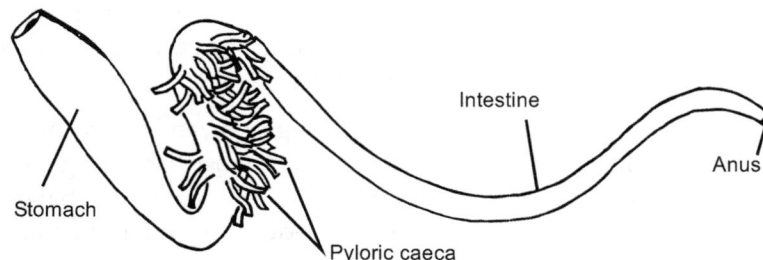

Fig. 1.10 : Pyloric caeca of salmon.

6. Digestive glands : The digestive glands comprise of liver and pancreas (Fig. 1.11). The liver has two lobes – the anterior and the posterior lobes. In a healthy fish, the liver should appear brown or dark red in color. There is a gall bladder which is situated on the ventral side of the right lobe of liver. The gall bladder is the repository of green-colored substance – called *bile* which is released via the bile duct into the intestine. Ordinarily, one hepatic duct orginates from each lobe of the liver and joins the cystic duct from the gall bladder to form the bile duct. The gall bladder in a feeding fish, therefore, empties at frequent intervals. If the gall bladder is distended, it indicates that the fish has not fed for several days. Liver function includes bile secretion and glycogen storage, in addition to several biochemical processes.

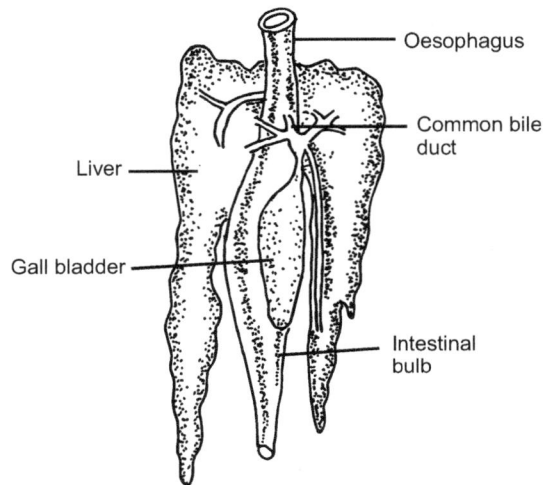

Fig. 1.11 : Digestive glands of a typical bony fish *Labeo rohita* showing the relationship of the liver and gall bladder with the alimentary canal.

Among the bony fishes, the pancreas is either remains in diffused conditions along the coils of the intestine or located in or around the liver or may remain within the spleen tissue. In advanced teleost fishes, the pancreas and liver are combined into a "Hepatopancreas". The pancreas is, however, a compact organ, usually consisting of two lobes.

The pancreas secretes several enzymes which are active in digestion. In addition to this, the pancreatic islets have the endocrine function of producing insulin.

Feeding and Digestion : Structure of Mouth and Feeding Habits

Fish that feed on small invertebrates have an "O" shaped mouth with contracting lips, while fish that consume larger animals usually have a wide mouth. The position of the mouth, however, provides a clue to the eating habits. Most of the fishes have the mouth at, or very near, the front end of the head, which opens forward. Bottom-dwelling fish have mouth that point downward, and surface feeders have mouths point upward. Upward mouths are possessed by relatively few fishes – those that capture food from the surface or those that wait at the bottom to catch prey passing overhead.

Except the family Cyprinidae, the majority of teleostean fishes are carnivorous. Predatory fishes have stronger jaws and sharp teeth such as those of the trout and pike (*Esox* sp.) and very many others. The teeth on the jaw serve to bite and catch the prey. The teeth on the wall of the pharynx – so called *pharyngeal teeth,* help prevent the prey from escaping if it is swallowed whole. Some fish are provided with a filtering system of branchial gill-rackers such as herring.

Herbivorous fishes have crushing teeth similar to those of the carnivorous. The cyprinidae have no teeth on the edge of the jaw and hence the name *'leather mouths'*. There are, however, teeth on the pharyngeal floor of the skull. These fishes are mainly vegetarians, but several species take mouthfuls of mud and nutrients are extracted from the natural food organisms it contains.

1. Variety of foods*

1. Variety of Foods : Most species of fishes are predatory, feeding on live animals or plants. But in the tropics, 10 to 20 per cent of the species may depend primarily on plant materials for food, and in freshwater lakes where soft bottom materials accumulate, there may be detritus feeders. However, a brief outline of feeding types in fishes is noted below.

(i) *Carnivores :* The benthic invertebrate fauna furnishes a significant portion of the food for carnivorous fishes. For the most part, organisms such as insects and their larvae, form food for most of the fishes. Some bottom-dwelling fishes hunt by searching and capturing individual organisms as they are confronted with the prey. Many bottom feeders swim above the substrate and search by sight for prey, and a few others supplement sight feeding by tactile sense, lateral line system, etc. Barbels and lips of many bottom-dwelling fishes carry organs of touch and taste aid in search of edible bottom organisms.

A number of species feed on zooplankton, some by straining the water through gill rackers and others by selecting individual organisms. Given the wide range of size in zooplanktons and in fishes, planktons are virtually detected and selected individually by tiny fishes, or fish larvae might be strained from the water by larger fishes. Fish of a given size might filter out plankton up to a certain size but select somewhat larger organisms individually.

(ii) *Herbivores :* There are few or no fishes that are herbivores throughout their life, because larvae and juveniles begin to feed on zooplankters. In addition, fishes that feed on filamentous algae or vascular plants also ingest animals (such as snails, insect larvae etc.) which are attached to the vegetation. Nonetheless, many species depend, throughout the adult life, on plant materials and exhibit suitable structural modifications for gathering, processing, and digestion.

Several species such as grass carp, certain *Tilapia* species, and *Puntius javanicus* browse on large filamentous algae or on vascular plants. Also, many species that live in freshwater habitats are adapted to scrape flims of diatoms and other

---

* For further study, see Hyatt (1979) and Gerking (1994).

periphyton grown on substrates (For further detail, see Chapter 13). This grazing is, however, accomplished by means of modified Jaw edges in some species and by teeth in others. Either only the modified lower jaws or both the jaws are involved in scrapping by a chiseling action. The best known examples of grazers and scrapers are various cichlids (*Pseudotrophenus* spp. *Gephyochromis* spp) parrotfishes, suckermouth catfishes, and chiselmouth (*Acrocheilus*).

(iii) *Omnivores* : Though fish species are specialized for a narrow range of diet, several species turn to a wider choice of food when the opportunity arises. A variety of herbivores can be caught with animal bait. Many of the cichlids of the African Grate Lakes have jaws, teeth, and pharyngeal teeth that are morphologically specialized for feeding on a narrow range of natural food organisms but are opportunistic nonetheless. They always take advantage of abundant food items.

Some species such as channel catfish (*Ictalurus punctatus*), common carp (*Cyprinus carpio*), and rainbow trout appear to be omnivorous and consume a variety of foods in their daily rations since they have a wide taste in foods.

2. Feeding : Structure, Function, and Mechanism : Fish having jaws exhibit a great variety in structure and function of the feeding apparatus. There has been a progression from the simple long jaws which are able to catch and held prey to suction feeding aided by protrusible jaws. The evolutionary sequence of protrusible jaws generally involved the development of movable maxillaries and pre–maxillaries, enlargement of pre-maxillaries, exclusion of maxillaries from the gape, and development of a long ascending process of the pre–maxillaries. However, the feeding apparatus of fishes can consist of more than 30 movable bony parts and over 50 muscles.

In opening the mouth of lower bony fishes, contraction of the body muscles elevates the neurocranium – lie just posterior to the premaxillary, and contraction of the muscles along the ventral aspect of the head lowers the mandible. In the teleosts, an additional coupling serves to open the mouth. Operculi muscles generate adequate energy to raise the opercular apparatus, which is coupled with the mandible either through the inter-operculo-mandibular ligament (in lower teleosts) or the inter-operculo-hyoid ligament (in higher teleosts). The ligamentous coupling between the mandible and the maxillary causes the maxillary to pivot on its connection to the neurocranium and swing downward and forward thus forming the lateral edges of the open mouth.

Generally, the mouth-opening sequence is accompanied by the lowering of the floor of the mouth by depression of the hyoid apparatus. During this time, the opercular opening is closed. This can lower the pressure in the orobranchial cavity and help in capturing of food. Opercular expansion essentially contributes to the suction but occurs as the mouth begins to close. Closure of the mouth is accomplished by the adductor mandibulae.

In cypriniforms and in higher teleosts, the mobile premaxilla makes jaws protrusion possible, and suction feeding is well developed. However, protrusion of the maxilla provides the velocity of approach to the food, allow the fish to reach for food, and help in forming an efficient funnel for suction.

The current of water passing into the mouth and out of the opercula serves as a feeding mechanism in teleostean fish. The food is drawn in by expansion of the mouth cavity assisted by protrusion of the premaxilla. The sequence of events during the feeding of fish consists of a preparatory phase which is followed by three phases. In the preparatory phase, the volume of both oral and branchial cavities significantly decreased. In the first phase, the opercular and buccal cavities also decreased which is followed by the closing of the mouth. In the second phase, the mouth is opened and expanded to full gape as the oral and branchial chambers are enlarged. And during the third phase the mouth parts reach the initial position which helps close the mouth again and compressed to force water and ingested material posteriorly, where water is forced out by compression of the posterior chamber and food is retained on gill-rackers.

Gill-rackers are of great variety, reflecting the diet of the species involved. The branchial sieve is not a passive filter in fishes that consume plankton. Space between adjacent gill arches can be controlled and gill-rackers and the grooves between them are controlled by musculature, which can govern the ability to retain particles of different sizes. However, food organisms are trapped in the basket-like array of gill-rackers. Pharyngeal teeth and the gill-rackers help in forcing prey into the oesophagus. In many cases, pharyngeal teeth are modified and specialized for crushing and grinding before food organisms are swallowed.

3. Digestion: In most of the fishes, the chemical digestive process starts in the stomach. Stomach secretes pepsin, protease, mucus, and hydrochloric acid. Hydrochloric acid maintains the pH of the stomach contents in a range suitable for the action of digestive enzymes on food organisms. The stomach generally acts as a storage organ and initiates digestion by mixing the ingested food organisms with the gastric juices.

As noted earlier, a variety of fishes lack a true stomach. Stomachs are absent in at least some of the members of the following families: Cobitidae, Cyprinidae (minnows), Belonidae, Cichlidae, Cyprinidontidae etc. However, stomachless fishes have an expanded portion of the intestine in which large morsels can be stored while undergoing digestion.

The pyloric caeca are considered to be sources of both carbohydrases and proteinases, but attributed to the secretion of these enzymes from the diffused pancreas. But a growing body of research indicates that the pyloric Caeca have no secretory function; rather, they have a function in absorption as well as increasing the area of digestive membrane.

Among the teleosts, the intestine varies in length and harmonization with food habits and with individual diet. Fishes that habitually ingest a large proportion of indigestible materials with their food appear to have the longest relative gut length, and in these fishes the gut is coiled in an elaborate manner. The added length increases the retention time of the food and permits for more efficient digestion of substances that are difficult to digest.

Digestion proceeds in the intestine in a neutral to alkaline medium. Enzymes involved are secreted by the pancreas, intestinal mucosa, and possibly by the pyloric caeca. The types and amounts of enzymes present in the digestive system of a given species are related to the general food habits of that species. Enzymes secreted by the pancreas include trypsin, chymotrypsin, and amylase. Intestinal mucosa secretes some carbohydrate-splitting enzymes. For example, maltase is known from the intestines of trout, salmon, and carps.

In contrast to carnivorous fishes, herbivorous species have a relatively rapid food passage. Examples include complete digestion of salmon and trout fry in 24h (at 15°C), tuna in about 14h (at 20°C), grass carp in about 8h (at 20-25°C), cichlids in 7-15h (at 25°C), common carp in 48h (at 25°C), and *Carassius* spp. in 60-72h (at 20°C).

Respiratory System

Respiration refers collectively to the physical and chemical processes in living organisms which results in energy liberation. Since every cell requires energy to carry out its living functions, all organisms respire. However, respiration in fishes depends upon a system of supplying oxygen and removing carbon dioxide as a waste product. For efficient respiration in fishes requires a large respiratory surface such as gills and an efficient circulatory system to transport of oxygen to tissues and carbon di-oxide away from them.

In fishes, gills are the most efficient organs of respiration. Gills arise as a series of paired pouches of the pharynx which extends outwards and finally open on the exterior by the gill-slits. Each gill pouch thus communicates with the pharynx by an *internal* and with the outside water by an *external branchial aperture*. It is separated by stout fibrous partitions, termed as *interbranchial septa*. The membrane forming the posterior and anterior walls of the pouches is supplied with blood through apertures.

In the higher bony fishes, the interbranchial septa progressively reduced to narrow bars enclosing the visceral arches, so that a double set of free branchial filaments arises from each visceral bar and constitute a single gill. In teleosts, the septa become greatly reduced to no more than small ridges along each gill arch, from which the gill tissue extends as long filaments. The gill apparatus is thus more compact than in other groups of fishes. The entire chamber containing gill arches and filaments is covered on each side by the bony operculum. Loss of the septa results in greater respiratory efficiency because the flow of water through the secondary lamillae is facilitated.

All teleosts have four gills lying below the operculum on either side of the head. Each gill is, however, composed of a curved bony gill bar bearing many gill filaments in a fringe-like formation (Fig. 1.12); the filaments bear smaller filaments along their length, and these in tum divide into still smaller filaments (Fig. 1.13). In general, there are two rows of filaments on each gill arch. In most of the teleosts, each gill arch contains one affarent branchial vessel and one effarent branchial vessel. But in some species such as *Labeo* sp., *Clarias* sp., *Catla* sp. *Heteropneustes* sp., *Trichogaster* sp., etc. two different vessels are present.

In the higher bony fishes, a complete gill is formed which is termed as *holobranch* and it is the morphological equivalent of two half-gills, called as *hemibranch* or sets of branchial filaments belonging to the adjacent sides of two consecutive gill-pouches.

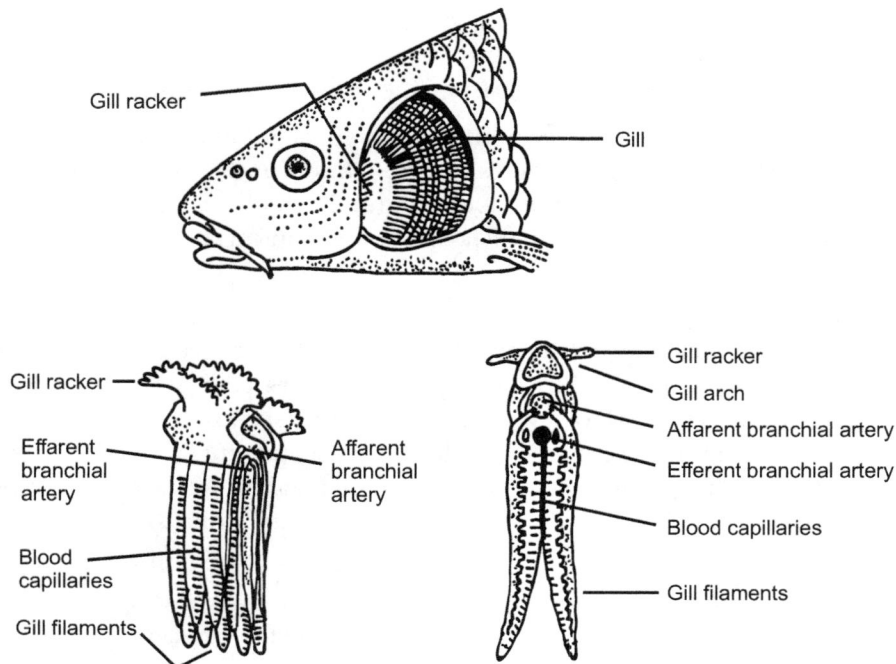

Fig. 1.12 : Structure of gill in a teleost fish : (A) Location of gill in the branchial chamber; (B) A few gill filaments; (C) Structure of a gill filament.

Fig. 1.13 : Tips of gill filaments of a teleostean fish.

The gills are borne on the first four branchial arches, the fifth arch being none. On the inner surface of the operculum is a comb-like body, termed as *pseudobranchia*, formed of a single row of branchial filaments, and representing the vestigial gill (hemibranch).

Bony fishes typically retain a pseudobranch at the site of the hyoid arch and holobranchs on each of the four branchial arches. Reduction of gills has, however, occurred in some sir-breathing teleosts and other groups of fishes. Some members of the family Synbranchidae (such as swamp eels) have only one well-developed holobranch.

Although the pseudobranch closely resembles a functional gills in some species, it receives only oxygenated blood and hence cannot function in respiratory gas exchange. Moreover, the pseudobranch of most of the bony fishes (for example, in the Family

Salmonidae) may be reduced to the appearance of a glandular structure located beneath the skin. This structure secretes oxygen to the retina of the eye.

1. *Branchial Sieve :* Effective exchange of gases in a gill-breathing fish principally depends on bringing the blood and respiratory water in close apposition to either side of a membrane through which the gases can diffuse. This system of gas exchange works best if the blood and water flows opposite to each other, and fishes have developed this counter-current system that requires a force to move the water and blood and finally divide, channels through which these fluids can flow. The channels are provided by the structure of the gills.

   Each gill arch bears numerous filaments. The actual number varies with the size and surface area of the fish and the general habits of the species. For example, sluggish and bottom-dwelling fishes generally have less filaments (numbers vary from 300 to 700) than active species such as mackerel (weighing 800g have about 2,400) and perch (weighing 30g have about 1,500). Moreover, less active fishes (such as *Coregonus* spp.) have less surface area of gills (One kilogram of fish possess gill area of 290 mm^2/g) than more active fishes such as herring (weighing 11g possess a gill area of 636 mm^2/g). These examples definitely indicate that the greater the surface area of gills the higher is the exchange of respiratory gases. In other words, a great surface area for gaseous exchange is the result of having numerous filaments bearing the small but numerous lamillae. The increase in number of filaments causes a very large surface area through which oxygen can diffuse rapidly from the surrounding water through their thin walls.

2. *Flow of Blood Through Gills :* The heart provides the major force to move the blood through the ventral aorta and into the gill filaments. The blood flows throughout the fine gill lamillae which appear bright red in healthy fish. The gills surface is so thin that it makes the fish particularly vulnerable to gill damage caused by infection, silt clogging or irritation and chemicals. There are two pathways that blood can follow in the lamillae : (i) the arterio-arterial pathway and (ii) the arterio-venus pathway. In the former, blood passes through the lamillae is oxygenated and proceeds to the efferent branchial and then to the supra-branchial artery which, in teleosts, develops into the carotid artery anteriorly and the dorsal aorta posteriorly. In the second pathway, blood can pass from the filamental arteries to the arterio-venus anastomoses and thence to the venus system.

   Ventricular contractions produce a considerable pressure that is attenuated as the blood is forced through the interlamellar channels and into the dorsal aorta via the efferent branchial arteries. The bulbus arteriosus maintains a positive pressure on the blood even through the diastolic ventricular pressure may drop to zero.

   It should be added as a note that although the heart provides a major force for maintaining blood flow in the dorsal aorta, systemic arteries, veins, and body musculature may patronise in the flow of blood.

In most of the teleosts, the mechanism of respiration is similar and respiratory valves are developed in the mouth cavity. But there are considerable differences in detail, particularly as regards the relative importance of the opercula and the branchiostegal membranes* in carrying on the movements of inspiration and expiration.

3. *Mehanism of Respiration /Breathing :* The respiratory pump that forces the water across the gills consists of the bucco- pharyngeal cavity along with all the mechanisms for opening, enlarging, and constricting it, and the parabranchial cavity (a cavity between the gills and the operculum) which can be expanded or constricted by the action of the operculum and the branchiostegal membranes. The free edge of the operculum is provided with a membranous flap called as *branchiostegal membrane.* The movements of the opercula and the mouth are, however, well co-ordinated to generate a stream of water over the gills so that adequate amounts of oxygen are constantly made available. Water enters the buccal cavity through the mouth, passes over the gills, and out of the opercula. The mechanism of generation of the water current involves the pumping action produced by a buccal pressure pump and opercular suction pumps, resulting from sideways movements of the operculum, and enlarging the branchial chamber. The details of the pumping mechanism vary in different species, but the following is a general description (Fig. 1.14).

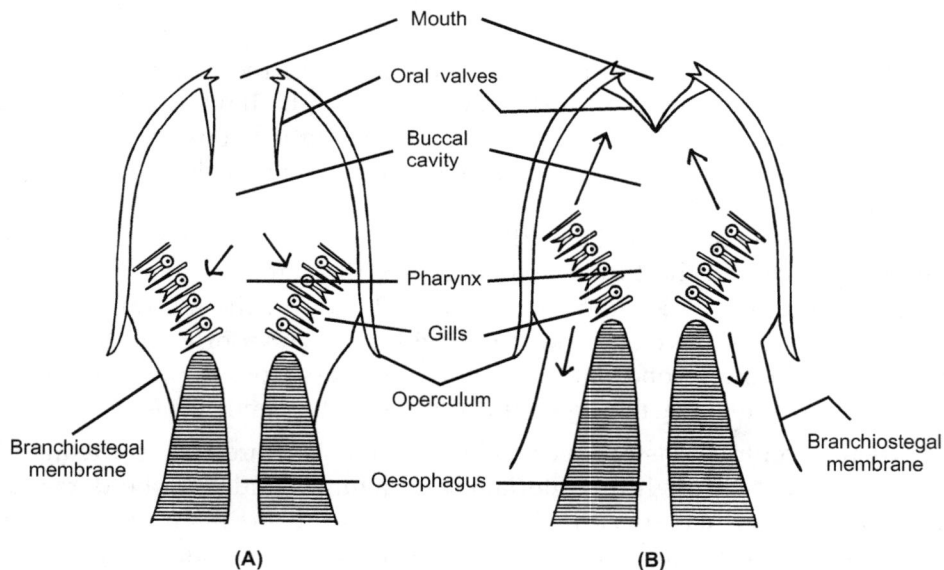

Fig. 1.14 : Schematic diagrams showing the physical mechanism of respiration in a teleost fish (*Lates* sp.). (A) Inspiration; (B) Expiration. The arrows indicate the direction of respiratory water currents.

---

* The free edge of the operculum is provided with a membranous flap called as *branchiostegal membrane.*

(i)   The pressure in the mouth cavity is reduced by lowering the floor of the mouth. As a result, water enters the mouth, while the free edges of the opercula are kept firmly closed against the body wall by the higher pressure of the water outside them. The branchiostegal membranes prevent the inflow of water from behind.

(ii) The floor of the mouth is raised, decreasing the volume of the mouth. The pressure of the water in the mouth brings about the closure of two intumed folds of skin along the upper and lower jaws, and forces water across the gills filaments. The movement of water is assisted by a simultaneous outward movement of the opercula which sucks water from the front to the back of the mouth cavity out over the gills.

(iii) The opercula and the mouth are closed. The branchiostegal membranes are forced outward by the pressure of the water inside it, and water escapes between the body wall and the operculum.

It should be recalled that inspite of the presence of more oxygen in air (20.6 per cent) than in water (4-6 per cent), a fish will suffocate in air . This is possibly because the muscular system of mouth and opercula which can work well in water will not function in air. In other words, the system of valves, although water tight, is not air tight. Moreover, the gill filaments cling together into a mat when out of water, because of the surface tension of the water which covers them so that the total gill surface exposed is very much reduced.

4.   *Oxygen Uptake by Fish :* In resting condition, the oxygen consumption for some familiar freshwater fishes is shown in Table 1.2. Under favorable conditions, fishes can remove about 90 per cent of the dissolved oxygen from water passing over the gills. Such efficient removal of oxygen occurs when dissolved oxygen is high and respiratory volume is low. In general, the range of oxygen uptake varies between 50 and 60 per cent; and under conditions of low dissolved oxygen, high temperatures, and increased respiratory volume, the uptake may fall to 10 to 20 per cent. Actual oxygen consumption in fishes, however, depends on several factors such as temperature, size, activity, metabolism, oxygen and carbon dioxide

Table 1.2 : Resting Oxygen Consumption in Some Selected Freshwater Fishes

Species	Temperature (°C)	Oxygen Consumption (mg/kg/h)
Carassius auratus	10	15.7
	22	30-160
	32-35	127-262
Cyprinus carpio	10	17
	20	48
	30	104
Salmo trutta	10	80
	20	128-228
Labeo rohita	30	30-120
Catla catla	25	45
	30	95
Oreochromis mossambicus	26	20-60
	32	35-85

pressure, salinity, and pH of water. For example, anaerobic metabolism acquires an oxygen debt, and the fish must rid the blood of excess lactic acid. When dissolved oxygen content of the water is lowered, the fish responds well by increasing the rate of gill ventilation. This added activity requires a greater utilization of oxygen. Consequently, the fish can be placed in the situation of trying to draw a greater amount of oxygen from a smaller supply. This condition makes the situation intolerable for fish.

Circulatory System

A circulatory system is a system of vessels that carries oxygen, nutrients, and waste materials around the body. In fish, however, the main transport fluid is blood and it is pumped around the body by a heart.

Fish, which obtain oxygen and eliminate carbon dioxide through their gills, do not have separate chambers in the heart for oxygenated and deoxygenated blood.

The circulatory system of different groups of fishes is composed of heart, affarent branchial vessels, efferent branchial vessels, and veins.The heart,however, lies in its own body just in front of the abdominal cavity below the oesophagus.

1. *Heart :* A typical fish has a four-chambered heart, with the chambers arranged in a row a thin-walled lightly muscled sinus venosus (into which the great veins empty), an atrium, a thick-walled ventricle, and an elastic balbus cordis (also known as *conus arteriosus*). Depending on the species, either there is presence of contractile conus arteriosus or an elastic thin-walled balbus arteriosus (Fig. 1.15a). The heart is contained within a pericardial cavity located below the gills. The various groups of fishes have, however, many features of the heart in common.

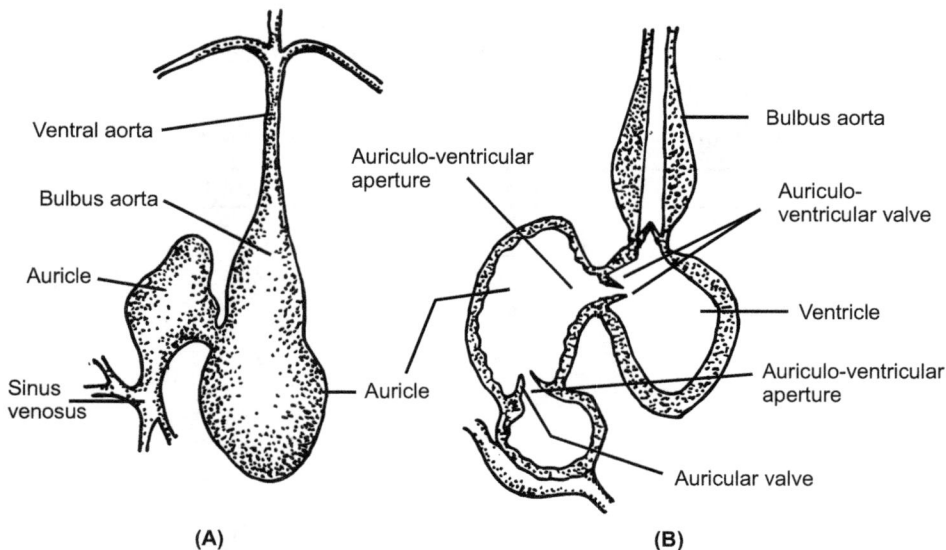

Fig. 1.15a : Structure of the heart of a typical bony fish (*Lates* sp.).
(A) External structure of the heart; (B) Longitudinal section of the heart.

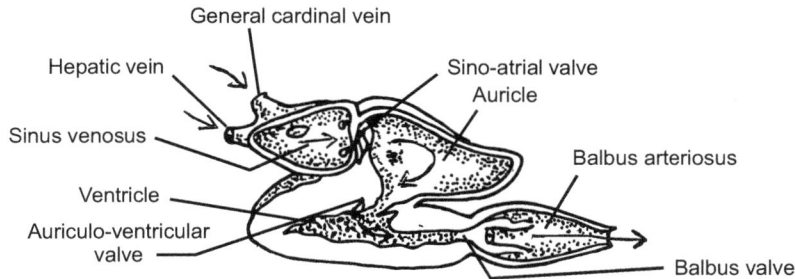

Fig. 1.15b : Diagramatic representation of the internal structure of the heart of a bony fish. Arrows indicate the path of blood circulation.

In most of the teleosts, the conus arteriosus is greatly reduced or absent, non-muscular, and bears only one set of valves between the ventricle and the non-muscular, elastic balbus arteriosus. The balbus arteriosus may appear at the base of the ventral aorta as a small but when the heart is pumping, the balbus aorta expands to almost the size of the ventricle (Fig. 1.15b). For example, the balbus arteriosus of a carp can expand about 700 per cent than the aorta of humans. Blood is prevented from flowing back into the ventricle by valves located within the reduced conus arteriosus. Since the balbus arteriosus is elastic and thin-walled structure, the walls of the balbus maintain pressure on the blood flowing to the capillaries of gills, maintaining an almost continuous flow, in contrast to the pulsatic flow from a conus arteriosus.

2. *Branchial Arteries :* The ventral aorta of fishes extends forward beneath the pharynx. In teleosts, there are four pairs of affarent branchial arteries which arise from the single ventral aorta and enter each of the branchial arches to supply

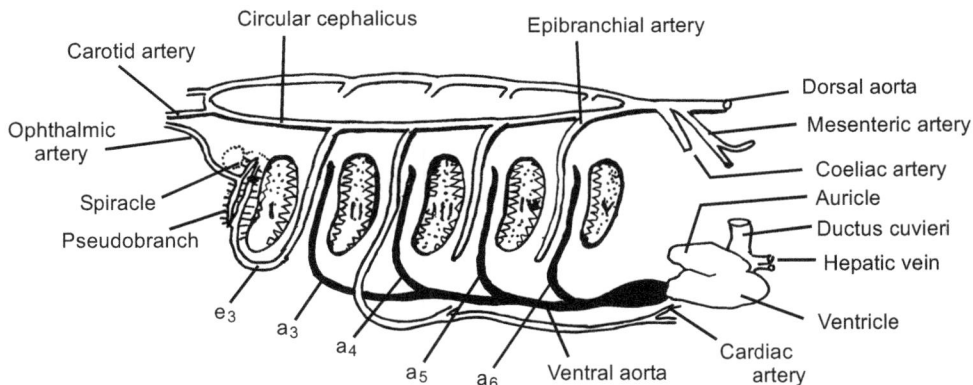

Fig. 1.16a : Diagram of the branchial circulation of a teleostean fish. a_{3-6}, four affarent branchial vessels from ventral aorta; e_3, efferent branchial vessel of first branchial arch; I–V, five branchial slits. (From Goodrich, 1909)

deoxygenated blood to the corresponding gills. The ventral aorta divides anteriorly into the first two affarent branchial arteries. The second, third, and fourth pairs of the affarent branchial arteries have separate and independent origin from the ventral aorta (Fig. 1.16a, b, c).

After oxygenation the blood from the gills is collected by the efferent branchial arteries and distributed to all parts of the body through venus system. The venus system includes the paired anterior and posterior cardinals, the unpaired hepatic and renal portal veins, and paired subclavian veins.

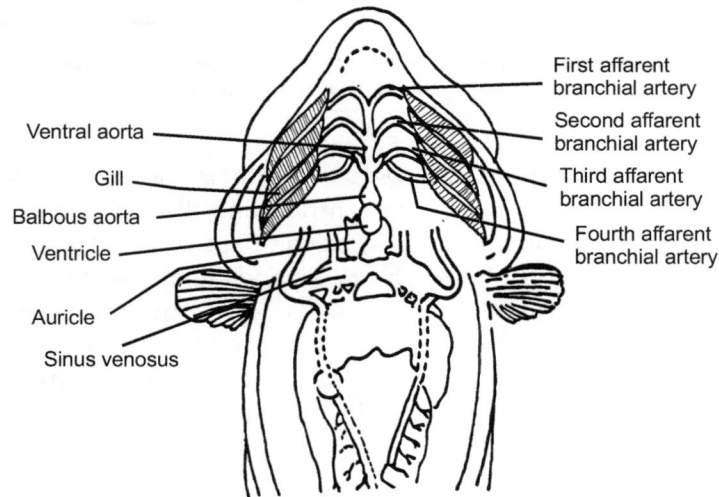

Fig. 1.16b : Affarent branchial system of a teleostean fish, *Ophiocephalus punctatus*.

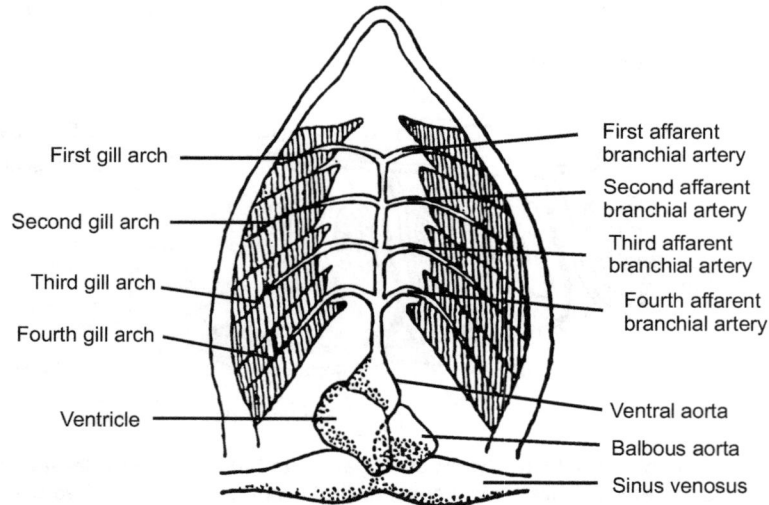

Fig. 1.16c : Affarent branchial system of a teleostean fish, *Labeo* sp.

Accessory Respiratory Organs*

The ability to extract oxygen from the atmosphere is a primitive feature of bony fishes and occurs as a convergent specialization in many teleost groups. The habit is encountered from the freezing Arctic bogs to the tropical swamps. Overall, there are many more air-breathing species among warm-water fishes than cold-water or temperate forms with the habit most common among the tropical swamp dwellers.

Many freshwater teleost fishes breath air, particularly those living in shallow and stagnant waters. These fishes develop certain specialized structures, generally termed as *accessory respiratory organs* or *branchial organs* – to meet the extra demand for oxygen since oxygen depletion is the common feature during summer in shallow waters. This situation has been foreced to develop some devices to live out of water for a short period of time. The evolution of accessory respiratory organs has a long history (*See* Panel 1.2)

Fishes are primarily aquatic vertebrates and they utilize the dissolved oxygen in water for respiration by gills. Those fishes which use atmospheric oxygen for respiration besides the oxygen dissolved in water. These fishes are called *air breathing fish.* Air-breathing fishes generally differ from others by virtue of their accessory respiratory organs, peculiar surfacing and hanging behavior, and feeding habit.

Freshwaters permanently low or deficient in oxygen can be inhabited by those fishes that can derive their oxygen from an alternate source. Tropical swamps of high organic content and heavy macrophytes cover support year-round populations of fishes, some of which are obligate air-breathers. Streams and swamps contain dissolved oxygen in appreciable amounts during part of the year, but stagnant and shallow waters contain negleglible amounts of oxygen during other times where facultative or obligate air-breathers inhabit. Fishes can cope with the drying of the water either by burrowing and aestivating or by moving overland to permanent bodies of water. Fishes that are able to aestivate during summers or dry swamps include the central mud minnow (*Umbra* sp..), the bow fin (*Amia* sp.), the walking catfishes (*Clarias* spp.), the swamp eels (*Amphipnous* spp. and *Synbranchus* spp.), channidae, the snakeheads, the climbing perch (*Anabas* spp.), and the mud fishes of New Zealand (*Neochanna* spp.). Some such as *Anabas* come onto the land when ponds dry up and others such as *Gambusis* take gulps of air for respiration.

1. ***Organs For Air-breathing :*** Fish gills are efficient for exchange of gas in aquatic environment. The lamillae of the gills supplied with blood vessels collapse under gravitational force in air. Consequently, air-breathing fishes had to develop certain specialized structures for efficient gas exchange in air. The accessory respiratory organs in air-breathing fishes are modifications of existing structures or these organs may be newly formed. The occurrence of air-breathing organ in fishes is followed by reduction in the surface area of gills.

---

* For further study, see Munshi (1992)

Air-breathing organs show diverse structural modifications. The adaptation for air-breathing was the development of elaborate air-holding chambers at the mouth (such as *Electrophorus electricus*) or throat (such as *Anabas* spp. and *Ophiocephalus* spp). Some species have developed gas bladder which acts like a vertebrate lung (such as *Notopterus* spp.). In many air-breathing fishes such as *Neochanns* sp., *Anguilla* sp., *Amphipnous cuchia*, *Periophthalmus* and *Boleophthalmus*, the skin serves as a respiratory organ. These fishes can, however, obtain from 20 to 70 per cent of their required oxygen through the skin. Cutaneous respiration is, therefore, of great importance mainly during the periods of low activity or of relatively low temperatures. A few species have modified stomach or part of the intestine for air-breathing (such as *Lepidocephalichthys* sp.). They swallow air and force it out at the mouth or the cloaca.

2. Structure of Accessory Respiratory Organs in Some Common Air–breathing Fishes

(i) The Common eel fish (*Anguilla*) : Though the fish has no special respiratory organ, it has vascular area in the skin by which it can breath both in water as well as on land. The gill chamber is closely covered with a rounded operculum which has a very small opening so that it can retain water in the gill chamber for longer period.

(ii) The climbing perch (*Anabas*) (Fig. 17 a) : There is a specialized structure called *accessory branchial organ* to carry out the major part of the respiratory process. This structure is situated in a cavity above the gills, consisting of a series of exceedingly thin bony lamillae situated one above the other and covered by vascular mucous membrane. The first epibranchial bone on each side of the gill chamber gives rise to three concentric rings – called as *dendrite organs*. Since this organ is highly vascularized, it is known as *labyrinthine organ* This fish is so dependent that even when in water, it comes to the surface to engulf air. The fish can survive for a long period on land.

(iii) The Indian Murrels (*Ophiocephalus* spp) (Fig. 17 b) : The accessory respiratory organs in these fishes are very simple, consisting of a pair of hollow pouches on the roof of the mouth on either side of the palate. In fact, these are pouches of the pharynx. The pharyngeal outgrowth that rests above and in front of the gill is highly vascular and can survive on land for a long time by engulfing air and respiring through pneumatic sac.

(iv) The Indian Cuchia eel (*Amphipnous cuchia*) (Fig. 17 c) : The fish has poorly developed gills but on each side of the body, there is a vascular sac developing from the pharynx and protrude into the gill chamber. This sac opens anteriorly into the first gill cleft through which they engulf air when they come to the surface of water.

(v) The Indian Catfishes:

» *Heteropneustes fossils* (Fig. 17 d) : This fish possesses a tubular air sac or pneumatic sac one on each side. This sac arises from the branchial chamber and extending laterally through the trunk muscles upto the tail. This tube is filled up with air for respiration.

» *Clarias batrachus* (Fig. 17 e) : This fish has a pair of complicated structure – called *suprabranchial organs* each lying under each gill and divided into two parts: one part is called the *arborescent organ,* which is highly vascularized, formed from second and fourth branchial arches and the other part – known as *vascular sac,* of the branchial chamber which encloses the arborescent organ. At the entrance of the suprabranchial chamber, the gill filaments coalesce to form the gill fan. Air is entered into the arborescent organ through the mouth ceaselessly on and can live outside water for a long time.

(vi)   The Loaches (*Misgurnus* sp.) : This species occasionally comes to the surface of water and engulf air. The air passes through the intestine – the inner wall of which is built by highly vascularized mucous membrane where gaseous exchange takes place and carbon dioxide is voided by the anus.

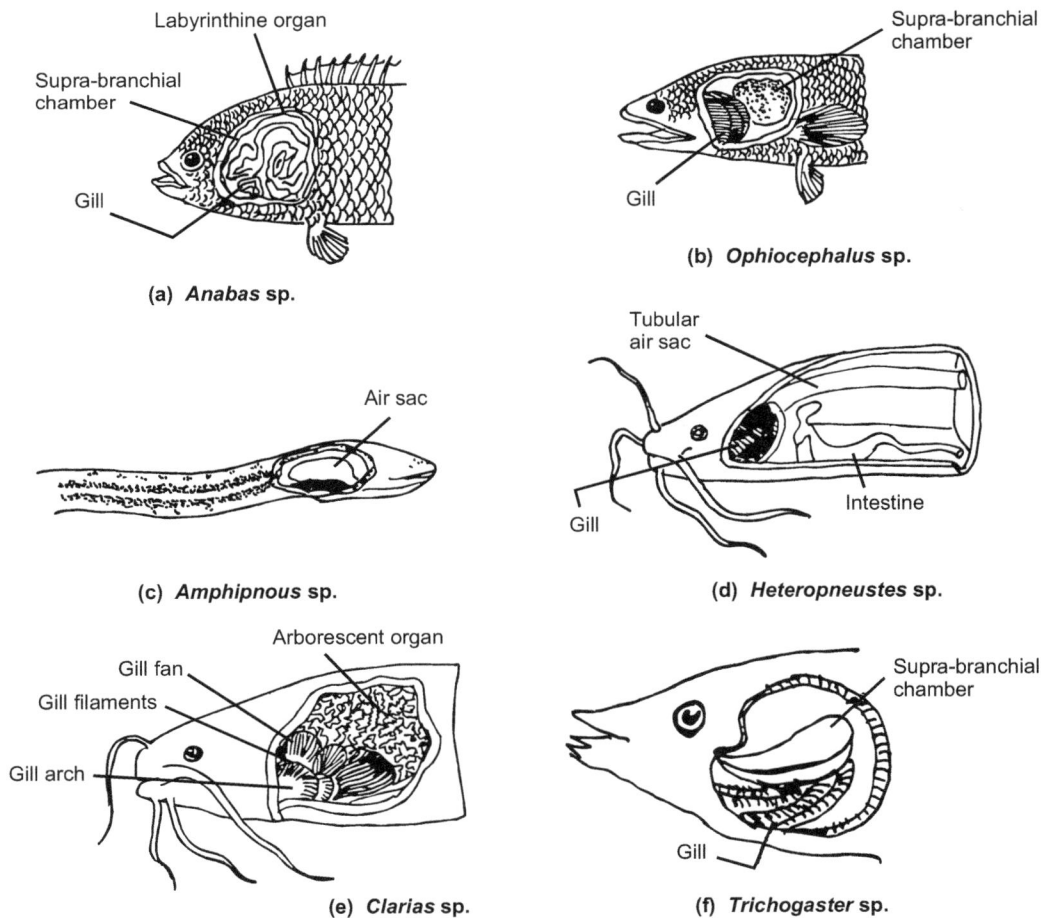

(a) **Anabas** sp.

(b) **Ophiocephalus** sp.

(c) **Amphipnous** sp.

(d) **Heteropneustes** sp.

(e) **Clarias** sp.

(f) **Trichogaster** sp.

Fig. 1.17 : Accesory respiratory organs in some freshwater fish.

(vii) *Trichogaster fasciatus* (Fig. 17 f) : The accessory respiratory organs in this species consist of a suprabranchial chamber, a labyrinthine organ, and the respiratory membrane. 'The suprabranchial chamber is situated above the gills on either side and as in *Anabas,* communicates with the pharynx by means of an inhalant aperture and with the exterior through opercular chamber by means of an exhalant aperture. The labyrinthine organ develops from the epibranchial of the first gill arch and the structure is very simple than that of *Anabas.* It is in the form of a spiral organ possessing two leaf-like expansions and is composed of loose connective tissue which is covered by a vascular epithelium. The respiratory membrane lining the air chamber and covering the labyrinthine organs, essentially consists of vascular and non-vascular areas, of which the former possesses a large number of 'islets' containing parallel blood capillaries. The islets are believed to be derived from the secondary lamellae of a typical gill filament.

3. Behavior of Air-breathing Fishes

Air-breathing fish exhibit two important behaviors that follow.

(i) *Feeding behavior:* More than ninety per cent air-breathing fishes are carnivores. They have large mouth, many pointed teeth, large stomach, and short intestine typical of a carnivore. The pressing temporal cost of surfacing very much restricts the time available for food procurement in air-breathing fishes. Consequently, they are forced to choice energy-rich carnivorous diet. It seems quite plausible that this type of feeding habit and hanging behavior have evolved in air-breathing fishes to make up for loss of their energy requirement and time-consuming surfacing behavior.

(ii) *Surfacing Behavior:* For breathing, these fishes generally come to the surface, break it and take a bubble of air which is then passed on to the accessory organ. They always avoid the problem of dessication and carbon dioxide accumulation in the blood by remaining in water, exploiting the atmosphere as a source of oxygen.

Air-breathing fishes are of two types such as (a) Facultative air-breathing fishes (*Heteropneustes* sp.) and (b) Obligatory air-breathing fishes (*Macropodus* sp. *Anabas* sp., *Ophiocephalus* sp). In the obligatory air-breathing fishes, only the interval between successive surfacings is affected by the partial pressure of oxygen dissolved in water. In the facultative air-breathing fishes, on the other hand, it is not only the interval between successive surfacings but also the surfacing itself that depends on the partial pressure of oxygen. Generally, decrease in oxygen content of water draw out increased surfacing activity in air-breathing fishes. When air-breathing fishes are kept in air-saturated water, the facultative air-breathers rarely come to the surface because their gills are very efficient to exchange both oxygen and carbon dioxide. But the obligatory air-breathing fishes have to surface at regular intervals even in air-saturated water because of their reduced gill size. Experiments on *Ophiocephalus straitus* (10g in body weight.) have shown that in aerated water (150 mm Hg), the individual comes to the surface 850 times per day, whereas in oxygen-poor water (40 mm Hg) the surfacing frequency increases up to 1,130 times per day.

Due to surfacing behavior of air-breathing fishes, a considerable amount of energy is lost which, in turn, decreases the growth of these fishes. It has been observed that obligatory air- breathing fishes reared in deep water (40-50 cm depth) consumed more food (220 g cal/g/day) but converted it into body substance less efficiently (growth efficiency is 13 per cent). This is due to the fact that during surfacing activity, more energy has been lost. In shallow water (3 cm depth), on the other hand, air-breathing fish consumed less food (140g cal/g/day) but converted more efficiently (growth efficiency is 20 per cent) conserving excess energy spent on surfacing activity. It is, therefore, assumed that the structure of obligatory air-breathing fishes in shallow water bodies is more suitable than in deep waters.

KIDNEY AND OSMOREGULATION

1. Kidney

Kidneys are the specialized organs in most vertebrates including fish that filter the blood and eliminate metabolic waste products. The kidneys also maintain the levels of water and salts in the blood which is inevitable for the survival of fish.

In teleosts, the kidneys consist of an elongated brown mass extending along the whole length of the visceral cavity. They are situated dorsal to the body wall above the swim-bladder and are separated anteriorly. The ducts of the two kidneys are united posteriorly. The ureters, one from each kidney, opens into a thin-walled urinary bladder which is situated ventral to the cloaca (Fig. 1.18). The bladder opens into the urinogenital pore.

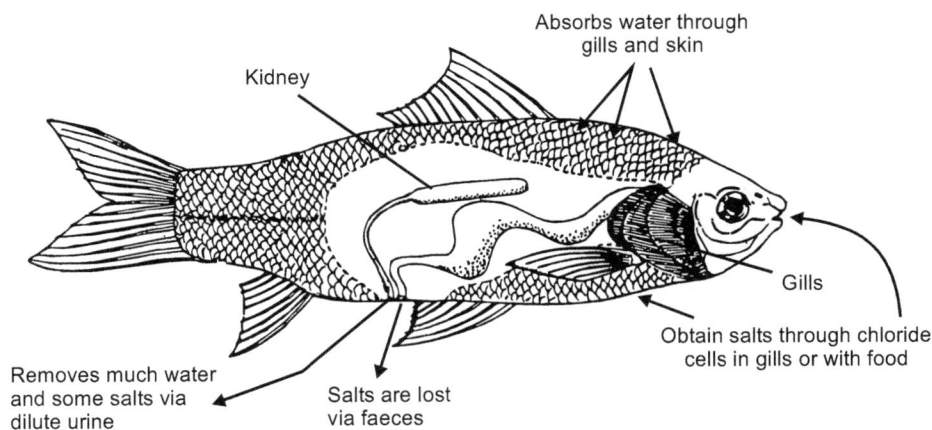

Fig. 1.18 : Diagram summarizing osmoregulation of a freshwater teleost.

Teleosts which are more diverse than other groups of fish and have a wide geographical and ecological distribution, have variety of kidney morphologies. The nephrons of freshwater fishes are generally composed of several parts such as: a glomerulus, a neck segment, a two-part proximal segment, an intermediate segment, a distal segment, and the collecting tubule. The only nephron structures that all teleosts have in common are the second proximal segment and the collecting tubule.

Most of the eels, cyprinids, and some catfishes have kidneys fused to each other anteriorly and posteriorly, but remain separate through the mid-section. These fishes have large glomerular capsules and all segments which are present in freshwater species, lack the distal segment.

In gobies, perch-like fishes, the snakeheads, and mackerels, the kidneys are found only posteriorly and the head kidneys are well differentiated from the opisthonephric kidney (see panel 1.3) . The glomerular capsule is relatively small and the distal segment is retained. In pipe fishes, however, narrow kidneys are connected only at the most posterior end and the head kidney is not developed.

> **(i) *Function of Kidney:*** Fish kidneys play their major role in osmoregulation, maintaining body fluids at the proper osmotic concentration. Freshwater fishes have large glomeruli that can filter much water from the blood. Some of this water is reabsorbed from the proximal tubule, but most is passed as urine. The urine of freshwater fishes is, however, much less concentrated than the blood plasma.

2. Osmoregulation

Simply defined, the term osmoregulation refers to the control of the water content of the body of an organism. However, similar to water the concentration of several dissolved substances in the body fluid and the blood must also remain within narrow range of tolerance because they regulate the movement of water as a result of osmosis. The regulation of these dissolved substances comes under ionic regulation. Since water and ionic regulation are inseparable activities, by extension the term osmoregulation is used to incorporate both the activities.

Some of the most interesting adjustments that fishes of all kinds must make in their particular environments concern the maintenance of proper water and salt balance in their tissues (Fig. 1.18) Failure to maintain the proper balance would results in lethal dilution of body fluids of freshwater fishes. The body fluids of freshwater fishes have a large osmotic concentration than their surrounding medium. This apparent osmotic disadvantage must be overcome for fishes to survive.

Because the osmotic concentration of a typical freshwater fish blood is in the range of about 265 to 325 mOsm/kg, fishes are hyper-osmotic to their medium and tend to gain water by diffution through any semipermeable surface. If unchecked or uncompensated, the inward diffusion would dilute the body fluids to the point that the necessary physiological functions could no longer be accomplished – a state termed as *"internal drowning."* Water-proofing the body would appear to be a means of preventing the diffusion but can be only partially successful. The gills, obviously, cannot be wateproofed and provide a great surface for diffusion of water as well as gases so, overall, water cannot be kept out.

If water in excess of the needs of the organism is driven in by inexorable osmosis, a balance must be maintained by driving the water out through some other means. The task of removing water from the body of fish is accomplished by the kidney. Blood from the dorsal aorta is led to the kidney by the renal artery, where it passes through the capillaries of the glomeruli and then through capillaries surrounding the kidney tubules before leaving via the renal vein.

The glomerulus is a filter that allows blood plasma containing dissolved materials to pass into the space between the walls of Bowman's capsule and thence into the kidney tubules. Blood pressure provides the force for glomerular filtration.

The urinary bladder appears to function as osmoregulation in freshwater teleosts. Water permeability of its wall is low, but Na^+ and Cl^- ions are reabsorbed through the walls. The resultant urine is diluted. The urine contains small amounts of uric acids, creatine, and ammonia. Ammonia is highly soluble in water and, at the same time, extremely toxic and hence must be eliminated from the body along with huge quantities of water. For this reason, freshwater fishes are genrally termed as *ammonotelic* and the phenomenon of ammonia excretion is termed as *ammonotelism.*

Although the concentration of salt in urine is low, the copious flow causes a significant amount of salt to be lost. Loss is also due to diffusion of salt from the body. These losses are balanced by salt intake along with food and by active absorption through gills. Uptake of Na^+, and Cl^- ions at the gills is involved in an exchange of. NH^+ and HCO_3^- ions. Some teleosts drink water, even though this places an added burden on the kidney and removing excess water.

Endocrine System

Endocrine system in all vertebrates including fish are made up of organs, tissues, and cells that control body functions by secreting hormones. Fishes, however, possess a well-developed endocrine system comparable to that of other vertebrate animals. The co-ordination of sorne of the functions of the body is effected by chemicals called *hormones* secreted by the glands termed as *endocrine glands.* These are glands which have no ducts or opening but their secretions enter the blood stream as it passes through their tissues, and are carried in solution in the blood all over the body. The effects of each hormone are much slower and more general in character. Many of the changes they control are long-term in their nature, such as the changes taking place during growth or in the development of the characteristic of sexual maturity.

Endocrine glands include the pituitary and hypothalamus, thyroid, parathyroid tissue, adrenocortical homologue (or inter-renal tissue), chromaffin tissue, and gonads. Among different endocrine glands present in teleosts, however, the pituitary gland is considered as one of the most vital organs so far the aquaculture production is taken into consideration. The pituitary varies considerably in structure from one group to another. The pituitary, in general, is a pea-sized structure that lies within the cradle-like space hollowed out in sphenoid bone of the cranial floor, called the *sella turcica*. It is attached to the hypothalamus – the hind part of the vertebrate forebrain by a stalk of nervous tissue. Part of the gland is composed of an up-pushing growth from the roof of the stomodaeum which unites and becomes incorporated with a down-pushing extension from the floor of the brain. The stomodaeal part is called as *adenhypophysis* and the innervated part is referred to as *neuro-hypophysis.*

The adenohypophysis surrounds the neurohypophysis. The former part in teleosts is divided in to three regions: (1) rostral pars distalis, (2) proximal pars distalis, and (3) pars intermedia which surrounds the pars nervosa. The pers nervosa is believed to produce oxytocin and vasopressin. However, the adenohypophysis and neuro- hypophysis are referred to as *anterior lobe* and *posterior lobe* of the pituitary gland, respectively. The anterior lobe is regarded as the main part of the pituitary gland as it contains six different hormones. Of them, growth hormone, follicle stimulating hormone, and leutinizing hormone are, significant as these hormones are involved in a wide range of physiological activities particularly growth and reproduction. In some bony fishes, thyroid occurs in a compact organ, but in many others thyroid follicles are dispersed in the region of the ventral aorta. It is suggested that thyroxine hormone or its homologue has a morphogenetic function. The islet tissue may occur outside the diffused pancreas in the region of pyloric caeca, gall bladder, and spleen. The pineal is regarded as a secretory function related to growth.

Immune System

The immune system is a collection of different types of cells and proteins that defends the fish body against foreign invaders. Similar to higher vertebrates, fish are susceptible to infections from bacteria, viruses, fungi, protozoa, and parasites. To fight these infections, mechanisms are required which recognize these foreign invaders of the body and deal with them. The immune system performs this task. The mechanism of immune system is controlled by the white blood cells or leucocytes which are present in the blood and in the lymphoid organs such as liver, spleen, and thymus. Leucocytes are of various types such as lymphocytes, macrophages/monocytes, and granulocytes. These cells are responsible for different defensive functions.

  1. *Non Specific Immune System :* In all vertebrate classes including fish, the first line of defence is the natural immunity. It rejects all foreign invaders, no matter what they look like. It is not specific and has no mechanism for recognizing an enemy previously encountered. It tends to be based on physical barriers such as health. The fish body concentrates several natural defences. Different secretions from fish body wash away microorganisms and contain enzymes to kill them.

If microorganisms manage to enter the fish body, they must be prevented from multiplying and causing harm. Certain types of white blood cells particularly granulocytes and macrophages ingest and destroy foreign invaders (such as bacteria, viruses etc.). Their reaction is essentially the same for any invasion, they are not able to differentiate between different disease-producing agents, and react in the same way on every encounter.

2. **Specific Immune System :** This immune system is the second line of fish defence. It is inevitable because some micro organisms manage to slip the first line of defence. However, the immune system can recognize all the antigens in different fish culture systems. It recognizes them using a set of receptors and proteins produced by cells – called *lymphocytes*. There are two types of lymphocytes generally recognized such as B-cells and T-cells.

Lymphocytes can distinguish between pathogens and furthermore can remember and react to individual pathogens which the fish has already encountered. This immunological 'memory' permits the immune system to ascend and effective defence thus giving immunity against the specific disease. This is the foundation stone of fish vaccination. By vaccinating farmed fish against a killed strain of the disease, the immune system is ready to react immediately the violent form is encountered. Vaccines are available against several fish bacterial diseases, and those against viruses (such as *Vibriosis* sp.) and parasites (such as *Icythyophthirius multifilis* – white spot disease and *Diplostomum spathecum* – eye fluke disease) have proved to be very successful.

(i) *B-cells or B-lymphocytes :* B-cells generate and liberate adequate quantities of antibodies which pass into the mucus and blood. The surface of each B-cell is covered with antibody molecules and when a pathogen is encountered, the appropriate B-cells bind to the surface of the pathogen. This recognition triggers generation of more of the same antibody molecules and thus the blood stream is charged with the specific antibody against the disease concerned. In contrast to mammals where five classes of antibody have been noted, fish only produce one type of antibody molecule. The foreign cell is lysed by a series of proteins in the blood serum which effect cell lysis in a sequential cascade manner. There may also be enhanced uptake and degradation by the non-specific leucocytes. Thus antibody itself does not destroy the pathogen directly but makes it susceptible to the rest of the fish's defense mechanisms.

(ii) *T-cells or T-lymphocytes :* B-cells often require some assistance to commence generating antibodies, even after they have encountered with an antigen. Unlike B-cells, T-cells do not make antibodies, but they have receptors in their membranes that recognize specific antigens. When an antigen is bound to the membrane of B-cell, the cell engulfs the foreign body and breaks it up into pieces. The B-cell displays the antigen pieces along with complex protein molecules. The T-cell then recognizes the antigen and helps release messenger proteins. These messenger proteins give signal to reproduce and produce more antibodies, and to appoint phagocytes to the area of infection and store them there until they complete the work of swallowing the pathogens. However, it is thought that T-cells have a variety of functions including the inspection of the body's own tissues for abnormal or potentially dangerous cells such as tumour cells or virus-infected cells.

Effect of Temperature and Hormone Levels on Immune System

The responsiveness of the immune system varies with temperature and hormone levels. In general, however, the higher the temperature within the fish's normal range of tolerance the faster the onset of immune response and the higher the magnitude. At low temperatures the immune response is delayed or completely ceased. The rate of multiplication of pathogens is also affected by environmental temperature so that at low temperatures when the immune response is inhibited, the bacterial growth rate is also inhibited. Several infectious diseases in fishes are predominant at certain times of the year and this may be caused by abilities of the pathogen and cells of the immune system to respond to fluctuate in temperature. Many diseases outburst in fishes occur in summer on a temperature rise when fungal and bacterial growth increases and there may be a lag before the immune system can counter the infection.

The immune system is also influenced by hormones which may vary seasonally. In fish farming, hormone levels are greatly modified by stress from handling and transfer of fish or by poor water quality. Stress results in reduced numbers of lymphocytes (lymphopenia) and monocytes (monocytopenia), and increased number of granulocytes (neutrophilia). Most possibly, the onset

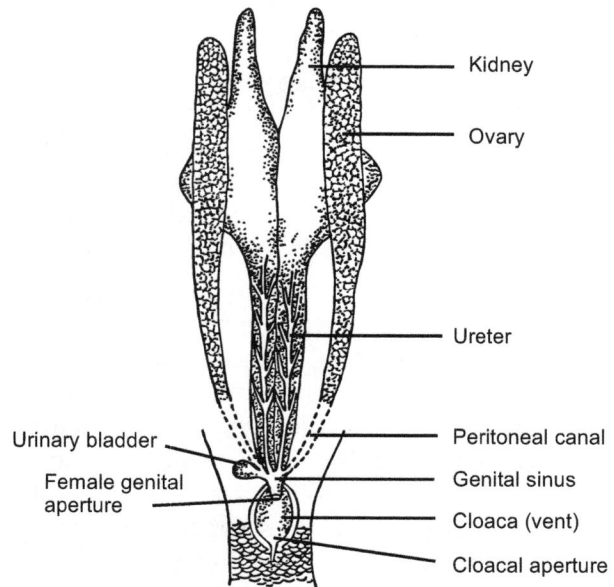

Fig. 1.19 : Male urino-genital system of a typical teleostean fish (*Labeo* sp.).

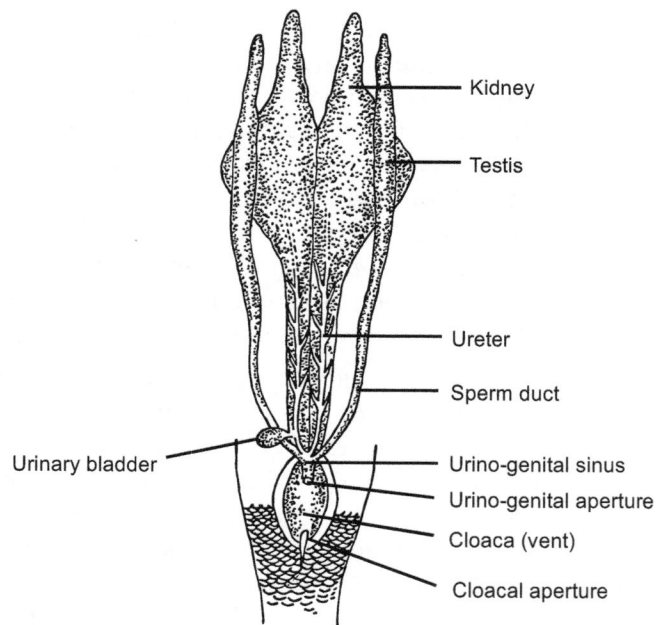

Fig. 1.20 : Female urino-genital system of a typical teleostean fish (*Labeo* sp.).

of disease occurs as a direct consquence of the reduction in the number of these cells. Secretion of cortical hormone as a result of stress is probably responsible for reduction in the number of circulating lymphocytes.

Reproductive System, Reproduction and Development*

Reproductive systems are the processes and organs that enable successful reproduction of a species. All organisms including fish reproduce, and in order to perform so, they require a specialized system that is best adapted to their particular reproductive needs.

Gonads are the main reproductive organs in sexual reproductive system along with some other associated structures (Figs. 1.19-1.24). The gonads of fishes are, however, generally elongated in structure and suspended by mesenteries from the dorsal aspect of the abdominal cavity. Their relationship to the kidneys and their associated ducts differ widely among different groups. In most of the bony fishes, the reproductive system is completely separated from the excretory unit in both sexes and there are separate openings to the exterior for the two systems, with the urinary pore posterior to the genital pore. In some forms, the sperm ducts connect with the urinary system in a urino-genital sinus located at the posterior end of the body cavity.

In general, teleost testes have a structure in which spermatogenesis

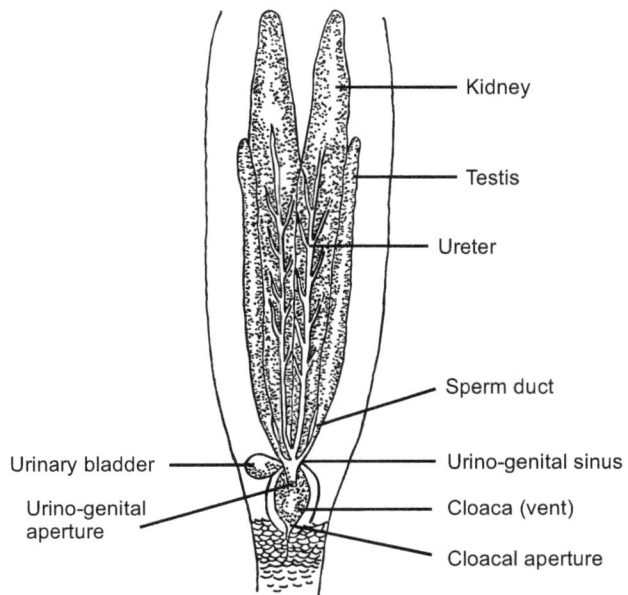

Fig. 1.21 : Male urino-genital system of *Oreochromis mossambicus*

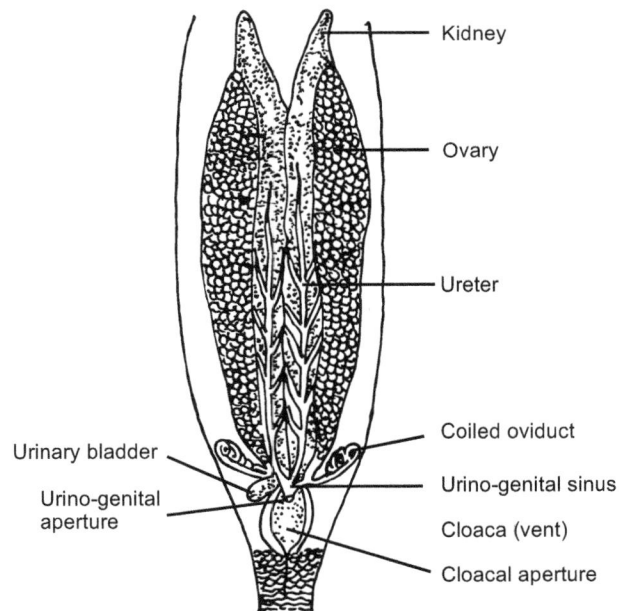

Fig. 1.22 : Female urino-genital system of *Oreochromis* sp.

---

* For further detail, see Hora (1969).

occurs along the length of a lobule, with a central lumen into which the sperms are shed. In females, however, oviducts are continuous with the covering of the ovaries so that the ova are not shed into the body cavity. This is called as *cystovarian condition.* On the other hand, most of the non-teleosts exhibit the *gymnovarian condition,* in which the ovaries open into the body cavity and the ova are conveyed through an open funnel to the oviduct. However, few teleosts have gymnovarian ovaries. These include Anguilidae, the loach, Salmonidae etc.

The testes, in general, are a pair of sacs that opens into the base of the urinary ducts. The ovaries are elongated in the carp and well separated in most fishes, but partial or complete fusion of the right and left organs can be seen in some bony fishes such as of the group percoids. In the large mouth bass, the ovaries join posteriorly to produce V-shaped structure. In the yellow perch, *Perca flavescens,* ovaries are so completely fused that they give the appearance of a single organ.

A pair of ovaries are situated in the abdominal cavity, one on either side of the abdomen. Each ovary consists of numerous ovarian follicles which are embedded in the connective tissue. Shortly before spawning, under the influence of follicle stimulating hormones, the ova are matured and are released into the body cavity. Considerable variations exist in the number of eggs generated in different species of fish (Table 1.3). The oviducts are, however, lacking. The anterior wall of the urino-genital sinus is pierced by a pair of genital pores through which the ova come out into view. In some bony fishes (such as Cichlids), the oviducts exist and it is thought that eggs are released through these ducts.

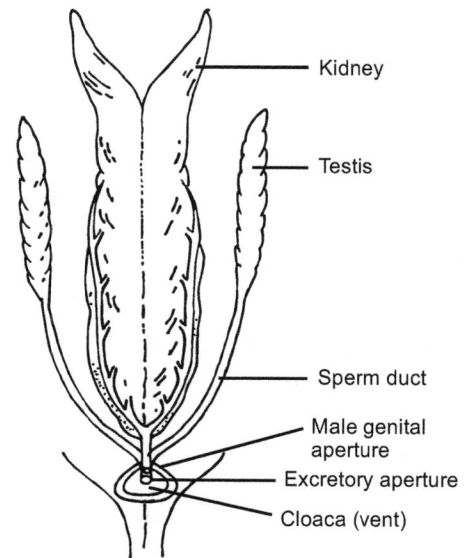

Fig. 1.23 : Male urino-genital system of *Ophiocephalus punctatus.*

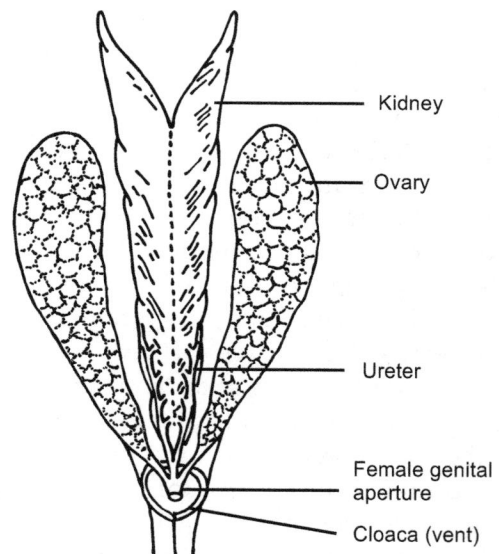

Fig. 1.24 : Female urino-genital system of *Ophiocephalus punctatus.*

Table 1.3 Number of Eggs Generaed in Some Species of Freshwater Fish

Species	Ranges of Eggs Produced per Female Fish	Body Weight (kg)
Heteropneustes fossilis	2,000-4,500	0.15
Anabus testudineus	4,000-35,000	0.15
Channa marulius	2,000-40,000	0.30
C. striatus	3,000-30,000	0.40
C. punctatus	3,000-25,000	0.25
Indian major carps (Rohu, Catla and Mrigal)	100,000-200,000	1-2
Cyprinus carpio	200,000-300,000	2-3
Ctenopharyngodon idella	300,000-600,000	3-5
Hypophthalmichthys molitrix	200,000-500,000	3.5
Tor putitora	5,500-6,500	1.0
Pangassius pangassius	200	0.001
Ictalurus punctatus	5,000-7,000	3.0

1. *Gonads and Sexual Maturation:* The two sexes of fish are not distinguishable in the early stages. The time of maturation of fish is highly variable depending on the season, environmental temperatures, hormonal influence, health condition, and the type of species considered for culture. However, as the maturity of fish progresses, the two sexes become distinguishable. The ovary develops a distinctly granular appearance as it swells with oocytes which are the precursors of the eggs. Attainment in sexual maturity of teleosts varies considerably between 3-months (as in Tilapia) and 2 years (as in most Cyprinids and Catfish).

Development of gonads is controlled by release of gonadotrophic hormones secreted from the pituitary gland. These hormones travel through the blood stream and act on the gonad. Release of gonadotrophic hormones is controlled by the pineal gland which is situated on the top of the brain and is highly sensitive to light. The pineal gland responds well to seasonal changes and is implicated in the synchronizing of spawning to a particular season of the year. Gonadotrophic hormones may also be released artificially by removing the pituitary gland from slaughtered fish, grinding them up and injecting the extracted crude hormones into other fish. This extract will induce fish to spawn – generally termed as *induced breeding* that has great practical significance in Asian fish culture.

The gonads contain hormone-secreting cells which generate the male (androgen) and female (oestrogen) sex steroids. These steroids help control gonad developnent and are also responsible for the development of sexual characters as maturation proceeds. Interstitial cells in the testis produce testosterone, androsterone, and progesterone as in mammals. The ovary produces oestradiol that regulates the growth, development, and nutrition of the female reproductive system.

2. Reproduction: The act of reproduction, by which a new life is brought into being, is associated with such activities as courtship and pairing of male and female individuals, and the actual reproductive act essentially consists of the fusion of two kinds of gametes such as sperm and ovum. In fishes, such pairing is the exception and in most of the bony fishes, the relation of the sexes at the breeding season are quite bewildered.

The time at which spawning occurs, varies in different species and naturally occurs at different seasons in various parts of the world. As far as our tropical fishes are concerned, fish breed in the first half of the year, the spawning season of the Indian major carps extends from April to July, that of common carp from March to July and November to January. Breeding season or spawing period of a particular species indicates that period during which some individuals may be found to contain mature ova or sperms and thus may last only for a few days or several weeks or even months. .

As the time for spawning approaches, the fish forms thick shoals in suitable habitats, and on some grounds they may be densely packed together at the breeding season. As a general rule of thumb, these shoals exhibit the females to be in greater numbers than the male partners.

During breeding season, both the paired testes and ovaries become greatly enlarged – the ovaries attain larger in size than the testes. Since oviducts are absent, the mature eggs are shed free into the coelom (body cavity) from where eggs emerge out through a pair of genital pore developed temporarily from the wall of the urino-genital sinus.

Most of the female fish (particularly carp) generate an enormous number of eggs at a time, which are fertilized and developed outside the body.Fecundity of Indian major and minor carps is fairly high. In general, fishes which do not exhibit parental care produce large number of eggs possibly to compensate for the loss of eggs and larvae as a result of unfavorable environmental conditions and predation 'The fecundity of the culvitated carps in ponds and in other aquatic ecosystems exhibits considerable variations. However, the fecundity of some carp and some other high-value fish is given in Table 1.4

In the majority of teleosts, the eggs float near the surface of water and are termed as *pelagic eggs*. On the other hand, eggs of some species such as *Labeo rohita, Catla catla, Ctenopharyngodon idella, Hypophthalmichthys molitrix* etc. sink to the bottom which are termed as *demersal*. In most of the fish species, eggs are non-adhesive and eggs swell up to about 10 times after coming in contact with water and drift downstream. Ova of teleost fishes are usually a few milimeters in diameter (Table 1.5) but in some forms it may be exceptionally large, especially in lung fishes. The sperms are produced in testes as a milky white substance called *milt* (or so called *spermatic fluid*) which is released over the eggs to fertilize them. Many fish leave their eggs since they have been fertilized. However, fertilization is external and the eggs of cichlids, trout, and some air-breathers are shed in small pits in the bottom. Indian major carps breed in flooded rivers or adjoining shallow and marginal areas where there is a good but slow current. Trout lay their eggs in the gravel beds where they themselves hatched. The eggs of trout and common carp are sticky and are attached to stones and aquatic plants, respectively.

Table 1.4 Ranges of Mature Eggs Generated in Different Species of Fish

Species	Range per g ovary weight	Range per g body weight
Catla catla	630-1,330	20-245
Labeo rohita	745-1,530	110-415
Cirrhinus mrigala	800-1,835	32-280
Hypophthalmichthys molitrix	320-1,770	45-320
Ctenopharyngodon idella	20-300	300-900
Esox reicherti (Amur pyke)	190-1,500	20-140
Oreoleuciscus pewzowi (Altai osman)	210-900	60-114
Coregonus chadery (Chadary whitefish)	210-790	30-400
Labeo calbasu	50-650	300-1,600
Cyprinus carpio	120-430	520-1,390

Source : Selected data from Chonder (1999)

Under optimum temperature conditions, Indian major carps and common carp are highly fecund and can be produced as eggs or young 4-5 times per year between March and September. This requires broodfish to mature before the start of the monsoon season and this can be achieved by selecting 3 year-old fish, stocking at low densities (1,000 kg/ha), and providing a well-balanced diet.

Table 1.5 : Diameter in Eggs in Different Species of Fish

Fish	Egg diameter (mm)
Catla	5.0-6.5
Mrigal	4.5-5.5
Rohu	4.0-5.0
Calbasu	3.5-4.5
Silver carp	4.2-5.0
Grass carp	4.5-5.5

The highly distinctive courtship and reproductive behaviors of cichlids (Tilapias) have been a consistent source of fascination of ethologists and fish biologists. Some species of tilapia begin to spawn at 3-months, when they attain the weight of only 30g, thereafter spawning as frequently as 4-6 times per year in nature and more in captivity. Spawning occurs in shallow waters, males establishing and defending a territory in which a shallow nest is excavated. 'There are differences among genera in terms of courtship and breeding behavior. By contrast with the polygamous (*Oreochromis* sp.) species, pair bonds formed among *Sarotherodon* and *Tilapia* species individuals may persist several matings and females assist in nest construction.

Egg numbers vary with species, age and nutritional status, and environmental conditions. Eggs of substrate-spawning (such as *Tilapia*) are laid in rows in the nest, where they are tended and guarded until they hatch. The males of some *Sarotherodon* species and both parents in others recover and incubate the eggs in their mouths, while among *Oreochromis* species the female picks up the eggs in her mouth and recedes to an area of safety to incubate them. Egg hatching takes up to 20 days, depending upon species and environmental conditions, and a further 5-7 days generally pass before the fry are actively swimming. Fry are released from the buccal cavity for increasingly long periods of time until they acquire independence, generally after a further 8-10 days. Fry of substrate spawners abdicate from the nest after 2-3 weeks.

Clarid catfish produces 20,000 to 1 million eggs depending on body size (60-70,000 number of eggs per kg body weight). The capacity to spawn clarids artificially under culture conditions is the key to the development of their culture in terms of the potential for stock improvement. However, hybridization of *Clarias gariepinus* should be considered under exceptional situations. The species has all the qualities of an aggressive and successful invader which adapts to a new habitat such as high fecundity, wide habitat preferences and environmental tolerance, rapid development, and fast growth. As a corollary, introducing *C. gariepinus* outside catchment areas can devastate many indigenous fish populations.

Several varieties of salmon ascend the rivers in the autumn and their reproductive act seems to be more exhausting, for after spawning the male and female drift helplessly downstream tail foremost and neither male nor female succeeds in returning to the sea. As soon as the reproductive act is accomplished, they die, and in some rivers, the dead spent fish can be observed for kilometres and sometimes piled to a height of several feet.

Two distinct types of eels may be recognized such as Silver eel (those in their special breeding livery) and Yellow eel (representing individuals in their ordinary feeding and growing coloration). Towards the autumn, a certain number of yellow eel assume their breeding livery and prepare to undertake the journey to the spawning grounds. Both male and femle fish cease to feed, the eyes become enlarged, the lips become thinner, the snout sharper, the pectoral fins become more pointed and blackish in colour, and the yellow or greenish coloration is replaced by a metallic silvery on the sides with a deep blackish back. These characters become more accentuated as the time for breeding approaches. They make their way down the rivers in the late summer and autumn. They, however, migrate across the Atlantic ocean to their breeding ground, which lies in the Western Atlantic and south of Bermuda. This stupendous journey, which may be as much as 5-6 thousand kilometers is undertaken because conditions suitable to the procreation of the species are to be noted. Having completed the release and fertizization of ova, the parents die, for it is beyond possible for them to return to their native place.

The eggs float for a time, hatched out and the young feed and grow at the surface of the sea, approaching the coasts of Western Europe when they attain about 4 inches long. Then they turn into elvers or glass eels. Millions of elvers commence the ascent of the rivers when about 3 years of age. In rivers, they feed and grow for several years until the time arrives for them to set off on their own breeding migration.

The members of the large and varied carp (Family: Cyprinidae) produce ova in large numbers, which adhere to weeds, stones, and other objects. When spawning is over, no further care of the offspring is taken by the parents. Under natural condition, the relationships between the sexes in the breeding season may be termed as *polyandrous*. In certain species, pairing occurs and in majority of the cases, each female is attended by 2, 3, or even more males. Males take part in the fertilization of the eggs when released. During breeding season, fishes develop hard, wart-like, nuptial tubercles over the head and skin. These excrescences (abnormal or morbid outgrowth) are used in the battles between rival males, in nest building, or assisting to hold the female and to facilitate the extrusion of the eggs by pressure on her body. These excrescences, however, disappear as soon as spawning has been completed.

All the carps and their allies breed in monsoon spring or early summer and even the most sluggish forms become intensely active under the stress of excitement and sporting at the surface of water or near the shore regions. As the time for spawning approaches, they congregate into shoals, and move into quiet, weedy/gravel shallows near the bank of the rivers/ponds/lakes/reservoirs. Whilst carps engaged in spawning they are forgetful to all danger, falling an easy prey to enemies of all kinds. The attendant males move above or round the female, and pressing her abdomen with their snouts to facilitate the release of the ova, which are produced at the rate varying between 200 and 800 in numbes at a time, each batch being timely fertilized by the males.

(i) *Fluctuations in Fish Fecundity:* The absolute fecundity (total number of mature eggs in the ovary of an individual fish which is related to the weight, length and age of the fish) in a single population may undergo considerable fluctuations in relation to the supply of adequate nourishment. A thin population having an adequate food supply generally exhibits high fecundity rates. In general, the fecundity significantly increased when intensive fishing is undertaken. The fecundity differences can be observed in fishes of the same length and weight if reared under different conditions. substantial differences are observed in the fecundity between closely related species.

(ii) *Factors Influencing Fecundity :* Biological factors such as stocking density, age, size at first maturity, gonado-somatic index, relative condition factor etc. exert a dominant influence on fish fecundity. Among all these factors, however, the gonado-somatic index and relative condition factor are very important that have great practical significance. These two aspects have been described at length in section 1.9.

* *Influence of Nutritional Requirements :* It is not difficult to differentiate between the effects of fecundity and nutritional requirements. A well-managed fish culture pond supplemented with balanced feed generally exhibit high fecundity rate than those in unmanaged ones. For example, if grass carps in addition to their normal food consume rice sprouts, peanut and soyabean dredy, generate about 58,000 eggs per kg body weight compared to the fish fed with rice bran (about 35,000 eggs

per kg body weight. Apparently, fish diet with different nutrient inputs and their better utilization will definitely encourage the rate of fecundity and thus a high production.

* *Influence of Parasites and Diseases :* Beside the broader aspects on biological factors and nutritional requirements just discussed, numerous parasites and diseases are involved in reduction of the fecundity rate of fish. Parasitic infections and diseases drastically affect the fecundity rate since the progress of growth and maturation is seriously affected and interaction always exist between fecundity and parasites/diseases. However, fish disease results in lower fecundities.

* *Influence of Environmental Factors :* Environmental factors especially temperature, dissolved oxygen, pH, and alkalinities of water which could affect the food consumption and food conversion efficiency and, at the same time, keep the fish under stress are found to be less fecund. Also, toxicants disrupt fecundity potential of farmed and wild fish, thereby decreasing the rate of fish production.

Environment-friendly and modern farming practices help maintain high fecundity rate compared to that of traditional farming. These practices ascertain to release mature eggs in high numbers that encourage fish farmers to generate robust fry/fingerlings for propagation.

The influence of these factors remind us on the importance of fecundity rate of fish in the maintenance of ecosystem productivity. Fishes that are kept in fertile condition by application of fertilizers, manures, lime, and balanced feed and by the choice of high-yielding fish varieties are likely to have more fecund.

3. *Courtship, Pairing and Parental Care :* A large number of fish species exhibit a definite courtship or at least a pairing of female and male during the breeding season, often exhibit marked differences in the two sexes. These may be of two types. Firstly, structural peculiarities are directly associated with the fertilization of the ova, usually taking the form of special male organs – termed as *claspers* (as in Shark) for introducing the milt into the body of the female and secondly, structural differences, peculiarities of colour and other features associated with the reproduction, having no relationship with sexual conjugation, but concerned with courtship and display, or with the battles that occurs between rival males.

The most important features between the sexes – the secondary sexual characters – are very common in many aquaculture species. These features have no correlation with actual union between male and female, may be present during the entire adult life of the male, but are frequently developed only as the spawning season approaches and ova disappeared as soon as this is over. In the great majority of bony fishes, the females are larger than the males and in some cyprinids, it is as much as six times as large as her mate.

Body coloration between sexes is also the most common feature. The males exhibit a brighter livery than their mates as in the case of Cyprinidonts, Cichlids, Damsel-fishes etc. Green, red, blue, back, and silvery white pigments are characteristic of the male, whereas the females generally exhibit dull or variously mottled hues. Many male Cyprinids become much brighter during the spawning season especially in the abdominal regions of the body and these colors may become very much intensified during the actual courtship. In the three-spined stickleback (*Gasterasteus aculeatus*) both sexes change their colors in the breeding seasons, the dark greenish color of the back, whilst the lower parts change from a silvery white to pale yellowish in the female and a red in the male.

In many Cichlids and Cyprinidonts, some of the rays of the dorsal and anal fins form fine streamers. Many male Cichlids develop a huge fleshy hump on the forehead which is gradually reabsorbed after spawning in some species and retained throughout life in others.

The males of many of the freshwater Gobies (Family : Gobiidae) are actively involved in hectic fights, rushing at each other and biting viciously, the victor then spreading his fins and showing his colors to the female.

In those fishes in which courtship and pairing occurs at the breeding season, the number of eggs produced by a single female is small or moderate and these are cared for to a greater or less extent by one or both of the parents especially the male. This parental care may take the form of consturcting nest for the reception of the fertilized eggs, varying from a simple hollow scooped out in the gravelly bed of a stream to a beautiful and elaborate structure or of some other precautions to ascertain the safety of the eggs/offspring until they are old enough to take care of themselves.

Fish as a group pay little parental care to their eggs and young. Most of them are content to ensure fertilization of their eggs but bestow little attention on them. This lack of parental behavior is correlated with the produciton of eggs in large numbers. There are, however, few notable exceptions that follow in which the eggs and young are guarded with utmost care.

(i) The nests of *Arius* (catfish) consist of circular excavations, about 20 inches in diameter scooped out in the bed of the river. Although the fertilized eggs are covered with a layer of large stones, the parents take no further interest in the fate of their offspring. The fertilized eggs are also carried in the mouth cavity by males.

(ii) The freshwater sunfishes (Family : Centrachidae) scoop out a shallow basin–like nest, from which all pebbles are removed, leaving a layer of gravel of fine sand to which the fertilized eggs are attached.

(iii) In many of the cichlids (Family : Cichlidae), the male constructs a shallow basin-like nest on the pond bottom at a depth varying between 30 and 150 cm. Both parents assist in the task of guarding the eggs. Members of the genus *Oreochromis* and *Sarotherodon* are substrate spawners. They remain in the spawning site as long as the embryos are developing and many of them tend the young fish, called *fry*, until they are independent. Many of the female cichlids protect their eggs by carrying them in their mouths, thus ensuring their safety. Later, the fry swim about in the water always keep near her head and when they fall in danger, they instantaneously return to their mouths. When fry are more developed, they move in small shoal accompained by the parents.

(iv) During the breeding season, the skin of the lower surface of the Brazilian freshwater female catfish, *Platystacus,* becomes sowllen and tender. When eggs are released and fertilized, she lies on them, presses them into this swollen tissue and all eggs become attached to the skin thus remaining fixed until hatched. Immediately after hatching, the skin shrinks to its original size.

(v) One of the freshwater mud-fishes, *Protopterus,* scoops out a hole in the mud of a swamp, surrounded by long aquatic weeds and grasses, the male being responsible for the construction of the home and for subsequent care of the eggs. The male fish not only defends encroachments of hunger fishes, but helps aerate the surrounding water by lashing vigorously with his tail. He keeps the eggs well supplied with oxygen essential for their development.

(vi) The three-spined stickleback, *Gasterosteus aculeatus,* is a freshwater fish of North American ponds and lakes. Each male fish takes up a small area of territory such as one among the stems of aquatic plants where the water flows with degree of regularity, digging out a small hollow in the bottom of the pond by biting movement and then roofing it with pieces of roots and stalks of aquatic plants or vegetable rubbish and cement them together by means of threads of sticky substance secreted by his kidneys. The nest is a hollow, somewhat rounded, barrel-shaped structure. The inside is made as smooth as possible by a kind of plastering system. The body of the male secretes mucus on to the inner sides on the nest and it hardens like varnish. It takes several days to complete the nest and then the male goes in search of a mate .and when selected, proceeds through an elaborate process of courtship.

When a stickleback female enters the territory of a male, he identifies her and swims round her in a special zigzag dance, trying to persuade her to enter through the circular aperture on the side. The female responds to male's dance and he turns and leads her to the nest, pointing out the entrance with the head. The female enters the nest, proceeds to deposit 4 or 8 eggs within, finally boring through the wall of the nest on the side opposite to the entrance and swimming away. While female is in the nest, the male swims round and round in great excitement, thrusting and rubbing his snout against her body. He then enters, deposits his milt on the eggs and absconds through the back door. Next day, he seeks out another female and repeats the entire process with her, with one after another, until adequate eggs have been deposited.

The male keeps watch over the nest for nearly a month and cares the young when they hatch. The male guards over the entrance and defands his charge with vigour against all invaders that attempts to approach the nest. From time to time, he repairs any damage of the nest. The two apertures of the nest permit a constant current of cool water to bathe the eggs within the nest, but this process may be carried out by keeping himself at the entrance and vibrating his pectoral fins. When the fry are able to swim about in the water actively, he gradually relaxes his activities, although still keeping a heedful eye on them until they swim independently.

(vii) There are over 30 species of murrels – so called *snakeheads* distributed in tropical Asia, including Northern China and in Africa. All species of murrels, however, exhibit parental care and spawn in nests built in shallow marginal areas with pieces of aquatic vegetation on similar material. Spawning lasts for 15-45 minutes and the eggs laid by the female in the nest are fertilized by the sperms shed by the male. The amber or golden-yellow coloured fertilized eggs float in the centre of the nest in a thin fIlm. The eggs hatch out in 20-60 hrs in temperatures ranging from 16 to 34°C, depending on the species. A four-day old free-swimming hatchling measures about 3.5 mm. Both males and females take part in carying for the newly hatched young for about 15-20 days.

Clarias makes their nests in the shape of holes from 20 to 30 cm in diameter and of variable depths. The holes are generally from 20-30 cm below the surface of the water, on the bottom of a rice swamp or in a bank with weeds and plants. The eggs are laid in the nest and stick to the roots of the plants or to the bottom of the nest. After spawning, the male watches over the eggs. They hatch out after 20 hrs at temperatures from 25 to 32°C. There can be as many as 2,000 to 5,000 fry to a nest.

The fish *Trichogaster pectoralis* spawns in freshwater ponds. Over a period of 1 or 2 days, the males prepare the nests which float and are made up of a mass of bubbles. The females lay their eggs under the nests where they float. Hatching takes place 2 or 3 days later. Each nest may contain more than 4,000 fry. The eggs and fry are protected by the male.

4. *Embryology :* In the preceding section, while discussing the matters involved in reproduction and reproductive system in fishes, we had stated that embryology and development/life history strategy will be briefly discussed separately in an exclusive section in view of the discussion which is important for fish culturists and for survival as well as growth of fish. This section provides a brief discussion on embryology and subsequent development as follows.

In normal fertilization, only one sperm is involved in fertilization of a mature egg. The sperm enters the egg through a tiny passage in the chorion of the egg known as *micropyle*. The great majority of fishes release sperm into the water in the vicinity of the eggs. Since the sperm of freshwater fishes live only a short time following release into the water, it is necessary to bathe the eggs in a heavy concentration of sperm. Fish farmers are interested in methods of long-term storage, and techniques are constantly being improved. Fish sperms have been successfully stored for several days or months at temperatures near freezing in suitable conditions. Sperm frozen in suitable media at the temperature of liquid nitrogen have achieved some success in fertilization of fish eggs.

The embryology and development of fertilized fish eggs follow more or less the same pattern as in other vertebrates. However, fish eggs in general contain an appreciable amount of yolk which in concentrated at one hemisphere called the *telolecithal egg*, but there are wide differences among fishes regarding the actual amount of yolk present in eggs. Some fishes such as sturgeons and lungfishes have very small amount of yolk material and in this case, holoblastic cleavage of fertilized egg is the usual rule. Some fishes have eggs with the cytoplasm thinly distributed around the large yolk, while others have cytoplasm concentrated at the animal pole. The cytoplasm forms a polar cap at the site of the egg nucleus following fertilization, and this begins to divide, forming the embryo on the surface of the yolk. This type of cleavage is termed as *meroblastic cleavage*.

Certain teleosts such as *Neothobranchius* and *Aphyosemion* deposit fertilized eggs in the bottom of decaying ponds. These eggs do not hatch for several months and complete development does not occur until the ponds are filled with water. The development of germinal disc, blastoderm, archenteron (rudiment of alimentary canal), and three germ layers (ectoderm, endoderm, and mesoderm) is assumed to be similar to that of any vertebrate which has been described in any text book of embryology.

The incubation period of fertilized eggs is governed by temperature. Within the optimum range for normal development, the period shortens as temperature increases. Trout and salmon eggs, for example, will hatch in about 50 days at 10° C, but incubation period requires about 6 months at 2° C. The eggs of the common carp incubate at temperatures ranging from 15 to 30°C, and hatching occurs in a week at lower temperature and few hours less at the higher temperature. The eggs of the Indian major carps hatch out in about 14-18hrs after fertilization at temperature ranging from 25 to 31° C. The incubation period of different species of fish is, however, variable.

5. ***Early Life History**** : Fertilized eggs are hatched out, the freshly hatched young fish is tiny, unformed creatures destined to undergo considerable additional development. The young fish is, however, termed as *hatchling* may still sustain the yolk sac and obtain food from it for some time. Generally, early development is divided among the egg, larva, and juvenile developmental stages. Following hatching, there is a stage – the *yolk sac larva* or may called as *prolarva* which ends with the absorption of the yolk sac. Immediately after the absorption of yolk sac, another developmental stage, called the *postlarva*, is developed but is still unlike the juvenile stage. The upright-swimming larva of eel is a good example of post-larva. If yolk-sac bearing larvae are transformed into juveniles, these larvae are called *alevins*. However, absorption of yolk sac results in the development of the young ones – called *fry* (2-3 mm in length) which begin to consume natural food organisms particularly zooplankton. When the fry becomes 5-10 mm long, it is called *fingerling*.

Aquatic plants provide shelter to many predators such as water bugs, water beetles, dragon- fly nymphs, snakes, tadpoles etc. They attack hatchlings and fry and prey on them.

---

* For further study, see Kamler (1992) and Hora and Randall (1988a b).

Ponds with an overgrowth of aquatic plants would obviously give a low survival of fish larvae. Filamentous algae and algal blooms are also harmful to larvae. Aquatic pollution may seriously affect the survival and growth of fish larvae and early fingerlings and therefore, appropriate control measures must be adopted. However, pollution may be caused by domestic, agricultural and industrial wastes.

Oxygen depletion, increased turbidity and other water quality changes may also affect their survival and cause large-scale mortality of fish larvae.

(i) Ecology of Early Life History Stages of Fish : The ecology of early life history stages of fish is complex which depends on many factors such as positive and negative phototactic movements, consumption of natural food organisms, density of prey, vulnerability to predation etc. Larvae of Walleye*, (*Theragra chalcogramma* and *Stizostedion vitreum*) and Pollock** exhibit positive phototaxis by moving horizontally from a darkened area to an area of low light intensity but negative phototaxis under high light intensity. They exhibit a daily periodicity in response to light, moving toward the surface in the dark and toward the bottom in the light.

Food for fish larvae at the initial stage of feeding must be small. Initial food of larvae may include zooplankton (particularly copepods, ciliates, cladocerans, rotifers etc.) and phytoplankton. Zooplankton make up the most important segment of the food of larvae.

Density of natural food organisms (prey) is of great importance to the survival of fish larvae at the onset of feeding. Higher the density of natural food organisms, the greater is survival of fish larvae. In ecosystems where the distribution of plankton is uniform, and fish larvae that are found in dense patches of plankton are generally well fed.

Many species of fish guard demersal eggs until hatching and some guard demersal young. This behavior reduces the predation to a large extent. The alevins of trout and salmon, for example, remain buried in the gravel nests until the yolk is used up and they are able to swim and seek food.Many predators are not able to consume larger eggs and larvae. Therefore, predators will never consume rapidly growing larvae in natural and controlled ecosystems. Development of sensory organs help to detect the presence of predators in ecosystems. Escape from predators is facilitated by development of the locomotory organs which help larvae to swim rapidly and to travel long distances.

1.7 Hybridization

In short, hybridization can be defined as a process of heterospecific insemination. Heterospecific insemination, on the other hand refers, to the process of crossing two different species. In nature, however, fish hybridization has been recognized for many years and fish hybrids have long been produced in hatcheries and laboratories. At the same time, natural hybrids can occur by two or more species spawnning in proximity.

---

* A north american pike perch with large, opaque silvery eyes.

** An edible greenish-brown fish of the cool family with a protrading lower jaw.

Hybridization involving pairing of different species is known, but there are reproductive barriers such as behavioral, temporal, or physical – that will preserve the integrity of species.

Since 7 species of carp account for almost 90 per cent of cultured cyprinid production, many hybrids of different freshwater carp such as common carp × mrigal, *Catla × Labeo fimbriatus, Catla catla × L. rohita, L. rohita × Catla catla, L. rohita × L. calbasu Catla catla × L. calbasu, Catla catla × Cirrhinus mrigala* and others have been in nature and it has also been paved the way for producing them through induced breeding and the results are highly encouraging. These hybrids are, however, better than one parent in some traits. Better quality of hybrids for different economic traits has great practical significance because hybrids produced have considerable benefit in respect of better adaptability to adverse climatic conditions and disease resistance. As with common carp, little selective breeding of Indian and Chinese carps have taken place, there have long been a marked dependence on wild-caught broodstock and consequently, evidence of inbreeding of hatchery stock is apparent.

Although the majority of common carp is not currently the result of selective breeding (to be discussed later on), varietal cross-breeding is used commercially to produce carp with heterosis (hybrid vigor) for growth and cold tolerance. Most interest in Chinese and Indian major carps which lack distinct races or varieties, focuses on inter-specific and inter-generic hybridization (see panel 1.4). several promising hybrids have been produced in Israel and China and the success of hybridization has been greatest among species with similar karyotypes such as grass carp × silver carp etc. There are, however, risks of infiltration of pure species broodstock by hybrids: grass carp × bighead carp hybrids exhibit incomplete triploidy, whilst silver carp × bighead carp hybrids are fertile.

Interspecific hybrids of the family Claridae have been produced between male African catfish and female Asian species: *Clarias gariepinus × C. macrocephalus, C. gariepinus × C. batrachus* and *C. gariepinus × C. fucus.* In each case, the maternal features persist in terms of appearance and taste and the improved growth rate is inherited from the male African catfish.

1.8 Selective Breeding*

Fish culturists are constantly seeking to maintain superior quality in the product or to improve the quality. Maintenance of genetic variation in captive wild fish requires alertness to genetic principles and requires use of brood fish in sufficient numbers and reduction of certain types of inbreeding. This helps practice to select breeders.

Selection means allowing the best stock to breed and produce the offspring. The principal effect of selection is to alter the array of gene frequencies. Selective breeding with fishes is accomplished in situations that allow the retention of a captive broodstock from which selected mating can be made. Objectives of selective breeding in domesticated stocks are varied such as rapid growth, high survival rate, hardiness and resistance to diseases, body forms, delayed or accelerated sexual maturity, excellent quality of flesh,

---

* For further study, see Kirpichnikov (1981) and Gjerde (1993), Gall (1983) and Kincaid (1983).

and many other attributes that could make fish culture easier, increase yield, and increase acceptability of the product in the market. Fish are selected for resistance to adverse environmental factors encountered in artificial surroundings including diseases extremes of pH, dissolved oxygen, temperature and other physico-chemical factors stress due to crowding and build-up of toxic metabolites in culture systems.

Selection can be made by scrutinizing prospective breeders for acceptable phynotypic attributes, and selecting those that appear to have the best balance of the characteristics demanded. And selective breeding requires that stocks are subjected to the same environmental factors in order that differences among individuals will have less chance of being environmentally induced.

It should be added as a note that in contrast to agriculture where 90 per cent of the food products come from genetically improved varieties, only 5 to 10 per cent of the yield is derived from genetically improved strains of finfish and shellfish. In this regard, the potential for genetic improvement of fishes must be realized and these weuld be of immance important for future, programs to increase fish production. A number of selective breeding programs have been taken up following application of the state-of-the-art technology coupled with a considerable economic potential in dissemination of improved fish seed to farming sectors. These programs would, however, further commercialize the genetic improvement work at large.

1.9 Relationship With The Environment

The most important aspects influencing the biology of fishes are those that has distinct relationship with different environmental factors. Environmental factors definitely affect fish biology to a greater or less degree and, at the same time, fish production potential. Dissolved oxygen, temperature, salinity, algal blooms, ammonia etc. are some of the factors that help estimate the relationship between different aspects of fish biology particularly growth, condition factor, gonado-somatic index and fecundity and the ecosystem in general. In most practical sense, this section is of greatest importance to fish farmers in dealing with fish production. Although frequent discussion has been made regarding environmental aspects and fish production in two volumes of the book 'Freshwater Fish Culture', attention must be given to the relationship between environmental factors and some important aspects of fish biology that follow and hence will now receive our attention.

Condition Factor

Simply defined, the condition factor of a fish is expressed by the ratio of length (in cm) to weight (in g) of the fish at any given movement. The condition of fish is expressed by the following formula :

$$K = \frac{W \times 100}{l^3}$$

Where K = Coefficient of conditions
W = Weight of fish (g)
1 = Length of fish (cm)

As the mane suggests, K is a good measure of the condition of the fish. K values less than 1 indicate poor condition. The fish is long and thin and has been starved or under-fed. K values greater than 1 indicate that fish is well-fed and in good condition. Fish approaching sexual maturation and swelling with eggs have a very high condition factor, (K value 1.8 or 2.5), and this falls to well below 1 after spawning. Tilapias and Indian major carps have a condition of about 1.0 and 1.3 respectively, and this can rise to 1.5 and 1.8 respectively in well-fed rearing and stocking ponds. Polluted ecosystems have, however, very low condition factor and the values vary between 0.5 and 0.8. The condition factor is very useful to compare different batches of fish and monitoring their progress. The conventions suggested here, although wide-spread, are by no means universal.

Growth

The weight of a fish in a well-managed farmed pond during the course of its life-span takes the form of a 'sigmoid' curve (Fig. 1.25) As the day passes, the weight of fish increase accelerates until the fish attains maturity and the growth rate decreases reaching a maximum weight beyond which little growth occurs. In fish culture, maximum returns in terms of growth take place during the early part of the growth curve. It should be kept in mind that the growth of fish is never the same from one day to the next and it requires a simple number which denotes growth rate of the fish. In fisheries, many mathematical formulae have been proposed to exhibit fish growth, whilst for commercial farmed fish it is best explained, in terms of specific growth rate. Simply defined, the specific growth rate (SGR) is the percentage of daily increase in weight and is best described by the following exponential equation:

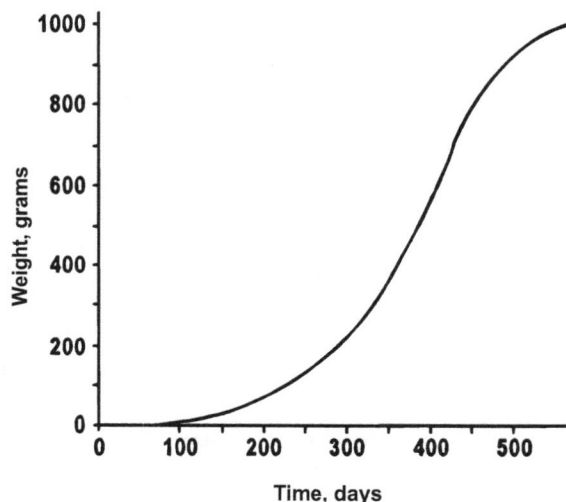

Fig. 1.25 : A typical fish growth curve. The increase in weight accelerates in the early part of life and then slows down to show the sigmoid curve.

$$Wt = W_O e^{Gt} / 100 \qquad \qquad \text{...(1)}$$

Where   Wt =   Weight of the fish after t days

Wo =   Initial weight of the fish

G =   Daily SGR as per cent per day

t =   Number of days of growth

1 =   Exponential constant = 2.718282

Hundred converts from percentage specific growth to actual fractional growth per day. This equation can be used to predict fish weight. An example given below will clear the matter concerned. Suppose if a fish grows for 84 days at a SGR of 1.5 per cent per day, what would be the weight of a 10g fish?

$$W = 10 \times 2.718282^{(1.5 \times 84)} \times 100$$

$$W = 10 \times 2.718282^{1.26}$$

$$W = 10 \times 3.525$$

$$W = 35.25g \text{ weight of fish after 84 days}$$

Whilst the SGR is defined as percentage weight gain per day, it is beyond possible to weight fish every day. Therefore, a random sample of more fish is weighed at an interval of 15 days. The SGR is then given by the following formula.

$$G = \frac{In\, Wt - In\, W_O}{t} \times 100 \qquad \qquad ...(2)$$

Example :

Over a period of 15 days, a fish increases in weight from 15 to 22g. What is the SGR?

$$G = \frac{In\, 22 - In\, 15}{15} \times 100$$

$$G = \frac{3.0910 - 2.7081}{15} \times 100 = \frac{0.3829}{15} \times 100$$

$$G = 0.01824 \times 100 = 1.82 \text{ per cent per day (SGR)}$$

This figure can then put into the equation (1) to predict future growth assuming the same conditions prevail.

The SGR generally depends on several factors such as fluctuation of temperature, food intake, degree of toxic effects on the environment etc. and these factors should be kept in mind. Since production of quality fish depends considerably according to the growth and feeding regime, adroit husbandry brings fish to marketable size at a steady rate or at seasons demanded by the market.

Gonado-Somatic Index (GSI)

Similar to the condition factor, the GSI of a fish is expressed by the ratio of length (cm) to weight (g) of the ovary or testis of the fish. The GSI of fish is expressed by the following formula.

$$GSI = \frac{W \times 100}{l^3}$$

Where W = Weight of gonad (in g) and l = Length of fish (in cm)

The GSI value less than 1 indicates poor condition of the gonad. GSI values higher than l indicates well-developed gonad and the fish has attained reproductive capacity and, at the same time, is ready to release mature eggs during breeding season. As the breeding season approaches, gonads tend to mature and when environmental situations permit, fish begin to release eggs in water.

1.10 Conclusion

General organizations and biology of commercially important species of fish have profound significance not only for their production but for marketing as well. Understanding the fish biology is most critical in polluted areas and areas of human interference and the management as well as sound knowledge of such ecosystems offers the most effective means of studying the biology of fishes. This study has, however, become increasingly popular in many fish producing countries of the world and at the same time is undoubtedly the most significant from both theoretical and practical points of view that must be kept in memory for high aquatic production.

Along with the progress in the study of fish growth and biology in the culture conditions, the understanding of the species has definitely increased. The biology of different freshwater farmed fish is now better understood than the biology of any other fish. This has stemmed from research studies inevitable for fish culture. Despite the success of freshwater carp, catfish, and murrel culture, the domestication of these species is a very recent issue and their successful culture should be based on a sound rise in value of the biological integrity implicated.

References

Bond, C. E. 1996. Biology of Fishes. Saunders College Publishing, New York.

Chonder, S. L. 1999. The Biology of Finfishes and Shellfishes. SCSC Publishers, Howrah, West Bengal, India, 514 p.

Gall, G. 1983. Genetics of fish: A summary of conclusions. Aquaculture 33 : 383-394.

Gerking, S. D. 1994. Feeding ecology of fishes. Academic Press, San Diego.

Gjerda, B. 1993. Breeding and selection, p. 187-208. In : Salmon Aquaculture. K. Heu, R.L. Monohan, and F. Utter (eds.). Fishing News Books, Blackwell, Oxford.

Goodrich, E. S. 1909. Vertebrata Craniata: Cyclostomes and Fishes. Part IX. A treatise on Zoology (ed. Sir Ray Lankester). Black, London.

Hyatt, K. D. 1979. Feeding strategy, p. 71-119. In: Fish Physiology, Vol. VIII. Bioenergetics and Growth, W.S. Hoar, D.J. Randall, and J.R. Bett (eds), Academic Press, New York.

Kamler, E. 1992. Early life history of fish: An energetics approach. Chapman and Hall, London.

Kincaid, H. L. 1983. Inbreeding in fish populations used for aquaculture. Aquaculture 33 : 215-227.

Kirpichnikov, V.S. 1981. Genetic basis of fish selection. SpringerVerlag, New York, Heidelberg.

Munshi, J.S.D. 1992. Air-Breathing Fishes: Their Structure, Function and Life History. Oxford and IBH Co, New Delhi, India.

Thomas, P.C., S.C. Rath and K.D. Mahapatra. 2003. Breeding and Seed production of Finfish and Shellfish. Daya Publishing House, Delhi, India, 402 p.

Questions

1. Define fish biology. Why the study of fish biology is important for fish culturists?

2. Discuss how environmental factors are related to fish biology .

3. What are the most important morphological features of fishes? discuss.

4. Why morphological and anatomical studies of fish are important in fish culture?

5. Briefly discuss how the swimming movement of fish is accomplished?

6. Briefly discuss the mechanism how movement of fish is performed.

7. What is respiration? With diagram discuss the structure of a teleost gill and explain the mechanism of respiration in fish.

8. What is circulation? Which structures constitute the circulatory system in fishes?

9. Discuss the structure of the alimentary canal of a teleost fish. What are the structural differences of alimentary canal between carnivorous and herbivorous fishes?

10. Briefly discuss the feeding mechanism and digestion of teleost fishes.

11. What do you mean by extra-branchial organs? Why it is so important in certain fish species? Name a few species of fish where these structures are found.

12. With diagram discuss the structure of extra-branchial organs in some air-breathing fishes.

13. Illustrate the behavior of air-breathing fishes observed in a culture system.

14. Define osmoregulation. Give an account of the mechanism of osmoregulation in a freshwater teleost fish.

15. What is immune system? Why this system is essential for fish life? Distinguish between non-specific and specific immune systems.

16. Discuss the changes that take place in fish at the onset of breeding season. Why fish lay huge quantities of eggs at the time of reproduction?

17. Discuss the structure of reproductive system in a teleost fish. Mention the importance attached to this system in the context of fish culture.

18. How does a freshwater fish controls the water and salt concentration within its body.

19. Discuss how fishes protect and care for their young in various ways until they are old enough to move about efficiently

20. Trace the development and life-history strategy of a teleost fish.

21. Discuss why hybridization and selective breeding are inevitable in commercially important species of fish.

22. Write notes on the following:

    (a) Pyloric caeca (b) scales, (c) swim bladder, (d) white muscle, (e) branchiostegal membrane, (f) labyrinthine organ, (g) pituitary gland, (h) beta cells, (i) T-cells, (j) parental care, (k) hybridization, (l) selective breeding.

PANEL - 1.1
ANATOMY IN THE COMPUTER AGE

Many years ago dissection was the principal means of anatomical study, but at present it is becoming increasingly obsolete, as dissecting instruments are replaced by keyboards and computer screens. Computer software technology has tremendous practical significance in relation to the study of anatomy. Students object to the dissection of toads, fishes, and other animals in many educational institutions of developed countries. This technology encourages the dissection. Using this software, however, students are able to dissect animals with a computerized scalpel to observe the internal organs.

The study of anatomy using computer software continues to be important in biological fields and different aspects on the subject concerned rely on such technology to gain clear-cut information. One of the best-known technologies in the 20th century is the introduction of computer software across the world that would definitely help anatomists to observe the structure of dissecting animals.

PANEL 1.2
EVOLUTION OF ACCESSORY RESPIRATORY ORGANS IN TELEOST FISH

During the early Devonian period (duration of the period was 50 million years) of Palaeozoic era (duration of the era was 395 million years), the oxygen content of air was more or less similar to that of the present day. In the late Devonian, this situation gradually declined due to drying up of lagoons, swamps, and freshwater impoundments. Such adverse environmental conditions led to the development of lungs from branchial/pharyngeal pouches in the Crossopterygeans. In some extinct forms of fish such as *Chelrolepis* sp. (the ancestor of the group Palaeoniscoidea), in addition to lungs, gills were developed and started aquatic respiration.'This group gradually gave rise to the Holostei and Chondrostei groups of fishes.

During the Tertiary and Quaternary periods (duration was respectively 2 and 63 million years), of Coenozoic era, both sea and freshwaters were dominated by the Teleostean fishes. At that time, the level of oxygen particularly in freshwaters drastically declined. Consequently, the gills were not able to absorb oxygen from waters. As a result of this adverse situation, some teleostean fishes such as *Ophiocephalus, Clarias, Anabas, Heteropneustes, Amphipnous,* and *Trichogaster* had to develop some other types called *accessory respiratory organ* by the modification of gill structures.

To sum up, the adaptation for air-breathing in fishes was possible in response to situations where aquatic respiration becomes impossible. It is emphasized that deficiency of oxygen in water must have responsoble for evolution of air-breathing habit in these fishes. High carbon dioxide levels in water could also have been an important factor in the evolution of air-breathing. It is important to note that about 85 per cent of air-breathing fish occur in tropical and sub-tropical shallow stagnant freshwaters where low oxygen and high carbon dioxide levels are common.

PANEL - 1.3
TRIPARTITE CONCEPT OF KIDNEY ORGANIZATION

The concept of kidney organization was based on the development of kidney tubules from the nephric ridge and their structural differentiation. According to this concept, the kidney tubules are developed from three distinct regions of the nephric ridge. The nephric ridge is derived from the intermediate mesoderm present in the dorsal side of the embryo. As development proceeds, the nephric tubules (which are formed from the nephric ridge) undergo gradual disintegration, addition, and replacement and consequently, different types of adult kidneys are formed.

In embryonic condition, the urine-secreting or urine-producing tubules of the kidney – commonly termed as *uriniferous tubules*, develops from a special mass of mesoderm that lies between the dorsal somite and the ventral unsegmented lateral plate. However, the uriniferous tubules develop first on the anterior segments of the body and continue to grow backwards. The kidney tubules formed in the first phase, are termed as *pronephros* or *head kidney*. In craniates, this type of kidney becomes functional in the embryo but functionless and often absent in the adult. Part of the kidney of vertebrates, arising later in development than, and posterior to the pronephros, and discharging into the Wolffian duct, are termed as *mesonephros* and becomes the functional kidney in adult unamniotes. In elasmobranchs, when the mesonephros develops, the pronephric duct divides longitudinally from its hind end as far forward as the anterior end of the mesonephros. One of these ducts retain connections with the mesonephric tubules (the tubules that appear later in the more posterior segments) and forms the excretory canal. It is termed as *Wolffian duct*. The other is *Mulierian duct* which retains connections with the pronephros and ultimately forms the oviduct in the female. In teleosts, however, the pronephric duct does not divide and the oviduct is formed differently.

In amniote vertebrates (such as reptiles, birds, and mammals), the second set of mesonephric tubules (the mesonephros) is developed later in development, and posterior to the mesonephros which becomes the functional kidney – termed as *metanephric kidney* with a special metanephric duct. This duct is termed as *ureter*. Since the metanephros is an individually separate and distinct structure, the adult amniote kidney has termed as *Opisthonephros* which, in fact, is a combined structure of both meso- and metanephros.

In general, some additional mesonephric ducts are developed from the posterior part of the nephrogenic mesoderm. Such extended mesonephric kidney with additional mesonephric ducts are termed as *opisthonephros*. However, functional mesonephric kidneys are developed in adult fish and amphibia.

PANEL- 1.4
FISH HYBRIDIZATION

The main objective of hybridization and selective breeding of fishes is to evolve a strain superior in quality to those of parent species with higher production potential. Hybrids in fishes may manifest in various ways. It may result in resistance to diseases and parasites, resistance to unfavorable environmental conditions, better food conversion, and acclimatization to new habitats leading to increased production.

Significant progress in this field among various species of fish has been made and several inter-specific and inter-generic hybrids have been successfully produced.

Several species of major carps and tilapia have been employed to obtain inter-specific hybrids (see below) and several valuable features have been noted among hybrids. It essentially involves cross breeding between two different species of fish. Hybrids are, however, intermediate in general appearance to their parent species. The growth of hybrids is faster and they attain sexual maturity within two years.

Male species	Female species
Labeo rohita	Labeo calbasu
L. calbasu	L. rohita
L. bata	L. rohita
L. bata	L. calbasu
L. calbasu	L. gonius
Oreochromis mossambica	Oreochromis nilotica

In contrast to inter-specific hybrids, inter-generic ones are the result of cross breeding of two different genera of fishes (see below) and, in several cases, the results are highly encouraging. Hybrids have several features of farming value. *Catla catla – Labeo rohita* hybrids, for example, develop a broad body like *C. catla* and small head as in *L. rohita*. Also, the hybrids contain more flesh than either of the parents.

Male species	Female species
Catla catla	Labeo rohita
C. catla	L. Calbasu
C. catla	Cirrhinus mrigala
L. rohita	Catla catla
L. rohita	Cirrhinus mrigala
L calbasu	Catla catla
Cirrhinus mrigala	C. catla
C. mrigala	L rohita
C. reba	L. calbasu
C. reba	L. rohita

Hybrid fish produced from cross breeding between *Catla catla* – *L. rohita*, *L. calbasu* – *C. catla*, and *C. catla* – *Cirrhinus mrigala* have satisfactory growth.

The hybrlds obtained by crossing between Indian and Chinese carps are noted below.

Male species	Female species
Ctenopharyngodon idella	Aristichthys nobilis
Catla catla	A. nobilis
Hypophthalmicthys molitrix	A. nobilis
Labeo rohita	A. nobilis
C. idella	L. rohita
H. moltrix	L. rohita
C. idella	C. catla
H. molitirx	C. catla
L. rohita	C. idella
C. catla	C. idella

Characters such as enhanced growth rate, greater hardiness and a sex ratio to 100 per cent of one sex or sterility are very much desirable skewed to fish farmers who practise planned pond stocking. Hybrids with any unwanted characters should be eliminated forthwith, otherwise it may endanger normal fish breeding program.

2

History and Development of Fish Culture

Fish culture development strategy has become an important issue for sustained production. Progressive destruction of capture fishery resources and depletion of fish catch have been forced to draw much attention to farming and aquaculture all over the world. Development and planning of freshwater aquaculture sector is responsible for the marked expansion of fisheries science and production of fish at large. This expansion is further compounded by momentous development of biotechnology in fisheries sector and establishment of various research organizations.

The study of fish culture development is important for several reasons. Firstly, extensive development of fisheries sector is necessary to supply the requirants of growing populations. Thus development technology must be available when fish farmers need it and it must come from experts, planners, and technocrats. Secondly, developmental strategy is the criterion that increases the standard of living of the farmers' community to a great extent. Thirdly, for better quality of farmers' life, rapid and intensive development of myriad types is inevitable, but it is necessary to frame and implant suitable plans for development depending on the local conditions. Fourthly, during planning it is necessary to examine the objectives of the development very candidly and critically and take the possible steps to ensure that resources are damaged to the minimum. Fifthly, successful development requires four other important items essential for high production — financial assistance, skilled personnel, important guidelines to fish culturists and growing awareness among fish culturists. And last but not the least, haphazard development of freshwater ecosystems for fish culture determine to a large extent the incidence of water quality deterioration and decrease in production potential.

2.1 Rationale*

Fish culture is considered as an aquatic counterpart of agriculture and the origin of agriculture extends back at least three thousand years. In contrast to agriculture, however, which has been the important way of obtaining food from land for several thousand years, fish culture has until recently contributed little in real terms to world fish production. In lieu of evolving towards cultivation, hunter-gatherer methods of procuring fish from aquatic ecosystems evolved along a different path by the continuous progress of tracking

---

* For further detail on the subject, see FAO (1997 and 1999).

methods and increase in killing power. Farming of the aquatic ecosystem was almost non–existent.

There are several reasons why aquaculture and agriculture did not develop in the same manner. Food in the lakes, rivers, and seas has, until recently, been luxuriant, and there was thus little demand to acquire the farm. Also, much of the aquatic ecosystems emerged unfriendly. It is not possible that a strategy which could support fish securely will grow rapidly for the market. Whilst the breeding of animals, harvesting, and planting of seeds was easily possible to obtain on land, it has been found arduous to breed many aquatic animals and to hatch their eggs and successfully rear their offspring. These problems were in part due to the fact that man was dealing with animals which were different from himself and in an environment which he apprehended and until recently did not understand.

World demand for fish has grown at a steady space since the end of World War II and has been met by the development of capture fisheries. Growth in the aquaculture industry, in general, was accelerated at a rapid rate during the 1960s when yields increased by an average of 6 per cent per year. During the 1970s and 1980s, growth rates were tardy and irregular, due extensively to the decline of fisheries utilized for fish meal and production of processed fish, and although same period has observed some increase in catches of fish for human consumption, there has been a decline in the overall growth rate.

The main reason of the decline in capture fisheries production is the drastic reduction of common stocks in large numbers which can sustain further increases in catch, and the position has been provoked by sharp increases in the development of freshwater aquaculture sectors. The latter factors have strike the major aquaculture nations such as China, Israel, Japan and USA which was dependent to a considerable extent on well-knit strategy for fish production.

Hopeful prophesy for the future suggests that a sustained growth of around 1 per cent per year is possible, providing adequate management of stocks, developnent of new fisheries, and marketing of novel fish products can be achieved. Demand of fish for human consumption is, however, projected to increase by around 4 - 6 per cent per year and by the end of 2020 capture fisheries will be inadequate to meet this demand. If fish meal based fisheries production is assumed to remain static during the next 18-20 years, then by the year 2020, 78 million tonnes of fish from capture fisheries resources will be available for human consumption.

Thus there has been a growing awareness that the stocks of world capture fisheries are exhaustible, and attention is increasingly being turned to the possibilities exhibited by aquaculture.

The growth and development of fish culture/aquaculture is thus a phenomenon of the 1970s and 1980s. In the following sections, the strategy/criteria that are responsible for the marked developnent of culture-based fisheries and current status as well as other developnental activities have been discussed at length.

2.2 Fish Culture Developnent Means Farmers' Welfare

The basic objective of fish culture developnent is to bring the farmers' group into the better standard of living. More than 60 per cent of the farmers in Asia are landless with no productive assets and are devoid of sustainable employment and minimum wages. Even the involvement of women in aquacultural activities which is now being gradually expanded in some developing countries, are denied both minimum and equal wages. The government has taken the initiative to increase the pace of economic development by setting and implementing viable aquaculture projects so that the farmers become self-reliant.

A number of financial institutions have set up to accelerate the economic growth and developnent among farmers. They provide financial assistance at a low rate of interest for income-generating programs and training in skills. At the same time, the government is providing momentary assistance to Non-Government Organizations for organizing developnental activities, income-generating programs and infrastructural development so that socially and economically deprived farmers may develop their standard of living to some extent and become self-reliant.

In the scale of priority, education and training to farmers are considered as a major component. It is the key to rural economic development. The low level of education and training are the source of all the problems they face during farming programs.

There is no denying the fact that government policies and affirmative action have had a positive impact on aquaaculture developnent. Inspite of this, malnutrition, poverty, and unemployment are the main endemic problems. It would, however, not be exaggerated that the implementation of policies and programs suffers severely from certain deficiencies such as deprivation, administrative apathy, and negligence. A pertinent question is often raised: can anybody say where crores of rupees earmarked for rural development in general and farmers' welfare in particular have gone? The current efforts of the government should now be intensified not only in the interest of the farmers but also in the interest of their communities. At the same time, a radical change in the seat of consciousness/feeling is the need of the hour.

2.3 Service to Farmers Through Development

Developing nations have failed to provide the institutional infrastructure required for a sound economy. It is important that the nations undertake their key responsibility more effectively and efficiently. The government machenary besides ensuring that it performs its own functions well, also has a role to monitor other institutions in the country.

Political and social systems are functioning with many weaknesses and inadequacies, which providing satisfactory services to the farmers' community. However, a number of principles should be given top priority for sustained development in the sector concerned: integrity in farmers' life, selflessness in service, objectivity, responsibility to fulfil obligations, honesty, and planning for sustained production. Planners/experts/ technicians are expected to be more constructive than corrupt. They are expected to become popular by their easy access to the farmers and their performerce. In many developing nations where the manpower is easily available, the organization is adequate, and the procedures are intelligently devised, are still running so persistently into severe crisis. For various reasons Indian planners, for example, have never treated the core function with sufficient depth and respect. If we are, however, serious about development in the sector we have to mobilize resources such as natural, financial, and technological for sustained production. Some developing countries have developed technologically to a greater degree and the speed of development must be sustained. Poverty reduction among farmers' community is also possible with structural changes.

2.4 Aim and Evolution of Fish Culture

Both fisheries and fish culture have a similar aim at maximizing the yield of useful organisms from the aquatic environment. The size of exploitable stocks is determined by four factors such as growth rate, recruitment rate, natural mortality rate, and fishing mortality rate. Capture fisheries try to maximize yields by increasing the fishing mortality rate, at the expence of natural mortality; although if too many fish are killed, then recruitment and growth are unable to compensate for the loss and the stock dwindles. Fish culture tries to achieve increased yields by manipulation of growth, reproduction and recruitment, and natural mortality rate.

Fish culture possibly evolved by initially achieving control over natural mortality, through capture and temporary holding of organisms in man-made facilities. The simplest facilities to construct would have been earth ponds. In some parts of the world, these would have been little more than mud walls constructed to temporarily hold water and fish following the seasonal flooding of a river. Growth manipulation through proper management schedules would have been an easy next step. Control of spawning and recruirtment is a recent achievement, as it is difficult to persuade many species to breed in captivity. There are also many technical problems involved in the hatching of eggs and the maintenance of feeding of larval and juvenile stages.

Fish culture has greatly evolved to gain control overall three of these population-determining processes. Great progress has been made in the fields of nutrition, genetics, engineering, physiology, and biochemistry which have all contributed towards improved yields. Since inputs of both energy and effort are involved in achieving total control, several types of fish culture have flourished by opting for lower yields at a considerable savings in human resources and expense. The release of juvenile fish into the wild for example, involves little more than increasing recruitment to the natural exploitable stocks, and does nothing to reduce natural mortality (due to diseases and parasites), or influence growth. There are considerable financial savings in terms of feeding and construction of holding facilities. Similarly, attention is once more being turned towards fish culture that reduces expenditure on feeds. Feed costs can account for as much as 60 per cent of production costs at a farm. However, feed costs can be reduced if advantage is taken from naturally-available foods. In ponds, fertilizers and manures will stimulate the growth of organisms at the base of the food web, whilst in lakes and rivers, cages and pens can be used to hole fish for growth.

To sum up, the farming of aquatic organisms is achieved through manipulation of an organism's life cycle and control of the environmental factors which influence it. Three main factors are involved in this regard: (1) control of reproduction, (2) growth control, and (3) elimination of natural mortality agents. Control of reproduction is essential, otherwise farmers must rely on naturally spawning stocks. The supply of fry from the wild may be restricted to a particular season and a particular area, and. there may also be shortages due to over-exploitation of wild stocks. For technical reasons, this step has not yet been achieved in the culture of many species. Growth can be increased through selection of broodstocks. As can be seen, the culture of carnivorous species is tied to the supply of high-protein and fish meal-based diets. There is considerable scope for reduction in feed costs providing the appropriate omnivorous/detrivorous/planktivorous species.

Last but not the least, rearing systems are essential for all types of fish culture. Such systems are designed to hold organisms in captivity whilst they increase in biomass, by minimi-zing losses through disease and predation and exclusion of competitors. Rearing facilities for fish, however, can either be land-based or water-based, the former type including ponds, raceway, tanks and silos and the latter one comprising cages, pens, and enclosures.

2.5 Origin and Growth of Fish Culture

The origin of fish culture is wrapped in mystery. However, small earthen ponds were most probably first used by the human society for generations as a convenient holding facility for fish, until adequate quantities were caught for consumption or marketing. Fish farming on commercial basis in ponds, lakes, tanks, and reservoirs is of very recent origin and have developed independently in a number of countries, all in Southeast Asia, Europe, Russian Federation, and Canada. Fish farming, however, is now widely practised in almost all countries of the world where potential water bodies exist. It is very difficult to pinpoint the origin of fish farming. In China, ponds were being first used to culture different species of carps in the 1100 B.C. At present, both marine and freshwater resources are being exploited in such a way that the conventional fish stocks have declined to a considerable extent. Fish culture industry has now become highly mechanized and commercialized, of course, such activity must have to be undertaken on a sustainable basis.

The history of fish culture is an old as the man's history. To search food and to satisfy the hunger, primitive men were caught fish with the help of harpoon from the seas, lakes, and rivers. Tao Chukung (500 BC), however, advocated that fish culture is and extremely profitable business. During the Stone Age (40,000 BC), the availability of fish refusals near the rivers and lakes clearly indicates the use of fish as food. In new stone Age (10,000 BC), fish were found to be preserved in a very primitive way. In the Bronze Age (3,500 BC), fish were preserved with common salt. So far the Indus Valley Civilization (2,500-1,500 BC) is concerned, it is evident that the utilization of fish as food was the most popular.

Among different fish species, carps are considered as the most profitable fish that can be cultured in many countries of the world. Carps are distributed widely in different rivers, lakes, and ponds around the world. It is very difficult to ascertain the original home for carp; of course, it is assumed that Asia is the native place of carp and were introduced from Asia to Europe in 1250; from Asia to Europe in 1496; from Asia to Denmark in 1560; from Asia to Russian Federation in 1729; from France to North America in 1830 and from Japan to North America in 1877.

Fish culture in ponds and lakes flourished in Europe during the Middle Age (500-1,500 AD). At that time, both capture and culture fisheries were more pronounced. Although most attention was paid to marine water resources, freshwater fisheries were to the forefront. Fish were kept for several years in ponds and lakes under controlled conditions where they grew and reproduced. It has been noted that carp, mullet, and bass have the capacity to adapt to a given condition enables them to grow well in rigid climate and the tropics. This is the main reason why the carp has played an important role since remote times.

In the 16th century, fish cultivation was most popular in many countries of Europe and Asia. Although the 30 years' war (1618-1648) in Europe caused a setback, fish cultivation were underwent a new boon through progressive pond owners in the 18th century. In the second half of the 19th century, fish were kept for a period of 4-6 years in ponds for the table. At the end of the 19th century, tremendous progress has been made through introduction of trouts (brook trout and rainbow trout) in Europe, while introduction of exotic species of fish in ponds of many tropical countries have given successful results in fish production since the turn of the 20th century. At present, fishery science has greatly contributed to further development of fish culture through extensive research in the field of feed, fertilization, manuring, induced breeding of fish, water management, and fish diseases. To day, fish culture is not an extravaganza but a strictly commercial proposition in many fish producing countries of the world.

Though China is assumed to be the first country where fish cultivation was initially undertaken, the most significant forward movement in freshwater fish culture probably started towards the end of the twentieth century in the United States. This is mainly due, in part, to the fact that as interest in freshwater sector inanimated in England, it flourished in the United States. The Federal Government established Bureau of Fisheries and began to breed and cultivate different species of fish and distributed them free of cost. Various State Governments, Universities, and Research Institutes were also established simultaneously with a view to introduce fish species in all possible waters, to propagate and protect fish and to improve the condition of water. Moreover, farmers have been encouraged to establish fish ponds both for supplying fish as food and give anglers an opportunity of good fishing.

In the last two decades of the twentieth century, much progress has been made in freshwater fish culture. At present, significant results from different countries are continuously forthcoming. Current research needs scientists to highlight the main problems on fish culture strategies. A number of branches on science such as ecology, physiology, embryology, biotechnology, pathology, genetics, chemistry, botany, and zoology are currently needed when fish culture operations in successive stages are contemplated. As fish culture gradually develops and civilization advances, fish farmers might face severe problems and to solve the problems, scientists and experts have become necessary. Although they are actively engaged in various aspects of fish culture problems and focus their research efforts to develop aquaculture industry, most of the farmers did not receive any technical and scientific informations. This is important because their results are based on laboratory and/or field experiments that can help fish farmers in various ways.

ROLE OF ROMAN AND GREEKS IN FISH CULTURE

The maintenance of the stocks of certain species of table fish by some form of fish culture is a very ancient practice, and it is known that in classical times the Romans and Greeks made a habit of cultivating fish in ponds and lakes for the table. Two main types of fish culture may be recognized. *Firstly*, the rearing of fishes in confinement until they are large enough to be eaten and *secondly*, the stocking of water with eggs or fry obtained from fish breeding. The ancient Romans carried on the first method, and were in the habit of admitting young fishes from the river/sea into enclosures, and then fattening them up for the table. The number of different kinds of fish which can be reared from egg to adult in captivity is very small, and the best known examples among freshwater forms include the Eel, Perch, Chinese carp, Common carp, Trout etc. are cultivated to a greater or less extent by fish farmers of Europe, but only the Carp makes a commercial proposition. Carp is the main fish used in pond culture throughout the world, and in Southern and Central Europe and in other oriental countries, carp culture forms a flourishing industry because the carp is a very hardy fish, can be bred and reared to maturity under all kinds of conditions, requires no costly food, consuming refuses, and has a delicate flavor. In some tropical countries, the rapid growth of different varieties of carps is more striking.

The growth of the second type of fish culture, the induced breeding of fishes for stocking purposes, was carried out in as early as the late twenties in different parts of Europe and America. However, Brazil was the first country to develop a technique for artificial propagation. In the early thirties, Von Ihering, Cardoso and others conducted experiments on induced breeding of fishes with success. Since then, Brazilian fish farmers have been employing, this technique to obtain fish seed for fish culture programs. Russian Federation is considered the next to adopt artificial propagation of fishes in 1938. In India, the first attempt to induce carp fishes was made by H. Khan in 1937. Thereafter, great advances have been made in this industry with marked growth of fish farming in some countries of the world. At present, artificial propagation of fishes is now extensively carried out in India and China through circular hatchery and now forms a regular part of fish culture program.

ORIGIN OF FISHERIES : INDIAN SCENARIO

It is believed that the history of Indian fisheries had its origin in the days before Christ (BC). Document referred by Kautilya in his Arthosastra (320-319 BC) clearly indicates that at that time fishes were grown in confined waters.

In the records of Pal dynastry in Bengal (810-850 AD) fish production in India has been mentioned. In 1300 AD, Srinathacharya and Vamadev Bhalla quoted in their slokas that fish should be used in the daily diet. In the official household manual of Emperor Akbar (11th century), the "Aine Akbari", a number of fish dishes for the royal dinner had been reported.

Russel, a Botanist with the East India Company, carried out a systematic study for about 5 years (1785-1789) on the Indian fish fauna. He listed about 200 species of freshwater fishes found in the neighbourhood of Calcutta. Mr. B. Hamilton — Superintendent of the Botanical Gardens, Calcutta, carried out an extensive study on the Biology of Fishes of the river Ganga for about 28 years (1794-1822). He prepared a catalogue of the Indian fishes. It is assumed that this was the first official systematic catalogue in India.

Evidences have shown that as early as 1780, there was overfishing of many Indian freshwater fish species from major river systems and their tributeries that led to dramatic decline in the fisheries of India by 1805. However, in 1839, MacLelland first suggested that commercial exploitation of Indian fisheries from different riverine systems should be judiciously manipulated.

In the nineteenth centuxy, Sir Francis Day, Surgeon Major with the British troops in Bengal carried out an extensive but systematic studies on the Indian fishes for over 20 years. After his death, his voluminous work was published in 1889 in two volumes named as "The Fish Fauna of British India" where 351 genera and 1,418 species of fishes found in Indian seas and rivers have been recorded.

So far the marine fisheries in India are concerned, it was evident that Mr. C. Alcock added 86 new genera and 200 species to the book published by Sir Francis Day. Since then, many Indian Ichthyologists carried out further work and have been incorporated into the book "The Fish Fauna of British India". In 1878, H.S. Thomas submitted a record of events for further use to the Famine Commission appointed by the Government of Madras. He advocated the development of Fisheries as one of the measures to fight against famine. In 1898, Sir Frederick Nicholson submitted a note on fisheries to the Govemment of Madras. But it is unfortunate that he did not able to attract the attention to the government.

In the beginning of the twentieth century, Sir F. Nicholson was appointed as Fisheries Officer in 1905 by the Government of India and the Government of Madras to investigate both freshwater and marine fishery industries of the Madras Presidency. At the same time, James Hornell and H.C. Wilson were appointed as a Marine Biologist and Pisciculturist respectively in 1908 by the Govemment of Madras. While Hornell was working with the Chunk and Pearl fisheries in Ceylon and Madras Fisheries Departments, Wilson first started fish culture operation by setting up a number of fish farms in the country.

The first Indian Director of Fisheries, Govemment of Madras was Dr. B. Sunder Raj. In 1929, the Government of Madras appointed a committee to evaluate the activities of Madras Fisheries. The Committee declared that it is necessary to close down the Fisheries Department as it did not serve any purpose.

In 1928, Government of India appointed the Royal Commission on Agriculture to investigate the activities of Indian Fishery industries. The commission reported that no progress has been made on fishery industries in India. At the same time, the commission also recommended to the Government that the usefulness of the Fisheries Department should not always be judged from the economic standpoint of view.

The 1960s were the dawn of expansion of fisheries and aquaculture developements. Since then, the concept of fish culture/aquaculture development has undergone a notable change and has become more comprehensive in different sectors. Fisheries-aquaculture-agriculture-rural developement chain has been set up by the national government in different plan periods to gear up the economic condition of the farmers. And in the new millennium, the development that has been described in succeeding chapters has reached a plateau especially in culture and export fronts by active involvement of expert personnel vis a vis by importing state-of-the-art technology and the country has chosen to equate it with rural and agricultural developments integrated with aquaculture in the widest sense so as to embrace different aspects of development.

1. Fisheries Development in Bengal Presidency: In Bengal, similar development took place in the country's fishery industry. In 1908, an ICS Officer first recommended the establishment of a Fisheries Department. Accordingly, the Government of Bengal appointed S.H. Ahmed to look after the Fisheries Department although he was a Revenue Commissioner to the Government. He carried out a fisheries survey of the Bay of Bengal.

The Government of Bengal appointed an ICS Officer, Mr. K. C. Dey in 1907, to investigate the fishing industry. Another person, T. Southwell, was appointed as the Deputy to K. C. Dey and he hold the post till 1919. In the same year, Dr. B. Prasad took over the charge of Fisheries Department and became the Director. In 1923, however, Abdul Karim Abu Ahmed Khan Gaznavi — a senior member of the Governor's council to the Government of Bengal — recommended that "there was no qualified person to take charge of the Department of Fisheries". In 1942, M. Ramaswamy Naidu of the Madras Fisheries Department was appointed as the Director of Fisheries of Bengal.

2. Fisheries Developement in Bombay Presidency: Similar to Bengal Presidency, two ICS personnels actively took part in drawing the attention to the Government for development of the fishing industry. In 1910, W. H. Lucas submitted a report to the Government on the improvement of fisheries in the Bombay Presidency. He made two important recommendations such as (i) "trawling off the Bombay coast had proved a failure and it would be premature to repeat the experiment until the bottom beyond 30 fathom limit was thoroughly examined" and (ii) "there was undoubtedly some scope, for the introduction of sardine canning and mackrrel pickling along the Ratnagiri coast but it must be left to the private enterprises".

In 1930, the Government asked another ICS Officer, H. T. Sorley, to suggest improvement in the fishing industry in India. He was collected a number of data from a large number of experts on fish curing by salting, shell fisheries of Bengal Presidency and various aspects of fishing communities. These reports were very much useful to sorley. Moreover, publication of annual reports on fisheries development at regular intervals by the Madras Fisheries Department also provided adequate support to H. T. Sorley to prepare reports on fishery industries. With the aid of these experiences, Sorley submitted a comprehensive practical report to the Government in 1932. This report covered various aspects on the development of fisheries industry such as fishing and fish curing methods, educational status of fishermen, research, and administration. His recommendations remain valid till today. Regarding research activities, for example, he said that "there is no need for the Government to indulge, in general, scientific research on fisheries. Research should be confined to certain questions likely to promise results on the economic importance within a reasonable time and it would be advisable to fix a time limit within which this research must be completed.

In 1944, the Government of Bombay appointed a committee and the committee strongly recommended the establishment of a separate department of fisheries. On the basis of this recommendation, the Government of Bombay appointed Dr. S. B. Setna as the first Director of Fisheries of Bombay Presidency. He was in that position till 1953.

3. The First all India Fisheries Conference : The first all India fisheries conference was held in 1948.This conference was very successful in getting recognition from the Government and did focus their attention on the need to develop fisheries industry in the country.

4. Tripartite Agreement for Fisheries Development : In 1952, a tripartite agreement between the USA, the United Nations, and Government of India was made for the development of the Indian fishing industry and a Technical Co-operation Mission (TCM) has been established. Through this mission, India received a variety of sophisticated equipments for fisheries industry at throw-away prices. Indian entrepreneurs realized that to develop the fisheries industry, modern equipments must be used.

DEVELOPMENT OF THE INDIAN FRESHWATER FISHERIES SECTOR IN 20th CENTURY

Although the progress in freshwater fisheries sector in the first half of the 20th century was significant, it was conspicuously pale to the development in the second half of the century. Generally, the development of freshwater fisheries in the first half of the century were restricted to high altitude fisheries for recreation with the introduction of some exotic fishes such as trout, tench, mirror carp, and golden carp by the British. At the same time, some hatcheries were also established in Nilgiri Hills (Tamil Nadu), Jammu and Kashmir, Uttar Pradesh, and Andhra Pradesh.

DECLARATION OF FISHERIES ACT IN INDIA

The Fisheries Act essentially defines fisheries management, development , conservation, protection and maintenance of fish stocks and matter connected therewith and prescribes penalties to the offenders. The coverage of the act is comprehensive in that it includes different inland water bodies over extensive areas. The existance of aquatic life depends by and large on the activities and performance of the humankind. Maintenance/development of fish stocks as a slipshod manner may take toll of thousands or hundreds of thousands of commercially important species of fish at a time. On the other hand, destruction of natural fish habitats and their breeding grounds as a result of pollution may prove fatal to fish life. We have barbarously damaged the existing fishery resources for the last 100 years or so. Therefore, some control measures must be adopted over limited or extensive areas so that fish loss can be reduced, if not totally feasible.

In India, there have been several Fisheries Acts (*See* Chart 2.1) both at national and sub-national government levels to prevent the loss of fish biodiversity. All these acts have been promulgated from time to time. These acts provide general powers to the government to take all necessary steps for the purpose of planning, prevention and control of fish stocks. These powers embrace (1) planning and execution of nationwide programs, (2) laying down about water and soil quality standards, (3) adoption of protective measures of game,

---

Chart – 2.1

Fisheries Acts

The dawn of the 20th century has witnessed the open declaration of Acts as follow.

1. Mysore Game and Fish Preservation Act (1901)
2. Jammu and Kashmir Fisheries Act (1903)
3. Punjab Fisheries Act (1914)
4. Game and Fish Protection Regulation Act of the Government of Travancore
5. Cochin Fisheries Act (1917)
6. Assam Private Fisheries Protection Act (1935)
7. Andaman and Nicobar Islands Fisheries Regulation Act (1938)
8. Hyderabad Fisheries Act (1946)
9. Madhya Pradesh Fisheries Act (1948)
10. United States of Gwalior, Indore and Malway (1950), Travancore-Cochin (1950)
11. Madhya Pradesh Fisheries Act (1948)
12. Bhopal Fisheries Act (1952)
13. Rajasthan Fisheries Act (1953)
14. Maharashtra Fisheries Act (1960)
15. Pondicherry Fisheries Act (1965)
16. Haryana Fisheries Act (1966)
17. Himachal Pradesh Fisheries Act (1976)
18. Kerala Fisheries Act (1980)
19. West Bengal Inland Fisheries Act (1985)

food and endangered fish species, and (4) regulation of fish stocks and many other related issues.

Similar to our country, many advanced nations have adopted fisheries acts to deter and punish the aquaculture enterprises violating the protection/conservation ethics. They have enacted Acts or amended their penal codes by introducing the laws with enough provisions to indict the offenders.

To sum up, all these acts have helped fisheries conservation and management strategy stances to recoup, but such activities have led fish to venture out to the humankind in search of food and water. Gradual shrinkage of most of the water resources certainly contribute a lot to this problem. Regular forays into aquatic resources, in many cases, have brought about actual changes in the productivity of ecosystems to a large extent. This occurs to be one of the important factors of the plunder of freshwater resources in yesteryear between the years 1980 and 2000. The outcome of such events is definitely the man-fish conflict posing threats to the long-term survival of high-priced fish species.

1. **Drawbacks of Fisheries Acts :** Though fisheries acts have many advantages, most of them have been plagued by one or more severe imperfections. Most of the acts have failed to consider the aspects of authentic survival, and literally have neglected the local community needs which has created repugnance towards the management programs. In fact, large lacunae that follows still exist both in terms of suitable Acts and their compulsion.

(i) Many of the existing Acts are updated versions of the earlier ones. They are basically specified significance to gear development and resource utilization in full potential and many of them were ordained at a time when ecosystem protection were not an important issue. None of these Acts have taken into consideration on the scientific and technological aspects.

(ii) Some of the Acts appear at times executing reciprocally frustrating social objectives. Where such potential ecosystems are shared by one or more states, acts ordained in one state may have serious implications on the other.

(iii) Many of the existing acts do not vividly indicate the social objectives. Consquently, the implementing authority have to reckon with their own explanation to enforce the Act which may not corroborate to the aim and purpose for which the Act was made. At the same time, alterations of national policies and situations have progressively rendered to many of these Acts antiquated.

FISHERIES DEVELOPMENT THROUGH RESEARCH INSTITUTES

The first half of the 20th century has witnessed the beginning of reservoir fisheries development, induced breeding, and pond culture in an organized way. In the mid 20th century, significant development in the management strategies of fish culture and fisheries has been made in such a way that next to China, India is one of the most important leading countries of the world in freshwater fish production. This has paved the way for introducing modern technologies through establisrunent of different research institutes, education, and development centres (Table 2.1) such as Central Inland Capture Fisheries Research Institute, National Bureau of Fish Genetic Resources, National Research Centre on Coldwater Fisheries and other agencies such as Fish Farmer's Development Agencies and Integrated Fisheries Project (successor of the Indo–Norwegian Project). These organizations have provided technologies to fish farmers for promoting fish culture. By 2008-2009, Inland fish production reached a level of about 34 lakh tonnes from 0.18 lakh tonne in 1950-1951.

Table 2.1 Establishment of Research Institutes in Different Parts of India for Development of Fisheries and Aquaculture

Name of the Institute	Place	Established in
Central Inland Fisheries Research Institute (Main station)	Calcutta (West Bengal)	1947
Central Inland Fisheries Research Institute (Sub– station)	Cuttack (Orissa)	1949
Central Marine Fisheries Research Institute	Chennai (Tamil Nadu)	1947
Central Institute of Fisheries Technology	Cochin (Kerala)	1957
Central Institute of Fisheries Education	Mumbai (Maharastra)	1961
Central Inland Capture Fisheries Research Institute	Barrackpore (West Bengal)	1987
National Bureau of Fish Genetic Resources	Lucknow (Uttar Pradesh)	1983
Central Institute of Freshwater Aquaculture	Bhubaneswar (Orissa)	1986
Central Institute of Brackishwater Aquaculture	Chennai (Tamil Nadu)	1986
National Research Centre on Coldwater Fisheries	Haldwani (Uttar Pradesh)	1986
Northeastern Regional Research Centre of Central Inland Capture Fisheries Research Institute	Guwahati (Assam)	2001

In the second half of the 20th century, a remarkable achievement was made on the production of freshwater fish culture through quality seed production. Methods of induced breeding of freshwater fish through hypophysation and application of synthetic hormones greatly spread throughout the country. At the same time, development of reservoir fisheries*, running water fish culture and wetland fisheries* followed by the establishment of more than 450 hatcheries (including major and mirror carps, freshwater prawn, Asian catfish and Trout), and fish genetic research institute for genetic upgradation of cultivable fish species in the last decade of the 20th century contributed to significant development of freshwater fish production in the country.

---

* For further detail, see Sugunan (1995) and Biswas (1995).

2.6 World Conference on the Development of Aquaculture

As early as the late sixties, many fish fanners and organizations believed that natural stocks could support fisheries industry and fish culture is not significant at all and that fish culture do not make any contribution to food production. Keeping this view in mind, attention has been given by National Governments for the development of aquaculture. In the history of aquaculture development, however, two important events have been observed. Two International Conferences on aquaculture were held so far: one in Kyoto, Japan in 1976 and the other in Bangkok, Thailand in 2000. The constructive effects of these conferences towards fish culture development is not unimportant. Not only is the freshwater ecosystem fosters overall productive to a greater extent but the pollution of aquaculture environment drastically reduced the productivity, provides sustained production of aquatic species and increases the profit margins. Kyoto conference, however, highlighted the opportunities in science, technology, manpower, network, and institutional strengthening for development of aquaculture. From 1976 to 2005, dramatic changes in aquaculture have taken place in several areas ranging from a small-scale household farming to a large-scale commercial one. The major development as a consequence of Kyoto strategy includes the following: (1) An increase in Governmental and institutional support to aquaculture which triggered technological advances of the sector, (2) Increase in planning and development effort by Governments that helped in increasing aquaculture to the level of capture fisheries, and (3) Technical co-operation for regional aquaculture development.

DEVELOPMENT OF AQUACULTURE SINCE THE KYOTO CONFERENCE

The main strategy of the Kyoto conference was to develop the status of fish culture in many fish producing countries of the world. Since the Kyoto conference, aquaculture has progressed gradually in many parts of the world particularly in Asian and Southeast Asian countries. At present, aquaculture has been the most important food production sector in the world for more than two decades. This sector, however, exhibited an overall growth rate of 12 per cent for terrestrial farm animal meat production and 1.5 per cent for landings from capture fisheries. The per caput availability of fish from aquaculture sector increased from 2.5 kg per annum in 1984 to 12.5 kg per annum in 1998 inspite of the increasing population and stagnation of fish production from capture resources. Moreover, development of technology and management systems during 1990's have substantially increased fish production. Cultivation of fish in submergent and uncultivable areas using more indigenous species of fish also exhibited significant production although different technologies face severe problems. Such problems include the environment, community participation in development processes, and resource allocation.

Although significant development has taken place in aquaculture sector of the world, a number of problems have threatened the sustainable expansion of aquaculture sector. For example, the most important constraint on the development of Asian aquaculture is the outbreak of disease that caused severe losses at 1.6 billion US dollars in the early 1990's. Furthermore, environmental degradations in regions with very high population densities also caused drastic decline in aquaculture sector to a significant extent which have serious socio-economic implications.

STRATEGY AND PLAN OF ACTION FOR AQUACULTURE DEVELOPMENT BEYOND 2000

The network of aquaculture centres in Asia-Pacific (NACA) – an inter-governmental organization – organized a global conference in Bangkok at the onset of the new millennium (Feb 20-25, 2000). In general, the NACA and the FAO (Food and Agricultural Organization of the United Nations) reviewed aquaculture developments in Asia, Latin America, Africa, Europe, North America, Near East, Russian Federation and South Pacific by organizing regional conferences and held meetings to evaluate trends in aquaculture developement. However, the Bangkok conference declared the followings:

Aquaculture practice should be pursued as an integral component of development such as sustainable livelihoods for poor sector of the community and promoting human development.

- Aquaculture development should be carried out within the framework of international, national, and regional agreements.
- For aquaculture development, active co-operation of the Governments, private sector, and other stakeholders is necessary.
- Aquaculture policies should encourage management and farming practices.
- Aquaculture sector should be developed in such a way that the economic growth, standard of living, and food supply may be assured.

STRATEGIES IDENTIFIED

On the basis of the above declaration, the following strategies have been identified.

1. Modern technologies should be adopted in aquaculture which are applicable to the local environment.
2. Development of environmentally sound technologies and resource-efficient farming systems.
3. Information flow and communication at national and regional levels should be improved. Information includes policy-making, education and training, planning, and adoption of procedures.
4. It is necessary to built the skilled personnel so that they may actively engage in the sector.
5. Investment in aquaculture research must be increased for use of resources more efficiently.
6. Culture-based fisheries should be improved since it has considerable potential for increased fish supplies and generating income in rural areas.

7. Application of genetics and biotechnology to aquaculture should be encouraged. These strategies enhance survival rate of fish, reduce production costs, and several other factors that should be considered for aquaculture development.

8. Development of marketing strategy and understanding consumer requirements are necessary.

9. Instiutional support should be strengthened to implement policy and regulatory framework. More attention should be given towards economic incentives in the planning and management of aquaculture development.

10. Aquaculture should be integrated with the related rural program.

11. For better production, it is necessary to satisfy the nutritional requirements of cultured species. Development of feed is necessary through efficient use of resources.

12. Involvement of women in aquaculture and aqua–product industries constitute one of the most important development issues (See Chart – 22). It is a major issue with far more momentous implications in aquaculture development.

Chart – 2.2
Role of Women in Aquaculture Development

In many tropical and sub-tropical countries of the world, aquaculture development has been accompanied by a large-scale involvement of rural women in different aquaculture sectors. These women live below the poverty line. This phenomenon has, however, come to be identified as a separate group of population who are involved in fish culture and post-harvest management strategies and are termed as *fisherwomen*. This tendency has been accounted for by the fact that the expanding aquaculture sector which is now based on the state-of-the-art technique offers more employment opportunities and consequently, a large part of the rural women have more serious attention toward aquaculture industry.

Aqua-industries embrace different phases of fish culture starting from pond preparation to the marketing of aqua-products, which are carried out by most of the women as a spare time or whole-time occupation. Aqua-based industries principally depend upon the preparation of fish culture ecosystems in right time, supply and use of inputs, farm management, harvesting, preservation, and marketing of products. The industries can, to a large extent, solve the overall problems of fisherwomen by providing employment to them at their own/leased water areas. At the same time, adoption of technology by the fisherwomen and providing technical advice as well as financial assistance to them *vis– a –vis* production- and processing-related training/extension/managerial programs which are often forgotten due to administrative

contd...

coldness must be geared up. These factors have tremendous potential in speeding up their socio-economic conditions to a large extent.

It seems plausible that aqua-industries definitely help improve the standard of living of rural women if Fisherwemen Co-operative society is established. The all-round development of fisherwomen hinges on the formation of such society. If co-operative forms can become the means for combining aquaculture with agriculture forms of work and making use of available fisherwomen in full potential and in activities requiring rising levels of skill, they can do much to increase fish production and marketing and hence their income as well as find avenues in productive work.

The woman is a crucial part of the aqua-farming industry In recent times, the profile and role of the fisherwomen have been undergoing significant changes. The percentage of working women both in agriculture and aquaculture sectors has actually been growing at a steady space. Their newfound purchasing power has fuelled to meet the demand of their family. The fisherwoman possesses a good awareness of the changes taking place in aquaculture sectors around the world. Her moderate educational background and the growth in media has substantially contributed to this development. Magazines/leaflets in the topic concerned written in regional languages carry a lot of information targeted at her. These reports also carry information on fish culture and processing techniques and message on a variety of topics of interest.

To sum up, the changing position of fisherwomen in aqua- farming industries has to reckon with. In the beginning of the new millennium, the position of fisherwomen especially in the middle- and/or poor-class segments of the rural populations is indeed changing fast. From the role of simple housewife, they have transformed into earning members of their nuclear families by involving themselves in different agro- and aqua-industries sharing the responsibility with the fishermen.

BENEFITS OF STRATEGIES

Benefits of these strategies promote several conditions favorable significant production of fish upto the carrying capacity level and natural organisms used as food by fishes and provides greater stability of fish culture industries. Pollution — the bad effects of fish culture ecosystem due to accumulation of toxic substances — is alleviated. Pollution undoubtedly breaks production strategy and seriously disrupts the farming systems. By adopting technologies, fish production maintains a sufficient and effective level. Also by different magement options, the volume of production from which economic condition is strongly established among farmers' community is maintained to a higher level.

The greatest benefits of fish culture development strategy relate to technology. Good technology which is suited to local conditions must be enforced that promotes the overall yield of fish biomass. The availability of the technology greatly influences the entire management system in fisheries sector. Moreover, the production potential increases if adequate training and education to fish farmers are given in times of need and skilled personnel is available to meet the farmers' demand.

A well-functioning aquatic ecosystem permits the farmer to start fish culture operations effectively following suitable management systems in due time. And this is highly significant in modern fish farming.

2.7 Evolution of Fish Culture

The evolution of fish culture can be briefly traced as follow:

THE STAGE OF CAPTURE

The pre-aquacultural revolution world was mainly characterized by simply capture of fish from different water bodies through traditional system of fishing. People captured kaleidoscopic variety of aqua-products and consumed. There was no elaborate culture system as the needs of the people and the prevailing technology did not demand such a system.

THE STAGE OF CULTURE REVOLUTION

This is the most important stage so far the production and nutrition fronts are taken into consideration. Due to gradual reduction in fish population in natural resources as a consequence of myriad factors stated elsewhere in two volumes, far- reaching changes took place in this stage. Both cultural and technological revolutions bore the germ of new farming strategy. It introduced new culture systems and new technologies to suit the local environment and brought about sweeping changes in the economic environment of the farmers' community. Sustained development of fish culture industry following the application of biotechnology, induced breeding, and culture strategy for mass production became the order of the day. A variety of fish species of commercial importance became available in great abundance. The fish culture revolution also generated the income revolution, giving a great deal of disposable income to a large mass of rural people. And it was this income revolution that sustained the mass production and mass consumption unleashed by the cultural revolution.

THE STAGE OF COMPETITION

The mass production and mass consumption brought by the culture revolution soon led to the stage of competition. The ever-increasing number of aqua- farms generated the phenomenon of competition. In early times, the main task of farmers was to stock fish in enclosures for certain periods until they grow to table size. Due to production of huge quantities of edible fish following adoption of current farming technologies, competition became the main issue. The situation demanded a conscious effort on the part of farms to ensure that their products were preferred to those of their competitors.

THE EMERGENCE OF CULTURE STRATEGIES

In the sixties and .seventies, fish culture strategies in many countries of the world changed enormously. There was a substantial increase in population, new aquacultural concerns sprang up rapidly, a great variety of new aqua-products strengthened the rapidly developing consumer market and selling of products became difficult because of the high intensity of competition. However, different culture strategies were made available to the farmer and the farmer began to occupy a place of unique importance. The progressive fish culturists soon realized that it was not enough if they somehow culture fish in their own or leased ponds, they had to ensure that the consumer who purchased their products once, came back to them again and again whenever he needed the product. They also had to ensure that the aqua-product should be healthy, zero defect, and in good condition. If a well-knit strategy is adopted during fish cultivation, quality fish will obviously be made available at the market to the consumer. However, the emergence of closed recirculatory system in fish culture, genetics, biotechnology, and conservation of farmed fish has further extended to the production of aquatic species of commercial importance.

2.8 Domestication of Farmed Fish

Fishes are the first aquatic vertebrates which developed before man in geological time and they have therefore always represented and important part of the entire aquatic environment on which man is entirely dependent for food and nutrition. Since time immemorial men had hunt fishes in order to procure food. This is still true today in almost all parts of the world, but in most moderately and densely settled areas, a number of important food fish have been eliminated by continual fishing over decades and by the pollution of waters which they live.

Not all fishes are important for human consumption. About 35 per cent of the total fish species are used as food. Though hunting of some terrestrial animals remains a major source of food, hunting techniques are also found in the fishing industry despite the great advancement made in other aspects of food supplies.

At some stages in the history, probably in the Stone Age, though the exact time is not known, man began to cultivate fishes. The first country to be cultured was probably China. Subsequently, man soon realized the importance of fish as food in daily life.

Man has not only used as food but has also taken drastic steps to ensure their survival and multiplication in order to produce increasing supply of food and other products through traditional culture which were then followed by advanced methods of farming. This has been done following the application of recently developed farming technologies to a large extent, by the treatment of fish diseases, by preventive measures against predatory animals, and by the application of genetic engineering in fish culture. In modern times, the importance of fish to science, culture, education, research, food, and profit has been fully realized and they are now cultured for the interest of the people and protect them from human interference in nature reserves and sanctuaries.

Fisheries and fish culture have also been improved from commercial viewpoint during the course of time. They have been specially bred following the application of induced breeding techniques to possess the quality of fish stocks which are widely used by fish farmers. Yields of fish and fish products have been greatly improved and are still being increased; resistance to diseases as a consequence of hybridization has also been achieved. Moreover, farmers themselves have learnt the best ways of rearing fish in farmed ponds or in other enclosures and managing the fish to obtain the best possible returns from them.

2.9 Two Approaches – Commercialization and Mechanization in the Development of Fish Culture

Developmental aspects of fish culture and fisheries suggest that two concepts have evolved through 3-4 decades of scientific research. The first concept treats the development as the improvement of fish culture and fisheries sector — a more profitable and labor-intensive business as a commercialization. The other treats as a mechanization. Moreover, because studies on the commercialization and mechanization contribute equally to development of aquaculture in general and fish culture in particular, these approaches cannot be fully segregated. In the following sections (sections 2.10 and 2.11), these two approaches have been briefly illustrated.

2.10 Commercialization of Fish Culture/Aquaculture

The term commercialization can be defined as a process wherein the productivity of fish farm is improved by using higher input-output ratio and market-oriented strategy for high economic returns. However, commercialized fish culture activity would have to be carried out on a sustainable basis. Since the 1970s, the concept of commercialization of aquaculture and fisheries has undergone a notable change and has become more comprehensive. Since this concept is of recent origin, there has been a significant breakthrough in the developnent of various technologies at different research institutes although only a few of them could be adopted by entrepreneurs. Some other refined technologies developed in developed nations may also be adopted for commercial application in India. Commercialization involves the adoption of technologies for development of aquatic resources on various aspects to all categories of farmers who seek livelihood in the rural areas. The category includes small/marginal farmers, tenants, and the landless.

COMPONENTS OF COMMERCIALIZATION

A developmental strategy adopted to bring about aquacultural production has been incorporated the following components.

1. Introduction of circular spawning and hatchery tanks based on Chinese model for mass production of fish seed and the technology is being used most widely and successfully.

2.   Composite fish farming (for further detail, see Vol. 1) should be considered as the second component of commercialization. Among the means of production, the most important in fish culture establishment is clear and unpolluted waters with good management schedules. This farming system has resulted in increasing fish production to the tune of 6,000 - 10,000 kg/ha/year depending upon the type of farming. And many values in between these figures have been reported in many parts of the world.

3.   The third important component of commercialization is integrated fish farming. Fish farming has been integrated with other components like poultry, livestock, agriculture, horticulture, and the like and several medium/large farm enterprises have successfully adopted.

4.   Shrimp/prawn Farming: The exploitation of both freshwater and saline water to their maximum capacity of these lucrative and value-added species generally depends upon the technical know-how employed in terms of production of prawn/shrimp seeds and juveniles. One of the major developmental constraints in shrimp farming is the pollution hazards as a result of intensive shrimp cultivation, salinization of land adjacent to shrimp farms and outbreak of white spot disease (a virus disease of shrimp). All these have resulted almost all large shrimp enterprises to virtually collapse the farming activity with some of them still continuing to operate.

A number of shrimp hatcheries (about 242, data for the year 2009) and shrimp/ prawn hatcheries for dual purpose (about 35 in numbers) have been established in different regions of the country following the technology imported from UAS and France. This development has facilitated production and supply of shrimp/ prawn seeds to farmers.

5.   Processing and Value Addition: The development of processing of aquaproducts and substantial value addition to products introduced through Individual Quick Frozen (IQF), freeze drying etc. have great practical significance not only for increasing the supply of food fish but also for increasing foreign currency earnings. Of late, these aspects have attracted more attention to farms. During the 8th and 9th plans, efforts have been made to increase export markets of both finfish and shellfish through commercialization for increasing incomes of farming areas. For this purpose, research and developmental efforts as well as infrastructural support and services for improving the quality of output, processing technology and marketing strategy will have to be strengthened.

ESSENTIALS OF SUCCESS FOR COMMERCIALIZATION

1.   If commercialization is to achieve a success, it is necessary that the policy guidelines are first to be formulated. A number of agencies should be involved for commercialization of technology concerned. Common water resources owned by individual farmer or Governments or Non Government Organizations need to be utilized in a sustainable and in an ecofriendly manner.

                                     Freshwater Fish Culture Vol. 2

2. Prior to commercialization of any suitable technology, it is necessary that all the infrastructures required for their effective implementation is made available to the entrepreneurs so that the transformation of technology could be easy.

3. Suitability of a technology under any regional conditions is the most important criterion for its commercialization. This approach is more applicable in case of technology transfer from other countries.

4. The success of technology which forms the crux of the strategy of commercialization, hinges on its excellence at the field level and the result is successful. It is claimed that the technology is suitable for transfer to the farmer level. Generally, all experiments should be replicated several times on a commercial scale very precisely before highlighting the results as achievement. This will enable to adopt commercialized technologies with confidence.

5. The level of interest, skill, and information on the part of the farmers should develop to such an extent that they should seek new practices and methods of their own and the field technologies should be so strengthened as to cope with the new demand.

6. Farmers education and training programs are aimed at training farm managers who would guide and influence fellow farmers in the areas concerned for adoption of improved technology. In actual practice, farm managers do not necessarily pass on their knowledge to their neighbours and the required dissemination of knowledge and technique does not take place. In order to fill the gap, the transfer of technology from lab –to– land plays a significant role. Under this system, farming facilities are introduced to relevant technologies that would help in diversification of labor use and introducing of supplementary sources of income in all fields of animal husbandry, agriculture and horticulture in addition to fisheries as the main enterprise.

MARKETING

Marketing of aqua-products both for domestic and export needs to be emphasized for better returns from the commercialized aquaculture. Even entrepreneurs are aware of the market condition for his/her produce before implementing the scheme. Marketing problems can severely affect the project concerned in a variety of ways. Unless a robust marketing system is established, commercialization of aquaculture in general and fish culture in particular will be of no value.

CREDIT

The main aspect of aquaculture/fisheries credit schemes for successful commercialization of the concerned technology is the co-ordination between credit and other economic activities such as cultivation, harvesing, processing and marketing. It is indisputable that most of the farmers no doubt require credit and mere credit alone will not enable him to carry on better farming practice and realize good production of fish so that he may repay the loan. Since the farmer is in need of the supply of fish seeds, feeds, fertilizers, gears etc., co-operative credit societies should help them in all farming activities. Thus there should be full integration of credit with other economic activities of the farmer. For this reason, the credit societies should either be integrated with the marketing and processing units or completely integrated with them in order that the credit facilities extended to the farmer will become purposeful and productive. The best way is to convert all the credit societies into multipurpose ones to make the scheme workable on efficient directions.

At present, in addition to government, a number of financial institutions such as co-operatives, commercial banks, farmer's service societies, National Banks for Agriculture and Rural Development and Gramin Banks have acquired a significant position in aquaculture finance systems of the country. They disburse credit through a variety of fish culture schemes and meet the financial needs of the farmers. They are gradually involving themselves into multifarious activities of fish culture development and development of allied activities as a whole. Apparently, the financial institutions have to evolve new producers, principles, and practices which may be found workable and acceptable.

1. **Two Requirement for Implementing Credit Schemes :** More important among these are the principles that should govern a financial institution's lending policies. There is no differences of opinion as to the view that the institutions should finance only those schemes which satisfy the following two requirements: (i) Technical feasibility and (ii) Economic viability. The former implies that the proposed activity for which the credit is required is feasible under the available agro-climatic conditions and technical know-how. The economic viability means that the proposed project should produce at least minimum of monetary benefit that would create adequate resources to meet the obligations of debt in time and after meeting the above requirements leave adequate surplus to the borrower that should serve as an incentive to the continuation of the activity.

2. **Illustrations of Technical Feasibility and Technical Viability :**

   (i) Technical Feasibility of the Construction of a Fish Farm : The technical aspect of the construction of a fish farm should include the general description of the project such as the latitude and longitude of the project site, its distance from the processing plant and marketing places, whether connected by road, the number of farmers who would be benefitted etc. Similar technical feasibility study of a processing plant should concentrate on the following factors: whether adequate space and sufficient quantities of water are available to run the plant smoothly, whether the plant will receive huge quantities of raw materials for preparing aqua-products, whether the technical assistance/personnel is available, the project should provide qualified processing technologists and necessary administrative and processing/marketing staff. In case of crop loans, the technical aspects to be examined include the availability of quality fish seed, organic manures, fertilizers, feeds, water and technical help.

Any project formulated should aim at adopting a higher level of technology or achieving a higher level of productivity. Different technical aspects, however, include the availability of better technology, their applicability to local conditions, farmers' wisdom or ability to understand the technology, and existence as well as adequacy of the extension network to promote the dissemination of technology.

(ii) Economic Viability: Before constructing a fish farm, the expected farm income and farm cost should be estimated. The farm income refers to the value of gross produce from farm (fish in case of polyculture or fish and additional products in case of integrated farming) less the current cost of faming. In calculating the current cost of farming expenses in cash and kinds incurred on, the following items should be considered:

- wages paid to labors.

- maintenance cost of crafts and gears.

- expenses incurred on fish seed, fingerlings, feeds etc.

- water and soil improvement work.

- development work.

- watchmen/night guard.

We can further illustrate the economic viability of a polyculture system (one hectare water area) with the aid of summarized financial data (For the year 2007) is noted below.

	Amount (in Rs.)
A. Expenditure	
Rental charge of one hectare pond	5,000.00
Net and other equipments	1,500.00
Bleaching powder	1,500.00
Pond reclamation and management	3,000.00
Fingerlings (5,000 Number)	3,500.00
Lime (500 kg)	1,500.00
Organic manure (Cattle manure @ 10,000 kg)	2,000.00
Inorganic fertilizers (i) Urea (175 kg) and (ii) single super-phosphate (250 kg)	3,000.00
Fish feed (Fish meal, Mustard oil cake, rice bran etc.)	30,000.00
Salary of labors	13,000.00
Loan interest	7,000.00
Total Cost	**71,000.00**
B. Income	
Minimum sales price of 3,000 kg of fish (approx.) @ Rs. 45/- per kg	**1,35,000.00**
C. Profit (B – A)	
(Rs. 1,35,000.00 – 71,000.00)	**64.000.00**

Within these concepts of technical feasibility and economic viability of a proposed need, a financial institution need be governed by the principles that follows while considering the finance any activity in the aquaculture sector. *Firstly*, the nature of security is the important issue. It is an accepted principle that financial institution must secure itself by means of a security against the possible loss arising from the non-recovery of loan advanced. A farmer can submit his pond/tank to the financial institution as the best security. A fish culture pond/tank, at any time, is subject to the vagaries of nature and the net fish production of the most evil state may turn out to be nil. Likewise, a farmer cannot be estranged from his pond if he fails to repay the loan. Small and marginal farmers have hardly anything else to provide by means of security. In these situations, the financial institutions should not insist on a perceptible security. The integrity of the farmer, his purpose and character, and the productivity of the investment proposal should be the main criteria to determine whether the farmer is eligible to receive loan in due time.

Secondly, the loan sanctioned should be adequate. An inadequate amount has two consequences: (i) the farmer may not be in a position to utilize the loan in full potential to him and as a corollary, may waste away the loan in other activities thus rendering its repayment and recovery arduous, (ii) the farmer is compelled to go back to indigenous sources such as the money lenders who charge him at exotbitant interest.

Thirdly, the credit should be timely available to farmers. If the credit is not given in time, it is likely to result in imparting the repaying capacity either through failure of the scheme, impeding the implementation of the scheme or in total misutilization. A loan sanctioned for induced breeding programs, for example, should be made available just at the onset of breeding season. A delay in disbursement of the loan may make all the difference to the farmer. Similarly, a farmer can be redeemed from curing to distress sale of his product only if the amount is made available in times of need.

Fourthly, the loan should be disbursed in such a way that it assures a proper end-use. A loan, for example, should not be disbursed is cash to the farmer. Instead, it should be used for settlement of bills of different types of inputs supplied by local dealers. The diversion of the loan for which it is intended not only affects the repaying capacity but also debilitates the base of production and investment.

Fifthly, the loan should be characterized by an element of flexibility. The credit mechanism must be pliant enough to adjust for cost changes.

Finally, it is necessary to disburse loans within a region which is easily accessible to the financial institution, otherwise it becomes arduous to manage the loan. Non-observance by the institution may, in many cases, lead to disbursing of accounts over unmanageable extensive areas, making the scheme most difficult to put under care. This is one of the reasons that account for the hassle of recovery and overdues that most of the financial institutions are currently confronted with this situation.

3. **Problem of Overdues and Recovery of Credit :** The term overdues refers to the amount due by a certain date but not realized by that date. Defaulters, on the other hand, refers to the farmers against whom overdues persist. The analysis of overdues is, however, cardinal for aquaculture management and development. This also helps in bringing about improveness for smooth running of the credit institutions. Overdue analysis essentially involves two important factors such as magnitude and causes.

(i) Magnitude of the Problem Involved :

* The overdues may be measured both in terms of absolute amounts and as a proportion of total demand as well as precentage of the total amount unpaid.

* Both principal and interest amounts due will have to be calculated.

* The category of overdues should be identified, such as which category of farmer are more vulnerable to default, either small or large farmers.

* Overdues may be grouped according to age, such as how much has been overdue for more than a specific period. Overdues with short periods (suppose 6-months or so), have a better opportunity of recovery. Longer the period of overdues, on the other hand, confiscation of the property of defaulters by the financial institution are the chances of being recovered.

* Since overdues vary greatly according to the local conditions of the area concerned, it needs be analyzed separately for different regions and for different financial institutions. While institution-specific defaulters require liberal policy management action by the affected institutions, region-specific ones definitely require a regional policy action.

(ii) Causes of Defaulters : There are two prime factors for concern over the causes of overdues such as external factors and internal factors. The external factors include wilful breach of contract, natural disasters and infractuous investment which are beyond the control of any credit institution. The other sets of farmers such as faulty systems of procedures with regard to identification and selection of farmers, lending policies, disbursement procedures, and inspection of the end-use of credit.

Co-operative institutions face another set of important factors such as failure of co-operative institutions to take up steps to help farmers in times of stress and strain, political patronage to defaulters which is sheltered and not visible and carelessness on the part of co-operative organizations rather than innate inability of farmers to repay.

(iii) Consequences and Problems : Poor recovery performance results in hassles which have long-term adverse effects.

* The absence of recycling of creditable resources may slow down the pace of credit supply which may have deleterious effects on development.

* It renders the credit institutions not fit to draw refinance from the refinancing agencies and consequently, the working capacity of credit institutions is affected.

* It may result in increased cost of credit and keep the staff active in the task of loan recovery instead of devoting time for development of the sector concerned.

(iv) Writing off of Overdues : In case of small and marginal farmers, the problem of overdues is of a different type. Inspite of their best intention, they are not able to repay their loans in time. To help this section of the farmers' community, some sub-national governments have applied to the practice of writing-off of their overdues. But such practice suffers from serious limitations that follow:

* Middle and big farmers may try to get into the category of small and marginal farmers to take the opportunity of the concession offered to small and marginal farmers.

* Big farmers are likely to create pressure and demand the facility of write-off.

* The demand for relief may not be restricted to only aquaculture credit sector, it may spread to other sectors of the region.

* If this practice continues, farmers will adopt agitational methods and will reap other benefits.

* The practice of write-off may cause to trouble the genuine farmer who has to repay the amount received as loan with interest. This condition will invariably cause a perception that corrupted farmers are rewarded and blameless ones are punished. Consequently, genuine farmers will follow the wilful defaulters, their numbers will gradually increase thus heightening the hassles of loan recovery.

It should be added as a note that complete write-off of overdues is not desirable. Some facilities to the sincere and affected defaulters may be considered necessary.

4. **Fisheries Insurance Scheme :** Similar to the comprehensive policy devised in 1978 to offer protection to poultry birds against death due to disease outbreaks, comprehensive cover for inland fisheries has been devised since 1979 to protect fish farmers under Fish Farmers Development Agency scheme. The progress of fisheries insurance scheme has, however, been limited. The benefits availed only by those farmers who belongs to the financial institutions. For successful operation of this type of scheme, it is necessary to extend this activity on an individual farmer basis. Given the present position, we have to proceed a long way.

(i) Advantage of Fisheries insurance: Fisheries insurance has several advantages as follows.

* The scheme will stabilize the farm income. As a generalization, farm incomes oscillate violently because of natural calamities. This instability has a noticeable dent on the investment decisions. Hence this scheme would cater a useful aid in dealing with this hassle.

* Fish insurance not only encourage farmers in areas where high-risk exists but also may serve the farmers to adopt better farming methods more readily.

* Fish insurance increases credit-worthiness of farmers from institutional credit viewpoint, reduces the necessity of their having to incur 'distress credit' from private money-lenders to ascertain the continuation of farming even following a season of crop failure.

(ii) Benefits of Fish Insurance :

Farmers

* Stability of income.

* Protection of investment.

* Promotion of greater risks and capital investment.

* Stability of supply and prices of inputs.

* Supply of credit at very low interest.

* Accumulation of savings.

Credit Institutions

* Better credit worthiness of farmers.

* Greater liquidity of funds.

* Less decapitalisation.

* Lower loan default by borrowing farmers.

Government

* Reduced call for government aid in the event of crop failure.

* Chances of repayment of credit to farmers.

* Stability in collection of dues.

* Less liability of disaster to help farmers.

Others

* Promotion of co-operative vigor among the farming community.

* Improvement of information about farmers.

* Streamlining of administration in rural areas and a link between the farming community and the government administration.

(iii) Problems of Fisheries Insurance: In the implementation of the fisheries scheme, several practical difficulties have crept in; some of which are noted below.

* Absence of reliable long period data on fish yields and losses.

* Wide variety of aquacultural practices.

* Existing land record system.

* Ignorance and poverty of farmers.

* Lack of trained personnel.

* Limited financial resources.

* Development of a sound actuarial structure.

The sum and substance of this sub-section indicates that the different credit schemes–a major step towards aversion of risk in farming, needs welcome from every comer of the farming community. Although it is the first step towards aquaculture development to a large extent, in the true sense, certain limitations still exist nonetheless principally due to lack of experience on the part of implementing agencies. And the power of effective action of current technology of different aqua-farming strategies following the adoption of research reports in respect of aquaculture development vis-a-vis the active involvement of technically qualified personnel of the credit institutions will produce reverberations among progressive fish culturists. If different credit schemes are adequately monitored and evaluated and then modified according to local conditions, it will help increase farm income of individual farmer particularly small and marginal ones. .

2.11 Farm Mechanization

Aquacultural farms in most developed countries have long been mechanized. The introduction of machinery was also an important part of the program for farm modernization in China, Isreal, USA etc. But in most of the parts of Asia, Africa, and Latin America, manual labor and the use of traditional technology retains its traditional importance. At present, however, smaller equipments have been developed for use in small- and/or medium-sized fish ponds in several under-developed countries. Some developing nations have mechanized their traditional farms and thus overcome the hassles inventing smaller equipments.

Now that increasing yields and larger harvests are being obtained as a result of induced breeding and the introduction of aquacultural schemes, and fish producing countries are experiencing a rapid increase in farm mechanization to allow the farmers to cope with the extra works. Farm mechanization has several advantages as follow:

1. Farm mechanization enables the various activities on the farm to be done more efficiently.

2. In integrated fish farming, mechanization allows each work to be done rapidly so that it dose not interfere with other works.

3. The use of mechanization for harvesting, processing, and marketing of fish and fishery products can often lead to better quality. Fish are perishable commodity. If the harvested fish can be rapidly marketed through successive processes, the resulting product will be of better quality.

4. Mechanization of fish protection measures such as disease control and other production- related management schedules ensure that substantial quantities of robust fish will be produced.

Despite several advantages to be gained from farm mechanization, a lack of proper planning can lead to some problems which may partially affect the advantages in the short term.

In many small farms' the purchase of expensive equipments is simply uneconomic, because the farmer never makes enough profit to purchase equipments. Thus where farms tend to be small, as in most tropical countries, the best way of increasing mechanization is through co-operatives or farmers' associations.

UNEVEN MECHANIZATION

The initial cost of farm equipments tends to be very high. Thus only those with sufficient capital can afford to purchase equpiments for themselves. This is the main reason why the farmers of many developed nations employ so much more equipment than Asian or African farmers. Their standard of living is higher, their products are of better quality and fetch a higher price. Most of the fish farmers are too poor to buy production and marketing-related equipments. At the same time, most of the farmers do not know how to manage the farm, how to look after the farm or how to get the best results from different culture systems. Consequently, some farmers become more efficient than others to yield fish. This may lead to personal conflicts or to difficulties of implementing production-oriented schemes.

Summarizing the various advantages and disadvantages it is clear that if farm modernization is not carefully planned, or if it is introduced in areas where it has traditionally not been used, it can lead to certain problems. These can, however, be overcome by adequate planning in the long-term and farm mechanization will increase efficient out put and productivity.

MEANING AND SIGNIFICANCE OF FARM MECHANIZATION

The productivity of fish culture ecosystems and total aquacultural production can be raised substantially through improved fish seeds, feeds and other infrastructures. But these inputs become more effective and their potential better utilized if well-knit strategy and refined technologies are available to farmers. Most of the developing countries still broadly rely on the extensive culture system for fish production at minimum levels. According to an estimate, more than 6,500 kg/ha/year is possible if fish culture farms are highly mechanized. This dependence on extensive culture system does not go in harmony with the modern fish farming technology of pond lining or recirculatory system. In contrast to developing countries, many advanced aquaculture nations have reached the plateau of production where farms have been mechanized to such an extent that it would take time to convert traditional fish farms into highly mechanized ones. This conversion is termed as *mechanization of fish farm.*

Mechanization of fish culture farms in highly populated countries has been the subject of controversy. Those who have encouraged mechanization, has strengthened the controversy that it stimulates productivity and further that the increase of inputs particularly fish seeds and quality fish feeds are quite worthless unless these inputs are adequately utilized by the mechanized farms. Professional reviewers of mechanization generally focus to the unemployment potential of mechinary in a labor-surplus economy of under-developed countries.

CONSIDERATION FOR MECHANIZATION

1. A number of field studies have shown that mechanization promotes productivity. The difference between the productivity of a mechanized farm and a non-mechanized (traditional) farm may range up to as high as between 30 and 40 per cent. Depending on the nature of farming systems, other management options and harvesting can be done more efficiently. For example, the use of aerator, feeding and water exchange devices permit better production of fish than can be achieved in a traditional operation.

2. It is believed that mechanization leads to unemployment since it helps to work quickly and effectively and it will displace human power. But it should be noted that aquaculture and fisheries are labor-intensive and hence the increase in labor requirements can to a greater or lesser degree increase costs and lead to less income. But nonetheless, mechanization must be introduced so that farmers are not replaced before they can find alternative employment.

3. Mechanization of aquafarms is an important catalyst of fish culture development because many farmers still produce only as much as they have to yield and are not interested in getting the full output of their resource. The high cost of mechanized fishing crafts and gears have forced the farmers to maximize their output in order to meet the cost of maintenance. The experience in several regions clearly indicates that the farmers are not unable to do this quite easily as is clear from good repayment records of financial institutions from farmers regardless of their holdings, small or large.

4. It must be pointed out that mechanization generates employment through manufacture of farm machinery, fish feed, repair of equpiments etc. Therefore, at the present stage of development, the minimal loss in direct farm employment resulting from mechanization can be more than offset by off-farm employment in the secondary and tertiary sectors of the economy.

CONSIDERATION AGAINST MECHANIZATION

1. The constituents of mechanization point out that small sized water bodies makes it impossible to carry out activities without any degree of efficiency. Therefore, there is a strong argument that the small holding makes the use of farm machinery uneconomical.

2. It is argued that most of the fish farmers having small farms cannot afford to withhold capital out of their own savings by machines. Therefore, in some conditions it may not be possible to receive the advantages of farm mechanization simply by the defect because the farmer cannot support to purchase equpiments.

However, if we accept this argument it would imply that we are not ready to drive home the benefits of the new technology. The success of new technology hinges closely on the availability of improved inputs including capital.

3. It is contended that most of the developing countries have a scarcity of mechanical skill, but this argument does not seem true.

To sum up, regarding the question about mechanization, it can be said that in the present available situation a complete mechanization of fish culture is not acceptable. But this is a position when strategies have to be formulated for future, and a platform should be prepared for action. If the nation is serious about raising farm productivity and yield, at least a partial mechanization would have to be implemented right now, gradually building up the infrastructure that would ultimately paved the way for complete mechanization. It seems that there is no other alternative to mechanization. It can meet the aggregate demand of both fresh-water and marine water fishes for the posterity.

2.12 Extension of a Fish Culture Farm

Extension of a farm and extension education relate to the process of conveying the technology of fish culture to the farmer in order to enable him to make use of the knowledge in full potential for better fish culture and a better economy.

Extension services afford the necessary skills to the farmers for undertaking improved fish culture operations, to make available to them timely information on advanced practices in a digestible form suited to their degree of literacy and consciousness and to generate in them a friendly attitude for innovation. Moreover, it is also necessary to change the outlook of the farming community. The farmers have not only to be educated about the impact of the use of farm inputs, but also have to be shown the actual difference they make to fish production and farm returns.

Most of the farmers generally have little linkage with advanced technology. There is need for a massive education and extension effort to modernize the outlook of the common farmer and make him innovative, enterprising and willing to adopt readily to changing situations and new technology. Education of women on the farmers' household in fish culture innovations, subsidiary occupations and nutritional aspects should be an important part of the program of farmer's education.

METHODS OF EXTENSION

Three important methods of farm extension are as follow.

1. **Demonstration:** Demonstration is an important method of extension in upgrading the farmer and occupies an important position in the national extension program. With the introduction of the new technology, demonstration programs must be followed. A system should be developed to maximize return and to impress the farmers which require a package approach by the experts. It is necessary to

demonstrate the efficacy of new practices based on scientific outlook, information, and skill by bringing the research results to farmers.

2. **Farm Information and Communication Support :** Farm information and communication support are regarded as essential components of aquaculture extension program for supporting the production activities of the farmers. Both national and sub-national governments have been providing information and communication support to educate farmers. Also,Fisheries Universities and Colleges have established extension sections and communication centres.

3. **Farmers' Education and Training:** This program is aimed at training to farmers who would guide and influence fellow farmers in the adoption of new technology. But in actual practice, trained farmers do not spread their knowledge to other farmers. For this reason, Lab-to-Land Program has been implemented and the results of this program are highly encouraging.

2.13 Organization of Fish Farming

Types of farming have been considered in terms of the methods and types of fishes grown in cultivable water areas. It is also important to realize the effects of farming organization on fish culture development. Farming organization is most affected by the way in which the water of good quality is held; whether the farmer possesses his farm, is a tenant or merely a labourer. In many areas, recent changes have taken place in land tenure.

In contrast to agriculture where land reform policies have drastically changed the entire farming scenarios in most parts of the world, not much attention has been given to water reform policies. Water management, which is one of the important ways of improving fish production in any type of farming system, is considered as a good farming technique. However, in fish farming systems, two principal kinds of farming organizations may be adopted for successful fish culture development: (1) owner occupation and tenancy and (2) cooperative farming.

OWNER OCCUPATION

Many farmers possess their own water bodies and take rent or lease from pond-owners. Farms owned or rented vary much in size and range from large to small ponds, often almost too small to be worked economically. Income also vary much so that some farmers get good returns and have a high standard of living while others may be poor, or nearly in debt. Farmers purchase their own fish seeds, fertilizers, nets etc. and grow and harvest their products. In other words, they work individually. Owner occupancy and tenant farming may be modified in a number of ways. Profits may be enhanced by the farmers joining together in a cooperative or, in areas where farming returns are generally low and farmers suffer from a lack of capital and certain special tenancy arrangements may be made. In such case, the landlord provides the input because the tenant has no cash to purchase it, and the tenant then begins to work candidly. When the crop is harvested, the landlord receives a part, or the crop in lieu of rent and to pay for the input.

CO-OPERATIVE FARMING

Co-operative farming is practised in developed countries though the principles of co-operatives which are being employed all over the world. This is an advanced form of farm organization which has proved workable in rural societies which have a literate population, capable of understanding the co-operative principles and components in managing the business of the co-operative in the best interest of the farmers.

It is needless to mention that the aim of co-operatives is to eliminate the middleman, also generally affords a valuable service in collecting the farmer's produce and bringing it to market. Rather, the co-operative tries to perform this function itself, in addition to farm production, and thus enables its members to reap profits. At the same time, by acting as a buying as well as selling agency, co-operatives can also take advantage of economics of scale and buy seeds, nets, boats, fertilizers, feeds etc. in bulk, thus obtaining better terms.

The success of co-operative depends on many factors including the efficiency of the farmers or their paid executives, their ability to complete with long-established traders and the actual co-operation of the farmers themselves. Many farmers all over the world have attempted to run co-operatives but have failed because of their inability to complete effectively with the more dynamic businessman or because of the lack of confidence of the farmers in their elected executives. But in many advanced nations, the co-operative farming has been so successful that every farmer is a co-operative member. However, the capital resources of the co-operatives enable them to finance reclamation or farm improvements as well as setting up small-scale fish processing industries. Co-operatives help the farmers in the following ways:

1. **Purchasing :** Farmers can buy feeds, seeds, fertilizers, nets and can therefore buy at reasonable rates than an individual who only require small quantities of any item.

2. **Marketing:** After harvesting, sorting, storing, and packing, co-operatives are able to sell in bulk quantities and can dispose of their products at the favorable times. They can, therefore, obtain more favorable terms of trade than individuals. Co-operative also works together with government-run marketing boards which set certain standards for aqua-products guarantee fixed prices for such products. Maintaining of high standards in this way may assist the export trade.

3. **Processing :** Several fish processing industries are controlled by co-operatives. This enables the profits of such businesses, closely allied to farming, to accrue to a large number of co-operative members.

4. **Finance:** Emergence of financial services such as Grameen Banks, Co-operative Banks, Nationalized Banks etc. as a major development of farming community is one of the major offshoots of the reforms. These institutions provide loans to needy farmers on easy terms when they need to purchase ingredients necessary for fish culture. Such financial services have become an attractive new development proposition for the farming community. When most of the farmers as a class readily

started availing of the emerging financial services, quite a few farmers have gone to a step further and floated their own financial services and pursued financial service as one of their developments.

5. **Advice and Research :** Co–operatives are engaged in aquaculture research and farmers are entitled to advice from experts on the hassles associated with their farms. They are thus able to improve the efficiency of farms and the quality of products in order to get better returns from their ponds. Farm equipments can also be borrowed from co-operatives.

The co-operative societies have thus benefitted the entire farmers' community particularly the small and marginal farmers and may given high standard of living to farmers. The farmers not only earn a comfortable living on their farms, but receive income from their co-operative societies, as all profits made by the co-operatives are either ploughed back to expand the co-operatives or go back to the farmers. A farmer is thus encouraged to join as many co-operatives as he can and take part not only in production but in handling, processing, and exporting fish and fish products.

2.14 Appraisal of Fish Culture Development

Most of the tropical and sub-tropical countries have achieved substantial advancement of fish culture industry. A solid aqua industry has been created and a considerable degree of diversification and sophistication has also been accomplished on this sector. The growth of fish culture sector has been particularly striking in segments such as prawn, shrimp, cuttle fish and their preservation as well as export, production of carp etc. The traditional aquaculture and aqua-industries have also undergone rapid modernization during the latter part of the twentieth century. In recent years, fish culture industry has also undergone a qualitative change in addition to the quantitaive expansion. It has enfolded the impression of optimum scales of production, state-of-the-art technology, internationally comparable cost-effectiveness and level of productivity. Though it still has a long way to move in this respect, a good inception has already been made and it foreshadows well for the posterity.

Like agriculture, aquaculture is also an encouraging sector of the rural economy. It has always been the backbone of the national economy to a greater or lesser degree. Fish culture also supplies raw materials to the industry. substantial portion of the nation's exports is also provided by this sector. The future growth of fish culture industries will increasingly depend on rural prosperity, which, in effect, means fish culture prosperity. In fact, to day, export of aqua-products in several Asian countries seems to agree that fish culture constitutes the foundation of their overall development process.

As noted earlier, the 'fish revolution' relies on the state-of–the-art technology as the main source of increase in fish production. The fish revolution was a great turning point in many regions of the tropics and more than anything else, they explored the age-old myth about the tropical farmers — that of their total traditionally; it laid to rest the vision of the tropical farmers as a robot, changeless, obdurate, and despicably following their inherited fish culture traditions, from a home need-based land-use system to a market-oriented land-use system. And the progress in the farm front has provided a thrust to the overall economic activity in the area concerned. Fish culture has improved the economic condition of the vast masses of farmers in the rural areas; it has created additional employment and addi-tional purchasing power among the farmers and brought them into the consumption community. However, the fish culture scenarios that we observe in the rural areas of developmng countries are the direct outcome of the advancement brought in by the revolution on the farm front.

2.15 Development Trends in Fish Production

During the last 25 years, fish culture has developed as one of the most encouraging industries around the world due to gradual stagnation of riverine fish production. At the same time, development of biotechnology has stimulated the production potential of finfish and shellfish through adoption of intensive, extensive, and super-intensive culture systems.

NATIONAL SCENARIO

During the last forty years or so, Indian aquaculture production (both inland and marine) has increased several folds. During 1950s and 1960s, the increase in production came from the sea. During 1970s, the development of freshwater fish culture accounted for much of the increase. For statistical evaluation of fish culture growth, the country has been divided into three regions such as (1) Western (states include Maharashtra, Gujarat, Kamataka, and Kerala), (2) Eastern (states include West Bengal, Orissa, Andhra Pradesh, Tamil Nadu, and Andaman-Nicobar Islands), and (3) the land-locked (states include Bihar, Madhya Pradesh, Uttar Pradesh, Punjab, Jammu and Kashmir, Himachal Pradesh, Haryana, and North-Eastern states) regions. Among these three regions, the first two ones are equally important for the development of fish culture (fish production accounting for more than 40 per cent of the total) while the land-locked regions contribute more than 15 per cent of the total catch.

Since the latter half of the sixties, the growth of Indian aquaculture is on the increase. The improvement in fish seed production sectors has been particularly noteworthy. Over the decades, the country steadily increased fish production. Production of finfish has been hovering around 17 million tonnes and shellfish accounting for around 19 million tonnes (data for the year 2009).

INTERNATIONAL SCENARIO

Having expanded significantly, fish culture sector has provided about 26 million tonnes by 2009 compared to about 7 million tonnes by 1984. The value of freshwater aquaculture production in 1990 on global basis was nearly one thousand million US dollar. Global aquaculture expansion was, however, not encouraging during 1985-1992. During 1984-1989, the compound growth rate was 25 per cent by value and 15 per cent by volume. But it should be kept in mind that significant production has taken place in the developing countries (particularly in Asia) and the yield has increased from 76 to 87 per cent by volume during the period.

Freshwater fish production trends during 1990s indicate that there always exists regional differences in volume. Production trends in China and India, for example, varied between 6 and 9.5 and 1.5 and 4.7 million tonnes, respectively. One Latin American country such as Equador produced more than one lakh tonne and in Africa, only two countries such as Nigeria and Egypt produced more than 15, 000 tonnes. Production trends in Asian regions are dramatic — trend varied between 1.2 and 3.8 million tonnes. Though substantial amounts of carps are produced in most Asian countries, the production trends of catfishes, tilapias, trout, and salmon are also noteworthy. Whilst aquaculture production contributed to more than 25 per cent of the total fish supply in Asia, it is very less in Africa (1 per cent of the total fish supply). Most of the African countries produce fish less than 0.25 kg/caput. Although aquaculture could contribute significantly to food security in Africa, the potential is yet to be realized.

2.16 Shift From Capture to Culture Fishery Towards Development

Due to gradual destruction of capture fishery resources and depletion of fish catch, there is a shift from capture fishery to farming and aquaculture across the world. This shift has been considered unanimously due to rapid growth of fish and higher production in different culture conditions. However, the success of development in different aspects of fish culture includes (1) prevention and control of diseases, (2) construction of hatcheries, (3) use of probiotics, (4) development of quality feeds, (5) development of closed recirculatory system, (6) integrated fish farming, and (7) application of genetics in fish culture.

A number of fish species have been considered as the main object of fish culture and their commercial cultivation has flourished as an industry in the context of decreasing rate of capture fishery. The development of methods for mass culture of different species of edible fish such as carp, tilapia, catfish etc. in freshwaters with subsequent improvement in rearing techniques paved the way for increasing fish production to a marketable size within very short periods. A typical example will show how change from capture to culture fishery has given an encouraging result. In Japan, the culture of freshwater eel (*Anguilla japonicus*) is a very lucrative industry. The production of eel far exceeds that of other important freshwater fish. When the culture-based production varied between 40,000 and 50,000 tonnes in ninetees, the annual production from capture fishery is very low (500-1,000 tonnes). Recent advancement of eel culture is characterized by the construction of green house type ponds with the facilities of simple method of feeding, size-wise seggregation, and removal of diseased fish.

2.17 Quality Feed Ingredients, Disease and Health Management and Biotechnology

FEED INGREDIENTS

The use of fish meal for aqua-feeds is gradually increasing and it is assumed that it will rise further, while supply of marine feed ingredients is declining mainly due to expansion of the culture of carnivorous fish. Techniques formulated to reduce environmental damage by developing low-pollution fish feeds tend to induce greater utilization of fish meal. It is imperative that fish culturists promote the culture of harbivorous and/or carnivorous species wherever economically feasible, to reduce dependence on high-quality protein-rich feed inputs and to take the advantage of farm-made supplementary inputs. Research on the use of feed ingredients, attractants, , and probiotics is badly needed.

DISEASE AND HEALTH MANAGEMENT

The most important constraint to further development of fish culture in many regions of the world particularly in Asia is the emergence of disease problems. These problems not only cause economic losses but also public concern about the use of antibiotics and other therapeutics. Most of the diseases are closely associated with the environmental degradation. A number of viral diseases have no control measures and hence prevention rather than control must be encouraged. In tropical conditions, environment-host- pathogen interactions have not been properly evaluated and extensive studies towards sustainable development are necessary.

Scientific attention should be given by research institutes to improve the practices associated with the production of robust fish to check the spread of diseases. Although disease problems are widely known, the guidelines designed to alleviate them are not being implemented in many countries mainly due to lack of physical facilities and expertise.

BIOTECHNOLOGY*

Biotechnology has been the essential tool for fish production. Significant development of biotechnology in fish culture has been made in many countries of the world. Due to the technological development, fish stocks have been manipulated in such a way that production can be enhanced up to the carrying capacity level. The quality of fish feeds has also been improved to a significant extent and are subject to better quality control. Also, the use of probiotics, closed recirculatory system in fish culture, preservation of fish gametes, bioremediation, processing and preservation of fish and fish products, and development of quality aqua-products for export are some of the most important and significant strategies for biotechnological development in aqua industries across the world. These aspects have been discussed at length in separate chapters of this volume. The increasing importance of high-tech aqua-industries and growing awareness among progressive fish farmers have, however, translated the efficiency of the best technology

---

* For further detail, see Ranga (2002).

into action to increase opportunities for further investment in the sector concerned and improved performance has resulted in reducing the cost of production.

2.18 Rivalry for Water and Land

Along with the increase in population, the rivalry for water and land will increase and as a corollary, requirements for food will also be increased. Although it is possible to increase the land utilization for boosting agriculture production, the rivalry for water has serious consequences for freshwater fish culture.

It has been estimated that about 70 per cent of world, managed water resources are effectively utilized for agriculture production. Inland aquaculture, however, requires significant quantities of freshwater. At present, about 60 per cent of total aquaculture production comes from freshwater and the production has been doubled through polyculture with the utilization of freshwater resources in full potential.

In most of the tropical countries, carps and other cyprinids are produced through semi-intensive and extensive culture systems. For boosting up production of aquatic food items, fish culture must be integrated with the development of agriculture. Effective utilization of land and water is very important while selecting the pattern of culture systems for future development. It should be added as a note that fish production can be substantially increased through different forms of enhancement (See Panel 2.1).

2.19 Genetics and Biodiversity With Respect to Fish Culture Development

It is widely known that although most of the wild varieties of domesticated terrestrial livestock have been lost due to lack of proper management of the wild stocks, most of the aquatic genetic diversity is still found in wild and undomesticated species. Therefore, it is necessary to conserve this diversity to form the basis for future genetic improvement programs to sustain capture fisheries. Cross-breeding between wild and farmed stocks constitute a potential danger in fish culture. Depletion of wild stocks by predation, loss of indigenous breeds due to hybridization, competition for food in breeding grounds, changes in aquatic habitats, and transfer of diseases are the most important problems. Several methods such as cage and pen culture of commercial species in natural or culture-based fisheries form greater risks to wild species. ,

Although livestock breeding programs have been practised for centuries, reproduction of the Indian major carps and Chinese carps under controlled conditions has become possible within the past three decades. Selective breeding and hybridization and manipulation of chromosomes and genes have now been developed. These programs are definitely important in culture-related fisheries to follow the current farming strategy. Transgenic catfish, tilapia, coho salmon, ,rainbow trout, and carp, though in experimental basis, have been produced using growth hormones from another species to make them growth at faster rate.

Among fish culturists, scientists, and the public, there is a growing awareness that the diversity in fish forms harbored in different ecosystems generally termed as *aquatic biodiversity*, is an immensely valuable resource that will be cherished by future generations. One important region for preserving as many commercial fish species as possible is the existence of endangered species. If human activities inadvertently reduce biodiversity significantly, there is more risk of unexpected and unintended human efforts on aquatic environment.

PRESERVATION OF FISH GAMETES*

The most significant technique that has already been developed in culture fisheries development program is the method of preservation of fish gametes. Due to shortage in fish seed production arising out of asynchronization of gonadal maturation or of non-occurrence of simultaneous maturation of gonads, the preservation of fish gametes (Sperms, ova, and embryos) has emerged as a very successful technique to carry out induce breeding. Generally, twe methods are involved in the preservation of fish gametes such as short-term and long-term. In the former method of preservation, sperm are stored for several hours or weeks in an unfrozen condition. In the latter method — called as *cryopreservation*, in which gametes along with suitable medium (such as glucose, sodium chloride, and fish Ringer's solution) and cryoprotectant (such as glycerol, methanol, and ethylene glycol) are preserved for prolonged periods at very low temperature ($-196°$ C).

2.20 Management of Water in Relation to Fish Culture Development

Better growth and health management of cultured aquatic animals are dependent on the primary production of pond water. In farming system research, the interactions between fish biomass and environmental factors should be clearly assessed. In the context of fish culture development, the system of management of water must be adopted in a way which provides conditions such as (1) least favorable for the occurrence of disease and (2) suitable for cultured finfish and shellfish. The question of water management was discussed at the conference of experts of SAARC (South Asian Association of Regional Cooperation) countries held at Rai Bareli, Uttar Pradesh, India, in February 1991 and it was concluded that for sustained fish culture development several measures should be taken into consideration. These measures include: (1) biological methods of treatment of pollutants, (2) recycling of municipal and solid wastes on the outskirts of the city, (3) handling of effluent treatment plants with utmost care, (4) use of effective detergents having low phosphate, (5) effective implementation of water act to prevent water bodies from pollution, and (6) control of population explosion. All these measures cause severe pollution of freshwater bodies but it is unfortunate that nobody is interested in this regard.

The international water management Institute, Washington, USA, has cautioned that more than thirty countries which are situated around Asia and Sahara Deserts, North and South America will suffer from freshwaters for use in agricultural and aquacultural activities

* For further study, see Bayness and Scott (1980) and Stoss (1983).

within 2030. The amount of water to be used for the purpose will be greatly reduced. Consequently, farmers would be detached from farming activities. The institute has reported that not only drought-prone and semi drought-prone areas but also the areas where the crisis of water is not severe even today would be affected in the near future. The use of water is increasing day after day. If this trend continues without adequate management of surface and ground waters, trends in fish production will definitely be hampered.

2.21 Assistance for Development of Fish Culture

Financial assistance need be borrowed for fish culture purposes in order to meet farm and non-farm expenses. Aquacultural borrowings are, however, divided into two parts on the basis of the use to which the credit is put : *Firstly,* production and investment loans and *secondly,* consumption loans. The first group may be classified the loans contracted for different purposes associated with different cultural operations varying from the purpose of inputs to the marketing of produce. These can be listed as follows: (1) purchase of quality fish seeds, fertilizers, manures and feeds, (2) purchase of aquaculture implements, (3) repairs of aquaculture implements, transport equipments and farm house, and (4) other current farm expenses. All these purposes are, however, productive in the sense that they are used only for production-related activities. So far the consumption loans are concerened, fish farmers burrow to fulfil his consumption expenditure such as (1) education, medical and other family expenses, (2) payment of old debts and (3) purchase of domestic utensils. And all these forms of consumption expenditure do not bring any income to the farmer. Therefore, once contracted, farmers become a permanent liability for the farmer in the sense that they fail to manage excess funds to pay off these contracted loans. Consequently, the pressure of indebtedness goes on and abetted by the deceitful conduct of money lenders.

SOURCE OF AQUACULTURE FINANCE

Different sources from which fish farmers procure loans can be classified into two groups such as institutional agencies and non–institutional agencies. The former group is further classified as co–operative societies, commercial banks, regional rural banks, farmers' service societies, National Bank of Agriculture and Rural Development (NABARD) and the like. On the other hand, the non–institutional agencies include the local money-lenders and their agents and landlords. Most of the farmers entirely or partially depend on the non-institutional agencies because most of the institutional agencies are, at present, reluctant to provide assistance in the form of loan for consumption needs. Obviously farmers have to fall back upon the non–institutional agencies for that type of loans.

The institutional agencies have made a dent in the aquaculture finance situation. Attempts to build up the institutional assistance system for fish culture concerned with the adoption of credit societies. These societies relieve farmers from the traditional burdens of debt. However, different institutional agencies together have become a viable system of assistance for the development of fish culture. At present, the growth in the total institutional credit indeed is impressive.

The development of competition from institutional improvement in the state of fish culture in terms of technology and profitability helps improve the position of fish farmers *vis-a-vis* the money-lender and also helps to bring down the rate of interest charged by private credit to acceptable levels. It is possible that the importance of private credit organizations will be less and not satisfactory from a policy point of view.

Considering the state of being immoderate of the problem, the importance of private source of credit will continue to be important for quite some time. Besides the attempts to mobilize resources from private sector, the government should also establish some mechanism for smooth functioning of the private credit sector.

PATTERNS OF ASSISTANCE FOR DEVELOPMENT OF FISH CULTURE

The government of any country is principally responsible for evolving a suitable rural credit policy through different systems and the governments have also the overall responsibility to develop the organizational framework for institutionalization of credit so that the exploitative system of providing finance by money-lenders is replaced by a system of facile credit responsive to the needs of the farmers.

Keeping this point in mind, governments have been forced to implement a vast variety of schemes on the development of freshwater fish culture. Different items in this regard have been divided under two heads such as (1) ongoing components and (2) new components. The ongoing components include (1) construction of new ponds and tanks in farmers' own land with screened outlet, inlet, and shallow tube-well, (2) reclamation of existing tanks/ponds, (3) first year inputs such as seed, fertilizers, feed, manures, and preventive measures for a variety of diseases, (4) running water fish culture in mountain regions, (5) integrated fish farming, (6) pumps/aerators, (7) freshwater fish seed hatchery, (8) purchase of vehicles, and (9) training of fish farmers. The new components are characterized by the (1) establishment of laboratories for fish health and water quality monitoring, (2) establishment of hatcheries, (3) provision of water and soil testing kits, (4) setting up of integrated units, and (5) transportation of fish seed.

The assistance under different schemes is given in the form of subsidy. The subsidy is also available for development activities to the Fishermen Co-operative Societies. It should be added as a note that if the subsidy system is withdrawn, it may cause the precarious condition of fish farmers' community.

FOREIGN ASSISTANCE

During the 1990s, several fish culture projects have been funded by several countries of the world and these projects have already been implemented successfully. Trout farming on commercial scale in Himachal Pradesh with Norwegian assistance, pilot project on village trout farming in Jammu and Kasmnir with European Economic Community assistance, and. reservoir fisheries in Kerala with German assistance are apt examples. These clearly indicate that the development of freshwater fisp culture sector has lead to virtual expansion not only in culture sector but in fish processing and export as well.

The implementation of the World Bank assistance for fish culture project is also equally important to increase fish production to the tune of about 10, 000 tonnes/year.

2.22 Weaknesses in the Development Programs

Fish farmers are fortunate that different financial institutions are trying their best to work out and implement an effective strategy to increase the production of farmed fish. But it is unfortunate that crores of rupees allocated for fish culture projects on different aspects are not properly utilized for the purpose for which it was sanctioned or not even spent; it is diverted to other and non-priority purposes. Neglect of water management and disease control, lack of skilled manpower in rural areas and entrepreneurs, lack of awareness etc. have compounded the problem manifold. Establishment of government and non- government organizations are expected to help to prepare a master plan for production of robust fish. Crores of rupees spent annually to uplift the standard of living of fish farmers' community can be saved if the organizations do so with swiftness.

Before implementing most of the programs, inagural addresses are delivered by subject specialists to encourage fish farmers. But it is unfortunate that the follow-up action is not taken in-depth during the entire project plan period whether the project is running without any difficulty. Therefore, regular monitoring and survellance should be done for effective implementation of different projects pertaining to fish culture.

2.23 Freshwater Culture Fishery Projects in the 21st Century: Indian Scenario

Indian farmers have entered into the new century. At this juncture, much attention to the development of some aspects of freshwater culture fisheries has been given by scientists, planners, and technocrats. Different projects have been taken up by the national government. Some of the important sectors in this regard are noted below :

SUSTAINABLE DEVELOPMENT STRATEGIES FOR INLAND AQUACULTURE

Capture fisheries are related to open water systems where fisheries exploitation is highly limited due to variability in fish habitat. Management strategy circumscribes the sustainable exploitation of natural populations. On the other hand, fish culture in ponds and tanks is widely practised where there is scope for human intervention during the entire culture period.

Sustainable development strategies for inland aquaculture and fisheries essentially fall under four categories such as (1) Capture fisheries in rivers, lakes, estuaries, and backwater lagoons, (2) culture fisheries in ponds, tanks, wetlands, and in space partitioned off within large lakes, (3) Culture-related capture fisheries in reservoirs, floodplain wetlands, and (4) Stock and species enhancement in medium and large reservoirs.

According to the Food and Agricultural Organizations (FAO), development of sustainable fisheries refers to "the management and conservation of the natural resource base and orientation of technological and institutional charge in such a manner so as to ensure the attainment and continued satisfaction of human needs for preset and future development." This statement clearly indicates that fisheries development must be environmentally, socially, and economically sustainable. At the same time, 'low external inputs for suataitable aquaculture system' must be adopted. Therefore, sustainability in fisheries is relevant to both capture and culture systems and quality aqua-products which are useful to the humankind. Different inland open water fishery resources that follow (See Chart 2.3) afford different management norms more obedient with alternatives for sustainable development. Most open water resources are managed in most special events through different improvement procedures which are intermediate to capture and culture strategies. These strategies are definitely eco-friendly, non-degrading, and technically feasible.

Chart - 2.3
Inland Open Water Fishery Resources of India and their Strategies of Fishery Management

Resource	Resource Area	Management Strategies
River	29,000 Km	Capture fisheries
Mangroove	3,56,000 ha	Subsistence fisheries
Estuary	3,00,000 ha	Capture fisheries
Estuarine wetland	40,000 ha	Aquaculture
Backwater/Lagoon	1,90,600 ha	Capture fisheries
Large and Medium reservoirs	16,67,810	Stock and species enhancement
Small reservoirs	14,85,550 ha	Culture-related fisheries
Floodplain wetlands	2,02,210 ha	Culture-related fisheries
Upland lakes	1,50,000 ha	Capture and Culture fisheries

DEVELOPMENT OF BEEL FISHERIES

The term 'beel' refers to the floodplain wetlands (See Chart 2.4) which possess their permanent or temporary association with the parent river systems. They ceaselessly shift their beds when riverine systems are changed from one position to another. They make perfect example of foodplains such as Ox-bow lakes, sloughs (hopeless depression), back swamps, and meandering scroll depression of residual channels. They are either shallow depressions or dead river beds and particularly receive huge quantities of water during monsoon season. Consequently, they are termed as *open* or *closed* depending on their connection with the main river. Beels are repository of diverse species of fish and fish production potential is, therefore, great (See Chart 2.5). Unfortunately, most of the beels are on the verge of disappearance due to several factors such as anthropogenic tampering, siltation, etc. Due to excessive growth of macrophytes over extensive areas most of the beels have become good-for-nothing.

Chart - 2.4
Distribution of Floodplain Wetlands in India

State	Distribution	River basin	Area (ha)
Arunachal Pradesh	East Kameng, Lower Subansiri, East Siang, Dobang Valley, Lohit, Changland and Tirap	Kameng, Subansiri, Dibang, Lohit, Dihing Tirap	2,500
Assam	Brahmaputra and Barak Valley Districts	Brahmaputra and Barak	10,000
Bihar	Saran, Champaran, Saharsa, Muzaffarpur, Darbhanga, Monghyr and Purnea	Gandak and Kosi Dhar	40,000
Manipur	Imphal, Thoubal and Bishnupur	Iral, Imphal and Thoubai	16,500
Meghalaya	West Khasi hills and West Garo hills	Someshwari and Jinjiram	210
Tripura	North, South and West Tripura District	Gumti	500
West Bengal	24 Parganas (North and South), Hooghly and Nadia, Malda Murshidabad Cooch Behar Burdwan Dinajpur (North and south) Midnapore	Hoogly, Ichhamati, Bhagirathi, Churni, Kalindi, Dharub, Dharala, Pagla, Jalangi, Behula, Torsa and Mahananda	42,500
Total			1,22,210

Source : Nath and Das (2004)

1. **Suggestion to Improve Beels :** Several suggestions that follow may be reccommended to rejuvenate the damaging beels/wetlands.

* Adoption of Geographic Information System related inventory survey of culture-related fishery resources.

* Removal of macrophytes to the possible extent to restore the pristine condition of the region concerned and to save them from disappearance.

* Phase-wise excavation of littoral areas with the available labor inputs of the fishermens' community.

* Species enhancement following stocking and harvesting with degree of regularity.

* Adoption of pen culture in potential areas.

* Removal of weed fishes as far as possible.

* Potential beels should be declared as fish sanctuaries.

* Use of limestone at the rate varying between 300 and 400 kg/ ha/year to make the decomposition rate of organic load more effective.

* Forestry should be practised around the catchment areas to provide shade during summer and to make the soil productive.

* Introduction of exotic species into the beels should be prevented.

* Establishment of nurseries for *in–situ* seed rearing.

Chart - 2.5
Fish Production Potential in Some Beels of India

State	Name of the beel	Average Prdocution (kg/ha/year)	State	Name of the beel	Average Production (kg/ha/year)
West Bengal	Bhroma	1,230	Uttar Pradesh	Nainital	815
	Patari	740		Salontal	825
	Kola	335		Dahital	795
	Baloon	920		Rewati	1,290
	Dekole	1,135			
Assam	Boka	1,550	Bihar	Manika	1,640
	Na beel	1,300		Kanti	1,450
	Dighali	1,390		Sirsa	985
	Gala beel	480		Matwali	1,700
	Sone	1,310		Motipur	
	Samaguri	840		maun	975
	Mandira	1,060			

DEVELOPMENT OF DEEP WATER TECHNOLOGY

Placing the country into the league of select countries, Indian scientists have developed state-of-the-art technology for underwater lakes/reservoirs/sea exploration for activities such as lifting samples from different depths and analyze soil conditions. The technology has been developed by the National Institute of Ocean Technology, Chennai. Under the initiative, scientists have developed a world class deep water mining machine generally termed as *under-water crawler,* remotely operated vehicles and a deep water *in situ* soil testing machine. This method is propitious for reservoirs and large lakes where ordinary testing equipments fail to operate efficiently. Development of deep water technology was a frontier area for harnessing non-living resources on the reservoir/lake/ sea bed such as nodules that contain heavy metals apart from traces of other minerals. This technology will help measure the extent of contamination/pollution of aquatic ecosystems over unmanageble extensive areas.

RESERVOIR FISHERIES DEVELOPMENT PROJECT

The most important area that has been considered by planners is the development of reservoir fisheries. Experts know that without a bunch approach, an effective organizational mechanism, activities in respect of supplies principally associated with the supply of fish seed and regular operations of reservoirs to maintain the balance between capture and culture, the development of reservoir fisheries is not possible. The need for adoption of phased harvesting, establishment of cold storages, transport systems for marketing, and export-oriented development must be taken into consideration. Inspite of the availability of these components in hands, the acting arrangement of effective development of reservoir fisheries is a threatening study. This has encouraged experts to enforce the project for high production.

It is generally considered that reservoir fisheries development is the main plan of the fisheries plan document and therefore this plan should be implemented right now in suitable manner.

DEVELOPMENT OF POND FISHERY

It is important to note that the strategy of the present pond fishery operation must be improved. Unless export-oriented culture techniques of carps are planned, it is not possible to improve the production capacity of pond culture systens. At present, however, some North and southern states of India have reached a plateau of carp production. This is due to the fact that the adoption of new technology and use of quality fish feed in pond culture sector in the regions concerned have triggered to produce more fish seed and table fish. This situation causes a trend to reduce the cost of pond fishes in the market. To solve this problem, progressive pond fish culturists have given more attention towards freshwater prawn and also tiger shrimp which is now being cultured in freshwater ponds. Production technology of seeds of tiger shrimp and scampi should receive more emphasis for incorporating an economically viable scheme that involves supply of adequate but quality seeds to pond culturists through a networking system to facilitate production, processing, and preservation not only for domestic consumption but for export of the product. It should be added as a note that the development of the present pond fishery economics will not be improved unless export-oriented marketing system is considered and developed.

DEVELOPMENT OF FISH CULTURE IN CAGES

Many freshwater areas are known for better utilization of their potential for cage culture. Apert from serving as an ideal site for fish culture in cages, water bodies also serve as the main source of fingerling fishes for the culture industry. A variety of species particularly Pangasius catfish, snakehead , tilapia, Asian catfish, carps, and perch are widely cultured. However, in some tropical countries such as Vietnam, cage culture has expanded into an intensive fish culture industry and the production otained from the cages has been estimated to be around 60,000 tonnes/year. To increase the catfish production from inland waters, Vietnamese cage culture technology may be followed in India and adjoining countries because of the similar agro-climatic conditions and grow-out characterstic features of culture ecosystems. Uptil now, it has not been possible to adopt fish cage culture systems. Training and extension programs on different inland aqua-agriculture systems would be useful for the farmers to make suitable modifications in their farming systems and to consider the locational advantages of the site available.

Though Indian sub-continent is endowed with many rivers, lakes, and reservoirs, the cage culture potential is yet to be realized. However, the adoption of Vietnam type of floating cages of 200 to 1,000 m^3 in these water bodies may definitely lead to a breakthrough in the field of inland cage culture sector of the country. Because of the vast potential inland water areas in some tropical countries, the total quantities of fish produced by cages are likely to be very high. The fish cage culture may be considered as an exclusively another topic of great importance and will be discussed at length in Chapter 9.

INTEGRATED FISH FARMING

Development of finfish culture in most of the Asian and Southeast Asian countries comprises stocking of silver carp, grass carp, Indian major carp, tilapia, some species of catfish (such as *Pangasius* and *Clarias*), snake-head fish, and gourmy. In these regions, freshwater culture sector is dominated by many aquaculture systems such as rice-prawn/fish, fish-livestock, fish/prawn-plant crop, paddy-fish-prawn, garden-fish- livestock, and vegetable-fish/prawn. Production may, however, vary between 0.4 and 12 tonnes/hectare depending on the system followed. Substantial amounts of fish are produced from cage culture systems. Yield has been recorded upto 150 kg of fish /m^3/year. Different types of farming systems must be adopted for significant production. Expansion of integrated fish farming is being made in extensive manageable areas where the potential of this farming system is resplendent.

So far the integrated fish farming system is considered, it is evident that great strides have been made by progressive fish culturists in several states of India such as Punjab and Haryana. It is, however, expected that different plan proposals will recommend and implement a scheme on integrated fish farming by the planners to other states of the country.

Integration of cash crop (such as vegetable, banana, coconut and others) paddy and livestock (such as duck, poultry, cow and pig) with fish/prawn would be desirable proposition for effective utilization of the embankment and adjoining water areas. Such developmental efforts may result in generating regular additional income to the farmers. Therefore, such system of fish farming must be widely practised in areas where the culture possibility exists. In a typical integrated system, garden-pond-cage, prawn-cum paddy, and livestock-garden crop practices are undertaken. In an area of one hectare, for example, 0.6 hectare is used for fish culture and the rest for raising garden trees. Another integrated farming system involves the culture of cage-paddy-garden-fish of the total area of 3 hectare available for the purpose, the farm utilizes 2.4 hectare for paddy-prawn/fish integration and only 0.6 hectare is used for rice culture. Integrated fish farming systems have been described in Volume 1 (*see* Chapter 11).

DEVELOPMENT OF WETLAND FISHERIES*

Wetlands are called as different names in different places. In some parts of the world, for example, a marsh may be called as *bog*. But three factors such as soil type, vegetation, and hydrology are responsible whether or not a piece of land is a wetland. Wetlands are, however, often called as *swamps, flood plains of rivers* and *littoral area of lakes* and *ponds, paddy fields, tanks,* and many other names. Wetland ecosystems are composed of multiple-linked components that shuffle components through energy flow, water flow, and chemical as well as physical transfers.

1. Importance of Wetlands : Wetlands are fragile and intricate ecosystems that are very vulnerable to the activities that occur around and within them. Inspite of this, wetlands serve many functions for humans. Millions of people depend up on wetlands for sustenance activities such as fishing, farming, hunting etc. Water of wetland is used for aquaculture and agriculture. A number of fish species (both finfish and shellfish) hatch and develop in to maturity in the wetland swamps and when they become adult, they migrate into the ocean.

Wetlands contribute to the food chain by producing large quantities of food for fish and other animals. Since the young fish spend their early life stages in the wetlands, these areas are necessarily important for commercial fishing.

The total wetland area of the world has been estimated to be around 85,60,000 square kilometres which is about 7 per cent of the total area of the earth. India, however, has about 2,175 and 65,000 recorded natural and artificial wetlands respectively, spreading over an area of 1.6 and 0.3 million hectares, respectively. In India, about 45 per cent of the edible finfish and shellfish are collected from different inland wetland ecosystems.

Tropical wetlands are very fertile. According to one estimate, one hectare of fertile wetland is worth Rs 12.5 lakhs for the benefit derived. One cannot help but wonder why wetlands of the tropics have not been more effectively utilized for capture and culture fisheries. Wetlands seemingly have many advantages over other water areas. The fish growing season is usually year round. And some wetlands have physical characteristics far superior to the wetlands of the temperate zones.

2. Major Development Constraints : A relatively high pressure condition prevails over different types of wetlands. Pollution, weed infestation and eutrophication are the most significant world-wide constraints. These constraints are most notable in tropical regions and Africa. Siltation also accounts for this constraint. However, management efforts must be paramount for fish production on the tropics.

While most wetland development authorities are implementing several strategic plans/ schemes and state-of-the-art techniques to prevent wetlands from degradation/extinction,

* For further detail, see Biswas (1995).

some of the wetlands in the tropics are now on the verge of disappearance due to human interventions. This ubiquitous phenomenon has already pervaded the aquaculture sector. It is essentially a case of extinction not being considered seriously to keep the growing demand of these lucrative water bodies. The 21st century is a century of urbanization but with far more serious implications on wetland development programs. It is unfortunate that several wetlands are being considered to convert them into residential complexes. If this continues, a time will come when many economic species of aquatic animals will have to be recorded in the 'Red Data Book' and consequently, most of the wetlands will be wiped out from our environment. Already many species of finfish recorded from different wetlands have been declared as endangered or vulnerable.

Wetlands when polluted are essentially worthless from aquaculture point of view. Unfortunately, such wetland areas are extensive. Implementation of intensive and effective management/development programs is badly needed, however.

3 Research on Wetlands: Unfortunately, too little is known about wetlands and their management protocols. Research on these aspects, their properties, and their potential for fish production are not significant compared to that on other water bodies of commercial importance. Knowledge on their characterstic features makes it possible to identify only their productivity. And knowledge on the nature of the wetlands following the adoption of management practices will definitely prevent dramatic loss of fish from wetlands.

There is a good likelihood that more intensive study is badly needed that will help to identify the complexities of different wetlands.

TRAINING, EDUCATION AND THE ROLE OF GOVERNMENT

Culture and capture fisheries development aspects include fisheries training and education to all categories of fish farmers which should receive much attention and these aspects should be brought under the scheme because successful implementation of the scheme entirely depends upon the availability of well-trained personnel having a sound background on culture technology as well as management with regard to production, processing, and marketing.

Farmers' education and training are the two important parameters to be reckoned with as it calls for greater devotion in the area concerned. It also demands a positive outlook towards the farmers' community who, in this age of technology, often feel disregarded. In many regions, however, attempts to bring successful results have been less than satisfactory. The planning has been lop-sided, follow-up actions on the implemented schemes have been totally ignored and the administrative callousness has set-back the strategy of fish farmers' development programs to a considerable extent.

The exultation of aquaculture exists in the mind of every fish culturist. A productive fish culture pond is the life and soul of any progressive fish culturist. Hence, a well-planned and down-to-earth development network should be established in different farming communities to reap the benefit derived from the region concerned. A haphazard development can stifle his endeavor. And as a corollary, farmers fall into despondency after their failure in farming activities – it is at heart a cruel irony in the name of development strategy.

Developing countries are varring round to the view that the government should not be running aqua-industries except in rare cases where strategic interests compelled it. Millions of hectares of water areas are now being opened up for private enterprises.

Ascendency of the private sector over the development of aqua-industries is one of the major features of the developmental change resulting from the adoption of modern farming strategies. Developing countries are now discarding the traditional moorings of the past and are embracing the concept of new technology. In the new milieu, the private sector is beginning to hold commanding heights in the farming economy.

The government should rely on private sector for effective implementation of fisheries schemes to sustain production of inland resources. To enforce different schemes, private sector enterprises would have to be provided with financial facilities by setting up an aquaculture development fund. Moreover, the private sector enterprises must be allowed to enter into the joint ventures with overseas companies so that development strategies may proceed in an unhampered manner. Bilateral agreements between Indian and foreign concerns would undoubtedly provide for the provision of financial assistance, export of quality products, and transfer of technology.

Recently, the foreign agency onslaught regarding development of aqua-farms and processing units in some states of India is striking. Setting up an agreement between foreign agencies and the sub-national government (s) though significant from development point of view, but at the same time, may have serious implications (*see* Chart - 2.6).

2.24 Development of Pond Lining Technology

In several countries of the world, the bottom and sides of fish culture ponds are screened with high density polythylene sheets (HDPS). The reasons for this are not complex. It becomes very difficult to remove the accumulated sludge from fish culture ponds. This condition often results in disease spread during the culture period. Moreover, sludge removal at the time of pond preparation not only takes considerable time but also the work for the next crop is not completed. Though pond lining technology is a recent issue in fish culture management protocols for high production, the technology is circumscribed by several advantages and disadvantages as noted below.

Chart – 2.6
Foreign Agency Onslaught *Vis-a-Vis* Aquaculture Development

It may be mentioned here that the economic reforms and liberalization are being gradually altered the aquaculture business and marketing scenarios of India. Recently, the entry of several foreign agencies in Indian aquaculture sector forms one of the most important strategies for high production, processing, and marketing of aquatic food species in the years to come. Several state governments are trying to hand over extensive water areas to some foreign agencies for setting up aqua-farming industries and subsequently expanding their activity. It is, however, a major phenomenon with several dreadful ramifications. Intensification of competition is one facet of this phenomenon. There are several others with more serious implications. It may result in the regional aquaculture enterprises of the foreign agencies gaining a new strategic edge in the regional aqua-industry. And in joint venture, it may alter the power equation between the regional and the foreign partners in favor of the latter. In sum, however, it means a grossly unequal battle, a battle between players of unequal strength.

If this trend continues, other multinational agencies will try to enter into the Indian aquaculture business sectors in future with superior technologies. The financial plight of regional aqua–farmers is so poor that it would never be feasible for them to compete with the foreign agencies. It also may not be feasible for the regional farmers to rebuff the demand of the foreign agencies and consequently the farmers would be the bigger losers if the alliance is broken. In other words, the farmers did not have a choice they had to compromise majority control to the foreign agencies. And as a corollary, they also had to relinquish the managerial autonomy.

ADVANTAGES

The capacity of a pond to yield fish is determined by the type of technobogy to be used and factors that affect the farmer's incentive to produce. Lining of fish culture ponds has the following advantages.

1. The presence of HDP sheets between the superjacent water column and the pond bottom soil obstructs mixing of the waste materials with the bottom soil.

2. It prevents possible adverse composition of bottom soils.

3. This technology helps reduce the number of aerators for use during culture phase (generally ten aerators are sufficient for one hectare pond).

4. The use of HDP sheets has been estimated to last for at least 15 years and the additional expenses involved for lining will prove to be economically beneficial.

5. Feed costs are saved to a great extent.

6. The pond environment remains clean and consequently fish growth is accelerated.

DISADVANTAGES

Though lined ponds have several advantages, some disadvantages also exist. In lined ponds, for example, the transparency of water is very high due to the separation of water from bottom soil. Consequently, growth of phytoplankton and zooplankton tends to be reduced. This problem can, however, be solved by fertilization and manuring programs to lined ponds so as to encourage the growth of plankton populations.

Fish culturists in several countries such as Japan, USA, and some Asian countries have adopted pond lining technology. A survey carried out by Global Aquaculture Alliance disclosed that fish farmers who have accepted the pond lining technology received encouraging results. In most areas of the world, pond lining technology has been considered as an immediate method of increasing fish production. In time, however, as technology of fish culture management progresses, expansion of pond lining technology will likely to occur. As awareness among fish famers increases, use of lined ponds could aquire greater reception.

Special problem relates to sandy and sandy-loam ponds, which are characterized by high leaching phenomenon. These ponds are extremely deficient in inorganic materials and nutrients. If ponds are lined, appreciable quantities of nutrients and organic matter will be increased soon after fertilization. In ponds where inadequate water creates a serious problem, water conserving efforts must be paramount for fish production. Implementation of pond lining technology is, therefore, important in intensive and semi-intensive culture systems. Pisciculturists are challenged to devise this technology that will alter the fish culture scenario.

2.25 Satellite Technology Versus Aquaculture Development*

Use of satellite technology in aquaculture development is a very recent issue. Satellite is a powerful communication tool that has the capacity to exchange of infomation on a global basis which, at present, has an increasingly important role in the global information infrastructure. With the development of orbiting earth satellite-carrying remote sensing systems, the remote sensing has expanded into a major branch of various research aspects. Because orbiting satellite can image and monitor extensive geographical areas or even the entire earth, global and regional studies have become possible that could not have been carried out in any other ways.

Application of satellite remote sensing technology in India and other developing countries has recently been a subject of aquaculture and fisheries development. Those who consider this technology more suitable than others, argue that it undoubtedly promotes productivity but not suitable unless this technology is apprehended by fishermen.

TOOLS USED IN SATELLITE COMMUNICATION

Satellite technologies have further been developed by installing several tools which has increased the capabilities of satellite communication both in space and on land. Some

---

* For further study, see Brenard and Moksness (1996).

common and important tools that are used in satellite communication include Very Small Operation Terminal (VSOT), Global Positionting System (GPS), Geographical Information System (GIS), and Remote Sensing (RS). These technologies have several applications in aquaculture development. The aim of the ongoing research and development activities between Fishery Survey of India and Indian Space Research Organization is to incorporate this knowledge with satellite data to predict the fishery scenario of the country.

1. VSOT and GPS : These are the abbreviations for Very Small Operation Terminal and Global Positioning System, respectively. While the VSOT technology has a significant role on independent communication networks connecting large number of geographically dispersed sites and provide value-added satellite-based services capable of supporting database enquiries, internet, E-mail, voice communication, sale monitoring etc., the GPS has numerous use in civilian and commercial applications. The global positioning equipment is a satellite navigation system and uses the satellites in space to locate the position anywhere on the earth. This system is an aggregation of 24 satellites that orbit the earth twice a day, transmitting precise informations to anywhere on the earth. The GPS has, however, the following applications in aquaculture.

(i) This system measures atmospheric factors that affect the aquatic habitat.

(ii) This system explores aquaculture potential regions in streams and rivers and also helps locate these areas on the map.

(iii) The system indicates the meandering nature of rivers and streams.

(iv) It helps locate the relative distance among potential and explored areas.

2. Geographical Information System : The Geographical Information System or GIS is a rapidly expanding system and is increasingly being used in both marine and freshwater ecosystems. The GIS are means of correlating spatial data obtained at different times, scales, and formats for evaluating reliable informations for future planning and development of aquaculture and fishery sectors.

The GIS has emerged out as a very powerful tool to store, collect, retrieve, analyze, and integrate various kinds of data – whether spatial or non-spatial, for planning and decision-making in fisheries sector with recent advancement of computer technology, hardware, and software. The sustainable development of fisheries will lead to improve the socio-economic conditions of fishermen and upgrading the activities of entrepreneurs.

3. Remote Sensing: Remote sensing refers to gathering and processing of informations from great distances and over broad areas, through use of instruments mounted on aircraft or orbiting space vehicle. The data acquired by these instruments called *remote sensors* is typically displaced as image-photographs or similar depictions on a computer screen or color printer – but is often processed further to provide other types of outputs such as maps, mountain features, graphs, and statistics. The usefulness of remote sensing technique to aquaculture and fisheries is summarized below :

(i) The technique provides an indication to temperature gradients of water for identification of potential fishing grounds in vast water areas.

(ii) The technique helps draw the diagram of surface waters and seasonal variations of water temperature.

(iii) It provides to explore the productivity of water bodies through measurement of sediment load and eutrophication.

(iv) It gives a correct dimension of lakes, reservoirs, and rivers.

(v) Watershed development and water resource planning are carried out. These include water and soil conservation and their development, wasteland development, water harvesting and reservoir siltation.

Thanks to GIS and satellite technologies, the fish production figures are always in constant changes with notable oscillations, even as animals and plants undergo evolutionary development. Technologies produce and maintain a variety of farmed fish that are too important to earn foreign-currency.

2.26 Appraisal of Satellite Technology

Though satellite remote sensing technology embraces various relevant processes such as biological and environmental, and the technology is of immance use for sea fishing, considerable development is taking place in the field of inland fisheries such as identification of suitable areas for aqua-farming and selection of appropriate dosages of different inputs for sustained production, planning for aquaculture development, plankton and nutrient dynamics, sediment fluctuations, and identification of potential fishing grounds for exploration. 'Therefore, great opportunities for developing aquaculture systems are gradually expanding. Satellite technology in the field of aquaculture has given high priority in many developed and developing nations and the success achieved is highly encouraging in recent years.

The application of satellite remote sensing in aquaculture and fisheries should not only relate to the short-term forecasts but should endeavour to have necessary impact of the advanced technology in long-term forecasting of fish stocks. The ultimate aim at developing the technology should be the exploration of both marine and freshwater resources for assessment of fish stocks, monitoring, control, and surveillance over sustainable fisheries management.

The application of remote sensing to aquaculture and fisheries has been developed quite late. Information on aquaculture within a short time through satellite communication technologies is very reliable and consequently, it has until recently contributed more in real terms to global fish production.

2.27 Problems of Freshwater Fish Culture Development

Preceding discussions remind us that the development of freshwater fish culture sector is an important issue subject to only systematic management in a particular region if different development processes proceed near normally. In seeking the cause of fish culture problems, one should keep in mind the conditions under which several factors in relation to development are likely to occur. Although problems regarding development of culture-based fisheries may reduce the ecosystem productivity and thereby decline in fish production potential, major focus should be on those managment that helps increase production potential. Also the most important reason for concern over the haphazard development of aquaculture is the harmful effects on the environment in general and ecosystem in particular (see panel 2.2). However, some of the notable hassles associated with fish culture development that follows are important.

INCREASE OF WATER VOLUME

A sudden and drastic increase in the water volume of ponds, rivers, and tanks during monsoon (See panel 2.3) due to gradual siltation of these waterspreads bring about considerable changes in the physico-chemical properties of the environment of fish populations. The conditions of most of the water bodies are very miserable. The destructive effects of siltation are very serious. Not only is the depth of water reduced but fish crops can be removed root and branch during monsoon, left to die with mud exposed or covered by the drifting debris. The load of sediment carried by flow-off waters to most of the fish culture ecosystems is enormous. Siltation problem is more severe in some other areas of the world. The life time of fish culture ponds is shortened dramatically by inputs of sediment from eroded lands.

BROODSTOCK AVAILABILITY

The fish culture ecosystem can hold water for more than five months in a year. To produce broodfish in large numbers, it is necessary to retain water in a pond throughout the year so that broodstock management is possible. Consequently, brood fishes are not being produced for seed production. It takes more than one year to produce a brood fish from a pond or tank. Therefore, retention of water in the pond is a very important issue for broodstock availability.

The act of fishing with hook and line, poisoning of fish culture ponds with certain chemicals, and political conflicts among farmers in many rural areas are the causes for drastic decline in production potential of most of the table-sized fish. As a consequence, fish farmers do not show any interest towards farming practices. For these reasons, fish farmers of several Asian countries are not reluctant to produce table-sized fish. However, if broodstock management protocol is not properly taken into consideration, seed production units (hatcheries) will cease to function. In the meantime, the function of some of the hatcheries has already been closed for ever.

POLLUTION

Heavy floods during monsoon and pollution of rivers and other water bodies by the effluents of different industries or through agricultural run-off takes a heavy toll of fishes every year. Reduction in the quality of water and soil due to pollution is a serious problem across the world. Pollution is most certainly a concern for all society. Productive capacity of ponds/lakes/reservoirs drastically reduced only when the pollutants exceed the tolerable levels. The ultimate influence of pollution or siltation on fish culture ecosystem productivity is determined by some factors such as management and water depth.

A well-managed and deep water pond, for example, may not lose its fish production potential even though it suffers from siltation and pollution hazards. In contrast, siltation and pollution of a shallow pond may bring about rapid decline in pond productivity.

WEED INFESTATION

The next important problem associated with pond culture of introduced fishes is the eradication of aquatic weeds. Excessive buildup of obnoxious aquatic weeds such as *Cardenthera, Chara*, Arrow head, duck weeds, *Hydrilla, Vallisnaria,* Water hyacinth, and *Utricularia* occurs over extensive areas of tropical freshwater ecosystems and consequently, waters are deficient in oxygen. Control of these plants is badly needed to guard ecosystems for fish culture, though their control is an arduous problem. While the growth of aquatic plants is serious, their long-term existance leads to fish culture problems. The control of aquatic plants through chemical and biological means have, however, been described in Volume 1 (See Chapter 4).

FLUCTUATION OF ENVIRONMENTAL FACTORS

Sudden and large-scale mortality of fishes is also a common problem. Fluctuation of physico-chemical factors of water (such as pH, dissolved oxygen, turbidity, toxic gases, and temperature) and also the lack of food as well as diseases (biological factors) are inevitable for fish mortality.

EUTROPHICATION

Gradual addition of domestic waters, and nitrates to intensive fish culture ponds through fertilization and feeding cause high fish mortality. Such ponds beccome rich in nutrients, especially nitrate and phosphate ions. Thus the water bodies become highly productive or eutrophic and the phenomenon is termed as *eutrophication*. Accumulation of nutrients stimulate luxuriant growth of algal blooms in pond water. Algal blooms not only compete with other aquatic plants for photosynthesis but also release some toxic substances which kill fish and fish food organisms. At the same time, decomposition of algal blooms in water also depletes the content of its oxygen with disastrous effects on biotic communities. Thus, in a pond where oxygen concentration drastically reduced with higher carbon di-oxide levels, fish and other aquatic animals begin to die and consequently, the water body will emit an offensive smell.

Eutrophication is thus a limiting factor in supply of unpolluted water for fish culture. Therefore, it is necessary to reverse or stop eutrophication at any cost.

DISEASES AND PARASITES

The most important problem relates to fish culture is characterized by the occurrence of diseases and parasites. Freshwater fish culture industry in many tropical countries have, however, been experiencing changeable success and failure. 'The changeable trend can be converted into an uninterrupted successful one by prosecuting culture practices in such a way that bacterial, fungal, viral, and protozoan infections could be eschewed to a large extent and consequently sustainable and disease-free culture conditions are established. Changes in the pond bottom conditions and water quality generally alter environmental conditions which ultimately affect the health status of fish. Fish diseases and parasites have become a great pitfall and must be cured by suitable treatment immediately after their occurrence. Disease problems can be solved by adopting the following prerequisites such as (1) careful preparation of pond bottom by the use of lime/zeolite/soil probiotics, (2) regular monitoring and management of healthy pond bottom and water quality, (3) pumping of filtered chlorinated water to the culture ponds, (4) regular use of feed probiotics, digestive enzymes, vitamins, and immunostimulants, (5) exchange of pond water at 20-30 percent level according to the condition of the pond bottom. Frequent exchange of water can, however, be avoided by using soil and water probiotics along with lime, and (6) frequent inspection of fish health is the most important aspect for healthy fish production.

Many countries have a problem of the availability of quality water. Inadequacy of quality water is a serious limiting factor for much of Asia. Monitoring efforts of water and soil qualities must be paramount for fish production on this continent. It is unfortunate that most of the fish farmers are not conscious about these problems and in most of the cases they are reluctant to solve the aforesaid problems inspite of providing technical and/ or financial assistance to farmers. Fishery scientists are, however, challenged to devise methods that will solve these hassles in the near future.

2.28 Dearth of Fish in Freshwater Ecosystem

In recent years, the scarcity of commercially valuable species of fish in many freshwater ecosystems is a very important facet as a result of floods, over exploitation, siltation, anthropogenic load, tampering of catchment area, pollution, habitat destruction, eutrophication, and human settlements around fish culture ecosystems. The fierce reality is that the human population has shown far too little resolve since the beginning of aquaculture development to rectify damage inflicted mindlessly on aquatic ecosystem over the past three decades. It is ironic that developmental activities have happened whilst agricultural and aquacultural growth have taken the output of marketable product for most of the countries in the tropics and sub-tropics. In the process, however, we have plundered the aquatic resources to an extent that have threatened numerous fish life support systems all over the world.

Several aquaculture problems can add to the precarious condition of the aquatic resources which are in any case being degraded by pollution and habitat degradation.

The problems of soil and water pollution (See Panel 2.4) and deplorable condition of fish farmers' community are much too familier to neglect. At the core of these problems is the inability of decision-makers to analyze the erroneous conflict between fish culture ecosystem and development

At present, it is not enough to emphasize the protection of aquatic resources. What is needed is to actually invest in the rejuvenation of natural capital. And the policy should target the much larger returns that would accrue from investment in the virtual expansion of natural capital such as fish biodiversity and unpolluted water as well as soil for sustained yield.

2.29 Research For Fish Development

Systematic research on fish culture/aquaculture (*see* Chart 2.6) is definitely an important issue for successful development in fish culture and sustained growth in aquaculture in general is feasible if investment in research is continued. Since fish production from capture fisheries resources is limited, aquaculture is being called upon to reduce the shortfall in food fish supplies in the market. A number of issues involved in sustainable development (See Panel 2.5) should be informed to farmers if current production levels are to be maintained. Applied research is, however, inevitable to evaluate the role of aquaculture in rural development and to elaborate effective development approaches for its use in this respect where opprotunities exist. Extensive research must be carried out for exploring the potentiality of aquatic resources and intensive use of inland areas for augmenting production. Many important informations concerning about the development of freshwater resources for high production in tropical and sub-tropical regions are lacking and should be gathered through research.

The national and sub national governments play a proactive role in the establishment of regional centres for aquaculture research. These centres continue to assist farmers through direct technical and financial supports and training activities.

Future expansion of aquaculture depends not only on continuous investment in applied research but also on suitable training for sustainability in an environment-conscious society.

It is fortunate that progress related to aquaculture research has been made. Moreover, further research programs focussing on development in integrated aquaculture system and a better understanding of the economic and social environments for further development have already been accepted. Research information about the existing problems related to development of aquaculture and fisheries and recommendation of reputed international organizations which are actively involved in applied research are also significant.

Chart - 2.6
Systematic Research on Fish Culture/Aquaculture

1. Classification of Aquaculture Research Activities : Aquaculture research activities can be divided in different ways such as :

• Routine problem analysis and research on non-routine problems.

• Research on short-term and long-term problems.

• Classification on the basis of the actual subject of the research.

2. Definition of Aquaculture Research: It is the systematic, objective, and exhaustive search for evaluation of all factors involved in farming operation and fish production, and study of the facts partaining to any problem in the field of aquaculture.

3. Steps Involved In Aquaculture Research: In applying aquaculture research for solving aquaculture problems, the aquaculturists have to go through several steps. Each stage has its own determinative role in the entire process. Right from beginning the culture activity down to the marketing and the preparation of research report, the aquaculturists has to proceed step by step that follows.

• Identifying the aquaculture research problems involved.

• Developing the research design and research procedures.

• Collecting the data/information.

• Analyzing the information and interpreting it in terms of the problem being tackled.

• Summarizing the findings.

• Preparing the research report and disseminating all reports to fish farmers.

While most aquaculture research assignments proceed through the above steps, in some cases, some of the steps are compressed, depending on the type of problem, costs, and benefits. There are several aquaculture research problems which can be handeled by certain simplified methodologies.

As a generalization, there is a lot of problems in any fish farming system. The problems belong to aquaculture farming. The aquaculture farming research problem is a derivation of the aquaculture problem. Hence, the farming problem involved should be clearly identified, defined, and conceptualized. Once the farming problem involved is accurately identified and defined, the identification of the farming research becomes a relatively easy task.

Often, research laboratories look at what the fish farmers/fishermen want to know about the fanning strategy so that production may be substantially increased. This may hamper identifying the farming research problem involved. It is enormous to think that the analysis of the problem can be made after the data comes in. The benefit accrues only when the problem is analyzed before the data is collected.

Development of research design and research procedure is the most important step to be considered. The choice of research design depends on the urgency of the work, and the time available for completing it. Research design is principally the blueprint of the farming project, and when implemented, it must bring out the information required for solving the identified problem. The research design indicates the method of information gathering, the instruments and chemicals required for the purpose, the method of sampling and the like.

Research procedure arises from research design, it spells out the plan for obtaining the information. Aquaculture survey is an example of research procedure. In most cases, however, an aquaculture survey may not be necessary for collecting the required information. Simpler procedures such as desk research of published data may be sufficient for the purpose.

Research equipments denote the equipments employed for collecting the data/ response/information. Equipment can be a mechanical or electronic device of observation on different aspects of ecosystem productivity.

To sum up, in many cases the problem of aquaculture research in relation to productivity has been causing serious concern and economic hurdles to farmers. Scientists must be exhorted the farmers to face the problems collectively and at the same time, ascertained that they always remain behind them. It is, however, assured that farmers' community would continue to strive to get the current technologies developed through research in full potential for high production.

The Food and Agricultural Organization (FAO) has been active in examining research priorities and the studies have been made in several regions of the world such as Latin America, Africa, Asia, North of the Sahara, and South of the sahara where a number of major stumbling blocks to further development of aquaculture have been identified, with several research fields which require immediate head. And the most important and common research themes have come into existence among regions. The following items should be given top priority for development of fish culture in particular and fisheries/aquaculture in general.

1. Use of conventional feed ingredients and replacement of fish meal.
2. Expansion of culture-based fisheries.
3. Genetic selection for increased productivity.
4. Improved hatchery technology.
5. Environmental, social, economic, and technical criteria for aquaculture planning.

6. Use of efficient commercial and farm-made feeds.

7. Improvement of pond design and construction.

8. Integration of fish culture with agriculture and/or livestock.

9. Selection of indigenous species for culture.

10. Specific attention to rural aquaculture.

11. Attention to fish health management.

12. Improvement in soil and water qualities.

13. Spreading of research findings in a form usable by farmers and planners.

14. Use of recirculatory system in fish culture.

Development of inland fisheries sector requires considerablere search with specific objectives and combining with several disciplines. Different types of freshwater fishery resources that need to be maintained in good condition should be brought under fish culture on commercial basis. To serve the purpose, different schemes based on regional conditions should be implemented. Progressive farmers who are interested in fish culture must come forward so that culture, harvesting, and marketing operations can be undertaken.

Introduction of commercially viable schemes for fish culture development require capital investment, application of technology, infrastructure, technical personnel, systematic research on culture fisheries sector, and market. Though technologies have been imported/developed within the country and technical manpower is, to some extent, available at present, research on fish culture is still limited. Extensive research and voluminous results have appeared in many journals and the results have not been arranged systematically so that fish farmers can have a vivid idea to implement the culture technology in their own water. Therefore, suitable planning should be enacted based on the establishment of different types of commercially-oriented fish culture industry. Since most of the tropical waters are fertile, development of advanced technology for carp and catfish culture and technology for economic reclamation of unutilized waters require specific research and financial support. Many neglected water bodies are abundant in some tropical countries and are not effectively utilized for culture due to multiple ownership and other reasons. These problems can, however, be removed if government adopts firm policy in these aspects. Lack of management of aquatic weeds, technology, non–availability of fish seed and quality feed, and pollution hazards are the main constraints towards development of inland fisheries sector.

To sum up, systematic research is badly needed so that culture programs can be successfully implemented according to the need of farmers. This involves the diversification of culture system, development of economic means of management at very low cost, refined region–specific technology for culture and effective feeds as well as feeding schedules for different types of farmed fish.

2.30 Importance of Farmers' Commission in Aquaculture Development

One of the major steps that should be taken into consideration with a view to improving the flow of credit among indigent farmers for the development of integrated farming sectors of the developing countries of the world is the establishment of Farmers' Commission. The government of India, for example, has established such a commission in 2005 headed by M.S. Swaminathan to suggest ways to make different projects viable. The report submitted by the commission in the month of October, 2006 has firmly indicated to form 'land use board' in all states of the country and to implement the farmers' principles. The land use board should also be established even at the district levels to identify the nature of different lands which could be utilized in full potential for agriculture production. The same principle may also be adopted for aquaculture production. The nature and type of different water bodies will be recorded by the board so that water bodies, whether unproductive or productive, may not be converted into other purpose. Inspite of the rapid development programs undertaken by different agencies, a large segment of the farmers' community is still beyond the reach of the organized sectors. This commission which is aimed at giving a new dimension to develop farmers' community by providing credit to the farmer for acquiring productive assets, may be operative in every states and would benefit millions of farmers. The commission will, however, assign to a key role in the comprehensive development of the farming sector and lay down clear guidelines for different agencies involved in commission functioning. Developmental aspects may include subsistance and credit facilities to farmers through nationalized/ co-operative/gramin banks, adoption of crop insurance schemes, farming on contractual basis, and other related issues. In short, although the commission is structurally sound, minimum infrastructure and higher degree of managerial skill and competence will be required to work in concert on the desired lines; otherwise, all efforts will be ended in smoke.

2.31 Conclusion

Mankind has entered into the new century. Optimistic farmers look forward to it with the hope that it would usher in a new era of optimum living standard and technological development towards an epoch of fellow-feelings among farmers of the nation. But such expectation can never come true unless farmers, technicians, experts, and planners bear the heaviest part of the attack on development-related issues. Fish culturists and experts must work with harmonious and locally-suited advanced technology for the development of the quality of life in rural areas. The technology which is inevitable for the development of different types of freshwater resources, should be harnessed for sustained production of aquatic food to nourish millions of people and devised means that can ensure to produce fish and fish products for the posterity.

The most important development features of the culture aspect has been the rise in value-added items which is a welcome indication. Culture of carps, catfishes, tilapias, and also freshwater prawns which are considered as important freshwater food species in tropical countries have increased several folds. Such endeavour has encouraged the farmers because of the stagnation in catches from many capture fishery resources due to pollution, disease problems, habitat destruction, and other problems related to fish culture.

Development and subsequent application of state-of- the-art technologies in aquaculture industry in the mid 20th century would definitely determine the fate and success of aquaculture strategies in the 21st century. Development strategies could serve as incubators for commercial production. There is beyond doubt, however, that the present century is an era of aquaculture and its related field which may be termed as *century of aquaculture*. As many as we would imbibe the knowledge on the development of aquaculture strategies, it will explore the limit of our knowledge, experience, and thinking in the interest of the human society. In fact, we have learned a lot on the subject concerned in the 20th century but have a lot to learn.

Adequate developnent programs may be taken for all-round development of both productive and unproductive water areas so that all potential resources are sufficiently developed that will accrue in increasing the income of farmers to a large extent.

Development strategies are easier dictum than done. To achieve the goal would require all the energy, intelligence, and selflessness of the noblest souls of the nation. Even at the risk of sounding sceptical, it can be said that it would be naive, considering the previous records, to have too rosy dreams about what the new century has in store for the farmers' community.

References

Baynes, S. M. and A. P. Scott. 1980. A review of the biology, handling, and storage of spermatozoa. *J Fish Biol* 17 : 707-739.

Biswas, K. P. 1995. Ecological and Fisheries Development in Wetlands. Daya Publishing House, Delhi, India.

Brenard, A. and E. Moksness. 1996. Computers in Fisheries Research. Chapman Hall, London.

FAO (Food and Agricultural Organization). 1997. Aquaculture Development. FAO Technical Guidelines for Responsible Fisheries, Fisheries Circular No.5.

FAO (Food and Agricultural Organization). 1999. Review of the State of the World Fishery Resources: Inland Fisheries, FAO Fisheries Circular No. 942.

Heidinger, R. C. 1999. Stocking for sport fisheries enhancement. In : Inland Fisheries Management in North America (Eds. C. C. Kohler . and W. A Hubert). American Fisheries Society, Bathesda.

Kershner, J. L. 1997. Monitoring and adaptive management. In: Watershed Restoration: Principles and Practices. (Eds. J. E. Williams, C. D. Wood and M. P. Domback). American Fisheries Society, Bathesda.

Monga, G. S. 2002. Environment and Developnent. Deep and Deep Publications, New Delhi, India.

Nath, D. and A.K Das 2004. Present status and opportunities for the development of Indian riverine, reservoirs and beel fisheries. Fishing Chimes 24 (1) 61 – 67.

Peter, T. 2000. Inland Fisheries Enhancement. FAO, Rome, Italy.

Range, M. M. 2002. Fish Biodiversity. Agrobios, Jodhpur, India.

Raymond, F. M.1992. Environmental Assessment and Sustainability at the Project and Program Levels. World Bank Conference on Environment and Sustainable Development, Washington, DC.

Stoss J 1983. Fish gamete preservation and spermatozoa physiology. In: Fish Physiology (Eds. W. S. Hoar, D. J. Randao and E. M. Donaldron). Vol. IX B. Academic Press, New York, pp. 305-350.

Sugunan, V.V. 1995. Reservoir Fisheries of India. FAO, Fisheries Technical Paper No. 345, FAO, Rome, 423 pp.

Questions

1. Discuss how better standard of living of fish farmers can be maintained through aquaculture.

2. What are the steps that should be taken into consideration for sound economy in farmers' community ?

3. Trace the aims and evaluation of fish culture.

4. Trace the origin and growth of fish culture.

5. Trace the origin and development of fisheries in India.

6. What are the principal strategies which were considered in the Kyoto Conference for the development of aquaculture sector? Why this conference was so important?

7. State what are the declarations that were considered in the Global Conference held in Bangkok, 2000, to assess the trends in aquaculture development. What are the strategies that have been identified for better fish production across the world?

8. What is sustainable development? How it is related to the growth of aquaculture sector in fish-producing countries of the world?

9. What are the factors that should be considered for sustainable development?

10. What are the indicators that are related to sustainable development in rural areas?

11. Describe how farmers are benefitted from co-operative society.

12. With reference to tropical areas, discuss how the productivity of fish culture ecosystems may be increased and outline the problems which are associated with the increase in productivity.

13. Farm mechanization is necessary for increasing yield — Discuss.

14. "The increase in aquaculture production is important for human nutrition but rural development means much more than that". Discuss this statement in the context of development efforts made in India and their effects on rural economy. Explain the answer with studies from the history of rural development.

15. Briefly discuss how the current technology has improved the mechanization/ modernization of aquaculture farm for sustained production of aquaculture food for mass consumption and export.

16. State the sequential steps involved in domestication of farmed fish.

17. State the essential components of commercialization of fish farms.

18. What are the indispensable factors for the success of commercialization?

19. Discuss how credit is related to commercialization of aquaculture.

20. What do you mean by extension? Why an ideal fish farm requires extension program? What are the methods of farm extension? Mention the significance of farm extension.

21. What are the developmental trends in fish production around the world ?

22. Why capture fishery strategies have shifted to culture-related fisheries?

23. Narrate how quality feeds, health management, and biotechnology are related to fish culture development.

24. Discuss why fish culture should be integrated with the development of agriculture for boosting aquatic food.

25. What is enhancement? Discuss how enhancement is related to inland fishery development. What are the importance of enhancement?

26. What are the geographical factors that largely influence fish culture?

27. Discuss how genetics and biodiversity are related to the development of fish culture.

28. Why management of water is so essential for fish culture?

29. Most of the tropical inland water resources have enormous fish production potential, but most of the developing countries are not able to develop their freshwater resources due to financial support. Discuss how financial assistance is involved in fish culture development strategy. Why such help from different funding agencies is needed for upliftment of the farmers' community? What are the weaknesses in aquaculture development programs?

30. What is wetland? State its importance. What are the major developmental constraints of wetlands?

31. What is pond lining technology ? What are the advantages and disadvantages of this technology in relation to fish production ? Is it possible to adopt this technology in different culture systems?

32. Discuss how satellite technology is related to aquaculture development.

33. State major problems which are related to freshwater fish culture development.

34. State how fish culture development can proceed through research activities.

35. Write notes on the following:

    (i) Meaning and significance of fish farm mechanization, (ii) Role of Roman and Greek in fish culture, (iii) Development of fisheries sector in the 20th century, (iv) Domestication of farmed fish, (v) Commercialization and mechanization, (vi) Co-operative farming, (vii) Fish revolution, (viii) Organization of fish farming, (ix) Enhancement versus sustainable development, (x) Impact of fish culture on the environment, (xi) Indicators for sustainability, (xii) VSOT and GPS, (xiii) GIS, (xiv) Remote sensing.

PANEL - 2.1
ENHANCEMENT VERSUS INLAND FISHERY DEVELOPMENT*

Enhancement is, in part, responsible for the marked expansion in inland fishery development specially which is being followed in the developing nations. Freshwater resources of any country are, however, under the threat of environmental degradation and consequently, it is not expected to play an important role in meeting the additional requirements of inland fish production. Sedimentation, water abstraction, dam construction, pollution, and habitat destruction of freshwater resources are responsible for the loss of fish biomass. Consequently, the eco-friendly option of developing culture-based fishery in ponds, small reservoirs and floodplain lakes coupled with species enhancement in large reservoirs holds the key for future development of inland fishery resources. Unless fisheries enhancement options are adopted in developing nations where protein supplies are already critical and where the technology for increased fish yield is partly/entirely not adequate, fish supply will continue to be among humanity's dangerous problems.

In a typical culture fishery, the entire operation requires a high degree of effective control over the water and soil quality variables. Management of most of the inland waters is exercised on the basis of culture-based fisheries (when the fish harvest depends on stocking) or in different forms of enhancement.

DEFINITION OF ENHANCEMENT

The term enhancement can be collectively defined as a range of management practices. However, the Food and Agricultural Organization (FAO) has given the definition of fisheries enhancement as "technical intervention in existing aquatic resource systems which can substantially alter the environment, institutional and economic attributes of the system". By enhancement, quantitative and qualitative improvement is achieved from freshwater systems through specific management protocols. Several types of enhancements which are relevant to inland water resources include species enhancement, stock enhancement, management enhancement, and enhancement through new culture systems.

IMPORTANCE OF ENHANCEMENT

The world contribution of enhancement to fish yield tends to be assimilated into the statistics of aquaculture production. It is apparent that enhancement of production is dominated by culture-based fisheries for freshwater and diadromous species. Annual production in this type is likely to be around 2 million tonnes which account for about 10 per cent of combined culture and capture yields; total production is 21.2 and 7.5 million tonnes respectively. Culture-based fisheries for protein requirement and recreation are well established components of aquatic resource used in Europe and

* For further detail, see Suguman (1997), Kershner (1997), Heidinger (1999), FAO (1997).

in North America. In these countries, fisheries departments expend an average of more than 20 per cent of their budgets for stocking in freshwater systems.

Contribution to fish yield of enhancements other than culture- based fisheries are less well documented. However, habitat enhancements (to be mentioned latter on) using indigenous technologies are more widespread in many African and Asian inland waters and the importance of such water bodies has been seriously considered. According to one estimate, very little amount of inland fishes is available from estuaries and rivers. Substantial amounts of fish are obtained from different open water systems such as reservoirs, small impoundments, and floodplain wetlands. The main objective of management in these waters is related to fisheries enhancement and culture-based fisheries.

OPPORTUNITIES AND CONSTRAINTS

Different techniques which are adopted for enhancement afford opportunities for poor sections of inland resource consumers. Enhancement not only commences an active management of freshwater resources leading to productivity, conservation, social benefits, but sustainable and efficient exploitation of resource as well. Enhancement techniques help to maintain abundance of community structure and ecosystem functioning in the light of environmental degradation and exploitation.

The most important and significant events derived from the Bangkok Conference which was held in Thailand in 2000, provided the importance of enhancement as a practice with important opportunities for poor sections of the human society. It is also necessary to afford facilities from aquaculture technologies that will definitely allow effective use of derelict and deceased resources. However, the conference has declared that to increase fish production potential from culture-based fisheries, the following three points must be realized:

1. Management of environmental impacts should be assessed.

2. Effective assistance and exchange of informations should be evaluated.

3. Adequate research and development supports must be assured.

Enhancement of fisheries necessitates investments in common pool resources and can be sustained under research and development institutions that permit direction of use and flux of advantages to those who are able to sustain the cost of enhancements. At the same time, technological interpositions are confined to some aspects of life cycle of stocks and the consequences are primarily dependent on natural conditions that are beyond the control of management. Obviously, management interpositions should be adjusted to local provisions and certain conditions may prevent felicitous advancement collectively. Both private and public sectors have an important role to play in expediting advancement inaguratives through setting up the institutional arrangements, management of environmental impacts, and suitable reseaarch and development support on fisheries advancement.

ENHANCEMENT VERSUS SUSTAINABLE DEVELOPMENT

Sustainable development, as defined by the FAO, is "the management and conservation of the natural resource based on the orientation of technological and institutional change in such a manner as to ensure the attainment and continued satisfaction of human needs for present and future generations". Such development should, however, be technically viable, socially agreeable, and environmentally non-degradable. Sustainable fisheries development is particularly related to both culture and capture systems which are intended to maintain constant productivity and beneficial to society. Culture and capture fisheries provide management discretions, which is an expression of dissatisfaction with the patterns of sustainable development. In tropical regions where most of the open water resources contribute to fish yield, are well managed on the ground of culture-based fisheries which are transitional to capture and culture fisheries patterns.

1. Types of Enhancement: In general, inland water-related enhancement involves several management options that follow.

(i)   *Improvement Fish Stocks:* Increasing the stock of fish biomass is the most important management option that is executed in vast open water systems to prevent undesirable fishes that may grow at the expense of important fast-growing species. The stock enhancement, however, not only involves the selection of species to be stocked but determination of the size at stocking and stocking density as well.

(ii) *Species Enhancement:* In vast open water systems, introduction of commercially important species of fish occurs either accidentally or deliberately with a view to establish fish populations for harvesting maximum fish. Stocking of introduced species in many freshwater bodies is a debatable issue because of their adverse effects on the indigenous stocks.

(iii) *Habitat Enhancement:* Since freshwater habitats are being gradually degraded due to human interference, improvement protocols of fish habitats must be carried out following recently developed techniques. An effective habitat enhancement in some tropical regions involves the construction of 'brush parks' which provide substrate for the growth of periphyton and defence from predators. Fish production from these parks has been calculated that varied between 2.0 and 5.7 tonnes/ha/year.

(iv) *Environmental Enhancement:* The fertility of inland water bodies is their capacity to supply adequate amounts of different nutrients in the balanced proportion for fish growth. This means that the fertility of freshwater bodies is concerned with the amounts of nutrients present in the soil. On the other hand, productivity of water bodies is the capacity of the soil and water to yield robust fish under specified systems of management. A productive inland water must be fertile and requires precise amount of nutrients.

Environmental enhancement is characterized by the improvement of the nutrient levels of water and soil by use of manures and fertilizers. Organic manures are added to culture systems to improve their productivity. Fertilizers are usually concentrated inorganic materials containing high amounts of nutrients in a form usable by fish through food chains. These nutrient carriers bring about certain amendments in the condition of certain disputable water and soil, just to make them more suitable for fish growth. Gypsum and lime, for example, are added to alkaline and acid waters respectively to control alkalinity and acidity accordingly. Although use of different nutrient carriers is a common management option adopted in small freshwater culture systems, adoption of this option in large water bodies needs careful consideration because of the possible danger on the ecosystem.

PANEL - 2.2
IMPACTS OF AQUACULTURE ON ENVIRONMENT

Aquaculture has both beneficial and harmful effects on the environment and concern about the impact of aquaculture has become an important issue. Although harmful effects have been recognized, their severity and range have often been exaggerated. A number of aquacultural operations possess minimal ecological risk but contributed significantly to fish supply and income. In most of the cases, aquaculture techniques are often constrained by increased pollution of cultivated water, degradation of habitat, and access to appropriate water and land resources is greatly reduced.

In contrast to developing countries, most of the developed countries follow, in aquaculture sector, regulations on farm discharges. In some developing countries, at present, environmental legislation is emerging although further information on this aspect is necessary to establish suitable regulatory system and to allow planning and pro-active management techniques. A number of informations available so far on the subject concerned relate to edible fish species used for culture in developed countries to those cultured in temperate conditions. Many developing countries have no suitable technical personnel, and facilities with which to carry out work due to limited financial resources. Therefore, well-equipped agencies should help national governments in many other ways such as establishment of guidelines for tropical regions, application of suitable technology for particular area, human resource development, formulation of policies, and development of applied research.

PANEL - 2.3
FLOOD CAUSES DESTRUCTIVE EFFECTS ON FISH COMMUNITY

Floods have been a curse of the fishermens' economy. We have probably seen enough media coverage of river floods to have a good idea of the appearance of flood-water and the damage inflicted by flood erosion and deposition of silt and clay.

We can define the term flood as the condition that exists when the discharge of a river cannot be accomodated within its normal channel. Consequently, the water spreads over the adjoining grounds which is normally cropland or small/large water areas used for aquaculture.

Floods bring out the germs or diseases or toxic substances and consequently, the fish crop is ruined and severe loss follows. Flood waters engulf fish farms, make the farmers oblivious and unprepared, resulting in loss of fish biomass. This damage burdens the farmers with additional investment for the following crops. If the policy of insurance companies is not insured for fish crops, farmers are compelled to mobilize needed funds on their own, mostly depending on money-lenders as the banking sector, by and large, is averse to extend financial assistance to farmers.

Floods take a heavy charge of farmers' community, fish culture, and fisheries. 'The intensity of flood damage in many countries of the world has been rising. A number of causes account for this situation, among which the more important causes are (1) construction of embankments and irrigation systems for agriculture have led to a fall in infiltration rate and in the seepage of rain waters into ground, (2) the flood control capacity of the dams has been corroded by rapid siltation, and (3) deforestation.

Though flood provides productive agricultural land due to addition of nutrients, recharge ground water under plains and refill wetlands, the most extensive damages of fish culture ecosystems and fisheries during flood are characterized by destruction of breeding grounds of fishes, siltation, severe pollution of water and soil and partial or complete elimination of endangered as well as commercially important species of fish. These situations entirely damage the fish culture industry.

Panel – 2.4
WATER AND SOIL POLLUTION

There are two prime reasons for concern over the loss of essential elements by leaching. First is the obvious concern for keeping these nutrients in the water and soil so that essential elements are available to plankton and fish. A second and equally important reason is to keep the nutrient out of streams, rivers, lakes, and ponds. Excessive growth of algae and other aquatic species takes place in water overly, enriched with phosphorus, nitrogen, and other nutrients. This process called *eutrophication* ultimately depletes the water of its oxygen, with disastrous effects on fish and fish food organisms. Also sources of drinking water have become sufficiently high in nitrates to cause health concerns for humans. Likewise, surface run-off waters may contain levels of substances toxic to fish.

PANEL – 2.5
SUSTAINABLE DEVELOPMENT OF AQUACULTURE FOR RURAL DEVELOPMENT*

During the last thirty years, the term sustainable development has been universally considered by experts and policy makers in all countries of the world as a goal to reach in their developmental efforts. The clear apprehension of this goal requires the presence of an adequate ecosystem to boost production which can be termed as *sustainable ecosystem.* The implication is that there is a close relationship between ecosystem and development. The present state of aquatic resources and ecosystem of a country places a limit on its capacity to achieve sustainable development. The reason is that the achievement of higher economic growth is inevitable to reduce the level of poverty in the developing world. The realization of higher economic growth, however, requires greater use of aquatic resources and ecosystem which in turn leads to their degradation and eventual decay. Moreover, increased population pressure on the declining supply of resources also contribute to the degradation of ecosystems. The alarming decline in the quality and quantity of resources makes it extremely difficult for a nation to achieve the goal of sustainable development. Since protection of aquatic resources is a key factor in the success or failure of all development programs, the central issue to be addressed in achieving sustainable development, therefore, relates to the measures to protect the ecosystem.

Concept of SustainableDevelopment: The concept of sustainable development is very broad and has been defined by different experts to include different aspects of

* For further study on the subject, see Raymond (1992) and Monga (2002).

sustainability. Barbier's concept of sustainable developnent includes: (i) the biological goal of maintaining genetic diversity and biological productivity and (ii) the economic goal of satisfying the basic needs (reducing poverty). The former requires preservation of fish populations and this goal cannot be reached unless the ecosystem is preserved. Likewise, the economic goal of reducing poverty by enhancing the standard of living cannot be reached in absence of adequate supply of food, housing and others.

One may say that sustainability is a process of change in which exploitation of resources, orientation of technological changes etc. are made consistent with the present as well as the past. We may say that economic development in rural areas is sustainable if the total resources does not decrease over time. This approach may be illustrated in alternative ways. According to Raymond, the most potentially measurable criterion for sustainable development is the preservation of productivity and full funtioning of the natural resource base.

When the total developmental process is considered, the problem of ecosystem pollution and conflict between industrial development and ecosystem degradation become very serious. At this stage merely the technical side of production like industrial waste and depletion of natural resources cannot be blamed for pollution. The utilization pattern, awareness, and attitude all reinforce the situation of ecosystem degradation. For sustainable development, we have to take all these factors into account along with technology. The United Nations General Assembly through Agenda 21 has provided a vivid picture of the interlinks related to sustainable development and aquatic ecosystem and the priority goals of the economy for sustainable development. However, combating poverty, improvement in farmers' community changes in resource utilization, pollution control, treatment of waste materials, control of hazardous materials, and production-related strategies are some of the vital requiremnets for overall sustainable development of the sector concerned.

Indicators For Sustainability : The question may arise what are the indicators on the basis of which we ensure that our projects and policies are sustainable? The sustainable conditions can be derived from dynamic mathemetical models. Since these models very complex, use of simple indicators is necessary to assess economic and environmental sustainability of the aquaculture development process.

At present, per capita income of farmers is the most important measure of development in aquaculture sector. But it fails to allow for capital maintenance of natural resources of the ecosystem contribution to economic activity. Therefore, in the context of sustainable development it is highly inadequate and should be replaced by a set of indicators reflecting changes in the ecosystem and economic activities. The ecosystem indicators for measuring development in the sector concerned should be very simple and practical. These indicators should be compared with their sustainable limits. Some important indicators that are related to aquaculture development are given below :

1. *Farmers' Community:* Farmers' community in a country is not only a source of rural economic development but a source of ecosystem degradation caused by poverty and lack of awareness as well. Therefore, their standard of living, health and education needs to be improved.

2. *Human Resource Development:* This is a very complex and variable index. It combines several aspects of development such as education, health, and financial condition. The United Nations Development Program (UNDP) occasionally publishes the magnitudes of the index for different countries.

3. *Use of Water For Culture:* Availabillty of adequate quantity of unpolluted water for fisheries and aquaculture has to be ensured for sustainable development of the sector.

4. *Habitat Degradation:* Loss of ecosystem productivity and habitat degradation due to pollution should be reduced to maintain sustainable development. Habitat degradation contributes substantially to loss of fish population and hence the ecosystem productivity and fish harvest drastically reduced. Besides this, the demographic pressure directly exacerbates the problem of habitat degradation.

   High pupulation growth means growing demand for requirement of fish protein. To maintain per capita fish consumption, production will have to be increased.Though finfish and shellfish production has substantially increased across the world, per capita fish consumption rate has not been noticeably increased. This is due to the fact that a large proportion of population of several tropical and sub-tropical countries does not have enough purchasing capacity to meet their requirements.

5. *Institutional Weakness:* This indicator is more pronounced in aquaculture sector. In India, aquaculture sectors were managed for long time as source of food. The technical issues involved in aquaculture have still lagged far behind agriculture.

A body of research has noted that about 3,000 experts are actively involved in aquaculture sector of Asia compared to that of agricultural sector where about 10,000 experts are involved (Data for the year 2009). In India, however, expenditure on aquaculture and fisheries is much less. Consequently, policy which is based on inadequate understanding, has plagued the strategy of aquaculture to a large extent.

3

Importance of Fish Culture in Rural Development

So far the rural development is considered, fish culture in particular and fisheries in general have to play a crucial role. Fish-culture development is basic not only to develop the rural areas but the industrial sector also. Fish culture provides a number of materials on which a large sector of the industry is dependent. Fish culture also provides nutrition to farmers community. Development of rural community as such is a large concept and encompasses the entire development of rural economy. Aquaculture includes development of infrastructure such as construction of fish farms, establishment of hatchery units, banking and marketing facilities, developnent of fish processing industry, different management protocols, and development of traditional industries along with other improvements to farmers community, such as training, education, and health care.

3.1 General Considerations*

The idea of rural development concerned earlier is only the development in the context of agriculture. This concept has undergone a notable change since the seventies and has become more comprehensive. A comprehensive rural development program aiming at the socio-economic uplift of the rural areas cannot be based entirely on agriculture, but should encompass a large complex of activities, termed as *Intergrated fish farming*. The World Bank has, however, defined the term (rural development) as a strategy designed to upgrade the economic and social life of a specific group of people — the rural poor. Rural development essentially involves spreading the advantages of development to the poorest among those who live in the rural areas.

In formulating the rural development policy, following several components should be suggested:

1. The means of production and appropriate technology must be kept in mind. Aquatic resources are the most important means of producing aquatic food. Maximum exploitation of water will depend upon the technical know-how employed in terms of the production of fish, use of the right type of nutrient carriers suitable to the soil and water condition, farmers' force, water management resources, and availability of quality fish seed.

2. The building of institutional infrastructure required for mobilizing economic resources, and the management of material resources and money.

---

* For further detail, see Volume 1 (Chapter 11), Gupta (1983), George (1984).

3. The aim of farmers is to develop proficient 'personnel and adequate knowledge so that farmers can exploit their resources for quality of life.

4. Self-reliance and distributive justice.

During the seventies and eighties, various types of rural development programs were executed in many developed and developing nations. some of them have either been replaced by other programs or have been amalgamated into new ones. The most notable rural development programs in India, for example, that are related to food production include Small Farmers' Development Agency (SFDA), Community Development Program (CDP), Integrated Rural Development Program (IRDP), District Rural Development Agency (DRDA), and Cooperative Society. Although the objectives of these programs were all-round development of rural economy and farmers' society, it is unfortunate that farmers' initiative and participation have always been lacking. For success of rural development programs the following criteria would have to be kept in mind:

(i) An upward trend of production should be sustained till the specific program reaches the target point.

(ii) The level of efficiency, attention, knowledge, and information on the part of the farmer should be developed to such an extent that they may obtain new practices, methods, and programs of their own.

(iii) The program should aim at concomitant development in different directions in a fundamentally correlated fashion.

(iv) Farmers should take more liability and exhibit their acknowledgment of the program by their sharpness to sustain the program.

(v) The significant impact towards the implementation of different developmental programs should be strongly felt among rural people.

(vi) To make the program a success, various problems should be identified and necessary remedial measures should be followed.

3.2 Aquaculture Versus Economic Development*

Aquaculture progress is generally regarded as a prerequisite of economic development. In fact, rural economic development, at present, has come to be associated with the growth of fisheries sector/integrated fish farming system; however, it is generally regarded that the growth of fisheries sector can follow on the proper heal of both culture and capture fisheries, the aquacultural sector has to be looked after with utmost care. This is more aptly demonstrated different farming systems in countries where through various ways such as extension services, subsidized farm inputs , fish seed supply at a reasonable price, and training programs, aquaculture is maintained at a high degree of efficacy. In a developing economy, fish culture has to be given top priority in order to accelerate the rate of economic growth. The way in which culture fisheries contribute to development of rural economy and which makes economic development feasible are briefly noted below:

---

* For further detail, see Jolly and Clonts (1993).

1. Farming systems are the source of supply of fish. Excessive shortage of fish production and mortality due to pollution and disease outbreak creates imbalance in the rural economy. Drastic reduction in fish production and damage to fisheries sector due to flood and drought causes an increase in prices in the market. Consequently, demand-supply equilibrium is not properly maintained. However, high cost of fish production in many farming systems may sometimes create difficult situations for the farmers to make any precise cost estimates.

2. Farmers' capital is another important aspect of culture fishery industry. It is now recognized that fish consumption in requisite amounts must be considered which improves the nutritrional quality of farmers' diet. Malnutrition causes both physical and mental retardation that affect general health status. Fish consumption in low-income food deficit countries is very low not only in calories but in proteins as well, and therefore, farmers' community are less productive.

3. Inspite of the application of various technological know-how, it has not been possible for fishery industries to supply different inputs in due time for fish production. The failure of fish crop due to disease outbreak, cyclone, flood, and drought spells catastrophe for fishery industry. Likewise, the aquacultural sector comprises a majority of people in a developing economy.

   Most of the fish farmers are reluctant to cultivate fish in their own ponds or in ponds on lease-basis. They try to engage themselves in other activities. The increasing demand for fish culturists can be fulfilled by encouraging rural youth and to educate them so that they can take steps to perform fish culture. Consequently, fish culture productivity must be improved.

4. Income of fish farmers is generated in the aquacultural sector. If fish cultute is well-developed, it can make a net contribution to upgrade the standard of living of farmers.

5. Fishery industry also needs a strong, well-developed market to operate and function efficiently. Some industries require a minimum size before they can take advantage of current technology. The demand for such industry in developing countries comes from farmers directly engaged in fish culture. If farmers in this sector do not earn an income, they will not be able to furnish the market which the industry needs for continuous growth.

Increasing fish culture productivity makes an important contribution to the general economic development, and that is one of the necessary conditions which must be fulfilled before an economy gets itself ready for a process of self-sustained growth.

It should be concluded that a program of fish culture development can not be pursued independently of industrial development programs. In fact, fish culture development programs will come to a halt if these are not supported by adequate industrialization. Industry makes possible aquacultural progress in a number of ways. *Firstly*, inputs used for fish culture are made available by industry. *Secondly*, industry supplies the equipments essential for the farm. *Thirdly*, aquacultural engineering is a significant branch of industry. Fourthly, most of the research activities have been undertaken for the technical know-how and other ingradients provided by the industry and Fifthly, industry is to meet the growing demand for consumer goods in the rural sector.

3.3 Importance of Fish Culture in the Rural Economy

The main importance of fish culture arises out of the position the fish culture sector occupies in the overall economy of the country. Fish culture is the second largest sector of the economic activity and has a crucial role to play in the country's economic development by providing fish protein, employment to a very large proportion of population, and capital for its own development.

1. Fish culture is the second largest source of national income and is also the main source of savings and hence capital formation for the economy. The pace of development is conditioned by the rate of capital formation in the economy. As early as the late seventies, large investment has been made in fish culture. In areas where fish culture practices are traditional, investment has been on traditional lines. But the pattern of investment in developed areas, where modern technology has been adopted, has been predominant in ecosystem improvement, production-related farm equipments and other infrastructures. Both fixed and variable capital in progressive areas are of a different order since substantial amounts of capital are required for different infrastructures and inputs to expedite production.

2. Fish culture plays an important role in the country's international trade. The main fish culture commodities which are exported are processed and frozen shrimps and prawns, crabs, and some species of finfish. The country has emerged self-sufficient in terms of finfish/shellfish in the wake of bumper production during the period 1990-2004. Consequently, the fish culture sector has emerged as a net earner of foreign exchange which is essential for maintenance and capital imports required in non-aquacultural sectors.

3. During the process of development, interdependence between industrial and fish culture sectors has become stronger. This interdependence is shadowed in various ways such as supply of raw materials and inputs from fish culture sectors to the industry, supply of materials from the industrial sector to the fish culture sector, and supply of fish flesh to the rural population. However, while fish production has helped to develop different types of industries and diversification of employment, a shortfall in fish production in some years may have an adverse effect on industrial production, and in turn, adversely affect prices, resulting in all-round disbalance and affliction in the economy.

It can, therefore, be concluded that fish culture occupies a central place in the rural as well as national economy. A solid basis of fish culture is an essential prerequisite for a rapid economic and social developments. Fish culture can contribute substantially to the improvement of the rural sector. A further forceful role is, therefore, planned for this sector in future.

3.4 Role of Fish Culture in Future

For an industry to flourish, proper analysis of strength, weakness, opportunity, and threat along with suitable plan towards development of that particular industry should be kept in mind. These parameters must be taken into consideration in working out a proper plan. These analyses would, however, help us in evaluating (1) the future plan programs, (2) the decision-making process, (3) the standard of performance, and (4) the smooth operation of an industry.

Concerning the problems of development associated with the national economy, the prime attention should be given on the rural sector. At present, there is a tendency to transfer a large-scale population from fish culture to non-fish cultural activities. Consequently, the rural economy will be needed to sustain a vast and growing population. This burden has a greater accountability in supplying fish for nutrition to both rural and urban populations, employment, capital, and foreign exchange earnings.

In future, as planned efforts are made to generate employment opportunity, it will be necessary to yield fish biomass for supply to the population at reasonable price to confer steadiness to the economy for planned development.

It should be kept in mind that the interaction between fisheries industry and fish culture will definitely encourage and support the development of these sectors. A lack of growth in fish culture can be a serious limiting factor to the production of white meat. Scientific development of this sector will lead to more non-farm activities such as training, education, research, and extension as well as building up of sector concerned with providing supplies thereby expanding the scope for employment. The development of fish culture may, however, proceed in an unhampered manner if farmers and their cooperative societies have an aggressive attitude to advance their causes.

McDonalds can be ousted as the kind of fast fish chains if fisheries can be adequately developed for future generations in the country. All over the world, eating habits are changing from red to white meat. Fish is not only the elixir of life but also the healthiest of all non-vegetarian foods and if this can be developed through a fast food chain in the country, there would be little or no demand for others. In addition to agriculture, fish culture and fisheries are, however, considered as the perfect economic rejuvenator for the rural areas. In implementing the future strategies of development, these meditations would have to be preserved in memory.

The fishing industry has generated and strengthened a large number of ancillary industries and it will continue to strengthen them. Thus the development of fishery and allied industries has a special significance on account of the following :

1. This industry caters to the upliftment of the socio-economic status of the rural population.

2. Production and consumption of fish will help to enhance the nutritional status of both rural and urban populations.

3. There is an urgent need to increase exports of fish and fish preparations that will mitigate hard currencies.

4. To withstand the international competition, it is necessary to develop this export-related industry on scientific basis.

5. Produce of the sector provides marketable surplus for local and export markets. These constitute a good source of foreign exchange earnings.

3.5 Five-Year Plans and Fish Culture: Indian Context

Fish culture and fisheries are regarded as important sectors not only of the country's economy for various reasons but the rural occupation as well. The importance of this sector is better appreciated in the light of the fact that it provides healthy food containing animal protein in the diet. Despite achieving surplus in agricultural production, the problem of protein deficiency continues to be acute. India's share of protein yield in global output is placed at 4.3 per cent as against 15.3 per cent in case of China — the leader of fish production in the world. This indicates the necessity to increase fish production and thereby the national income. In this context, development of fisheries has been receiving considerable attention by the national government.

Development of fisheries and fish culture depends upon adequate financial resources. Processing of aquacultural produce and raw materials, purchase of equipments and inputs, arrangement of the means of transportation, marketing and many others require sufficient finance. The recent phenomenal growth of this sector has been the consequence of the large and easy credit facilities provided by the credit organizations.

Indian fisheries industry has made several-fold progress over the last two decades. Indian shrimp exports in terms of volume rank highest in the world. Fish production in 1980 was 24.5 lakh tonnes which increased to about 37 lakh tonnes in 2000-2001 and is further planned to go up to 55 lakh tonnes by the end of 2015.

THIRD FIVE YEAR PLAN

The third plan (1961-1966) was of great significance for the aquacultural sector. It held out a great promise for the future. The new strategy of fish culture development came to be introduced by the establishment of the Indian Council of Agricultural Research (ICAR) in the year 1947 to evolve a new technology suitable under Indian conditions. In 1963, the Agricultural Refinance Development Corporation (ARDC) was set up to provide adequate financial support to rural areas. This corporation now merged with the newly set up National Bank for Agriculture and Rural Development (NABARD). Thus, it would be seen that the third plan has provided a new structure of machinery which was triggered to meet the challenge expected to be posed by the new aquaculture strategy.

FOURTH FIVE YEAR PLAN

In fish culture sectors, the fourth Plan had two objectives. *Firstly*, to provide the conditions required for a sustained increase in fish production, and *secondly*, to enable as large a section of fish farmers as feasible to participate in fish culture. Accordingly, the priority programs of fish culture development during this plan period fell into several categories such as (i) to maximize fish production, (ii) application of new farming technology such as composite fish culture, air-breathing fish culture, and induced breeding. These have opened up the possibilities of a multi-fold increase in fish production and, at the same time, the employment generation for the rural poor.

FIFTH FIVE YEAR PLAN

The major breakthrough is being achieved by fishery scientists in the field of prawn and shrimp culture technology under controlled conditions. At tbe same time, this plan aimed at maximizing fish landing from different inland resources. Better deal for farmers, fish/ shrimp handling, processing, and marketing, and stepping up of marine exports have been developed.

SIXTH FIVE YEAR PLAN

This plan explicitly recognized that long-term prospects of growth in Indian economy depend on a rapid growth in fish culture and rural development. The plan, therefore, emphasized that the fish culture growth pattern should be taken into account as immediate need of aqua-products both for domestic consumption and export. An assessment of achievements and targets of fish yield indicates that the gap between achievements and targets has narrowed down to a great extent. The plan sought to modify and rectify the drawbacks evidenced in the strategy of intensive production.

SEVENTH FIVE YEAR PLAN

During the five years of this plan period, new heights are estimated to have been reached in fish culture production. The planners and the technicians have set their sight still higher and are projecting to generate larger agricultural surplus for development of the rural economy at large.

In order to attain th growth pattern, special efforts have been made for effecting a breakthrough in finfish and shellfish outputs for increasing the productivity of inland waters. Such pattern of growth can provide the necessary impetus to rural development through dispersal of technologies.

The most important thrusts in this plan period are (i) development of fish farmers and fishermen, (ii) water management for healthy fish yield, (iii) training and extension programs, (iv) credit institutions, and (v) farmers' participation.

EIGHTH AND NINTH FIVE YEAR PLANS

In these plan periods, it is recongnized that the need for providing fillip to the utilization of the inland capture and culture fisheries, fish seed production, cold water fisheries, reservoir fisheries development and others that are related to rural development was considered. Considering that the need for reviewing these schemes has been tapering out, a new orientation seems to have given to these plan schemes.

TENTH FIVE YEAR PLAN

The accent of the tenth plan fisheries schemes is stated to be on augmenting and extending inland culture fish production. Though development of inland culture fisheries sectors has progressed to a significant extent, development of cold water fisheries, catfish farming, reservoir fisheries, water-logged areas for culture estate, and utilization of inland saline land for fish culture has not been progressed all along at the same pace. Trout culture fishery development has received a remarkable fillip in very limited places of the country particularly in the state of Himachal Pradesh with the financial and technical assistance of the Norwegian Government. However, developmental activities on these aspects can be taken up by implementing well-drawn schemes designed under the tenth plan. This activity would be a blessing to the inland fishers for increasing their incomes and at the same time would promote export-related programs.

Another important development of pond fishery which has already been undertaken in this plan period, is the consolidation of the technology of shrimp culture in freshwater ponds. The related technology, developed by the farmers themselves, has become fairly well established in inland waters adjacent to the coastal regions and in the land-locked areas of the country. The shrimps grown in inland waters are now accepted in many countries of the world.

Considerable progress has been made by the farmers in respect of culture and production of the giant freshwater prawns and prawn seed.

This activity has spread not only in the cultivable waters adjacent to the coastal regions but in inland waters of land-locked areas as well. Polyculture of major carps with prawns/ shrimps has started gaining favor among fish culturists. Further, the culture of monosex/ GIFT (genetically-improved farmed tilapia) has become very popular in several Southeast Asians countries for the reason that the fish has not only a wide market in the form of fillets in USA and in several European countries but also it is an eco-friendly with practically no adverse effects on the environment.

To sum up, it would be seen that fish culture has been conceded as top priority in the programs of development. The results obtained have been none-too-discouraging. Indian fish culture has made a long step in terms of productivity in recent years. Inspite of this success, no major dent has been made on rural poverty. Over the years there has been deterioration in the areas concerned. Population explosion coupled with the pollution of capture and culture fisheries resources have compounded the problems manifold.

Decrease in fish production potential in most freshwater systems (such as rivers, many reservoirs, and wetlands) viewed in the context of extreme pollution are regarded as the most important conclusion. Therefore, some concerted efforts have to be made to increase fish production and productivity generally so as to expand employment opportunities in rural areas and also aim at achieving a new cluster of objectives in harmony with the rising preference and demands of the present and the future.

3.6 Long-term Fish Culture Strategy

A fish culture policy with a panorama till 2020 is being developed. This will aim at increasing fish production to adopt the country's demand and acquiring the sustained objective of self-sufficiency in fish culture products.

The most important aspect of the fisheries policy will be the elimination of imbalance in yield. Wide fluctuations in the production figures in many reservoirs of the tropics and sub-tropics indicate that there is need for stabilizing fish production. Development of capture-culture balance in various reservoirs is an integral way and with a gathering approach to cater the needed infrastructure convenience for production, harvesting, domestic marketing, and exports. While prawn and shrimp production has been increasing steadily from year to year, there are marked fluctuations in yield of other fishes such as carps and carnivorous fishes.

The strategy will thus pay special emphasis to the need of insulating fish and fishery products from drastic fluctuations in yield. It is expected that this strategy will definitely accelerate significant results by way of foreign exchange savings.

A long-term policy that can be undertaken involves agri-aquacultural techniques and development of wetland ecosystems. Agri-aquacultural techniques enhance ecosystem fertility and therefore are highly efficient in food production with plentiful opportunity for development in the third world. Fish culture and capture in its customary primordial condition can bestow robust fish in larger quantities for another ten decades.

To sum up, fish culture certainly has shown a remarkable progress. But in view of the annual growth of population of more than 2 per cent, the balance between animal protein requirement and population has become unstable. The package approach adopted as a part of new policy has opened up new magnitudes of the growth of fish culture sectors. Under the impetus of new policy, the farmer is responding to production and market incentives as well as fluctuations and is thus sharing in a purposeful way in the economic development of the farmers' community.

3.7 Components of Aquacultural Growth

An increase in aquacultural yield can results from an increase in water body under cultivation and from an increase in the productivity of waters.

1. **Area Under Culture:** The area under cultivation can be increased through adoption of mixed fish farming or by the reclamation of uncultivated water areas. Efforts

have been made during the last thirty years of the 20th century to increase the area of water bodies. The increase in area under crops paved the way for the fact that reclamation of water bodies and water management programs were launched during the last twenty years. The suitability of unutilized water areas made these endeavour a success. The availability of infrastructural facilities also made possible to adopt mixed fish farming systems.

Significant contribution towards development of induced breeding programe has been forced to increase the water area under cultivation. Though the area under cultivation has increased to some extent, there is further scope for increasing the program concerned. Out of a total freshwater area of about 16.5 million hectares, the gross cultivable area for aquaculture is being estimated at about 1.5 million hectares.

In other words, about 10 per cent of the freshwater area is under cultivation. Though this compares very favorably with the world or average of about 26 per cent, increasing popllation has drastically changed the entire fish culture scenario. At present, aquaculture production is not adequate in relation to needs. Since there is scope to increase the water area under cultivation, fish production may further be increased. At the same time, fish culture potential may substantially be increased to fulfil its increasing needs and in increasing the level of productivity.

2. **Production Per Hectare or Productivity:** More significant is the increase in the production per hectare of water. Where most of the water bodies are not fit for fish culture due to pollution hazards, stringent measures should be undertaken to increase the productivity of water areas through adopting better techniques of production. However, fish culturists have tried to boost aquaculture production during the last twenty years or so.

3.8 Productivity of Indian Fish Culture System

We have already discussed about the trend in aquaculture production; but inspite of marked improvements having taken place during the last 25 years, the productivity of aquaculture is relatively not up to the mark. The low productivity undoubtedly indicates the vast potential of growth to explore in future, but at the same time a proper diagnosis is inevitable. It is, therefore, necessary to explore the parameters that follow are responsible for low aquacultural productivity.

TECHNOLOGICAL PARAMETERS

Most of the farmers still follow the extensive system of farming. Availability of quality fish seed and fish feed is not adequate for the farmers. The widespread use of chemical fertilizers and pesticides in agricultural fields and industrialization lead to a considerable damage to the ecosystem productivity. If this situation continues or is not controlled the productivity of Indian Aquaculture will be considerably decreased in the years to come. And the total responsibility will rest on the policy makers and planners of the country.

DEMOGRAPHIC PARAMETERS

The increasing pressure of population is the most important parameter for low productivity in aquaculture. Most of the inland waters (particularly rivers, ponds, and lakes) are highly polluted due to human activities.

The increasing population has spread a number of evils such as diseases, parasites, habitat destruction, and lack of awareness among people. All these evils have been responsible for low productivity in aquaculture sector. Many reports have, however, shown that 'fish production has substantially increased'. Of course, this is particularly true in case of marine sectors. But so far as freshwater resources are considered, the entire scenario is not encouraging.

OTHER PARAMETERS

Multi-ownership of most of the freshwater bodies particularly large- and medium-sized ponds, tanks and bheries is the another important parameter responsible for low productivity in this sector. Interested fishers are unable to pay adequate attention towards improvement and production. The multi-ownership custom has, to some extent, been abolished by the formation of fish farmers' cooperative societies.

The credit finance system is not adequate though a good deal of work is being done by the commercial banks, the regional - rural banks, and the NABARD. Since farmers do not show any interest to repay their loans for which it was sanctioned, the banking sectors are restrained from providing further assistance to farmers.

In constrast to some other countries (such as China), Indian fisheries and aquaculture sectors are devoid of entrepreneurship. To increase aquaculture production, it is necessary to reinforce aquaculture with private business enteprises to obtain desired results in yield. Lack of awareness among fish farmers equally obstructs the productivity of ecosystems. Under all these parameters, it is not easy to observe any significant increase in aquacultural production.

3.9 Measures to Increase Aquacultural Productivity

The economic advancement of the farmers' community hinges on increasing aquacultural productivity. Therefore, the entire economic development of this sector should be recognized for increasing production. Such efforts must require a harmonious blending of traditional aquaculture policies, which sought to alter the structural and institutional arrangements in the existing system, and a modern pragmatic approach, which aims at introducing a technological revolution in the system. Technological innovations generally refers to parameters that help productivity. Quality fish and fish feed, use of probiotics and nutrient carriers, and others that are related to production are useful examples. Broadly speaking, technological improvements will cover such measures as integrated farming, provision of better inputs, organization of aqua-industries, and investment in human infrastructures.

It is also important to note that farmers not only require the production and income-related technological information but sound supporting structure as well that will take care of inputs such as services snd credits. In absence of these inputs, the farmers' capacity to implement more modern techniques for fish production will remain limited. Joint venture between the government and the farmers' cooperative societies with the administration providing the management support may undoubtedly become a key factor in rapid transformation of Indian aquaculture in the near future.

To restore the sick fishery industries and to augment fish yield from different freshwater resources, long-term policy should be strictly adopted. In the present infrastructure systems, in addition to the culture and capture of freshwater fishes, strong emphasis should be given by the farmers for management of fish culturing and capturing equipments. But most of the farmers face great difficulty to manage the financial liability for the purpose. Consequently, they are compelled to depend entirely on government or non-government organizations or money-lenders for loans. For this purpose, it is necessary to separate the efficiency of fish culture/capture from equipments necessary in fishery industries and the fishery industries need to be corporatized. Infrastructure facilities should remain as an asset of the corporate body. The policy of financial encouragement must be implemented and enforced by the government. This policy will definitely increase the export-related potential of both raw, but unpolluted and value-added fish not only to Japan, US and European markets but to African markets as well. At the same time, effort should have to be made to sell at least 30 per cent of the total export product in the national markets. Strong emphasis should also be given to increase culture and capture efficiency of edible freshwater fish. If freshwater fish can be sold in cleaned and unpolluted conditions, there would be an increased demand for the product in both national and overseas markets.

For the promotion of fishery industries, a treasury should be formed by releasing government-guaranteed bonds. The assistance will be provided from the treasury to the fishermen community as loans but not as subsidy. If no compulsion is imposed on any industry, there would be no responsibility to repay loans. However, unless attention to export-related fish culture strategy is given, there is hardly possibility to develop the rural economy in the country. Whether there is any sensational affair regarding culture strategy, this sentimentalism has only remained as romantic outburst of passion. Fish culture and capture industries are the areas which experts felt needed tackling to make fish accessible to the general populace. The other areas where the fishery industry needed a boost was in the ready fish market which could be provided in urban areas.

3.10 Suggested Aquacultural Strategy

In view of the fact that the aquaculture technology has opened up new dimension of growth, it is imperative that on the one hand steps should be taken to obtain high gains in productivity and on the other steps should be taken to ensure that these gains reach all the corners of the economy. Hence, the overall strategy for enhancing aquacultural production should consist of the following important elements.

1. An uninterrupted improvement in production.

2. Availability and use of scientific inputs and credit.

3. Better utilization of resources which improves growth of aquaculture sectors.

4. Attention to the needs and potential of growth in areas with different levels of development.

5. An adequate support by extension, training, education, and research for 1 and 2.

6. The marketing structure which sustains growth of this sector.

3.11 Infrastructure for Fish Production and Rural Development

Infrastructure necessary for the progress of culture and capture fisheries resources involves the availability of quality water, transport facilities, arrangement for fish marketing and storage of products, aquaculture clinics and clubs, flood and drought control, organization for supply of brood fish, raw materials and inputs, set-up of fish breeding centre and fish farm, and a host of other facilities which are necessary from pre-production level to the marketing of fish biomass. Training and education to fish farmers and fishermen and formation of cooperative societies also form an integral part of infrastructural development. Some of these infrastructures that are related to fish culture and production have been described elsewhere. In this section we will limit to an analysis of the following components : availability of quality water, measures need to be initiated for infrastructure development, science and technology.

WATER QUALITY

Good quality of water is more valuable for fish culture and renders large areas productive which otherwise will remain barren. Good quality of water also provides better production that brings about a virtual transformation of the rural economy. But unfortunately, most of the water bodies are becoming polluted that needs well protection. Phyto- and bioremediation are the two most common and cheap methods which helps, to a large extent, remove pollutants from waters thereby increases production of robust fish. These methods have been described in Chapter 19 of this Volume:

It can be stated that proper water management (see Chapter 10 of Volume 1) and the availability of unpolluted water for fish culture requires action at levels higher than individual farmer and action is also required at the plannimg and policy levels. Integrated detailed planning for availability of quality water has to be carried out and designs have to be worked out well ahead of time. Institutional reforms have to be carried out to develop capabilities of commitments for these achievements. However, better management of water and adoption of modern techniques to improve the quality of water is badly needed for fish culture.

MEASURES FOR INFRASTRUCTURE DEVELOPMENT

The following measures should be initiated for the purpose:

1. Understanding the type and nature of activities which could be taken up and to ensure a reasonable and reliable income from different water areas.

2. Formulating technologically-feasible and economically-viable projects in the concerned area. The scheme, for example, for fish production or water management should *inter alia* incorporate all important informations on the availability of needed infrastructure and gaps in the existing infrastructure which can affect the implementation of the projects.

3. Need–based credit plan with details of the activity–wise budjet should be prepared and the assistance should be provided by participating financial institutions. For viable projects NABARD, DRDA, IRDP are willing to extend financial help in the form of short, medium and long-term loans depending upon the repaying capacity of farmers and the viability of the projects.

4. Government and non–government organizations should provide management inputs for successful implementation of the project.

5. Farmers, extension workers and voluntary associations should be adequately provided a production-related training.

6. Training should be organized in the region where farming activities have been successfully conducted so that farmers can esteem the thought and benefits and cast aside their old and wrong ideas.

7. Supply of inputs and marketing systems should be considered as a part of the rural development program.

8. The organization should be well equipped with skilled and experienced personnel to implement and enforce the developement project and to monitor the working activity of the project.

9. Adoption of package of practices requires the organization of village level piolot projects and operational research. Though a number of parameters are involved in the success of different projects for significant production, the most important factors are the concerned farmers, the interest and involvement of farmers, and the constant attention of trained and resourceful extension personnel supported by production and management experts.

SCIENCE AND TECHNOLOGY

Though technology is a key factor in rural development, the available technology should be appropriated to the local conditions. Efforts to promote appropriate technology in rural sectors have produced encouraging results during the last twenty years. These cover different aspects of the rural set–up among which the more important points are as follow :

1. **Fish Production:** Technological changes that have been introduced in fish production areas can be divided into three parts: (i) pre-production technology, (ii) production technology, and (iii) Post-harvest technology. The main aspect of pre-production technology are the improvement of fish culture ecosystems, improvement of soil and water quality, preparation of fish ponds for culture, and farm mechanization. This technology facilitates adoption of production level technology and therefore indirectly improves farm productivity.

   The production technology involves use of fertilizers and manures in suitable combinations, fish feed, stocking of quality fish seed, disease control, management of water quality, and culture practices. Interactions among all these arrangements make the technology more effective. Though production technology improves production capacity of the farm, the technology varies depending on the type of culture system to be adopted and species to be reared.

   The post-harvest technology includes preservation, processing, and marketing arrangements. This technology is, however, crucial for economic returns to farmers.

2. **Intergrated Farming :** Combination of two or more farming components which become part of the entire system is termed as *intergrated farming* and referred to as *Intergrated fish farming* when fish is included as a component. The main features of the approach are (i) recycling of farm waste for food production and (ii) efficient use of farm space for production. Freshwater fish culture is largely organic-based, with inputs derived from activities of agriculture and animal husbandry, with plant and animal residues forming the major component of feeds and fertilizers in carp polyculture.

   Huge quantities of plant and animal residues are generated each year. In rural areas also activities of mushroom cultivation, apiculture, and silviculture provide huge quantities of organic materials, that become a resource in fish culture. Likewise, domestic sewage, and effluents produced from agro-based industries and food processing plants that could be recycled to fish culture. Since the scope of integrated fish farming is considerably wide in most Asian countries, the application of this technology holds a great promise and potential for increasing production, betterment of rural economy, and also improving the socio-economic status of weaker sections of the rural area.

3. **New Methods of Management :** Innovation in management science could improve the performance in implementation of the projects. New methods of planning, network analysis, systems analysis, and input-output analysis can be adopted in rural development. Innovation in optimal use of resources for farm management is very important for high production.

To sum up, it is beyond doubt that no development is possible without the adoption of technology. Significant technological development has, however, occurred that has dramatically changed the rural picture. But further efforts have to be made to apply the new techniques in diverse sphere of rural activity. Extensive use of new technology in rural areas along with need-based viable schemes are not unimportant for rural development. It is important to note that the 9th and 10th plans have proposed to give high priority to this program. Moreover, the private sector should also play a key role in this regard by providing infrastructural supports.

3.12 Role of Cooperative Society in Rural Development

COOPERATIVE STRUCTURE

A well cooperation is a special mode of doing both agricultural and aquacultural activities. The principal idea behind cooperative business is organized by a group of farmers for promoting their economic welfare through mutual help. In a capitalistic society this device is adopted by farmers of the rural community for protecting their economic interest against the oppression of the stronger section generally termed as *middlemen, large-scale producers,* and *distributors.*

The cooperative society is a voluntary and democratic association of farmers for the purpose of conducting aquaculture and fishery activities for their benefits. The key feature of a cooperative society is its non-profit motive. Farmers organize themselves cooperatives for the purpose of supplying their own needs in the form of goods and services and loans when required.

The membership of a cooperative society is unrestricted. Any farmer can be a member if he subscribes to the common objective and the capital. The capital is divided into as number of shares of equal value. The liability of members of a cooperative society may either be limited or unlimited. If the liability is limited, the word 'Limited' is used after its name. A cooperative society may also be registered or unregistered. A society registered under the central or state cooperative Societies Act enjoys certain privilages and they are subject to strict government control and supervision.

Similar to agricultural credit societies, fish farmers' cooperative societies constitute widespread network and operate fish culture and fisheries sectors, supply of aqua-inputs, and financial support to the member farmers dissemination of practical knowledge or skill, and marketing of fish. If a well-managed cooperative society is set up, member farmers may receive a lot of benefits. Otherwise, the society may collapse (see Panel 3.1).

Since the cooperative society is a democratic organization, each member has only one vote irrespective of his capital contribution, which ensures equal voice of all members in the management of the cooperative business. Though every member has a right to take part in the management but in general, a managing committee is responsible for its action to the general body of members and has to execute the policy enunciated by the members at the Annual General Body Meeting. However, with the establishment of the cooperative society, the activity has gained vigor and vitality as the society is entrusted with the activity of planning and promoting programs to culture and capture fisheries, processing, and marketing of aquaculture produce through cooperative societies. Owing to mismanagement of the entire cooperative structure and foolishness of the member farmers, most of the cooperative societies have gone to the dogs.

ADVANTAGES OF COOPERATIVE SOCIETIES

1. The most important aspect of the cooperative society is the arrangement of disbursement of loans. Along with financial institutions, the cooperative societies have been instrumental, genrally speaking, in destroying strong repressive influence of rural money-lenders. Cooperative societies have provided not only more safe means of borrowing loans but also the loans are sanctioned at very cheap rates.

2. Co-operative societies play a crucial role in fish yield and management of freshwater resources. This is feasible by encouraging the use of better farming methods. Marketing societies are also equally important in removing malversation that prevails in local markets. An association of fish farmers in a highly defective rival market bestows an organized glimpse and can buffet a good contract.

3. Co-operative societies help farmers to confer a more systematic outlook which encourage them to save their interest. Co-operative societies also help restrain the farmers from the habit of extravagent expenditure. The habit to use the banking media for transactions has, on the contrary, acquired the fillip among the rural people.

4. Co-operative societies help a great deal to eliminate objectionable social conventions. Co-operatives have induced the farmers to give up the habit of acting of carrying on a process at law and the practice of betting.

5. In a well-functioning co-operative, all the members take an active role in business methods.

Thus fish farmers' co-operative societies can be considered as the interlacement of primary schools in aquaculture finance.

LIMITATIONS OF CO-OPERATIVE SOCIETIES

1. Most of the farmers have no idea about the importance of co-operative societies. They feel that these organizations are under the control of government and provide financial assistance and other facilities at subsidized rates. Lack of awareness, bureaucratic attitude, and politics are the three most important factors that drastically affect to promulgate the information of co-operative development.

2. Inadequate attention is often paid to the needs of farmers or their repaying capacity at the time of financing. When loans are granted to farmers, they do not show any interest to repay their loans to the society. Consequently, many cooperative societies are on the verge of extinction. Non-recovery of loans, politics in management, lack of coordination among various divisions of the cooperative

structure, farmers' propensity to receive loans from other agencies, and lack of auditing are the most important weaknesses. These difficulties undoubtedly create barriers in the way of smooth functioning of cooperative societies.

3. Bureaucratic inteference in the activities of cooperative societies sometimes disturb the entire network. As a result, farmers lose their interest. Farmers should understand that the cooperative societies are dependent on government assistance. Once the assistance is withdrawn the societies will perish.

4. Lack of technical experts, inadequate funds in relation to needs, and improper management in the societies may fail to spread their network with the result that the number of cooperative societies have reduced to a few.

PLANS FOR IMPROVEMENT

Though a number of advantages have gained by farmers from the cooperative societies, a number of defects have also been observed in the cooperative structure. The following important indications should be given to overcome these difficulties.

1. Orientation of cooperative credit policy in favor of the farmers should be intensified.

2. To check the irregularities in cooperative loans, cooperative by-laws should be reformed. Specific policies should be framed to reduce the rate of interest charged from farmers.

3. Adequate financial support to fishermen and aquaculture sectors will help to sustain rural development.

4. An integrated program of fish culture operation, processing, and marketing should be implemented and enforced which would, on the one hand, enhance their contribution to employment opportunity in rural areas and, on the other, increase fish production.

5. Though debt salvation laws have been passed, the laws have created extreme pain to the farmer because of the want of suitable and efficient system of credit. Hence, a promt adoption of a multipurpose credit system with efficient management is imperative.

6. Attention should be given to make the cooperative societies autonomous for better service to farmers, conscientious, and efficient in the discharge of their duties and free from politics.

In brief, there are both advantages and disadvantages of cooperative societies. At present, however, several new technological factors for different forms of group organization include (i) better management of freshwater resources, (ii) brood-stock management practices, (iii) obtaining credit for planned development, and (iv) introducing integrated fish farming systems. Planned cooperative societies are not as strong as it is necessary to serve the purpose and are not always stable. Future plan for cooperative societies must be conducted by these contemplations.

3.13 Role of Non-Government Organizations in Rural Development

In a planned economy, the developnent of fish culture sector depends, to a large extent, upon the percentage of outlay envisaged for them and also on various projects concerned. On an average, the percentage of financial outlay on freshwater fisheries sector in the entire plan periods is not sufficient to meet the demand of farmers. On the other hand, loans provided by authorized credit agencies for fisheries sector are nor secured. Since most of the farmers do not repay their existing loans, these agencies are very reluctant to provide further loans to the needy farmers. In this situation, the non-government or voluntary organizations (The type of nomenclature of such organization varies from country to country and even place to place) play a key role for upliftment of the socio-economic conditions of farmers.

Rural development through non-government organizations is, at present, highly encouraging in many countries of the world for proper allocation and utilization of funds since these organizations are free from politics and exploitation and has opportunity to recover loans. Moreover, in the marketing and transporting systems, these organizations destroy the importance of middlemen and farmers are able to receive the full benefits through profitable prices. The various activities performed by these organizations include credit programs, fish markets, adoption of refined technology, savings scheme, granting of revolving funds for marketing activities, social activities, creation of relief fund to compensate for the loss of farmers' family at the time of drought and flood.

3.14 Recommendations on Rural Development Through Fish Culture

For successful development of rural economy, the following principles may be recommended :

1. Integrated fish farming for production, waste recycling, and for optimum utilization of land-water resources. The existing mechanism of FFDAs could be extended for the purpose.
2. Establishment of backyard hatcheries of carp and catfishes for seed production.
3. Aspects of fish seed rearing, fish feed production, and preparation of fish products must be considered for involving women in fisheries and aquaculture sectors.
4. Intensive training on rural fish culture strategy.
5. Freshwater fish culture practices should be considered as an integral part of the watershed development program.
6. Practice of inland fish production in water-logged areas of reservoirs.
7. Development of reservoir fisheries.
8. Integrated development of research institutes and implementing agencies should be undertaken to keep the economic interests of local people in perspective.
9. Technologies should be demonstrated at different places to bridge the gap between the field activities and the work of research institutes.

10. Rural youth with vocational education should be trained at fisheries institutes.

11. Development of manpower for management of the increasing and ever-expanding resources should be taken up.

12. Awareness amongst fishermen community should be brought through vocational education.

3.15 Conclusion

Rural development generally involves agricultural and aquacultural activities that should be strictly followed through adopting new policy and technologies. Apart from training, education, and housing to farmers, the process of rural development through fisheries and aquaculture must be maintained by the farmers. And if they read about fish culture strategies between the lines, production of food from freshwater resources through integrated fish farming systems will be increased several folds. Export-related fisheries and aquaculture are inevitable for development of the rural economy and employment opportunity. The commercial possibility of fish culture industries is no doubt predominant that should be properly evaluated when rural development programs are contemplated.

In many tropical countries, the problem of rural development and degradation of aquatic habitats is due to population growth and poverty. These two phenomena reflect two sides of the same coin. The farmers of developing countries who spend most of their life through poverty cannot possibly improve their lot in life if they have to cultivate fish in water which is highly degraded. Essentially, the first step in economic upliftment of the poorest of the poor is to ensure that they have the capacity to re-build and regenerate aquatic resources on which their livelihood depends.

References

Gupta, G. S. 1983. Freshwater culture fisheries — Present status, prospects and Policy Issues in Inland Fishery Resources in India (U. K. Srivastava and S. Vasistha eds), Concept Publishing Company, New Delhi.

George, K. M. 1984. Rural Development Programmes — Its strength and weaknesses. Indian Journal of Agricultural Economics, July-September, 1984.

Jolly C.M and H clowts 1993. Economics of Aquaculture. Haworth Press Inc. New york.

Questions

1. Explain the role of new aquacultural strategy for development of India.

2. What are the strategies required for exploiting aquaculture sectors without generating ill-effects of the development process.

3. Indicate the main problems being faced by aquaculture in rural area.

4. How will you account for low productivity in Indian aquaculture?

5. Discuss the role played by different development agencies in the development of aquaculture activity in rural areas.

6. What are the components of aquacultural growth?

7. Discuss the need and prospects of aquaculture development in India. Mention special schemes/programs launched by the government for promoting aquaculture development. Indicate the role played by institutional credit agencies.

8. Economic development is impossible in the absence of progressive fish culture. Discuss.

9. Discuss the role of fish culture in the Indian economy.

10. Access the importance of cooperative societies in fish culture development. What are the advantages and disadvantages of cooperatives? Give some suggestions for improvement of cooperative societies.

Panel – 3.1
RAJENDRAPUR FISH FARMERS' CO-OPERATIVE SOCIETY

The Rajendrapur Fish Farmer's Co-operative Society is situated in the district of 24 Parganas (north), West Bengal, India and is one of the dynamic and largest co-operative societies in India. This co-operative society has been established in 1985 with a view to improve the socio-economic condition of farmers in these regions. This society is highly popular so far its reputation and production-related activities are concerned. This co-operative society possesses extensive water areas measuring about 600 hectares where more than 1,500 hatcheries and 1,000 ponds have been constructed for breeding Indian major carps and cat fishes. In this society, more than 2,500 fishermen are actively involved in fish breeding and culture-related programs round the year with an average annual turnover of Rs. 2,000 crores. In addition to these, more than 1,000 trucks come each day from different states of the country for purchasing fry and fingerlings. These data clearly indicate that this vast freshwater ecosystem has good production potential and must be improved at any cost.

Although there is vast scope to improve the production of quality fish seeds and fingerlings in these regions, siltation of these water areas, lack of good will and financial assistance to farmers, political interference, and disturbances of some local people are some of the factors involved in gradual dwindling of the activity of this co-operative society and hence must be reckoned with. And the co-operative society is always confronted with these factors. These factors are, however, not desirable at all and hence must be removed root and branch for the interest of fish farmers.

4

Fish Culture and Rural Development with Reference to Economic Criteria

Previous chapter has discussed at length the importance of fish culture in rural development. Fish culture and rural development are closely associated with each other and one cannot be realized without considering the other. We now focus on fish culture and rural development with special reference to economic criteria — the most important characteristic which significantly change the entire economic structure of the rural community.

4.1 Objectives of Fish Culture Development

Fish production from aquatic resources is one of the most important objectives of fish culture develement strategy. It is important in regions where protein requirement is not adequate. Production of animal protein is energy-demanding. On comparing different sources of protein it is evident that fish protein is highly favourable in terms of efficiency of energy conversion. Fish are considered both as subsistence production and in market economics. The difference in the economic structure across the world and inflexibility of capital has a marked effect that resources are not fully utilized. Therefore, economical action of fish culture has to be executed. The objectives of fish culture are based on the following three main groups such as society (macro-economic point of view), producers(micro-ecomonic point of view) and consumers.

SOCIETY

This group includes optimal resource assignment to yield fish for human consumption, excessive production of fish and fish products for export, and encouragement of rural reconciliation through increased standard of living by the total profitability of the industry concerned.

PRODUCERS

This includes small and marginal farmers to maximize income on investment.

CONSUMERS

Supply of aqua–products at low or resonable prices. Other important objectives of fish culture include production of sport fish, production of bait for sport and commercial fishing, production of food and ornamental fishes, recycling of organic wastes, and production of fish on commercial basis for the table. However, political decision is the main criterion to understand long-range macro-economic objectives of fish culture for greater interest of the nation.

Society is not concerned in either under-production or over-production. In the former case, it may result in high prices, inadequate food supplies, and poor utilization of resources. On the other hand, excessive products can neither be sold in the domestic market nor be exported at profitable prices. However, if maximum profit and income are not achieved, other objectives can never be gained. Fish farmers always attempt to maximize income or to maximize return on investment. In the latter case, the fish farmer wants to obtain this objective by one or two ways. *Firstly,* the farmer attempts to obtain fish by using least possible resources. With a given yield, the farmer attempts to obtain fish with least possible costs and *secondly,* with a given resource, the farmer attempts to increase his yield upto the carrying capacity level.

To sum up, the objectives are principally concerned with the income and profit. If freshwater resources are effectively utilized for aquaculture in general and fish culture in particular, the industry should be set up very close to the production farms and be actively engaged in the production process. Fish culture industry will be viable only if both objectives of profitability and those of social welfare are fulfilled.

4.2 The Role of Fish in Rural Economy

The role of fish in the farmers' economy differs from one part of the world to another. In most advanced nations where commercial fishing/farming is the major activity, they have a very important role to play. The whole economy is predominantly dependent on fish under such circumstances. Fishes are also very important in those countries/regions where they are consumed fresh, preserved or processed. In integrated farming systems, fishes have a significant role and the farm income is generated from a number of varied sources such as dairy, poultry, and agricultural activities. In these farming systems, the raising of fish and animal/plant products are inter-dependent.

The role of fish in providing by-products is highly important as their role as food in the daily diet is substantial. Apart from industries which provide processed fish foods, there are many industries based on fish by-products such as 'surimi', fish sauce, fish meal, fish oil, protein concentrate, and glue. Fish which have no commercial importance are subjected to sun drying, the dried products are ground up to make fertilizers, and food for fish, cattle and poultry.

A variety of products made from fish particularly 'surimi' and fish meal are now being widely used in many countries of the world which can be cheaply produced in large quantities. These products have a luxury market.

Fish farming has several importance which boost the socio-economic status of the rural people. This is very important in developing countries where most of the people live below the poverty line. At the same time, it creates employment opportunity in rural areas. A well-knit strategy of the farming industry will help foreign exchange earning.

The production and consumption of fish food is a guide to the standard of living of a country. The greatest consumers of fish and fish products are Japan and Portugal (48 kg/year) followed by Denmark, Norway and Sweden (30-44 kg/year) and Asian countries (20-40 kg/year). Countries with the lowest consumption are UK, USA, and Australia. This is due to the long-established traditional and polyculture with varied farming systems in different regions. Integrated farming has been the main method of farming in some temperate regions of the world, whereas in the tropics the composite culture and capture of a variety of freshwater fish especially carp, tilapia, murrels, and catfishes and their high production potential for an ever-increasing population leaves little room for farming activities. At present, fish farming is not beyond the reach of farmers which helps expand the area concerned. The lack of fish protein in the diet reduces nutritional levels and the population of such countries are not only poor but under-nourished. The fish culture revolution has greatly improved production and supplies of fish in under-developed countries, but unless the culture-oriented industry is not adequately maintained, it will never solve the shortage of fish proteins. Most under-developed countries will have to adopt more fish farming through better utilization of aquatic resources to overcome this problem.

4.3 Economics of Fish Culture

FISH CULTURE IN THE RURAL ECONOMY

In both developed and developing countries there has been a constant and continuous slow movement of human resources from rural areas to town and metropolitan cities mainly due to lack of job opportunities in rural areas. Aquaculture along with agriculture programs in rural areas may constitute an alternative in this regard.

While the cause behind current development in this field is explained, question may arise why terms of trade* have inclined against traditional rural activities. Adoption of new technology, changes in political factors, and the relationship between natural resources and populations have drastically changed the rural activities. Specialization in aquacultural sector and division of labor have also changed rural subsistence economy to production economy in which existing prices of output and of input in the industries greatly influence the level of income in these industries. Inspite of the adverse effects, technology can hardly be arrested and in course of time it will reduce production costs and thus improve standard of living.

---

* The rate at which one country's products exchange for those of other is referred to as terms of trade. It expresses the relationship between the export price and import price of a country.

Some fish and fish products are expensive and enjoyable in the household budget and others form part of the daily diet. Therefore, demand elasticities in prices will exhibit great variations. Since the price of food products determines the level of living standards, the governments in many countries control food prices which influence the income of fishery industries.

Uneven distribution of land/water resources is one of the main economic problems. Substantial portion of the water is possessed by a few big farmers which might increase the problem of efficient use of water/ land resources and betterment of the distribution of rural income.

GROWTH AND DEVELOPMENT

Economic growth or development is measured in terms of Gross National Product (GNP) which largely influences both social and economic aspects of the society. It connotes a change from a primary situation to a forthcoming stage and it is in this awareness of the fact that the development of aquaculture and rural integration problems must be taken into consideration.

Several external factors may indirectly affect the process of aquaculture development. These external events are characterized by increased pollution of natural resources to a greater or less degree, over-exploitation of natural fish stocks, recycling of organic and industrial wastes, and political conflicts. All these factors have indirect conclusion that can be drawn from the growth and development potential for aquaculture, both in total terms and considered in relation to traditional aquaculture.

Development of fish culture as stated in growth potentials may occur in four areas: new products, new markets, improved quality of being efficient, and organizing present activities differently.

MODELS FOR ECONOMIC ANALYSIS IN FISH CULTURE

For economic analysis in a fish culture ecosystem, a simple model is proposed. The model contains production-related technical and economic data and it can be used for analyzing the type of production and the total production. This model, however, helps analyze different factors responsible for the economics of aquaculture. The added value is the total of wages, salaries, interests, and profits. Capital, labor, and environment are some of the most important factors influencing fish production which indicates profit. Different methods to evaluate profitability may be assessed. In some cases, the remuneration of one input factor is measured and all other factors receive a normal remuneration. The calculation indicates the production of the selection factor.

Other important economic factors evaluated in the model for economic analysis in fish culture include: yield per yearly man-hours measures productivity, yield in terms of kilogram per surface unit or volume of water how ponds are intensively utilized, and investment calculated per yearly man-hours and per kilogram yield of how the activity of aquaculture is capital-intensive. In the process of economic analysis, the variations and interactions of factors which influence the income and variable costs must be kept in mind.

1. **Factor-Factor Relations:** Different combinations of yield factors are responsible for fish production. These factors are natural resources, capital, and labor and may substitute each other to reach a desire level of yield. For example, the mechanization process adopted in aquaculture is the substitution of labor by labor. The replacement affinities do not apply both between products and between yield factors. When the quantity of fish yield is known, the question may arise how this quantity has been produced by utilizing the least possible resources. In terms of values the resources which provide the lowest cost per unit yield should be amalgamated. In general, the farmer has always the tendency to replace expensive yield factors by cheaper ones.

2. **Product-Product Relations:** In fish culture, farmers would have to be kept in mind that the selection of suitable fish species and the size of fish to be reared in their ponds are most fundamental to achieve better production. Product-product relations connote which types of products are to be produced. Farmers would have to be selected whether air-breathing fish or carps should be cultivated or whether fry, fingerlings or table-sized fish should be reared. This is due to the fact that the production volume of fish culture is limited by the availability of resources. Consequently, a selection has to be made between different products that can be produced.

   Highest fish yield can be obtained at a given state of technology and with a given resource. Since both technology and resource are always changing, production targets are extremely variable. The following example will illustrate this point. Following the application of improved techniques, derelict or marsh lands can be made quite productive, while pollution of water may drastically reduce fish production.

3. **Economics of Scale:** The use of a high level of technology and capital per man-year and a minimum working hours per unit produced is the most important feature of the economics of scale. Savings of large-scale yield may only be employed if potentialities for specialization of yield and division of labor exist. However, on the basis of the formulated targets for aquaculture development, economics of scale may either be accepted or rejected.

   Though the techniques of economic analysis have not been discussed in detail, it is assumed that elaborate research is badly needed in this line to arrive at standard figures of yield along with the increased use of the individual production factors. This will permit both planners and fish culturlists to prepare more dependable budgets so that fishery industries may be benefitted in various ways and at the same time freshwater resources may be properly allocated.

4.4 Aspects of Integrated Rural Development

OWNERSHIP OF RESOURCES FOR FISH PRODUCTION

In general, water, land, fertilizers, feed, and fish are the vital resources for sustained production and very few farmers possess such resources. These resources are, sometimes, not fully utilized efficiently due to lack of extension services and unavailability of research data. Moreover, the net income is not distributed uniformly in regions where population pressure is severe. In contrast to land, large lakes, rivers, wetlands, and coastal regions are in many cases common to all where landless farmers may follow some aquaculture techniques such as the use of floating cages and harvesting by using simple gears. By charging taxes or subsidies on the use of such natural waters, fish yield can, in many cases, be increased. Natural resources can be marginal due to several factors such as topographical and climatic conditions, structure of ownership and distance to markets and these factors might result in variable costs.

So far aquaculture strategy in different countries is considered, it is seen that several hassles exist for management and ownership. In some situations, fish cultivation can be operated jointly owned and run by farmers, with profits shared. Fish culture may also be alternative to agricultural use of the land and in some cases it is found that substantial amount of water is evenly distributed between fish farmers and agricultural farmers. Generally, the type of ownership in agriculture is more or less similar to that of aquaculture. Some business organizations are operated by small/marginal tenant farmers while others by big land-owners. It should be remembered that organized ownerships are considered as best for effective utilization of land and water for fish culture. At the same time well-organized small-scale enterprises are also effective form of management to fulfil socio-economic objectives of rural areas.

POLLUTION-RELATED PROBLEM AND RECIRCULATION

Any form of aquaculture can be a polluter and a purifier of any aquatic ecosystem. Intensive fish cultivation contributes to water pollution through massive input of feed, fertilizers, excretory products of fishes, and other pollution-related substances and consequently, depletes the water of its oxygen that makes the aquatic ecosystem unsuitable or less suitable for fish culture. Fish culture ecosystems are very susceptible to many forms of pollution that results in increase in the risk of failure in this industry. On the other hand, fish culture can be integrated with aquatic plants for extraction of toxicants from contaminated waters and utilization of aquatic resources for fish culture.

Prior to assessment of the profitability of various enterprises, social cost of pollution will have to be estimated as the production cost. Gradual increase in awareness of the social cost of pollution among most of the farmers has resulted in the propensity to treat water for better utilization in fish farms. Many polluting industries now discharge their waste products after clarification either through their own treatment devices or by paying special taxes.

Large-scale operations of sewage-fed ponds for raising fish have now been extensively followed in many Asian and European countries with great success. Very simple methods should be adopted in potential rural areas for recycling water that contain organic matter which could also be made available for use in fish production. Recycling of waste waters provides ample scope for the development of rural economy. Elaborate discussion on this aspect has been done in volume 1 (*see* Chapter 16).

INFASTRUCTURE

In addition to the rules and regulations adopted by the government that must be strictly followed for sustained yield, well-furnished structures which are needed for the operation of aquaculture industry, research, training, and extension services can greatly improve the activity of individual enterprises. In regions where small/marginal fish farmers are well-established or where farmers are organized through cooperative society, a number of advantages such as fish seed producing units, market facilities, technical and financial supports, and fish quarantine measures may be available to farmers. But where such facilities are not easily available to serve the purpose concerned or partly available, government and non-government organizations should come forward to make the rural economics strengthen.

EDUCATION, EXTENSION SERVICE AND RESEARCH

The advancement and support of research on territorial basis and education are most vital for encouraging higher growth and development of aquaculture in rural areas. Therefore, human resource is one of the most important assets in any industry. A large number of informations have shown that use of human resources is one of the factors which is highly effective in encouraging development.

Generally speaking, progressive fish farmers are able to utilize technical know-how and for this purpose adequate knowledge about different production, management, and marketing strategies is inevitable. Similar to agriculture, a positive correlation between education and introduction of new things in aquaculture exists. Education and extension service should be the connection between fish farmers and research units which is essential for introducing new technology, improved production methods, control of culture operations, and disease control.

For proper functioning of aquaculture industry in rural areas, research and training centres should be installed in the farm areas that helps maximize the best use of existing facilities and resources. To increase the awareness among farmers, several demonstration farms should be established along with the extension service and fish culturists. Research results obtained from laboratories and research institutes would be corroborated in these farms through experiments. Demonstration farms would undoubtedly provide economic and technical data essential to improve aquaculture farms in rural areas. Research programs for aquaculture development must embrace both socio- and techno-economic problems.

Socio-economic research has not been seriously considered partly because in several countries of the world, fisheries and aquaculture have recently been developed as an industry. Also, progressive shifting from subsistence economy to commercial production economy has originated socio-economic problems. Both technological and biological research activities are best adjusted for international cooperation since research findings and conclusion drawn are not greatly instigated by economic, social, and regional discriminations.

AVAILABILITY OF FINANCIAL RESOURCES

The source of finance can be to (1) cover depreciation, (2) retain net profits of operations, (3) increase loans and other liabilities and (4) reduce assets. The first two helps increase owner's capital and hence improves the situation of the availability of liquid assets. At the same time, the availability of liquid assets may be greatly improved by measures under both the increase in loans and reduce assets.

The working capital in an aquaculture farm should be used for various purposes such as stocking of (1) feed, fertilizers, crafts and gears, (2) broodstock, (3) fish in warehouses, and (4) claims on customers. It should be remembered that the size of the working capital depends on a number of factors such as sales (the requirement of working capital with increase along with the increase in sales), seasonal variations (variations in sales will involve increased requirement of working capital), profit per unit, capital turnover (increased capital turnover will lead to reduction in the requirement of working capital), and the relationship between fixed and current assets (farm mechanization may leads to a reduction in the requirement of working capital. On the other hand, the need for asset to finance fixed assets will increase).

THE COST OF CAPITAL

The main cost of capital is the interest charged on loans sanctioned for the purpose. The rates of interest are influenced by demand and supply of capital and in several countries, some rates are controlled by government organizations. The time of repayment of loans greatly influenced the availability of liquid assets.

CREDIT

In rural areas, finance is one of the most important problems which is often on the increase. Investment of credit schemes is necessary for farm construction, purchase of equipnents and other related items, major engineering works, and improvement of transport systems. The amount of credit to be provided for fish farming will depend a great deal on the social structure of the farmers' community.

Small and marginal fish farmers often face a great problem in obtaining loans from any financial institution due to lack of their security and consequently some arrangements have to be made to guarantee the loan for working capital and investment.

Productivity of small and marginal farmers may be increased if adequate financial assistance and an exhaustive package of informations are provided among farmers. Allocation of long-term loans to farmers through government institutions help control monopoly situations in the area concerned. Adoption of credit policy among farmers' community could definitely increase yield in their ponds and sell their produce to reap the benefits offered by the vast rural market.

4.5 Sale Price and Market

The sale prices of fish and fish products have a great influence on the economics of fish culture and it is highly dependent on the total cost of input for yield. Fish farmers may improve their profit by increasing the sale prices or by obtaining a greater share of the final sale prices. These are, however, not always a sensible and practical idea since there are potential contest among small farmers and they have limited possibilities of influencing the final market prices.

The prices of harvested fish vary seasonally due to fluctuations in demand and supply of fish. However, farmers can sell their products either directly to the consumer or through various distribution networks. In developed areas independent sale structures for fish and fishery products have been set up through which farmers sale their products in the market. In some cases farmers join the existing sale structures in aquaculture or in agriculture.

VERTICAL AND HORIZONTAL INTEGRATION

Different forms of vertical and horizontal integrations are used to obtain economics of scale in production and processing and to control market prices. Vertical integration, however, refers to an structural coordination of activities of two or more farms on different levels. Horizontal integration, on the other hand, connotes structural coordination of the activities of two or more farms on the same level. The main aim of integration is to maximize the profit from invested capital. However, integration may be commenced either by authorities or by farmers themselves.

Cooperative society is regarded as one of the most effective forms of vertical integration. From the production centres integration is initiated forward towards the market and backward towards the raw material for farm supplies. In a cooperative integration, farmers together act as a collective integrator and the principal target of which is to maximize the profit from fish farm.

By integrating fish culture into other development projects in rural sectors, both rural activities and fish culture are benefitted. The advantages of an aquaculture industry are (1) co-ordination of successive production processes, (2) reduction in risk and stabilization of income since opportunities are extended to commodity markets and the margin of production goes to the farmer, (3) supply of new materials, (4) scope for altering production methods and control over the quality of raw materials, (5) chances of securing a market for products, (6) chances to provide economics of scale, and (7) potentiality for obtaining capital.

VALUE-SUSTAINED PROGRAMS

The increase in fish prices is directly proportional to the increase in the supply of fish in the market as production in small and marginal farms will become profitable. This increase in yield potential may be jeopardized by a corresponding reduction in the quantity produced by those farmers which get favorable conditions for yield. Hence with higher fish values these farmers receive the same profit with less yield. To obtain the same profit by farmers with variable ecosystems, adequate measures must be taken to influence profitability in the farmers.

The most important measures considered to influence production and profit include (1) regulation of production volume at each farm, (2) financial support in the form of subsidies and grants, (3) Purchase-loan storage programs to assist prices, (4) objectives that aim at influencing the real cost of production or influence on the demand for aquacultural products. All these measures have, however, various effects on prices and quantities of yield, but it is important to note that it does not contribute to reduction in the real cost of production. If the cost of production significantly reduced, the real income from fish sale will be substantially increased.

4.6 Factors Influencing Real Cost in Production and Marketing

PRODUCTIVITY AND EMPLOYMENT

The productivity of fish culture ecosystem is definitely significant to the individual farmer and to the national economy. In the case of individual farmer, productivity provides the farmer with a standard of living, and for the nation, productivity is the basis for providing the consumer food products at judicious prices. The government and the individual farmer can influence the productivity of the farm, but how this should be harmonized with the fact that in many developing countries, inexpert labor is the main production factor which is more abundant in rural sectors and at the same time, expert personnel, management protocols, and capital are not easily available.

SIZE OF FARMERS AND YIELD

In fish culture ecosystems the category of farmer might influence the cost of production. Thus the total cost of production drastically reduced with increase in size due to higher productivity while interest and depreciation tend to increase with the size of the farmer.

Labor-intensive versus capital-intensive fish culture is of prime importance so far the rural development is taken into consideration since labor and capital may substitute each other. Therefore, if a given quantity of fish is produced in a few large group of farmers, a limited number of farmers will get employment opportunity than would be the case if the same quantity of fish is produced by many small group of farmers.

In rural areas where labor is plenty and relatively cheap, enterprises become labor-intensive. On the other hand, investment in labor-saving equipments may improve the productivity of the farm. In cases, however, where production techniques are very complex, generally encourage the large-sized enterprises. In fish culture, the complexity of yield may be specified with regard to the (1) type of species cultured, (2) steps in production and marketing, (3) technique of production and level of technology, and (4) degree of interaction with other types of production.

ACTIVITY OF YIELD

In fish culture ecosystems, a number of factors are used to estimate the activity of fish yield. The following factors may be considered for the purpose: (1) production of fish in kilogram/year/cubic metre of water, (2) production in kilogram/year, (3) production in kilogram/farmer/year, (4) production in kilogram/feed conversion ratio, and (5) investment/kilogram of fish/year.

Generally the intensity of fish production depends upon the following norms: (1) fish rearing in nurseries for stocking in lakes, ponds, rivers, and reservoirs, (2) the capture and confinement of the young, (3) production of larvae from eggs and their culture in enclosures with adequate management until they attain for the table, (4) management of broodstock, (5) rearing of young fish in ponds or other enclosures until they attain to marketable size. In general, however, these forms of fish culture involve supplementation of natural stocks, improvement of commercial harvesting of both natural and culture stocks, feeding of fish, pond fertilization to encourage the production of natural food and monitoring of water and soil quality variables.

In contrast to the yield of small-sized fish, the production of fingerlings is more expensive, but in general, the percentage recaptured increases with the size of the fingerlings stocked in ponds. Thus the intensity of the cost of harvesting is reduced to obtain a given catch. The time of stocking until harvesting influences the mean weight per fish and the number of fish recaptured. If the duration of time from the stocking of fish to the time of recapture, the average weight per fish will be increased with the concomittent decrease in the number of fish captured.

RISK / HAZARDS

The aquaculture is a high-risk activity and is greater than in any other form of animal husbandry. This is due to the fact that fishes are cultured in aquatic environments which are very much susceptible to various diseases, parasites, and pollutants. At the same time, aquatic environment is also difficult to regulate by farmers. As in the case of enterprises with ambiguously, profits must be lower than in ventures where the hazard is lower, since redemption of the hazard has to be provided as a part of the profits. In aquaculture industries where the hazards are very common occurrence, low or moderate profit margins are subject to declension.

The price risks are delineated by price oscillations and must be kept in mind when commercial aquaculture strategy is contemplated. Price oscillations in different seasons of the year may be greatly reduced by increasing the (1) price and future supply of fish and (2) quantity of fish stored and reducing the prices in future.

Since high stocking density in ponds/tanks/reservoirs represents prices equal to the total sale per year, it is necessary for small-scale enterprises to protect their stocks against damages. Insurance, to a greater or less degree, may be effective, but insurance of fish stocks in aquaculture is costly since frequency of losses is very high and the evaluation of risk involved in this sector is very arduous. In many countries, aquaculture industry has no magnitude to provide facilities for insurance cover is a serious hindrance to the development of aquaculture.

4.7 Conclusion

Implementation of various projects through authorized agencies have been successful in raising fish such as carps, catfishes, tilapias, snakeheads, and prawns that have created a positive impact on the economics of fish farming in rural areas. These fishes are cultured through adoption of state-of-the-art technology for maintaining production – consumption – marketing – export chain. This requires considerable skill, planning, and financial assistance which will definitely improve the socio-economic conditions of fish farmers. Although different projects have been successful in producing and marketing fish in large farms, they have provided very little benefit to small-scale producers. The receipients of technology are circumscribed by parameters such as lack of quality fish seed, shortage of investment capital, inadequate extension network, poaching, deliberate poisoning, and existence of strong chain of middlemen. These problems have plagued the strategy of fish culture. Adoption of well unit strategy in fish culture is, therefore, regarded as the nucleus of the growth of rural economics.

References

George, K. M. 1984. Rural Development Programmes — Its Strengths and Weaknesses. Indian Journal of Aquaculture Economics, July - September, 1984.

Jolly, C. M. and H. Clonts. 1993. Economics of Aquaculture. Haworth Press Inc., New York.

Ling, W. W. 1972. Aquaculture in Southeast Asia — A Historical Review. University of Washington Press, Seattle.

Monga, G. S. 2002. Environment and Development. Deep and Deep Publications, New Delhi, India.

Phillips, M. J., M. C. M. Beveridge and J. A. Steward. 1986. The environmental impact of cage culture on Scottish freshwaters. In: J.E. Slobe and L. G. De. Effects of land use on freshwaters, agriculture, forestry, mineral exploitation, and urbanization. Ellis Horwood, Chichester, 568 PP. 504-508.

Pillay, T. V. R. 1973. The role of aquaculture in fishery development and mangenent. J Fish Res. Board Canada., 30 (2) : 2202-2217.

Pillay, ,T. V. R. 1985. Recent trends in aquaculture development in status and prospects on aquaculture world-wide. pp. 61-64. Aquanor 85 Trondheim.

Questions

1. What are the objectives that are necessary to establish a profit-related fish culture industry in rural areas?

2. Discuss the role of fish in rural economy.

3. Discuss different criteria essential to analyze the economics of fish culture.

4. What are the features of integrated rural development? Aquaculture is the nucleus of the socio-economic structure in rural community. Discuss.

5. What are the features that influence the cost of fish production and marketing?

6. What are the standards that are required for the intensity of fish yield? Discuss.

5

Role of Environmental Factors in Relation to Fish Culture

Fish interact ceaselessly on with their ecosystem in two ways. *Firstly*, they have certain environmental factors such as temperature, pH, light, salinity, oxygen, and some other factors that follow, which differ depending on fish growth. And *secondly*, fish have an impact on their surroundings and the biota is directly related to the excretory products, wastes, diseases, and parasites. The indirect effects on farmed fish include the infrastructure requirements of culture sites. These interactions may be linked, for example, when the ecosystem is not able to buffer culture-induced changes in the sediment or water quality which are then dangerous to fish themselves.

The geographical factors that stimulate fish culture are of critical importance and are discussed elsewhere in Volume 1. The first section of the chapter therefore summarizes the general aspects of the geographical factors influencing the farmed fish. The second and third sections will cover environmental impacts on fish culture with respect to sources and effects of waste products from fish culture, chemical effects, impacts on wild fish and other animals and impact limitation strategies.

5.1 Geographical Factors Influencing Fish Culture

The distribution of fish and farm activities are everywhere and they are greatly influenced by environmental factors. In some environments, fish farming is favoured by climate, soil, and water so that very little effort is needed to raise fish. In others, farmers are at the mercy of nature; and great skill is required to alter the environment to obtain minimal level of existence. The followings are some of the major factors that influence farmers and their activity.

CLIMATE

Climate is the general weather conditions prevailing in an area over a long period. As a consequence, climatic factors exert the greatest control over different types of fish culture. Among different climatic factors, temperature is very important. It is essentially a question of climate when trout are not widely cultured in the tropics and the scale carp flourish across the world. Although different technological developments have been made in fisheries science, farmer can do little to control climate.

TOPOGRAPHY

The term topography refers to the arrangement of the natural and physical features of an area. However, the most intensively cultivated areas of the world are the plains. Plains have the highest activities in fish culture than the high altitude regions where fish culture is very laborious and cost–effective. In some Asian countries, carps are grown in rice fields, ponds lakes, and tanks. The hills are terraced to create ponds for fish cultivation.

Aquaculture have devised some new strains of fish that thrive well at high altitudes but efforts to overcome topographical restrictions have so far affected only in some areas. The annual increase in the world's fish production comes mainly from greater intensification of farming on existing water areas such as use of quality fish seed, use of feed and fertilizers/manures, and types of farming which improve yields per hectare. Expansion of water areas for culture is possible only in regions where potential water areas have not been fully utilized.

SOIL

Soil is the upper layer of earth and a black or dark, brown material typically consisting of organic remains, clay, and rock particles. The soil, which is composed of a variety of mineral and organic substances form the physical environment of aquatic ecosystems and is fundamental to fish culture. As soils are so varied in their physical and chemical compositions, being so closely related to their climatic environment that their suitability for the cultivation of different species of fish varies greatly. The soil which is vital for fish survival and growth, differ so much that unless the farmer has a sound knowledge on soil properties, he is not likely to gain a lot from his ponds. In fish culture ecosystems, the composition and structure are very important which determine the overall productivity.

1. Soil Composition: Weathering processes carried out by rain, sun, wind or ice/frost break down rocks to form soil. Since rocks are so varied in their chemical composition, including such materials as calcium, potassium, phosphorus, aluminium, silica, and soda, the soils which result from the weathering of these rocks will also differ greatly. For example, the weathering of rock rich in calcium and iron usually produces a soil with little lime and iron oxides respectively.

Apart from their mineral constituents, soils also contain organic matter derived from the decomposition of animals and plants. This is termed as *humus* (see Chapter 1 of Volume 1). The fertility of pond soil is often determined by the amount of humus present, since humus greatly improves soil structure, ecosystem productivity, and fish production potential.

Fishes require varying amounts of different mineral nutrients in the soil and these nutrients are extracted by fish biomass via plankton and other fish food organisms. However, it is necessary to replace worn-out soil nutrients from time to time if a high level of pond productivity is to be maintained. Fertilization, manuring, and other management strategies are some of the most important methods used by farmers to keep their ponds in better condition.

2. Soil structure: The physical texture of the soil, depending on the size of the soil particles, may be coarse, medium, or fine. All soils possess some particles of all these groups but the proportions vary greatly.

Soil structure is also affected by the presence in the soil of various materials which bind the individual soil particles together. On the other hand, the work of micro-organisms, soil bacteria, and worms are some of the most common and important factors that affect the soil structure.

BIOTIC FACTORS

Fish culture in farmed ponds is hampered by several biotic factors such as parasites, weed fishes, aquatic plants, aquatic insects, and other animals. They compete with the farmed fish for nutrients or destroy fish stocks either partially or completely before it can be harvested. Farmed fish around the world have from time to time suffered from these biotic factors and the magnitude of destruction of fish stocks depends on the degree of their activity.

Preventive measures against fish parasites, pathogens (bacteria and fungi), aquatic weeds, and insects are many and varied and has been described elsewhere in this book and in Volume 1. Farmers should, however, refrain from using stocks that are suffering from diseases or ill-health. Regular management practices have the advantages of supplying more robust fish throughout the culture period.

SOCIAL FACTORS

Social factors affect farming systems in a number of ways. In the first place, the type of farming practised affects the type of fish which can be produced and the production which can be obtained. Intensive farming, for example, gives far greater yields than extensive ones. The type of farming which is practised depends on the culture adopted by farmers and to some extent on the physical and topographical features of the area in which farmers live.

The ownership and inheritance of land/water is one of the most important social factors that affect aquaculture. In many parts of Europe and in much of Asia, the water area of a parent is divided between his/ her children. This leads to the breaking up of already small farms into smaller units which are often uneconomic to farm. Elsewhere, public or cooperative ownership of land/water may affect the culture system of the aquaculture methods employed.

Finally, social influence has profound effects on fish farming systems which are reared in sewage-fed fish ponds. The belief held by many Western countries that fish produced from sewage-fed fish ponds are unclean, contaminated with pesticides and heavy metals, definitely limits the culture of fish in many parts of Europe and Africa. On the other hand, such type of culture in some Asian countries, particularly in India, limits the full exploration of the sewage-fed fish culture.

ECONOMIC FACTORS

In addition to the factors mentioned above, the farmer has constantly to take into consideration many economic factors. Small or marginal farmers cultivate fish in their own/ leased ponds and is either consumed within the farm or the village boundaries may not be so seriously affectd in case of a trade recession. However, the farmer's income is so small that in hard times he has very little to fall back on natural hazards such as droughts, floods and diseases that ruin his harvests pose an even greater threat to him than market fluctuations. Three kinds of economic controls are, however, operative in all farming practices across the world, except in state-owned farms where there is no individual ownership and economic problems have to be dealth with by government agencies.

1. Operative Costs: The cost of running a farm can be substantial and is highest in areas where the greatest mechanization is required or where land values are high. In order to take a turn the farm for the better, improvement must be covered with profits derived from the harvest of crops. Where profits are small, farmers have great difficulty or are even unable to improve their methods. The operation costs are of several kinds. *Firstly,* he must purchase or rent his land. Even when he inherits his land (or add resource) he has to meet maintenance costs to keep his land and farm building in good condition. *Secondly,* there are recurring expenses such as cost of fish, fertilizers, manures, and feed. Modern farming also requires great quantities of freshwater for fish culture, light, and good quality of soil. *Thirdly,* there are occasional expenses such as the purchase of equipments for water and soil testing and other farm machinery. In many cases, farm improvements are often expensive. In the event of crop failures or when there is need to expand farm area, loans from banks or co-operatives have to be secured and repayment of loans with interest unless interests are exempted. Sometimes mortgages may also be necessary to overcome bad times or raise money for other purposes.

In addition to this, many governments impose taxes on farm products and in most cases the state levies special duties on products and export commodities to make up for budget deficits or to finance national development projects. This is particularly true for shellfish where there is possibility of a large export market.

2. Marketing Expenses : When fishes are harvested and ready for market, other expenses are involved. Freight charges by road or rail are dependent upon the type of commodity and the distance from the market. For fishes destined for the export market, the farmer has to meet wholesalers' commissions, insurance costs (if any) and government tariffs or excises. Unsaleable products require storage facilities, since fishes are perishable products, greater care has to be exercised and very speedy transport is necessary in the disposal of products. Conveyance by refrigerated vehicles is expensive and in many cases, the rate of spoiled fish tends to be high due to transport over long distances and rough handling. In the event of any trade recession, perishable products are the first to be hit. However, marketing costs are one of the main factors which determine whether it is economic to cultivate fish in certain areas. If markets are inaccessible or distant, it may not be possible to cultivate certain fishes even though geographical factors are ideal for them.

Price fluctuation in farm products can be very violent, being virtually controlled by the needs of developed countries. When demand is great, farmers are often in capable of bringing into the market more than their usual output. But when demand shrinks, over-production will cause a glut and fish will face low prices. Farmers will have great difficulty in disposing to their products. Farmers are then not only at the mercy of climate and other natural hazards but also the very unstable market conditions of the world.

3. Government Policies: In addition to the costs of running the farm and marketing his produce, the farmer must also consider factors of government policy. Some countries may encourage the productivity and efficiency of farming by a system of subsidies or guaranteed prices to farmers. In most countries farm production is so efficient that there is a constant problem of over-production of fish. As a result, the government has introduced a policy by which payments are made to farmers who do not use part of their land.

5.2 Environmental Demands of Cultured Fish

Cultured fish have a range of environmental requirements which alter depending on their life stages. For proper survival, growth, and reproduction, these requirements must be gratified. When wild fish are confronted with an environmental challenge, such as high temperature and low oxygen, they escape to more elemental conditions. This discretion is at best only available to fish in culture. In general, fish should be cultivated within their range of environmental tolerance, since they have no means of physical escape, although they may change their behavior to curtail the effects of the environmental stressor or to make some physiological adaptations.

WATER QUALITY*

1. Oxygen : The growth of farmed fish is affected by the feeding rate, feed quality, and environmental conditions. The metabolic rate of fish is increased when they are fed regularly, but the food consumption is suppressed when oxygen is in short supply. Generally, oxygen concentration (expressed as mg/l or ppm) and the degree of oxygen saturation (expressed as percentage) are considered in aquafarms, but when oxygen requirements are considered as the ability of fish to extract oxygen from water, it depends on the relative saturations of the water medium and their blood. Since the solubility of oxygen in water is a function of both salinity and temperature, concentration statistics in isolation may be misguiding.

Fish response to hypoxia is of great importance to farmers as the high stocking density which is typical of intensive farming systems can lead to hypoxia. Though mortality of fish species vary widely in their requirements for oxygen, but few species are capable of rapid growth at oxygen level much below 50 per cent saturation, with carps requiring considerably more. Absolute lethal limits are of academic interest as culture operations must operate routinely at levels far above these limits. Drastic hypoxia can occur in fouled waters when, for example, the oxygen concentration in a cage of catfish cultured in fouled

---

* For further study, see Jensen et al (1993), Konstantinov et al (1988).

environments was reduced to 20 per cent saturation with no mortality. Many cage farmers in areas are prone to low current continuously or routinely monitor dissolved oxygen levels so that feeding can be discontinued during hypoxic conditions.

In ponds where phytoplankton blooms are abundant, the water becomes supersaturated with oxygen as a result of photosynthesis in day time and this poses no problem for fish. On the other hand, the respiration of algal mass at night can lead to hypoxia. Hypoxic conditions are, therefore, most likely to occur at dawn and may be exacerbated if fish are fed before it is fully lighted.

The supply of oxygen to intensive farms and to cages is mostly by aerator and by water currents respectively, including those caused by the interaction of currents and fish themselves.

Respiration of fish in cages will cause hypoxia if the rate of supply of oxygen is less than the rate of uptake. In addition, polluted sediments cause an oxygen drain on the water column so that the maximum oxygen level within the water column can occur at mid-depth, i.e. oxygen is consumed at the top of the water column by the fish and at the bottom by benthic metabolism.

2. Temperature : The exact temperature for growth of fish is extremely variable and temperature range of different species of fish varies between 12.0°C (rainbow trout) and 38.5°C (common carp). Most of the species exhibit a rapid increase in growth as temperature increases, passing an optimum peak before falling rapidly as temperatures approach the upper lethal limit. At extreme temperatures, fish mortalities occur for a variety of physiological reasons and at low temperatures due to physical problems such as the growth of ice crystals in tissues.

Due to the pivotal nature of the interactions between temperature, feeding rate, and growth, studies have been carried out on these relationships to optimize growth and food utilization. A variety of cultured fish and fish with potential for culture have been examined at a variety of life stages. Table 5.1 represents the temperature requirements of some commercially-important species of farmed fish.

The concept of a single optimum temperature for fish growth has been severely criticized by many authorities. It seems plausible that both stenothermal and eurythermal fish grow at faster rate when the temperature oscillates within physiological limits of the fish, the amplitude of the oscillation and the optimum rate of temperature change being species-specific and related to the ecological adaptation of that species. The optimal magnitude and rate of the change are, however, lower for stenothermal than eurythermal fish.

3. Salinity : The salinity tolerance of fish is of special interest to the aquaculture industry. There is a general desire to cultivate freshwater fish in sea or brackish waters due to the unavailability of suitable freshwater resourses in many areas. In contrast, it is desirable to reduce the requirements for pumped sea water to hatcheries by exploring the potential for culturing early life stages of marine species in less saline water.

The tolerance of salinity is related to temperature. In addition, the response of fish to other environmental factors may be affected by salinity. Euryhaline fish may have an optimum salinity tolerance where survival and growth are optimized and energy expended on osmoregulation is reduced.

The osmoregulatory abilities of fish (particularly salmonoids) are well documented. Pre-smolting salmon can be grown well in fresh and brackish waters and post-smolting thrive in full-salinity sea water. The salinity tolerance of juvenile salmon is related to body size, with larger fish being more tolerant.

4. Carbon dioxide: Carbon dioxide is often used in high concentration as an anaesthetic but it may cause physiological perturbation and reduction in the growth of juvenile white sturgeon *Ascipenser transmontanus* at much lower levels. The levels of carbon dioxide in water may increase through an excess of organic degradation relative to uptake by photosynthesis or as a result of fish respiration. The capacity of fish to amend for the fall in blood pH due to elevated levels of carbon dioxide by the uptake of bicarbonate is species-specific but the process is more rapid in sea water fish. This is probably related to the greater availability of bicarbonate in marine ecosystems. In intensive fish cultivation where levels of carbon dioxide are elevated due to respiration, there is also likely to be a reduction in the level of dissolved oxygen and the two effects interact. Though fish can compensate for the chronic elevated levels of carbon dioxide in water, it leads to poor growth of farmed fish.

Carbon dioxide drastically affects pH by the generation of protons during its hydration and hence has the ability to influence the toxicity of the ammonia-ammonium ion couple:

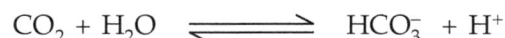

$$CO_2 + H_2O \rightleftharpoons HCO_3^- + H^+$$

5. Ammonia: It is a degradation product of protein metabolism and is excreted through gills. It is highly toxic to fish, particularly in its unionized form, and the equilibrium between ammonium ion and ammonia is pH-dependent.

$$NH_3 + H_2O \rightleftharpoons NH_4^+ + OH^-$$

The effects of ammonia toxicity have been noted in many literature for many farmed fish including carps, salmon, trout, channel catfish, air-breathing fish, and tilapias.

In fish culture ecosystems where re-circulatory systems are adopted, ammonia is removed each time, the water is recycled and this is accomplished by biological filtration process (for further detail, see Chapter 13). Though there is no report that ammonia levels are even likely to reach levels dangerous to fish in cage culture, severe problems have occurred in ponds. It is interesting to note that an effort to reduce ammonia in channel catfish ponds by cessation of feeding for 10 days led to little effect on total ammonia but an increase in the concentration of toxic un-ionized form due to increased pH of water.

6. Hydrogen Sulfide : Waste products developed in ponds can cause organic enrichment of the benthos with potentially detrimental consequences to fish health due to generation of hydrogen sulfide which is highly toxic to fish. Though significant correlation between fish growth and hydrogen sulfide concentration has been noted in many farm ponds, there are other factors such as poor husbandry which independently affect both benthic fauna and fish health.

7. Harmful Algal Blooms: When environmental conditions for fish growth are optimal, protistan algae can grow exponentially, leading to high local populations. This can occur in several species of dinoflagellates that contain poisons which are toxic to fish, depending on the species of algae and the class of compound they generate. The toxins are accumulated in fish flesh and when consumed by man cause paralytic fish poisoning. Many of these compounds are neurotoxic. Saxitoxin accumulated in shellfish, and when consumed, causes numbness of the mouth, lips, and face which reverses after a few hours.

Toxins like ciguatoxin from *Gambierdiscus toxicus* accumulates in fish flesh of grouper[1] and snapper[2] and when consumed definitely cause gastric problems, respiratory failure, and damage to central nervous system. This aspect has been death with in Chapter 9.

Some species of diatoms can produce domoic acid, which can accumulate in muscles and when consumed by humans, causes amnesic shellfish poisoning, a short-term loss of memory which can cause death.

Though optimum levels of algae are inevitable to perform several important functions (such as primary production, and symbiosis), the presence of algae in high populations, their dissolved organic matter and the presence of bacteria in high numbers that can supported by large amounts of organic matter lead to a condition known as *eutrophication*. Water is in a nutrient-rich state, biomass is high and oxygen demand exceeds than supply. Anoxic conditions rapidly develop, leading to the death of aerobic organisms.

Decomposition of dead material by bacteria leads to further demands for oxygen, and the entire ecosystem becomes anaerobic allowing for the growth of anaerobic bacteria and the generation of methane, hydrogen sulfide, and many other products of anaerobic metabolism. Eutrophication is commonly seen around untreated sewage outfalls and dairy farm waste flow-off, and freshwater streams are polluted by nutrient-rich agricultural field washes.

5.3 Environmental Impacts on Fish Culture

Recent increases in the production of farmed fish, driven principally by increasing demand as a result of general decline in many capture fishery resources, it is becoming apparent that fish culture as is currently practised will consequentially become limited by its environment. The regional scale of carrying capacity for a particular culture technique has not been fully explained.

The existing literature on environmental impacts is dependent on fish species in culture, culture methods and feed type and on the nature of the environment in terms of chemistry

1.   A large heavy-bodied fish of the sea bass family, found in warm seas (with many species).
2.   A marine fish for snapping its toothed jaws (with many species).

and biology; that the main impacts are on the benthos, usually within a localized region around the farm; and that food additives or chemical treatments may have a local impact, they were unlikely to pose a threat to the regional environment at current levels of use. Though this last view is widely considered, the question arises regarding the ultimate carrying capacity of aquatic ecosystems for fish before systematic effects have become apparent to the detriment of fish culture operations.

SOURCES OF WASTE PRODUCTS*

1. Carbon: The total amount of carbon present in fish feed and faeces has a significant impact on fish culture strategy. In intensive fish culture systems where fish are fed from external sources, lead to wastage of uneaten food. In open fish culture, fish faeces are lost to the ecosystem. Accumulation of uneaten food has a greater capacity than faecal material to impact on the environment both in terms of energy content and degradation rate. The loss of food might results in declining profits which compell farmers to use more efficient feeding strategies and also by the greater palatability of modern diets.

Intensive fish culture in developed nations has come to rely on the use of pelletised fish meal from high quality and hence, highly digestible sources. This has lead to lower feed conversion ratios and less waste. Intensive culture in other regions also relies on waste or trash fish of variable quality. Such diets tend to cause greater benthic impacts due to lower digestibility and due to wastage during the feeding activity itself.

2. Nitrogen : Waste materials from fish farms contain large amount of nitrogen, phosphorus, and silicon. However, between 65 and 80 per cent of the nitrogen added to different culture systems is lost to the environment, of which the majority (50-55 per cent of total nitrogen) is lost in dissolved form either directly from the fish or by benthic flux from solid waste. Modern fish diets contain more lipid and less protein. This results in a reduction in feed conversion ratios which approach 1 : 1, although more efficient feeding methods also play a large part in this case. The net effect is a reduction in nitrogen released to the ecosystem. Fish utilize dietary protein efficiently but a significant proportion is metabolized, thereby releasing large amount of nitrogenous wastes mostly as ammonia but some as urea. Excretion of nitrogenous wastes from different species of farmed fish is more or less similar. Nitrogen loss from fish is, however, dependent on temperature and on dietary protein content.

3. Phosphorus: Phosphorus has been the subject of considerable interest particularly with regard to freshwater systems where it is the limiting nutrient for plant growth. In intensive cage culture of rainbow trout, phosphorus outputs have been found to decrease from 16.6 to 8.9 kg/tonne of fish harvested due to combination of a reduction of phosphorus in feed (15 mg/g to 10 mg/g for grower diets) and a decrease in food conversion ratio (1.5 : 1.0 to 1.25 : 1.0 over the same period).

---

* For further detail, see Munday et. al. (1992), Kelly et. al. (1994, 1997), Hennessy et. al. (1996), Massik and Costello (1995), Gavine et. al. (1995), Tacon et. al. (1995).

4. Silicon: The element silicon has a key role in diatom metabolism. Almost all of the silicon added in food to the trout farm has been found to be lost to the ecosystem, but this accounts for about 20 per cent of the total silicon budget. The total loss per tonne of fish produced was 2.5 kilogram. Fluxes from the sediment to the overlying water were around 2.5 times higher than from local control sediment.

5. Hydrogen Ion Concentration: The role of water pH for the production of robust fish from ponds has been over-emphasized by fish culturists as one of the most key factors. There is a general consensus that moderately alkaline waters are highly productive while slightly acidic waters are less productive. Highly alkaline or acidic waters are not at all desirable for fish culture, however.

6. Alkalinity: The acid combining capacity of pond water — so called *alkalinity* — is a measure of its carbonate and bicarbonate concentrations that dictate the prime importance in the ecosystem of farmed ponds. The alkalinity of pond water is derived largely from contact with the soil and rock substances. In general, however, calcareous waters with medium alkalinity (50-100 mg/l) are highly productive, while ponds with very low alkalinity (below 50 mg/l) are poorly productive. In cases where high alkalinity of water is accompanied with a high concentration of calcium, phosphates in water are bound to calcium in a form that are not easily available to phytoplankton thus making the pond unproductive. This defect should be rectified.

7. Semi-intensive Ponds: Semi-intensive pond polyculture relies on the controlled eutrophication of the ecosystem by use of fertilizers and manures which stimulate the growth of a complex food-web which, ideally, is fully exploited by a variety of fish species. In contrast to intensive culture, these systems have little potential for polluting the aquatic ecosystem over extensive areas when adequately managed as they are essentially closed. This is taken further when they are integrated with an agriculture system involving paddy and cattle/poultry production, thus reducing the cost of external fertilization.

EFFECTS ON SEDIMENTS

Sediments are diverse environments supporting a range of fauna and flora existing in a complex matrix whose defining parameters include particle size, carbon (food) availability and redox-potential. In contrast to sea water, however, the effects on sediments in freshwater have received much less attention possibly because there is generally more interest in the effects of phosphorus emissions to the water column. And because of low concentrations of sulfate in freshwaters make the evolution of toxic hydrogen sulfide from polluted sediments unimportant.

Most of the high-latitude oligotrophic lakes have a large capacity for buffering impacts of nutrients and carbon. Such lakes are home to fish which are sensitive to pollution and hence ecological impacts may be much greater than that of an enriched ecosystem. In contrast to lake, a productive/fertilized pond sediment is generally not anoxic and hence benthic organisms are abundant.

The effects on freshwater lake sediment fauna are expected to be similar to those found in marine systems, i.e. a reduction in diversity and an increase in opportunistic animals. The oxygen demand of sediment plays a key role in biogeochemical sequences. This is important with respect to phosphorus under anoxic conditions.

In tropical ecosystems, benthic processes are accelerated by temperature. In several small lakes of Kariba, Zimbabwe, accumulation of organic material was found to be low inspite of a high sedimentation rate, possibly indicating that tropical systems may have a larger capacity to process wastes from fish culture than those in temperate areas.

IMPACT OF CHEMICALS AND OTHER AGENTS ON FISH CULTURE

A wide variety of chemicals are used in fish culture, many of which find their way into the ecosystem. In addition, some chemicals such as liming substances, fertilizers, zeolite, antibiotics, oxidizing agents, plant extracts, coagulants, osmoregulators, algicides, herbicides, pesticides, water quality enhencers, immunostimulants, and probiotics are known to be beneficial to the farmed fish. Of course, at recommended rates and the residues present in the flesh of harvested fish must confirm to minimum standards, which is highly variable regionally.

Aqua farmers throughout the world use a wide variety of chemicals and other agents as noted above to improve water and soil qualities, fish growth and control of fish diseases. Some chemicals such as inorganic fertilizers, liming materials, and substances containing trace elements are known to be beneficial to fish farms since these chemicals do not have any food safety concerns and do not cause ecosystem problems. On the contrary, use of human wastes and some organic manures or the contamination of aquatic ecosystems with agricultural or industrial pollutants definitely result in product contamination and hence food safety concerns.

SOIL QUALITY

As discussed in volume 1 (Chapter 1), different kinds of soils have several important roles in the fertility of any fish culture ecosystem. A fertile soil not only provides shelter and food to the cultured fish but also supplies essential nutrients to the overlying water. Soil also helps mineralize organic deposits in the bottom soil that is stored in the soil and subsequently released when necessary.

In contrast to land soils, the under-water soils differ in certain respects. *Firstly*, pond soils are formed as a mixture of different profiles that remain with permanently water-logged, having no gas phase. And *secondly*, large aquatic ecosystems receive dissolved nutrients in appreciable amounts and sedimentary particles drained by rain water. Nutrient and various suspended particles undergo sedimentation in changed physico-chemical conditions. The basic soil texture on which a farm is constructed exerts an influence on the productivity only for a few years. Older ponds, however, drastically modify their properties to a greater or less extent due to production and subsequent decomposition of plant and animal products.

Fish culture ecosystems which are not influenced by external factors, the physico-chemical properties of water are a reflection of the bottom mud. In general, production of a variety of nutrients is the result of various complex physical and/or chemical reactions in the pond soil. Hence a direct correlation between soil-water nutrients is not always observed. Different forms of nutrient carriers are generated from organic matter and mineral constituents of clay fraction of the soil through chemical and biochemical processes.

The pond soil has two general layers : (1) the upper loose and aerated layer that consists of colloidal organic matter and (2) the lower non-aerated and compact layer containing mineral layers in different compositions. The proportion of these layers play a crucial role in determining the physical, chemical, and biological processes in any fish culture ecosystem for sustained production. Some physio-chemical factors that follow are involved in fish production strategies.

1. Soil pH: Similar to pH of water, soil pH is also equally important and is dependent on several factors. In mudy water where there is deficient in oxygen, decomposition rate of organic matter gradually reduced followed by reduction in the amount of decomposition products or in some situations, substantial amounts are oxidized into fatty acids that make the soil more acidic. Unless the soil is buffered, the pond productivity significantly reduced. Soil pH greatly influences the transformation of soluble phosphates to unavailable forms and also control the adsorption as well as release of nutrients at soil-water interface.

2. Soil Nitrogen and Phosphorus: These two elements are very important for ecosystem productivity. While phosphorus deficiency is very common in pond soils and placing a constraint on its availability, the primary reliance has to be on the congenital phosphorus rank of the soil. The situation for soil nitrogen is entirely different since the supply of nitrogen comes from natural resources which is almost limitless. However, atmospheric nitrogen is fixed to the pond mud by four different processes such as Azotobacterial fixation, blue-green algal fixation, photo-chemical fixation, and atmospheric electric discharge. Different forms of nitrogen (such as ammonia, nitrate, and nitrite) exhibit a well-known cycle, termed as *nitrogen cycle* (for further detail, see Volume 1, Chapter 17) mediated by bacterial activity. When nitrogen fertilizers are added to this cycle, the concentration of different forms of nitrogen significantly altered resulting in the substantial loss of added nitrogen. Hence, efficient use of nitrogen carriers should be kept in mind. In general, soils with near neutral reaction, the loss of ammoniacal nitrogen is comparatively less than in highly alkaline pond soils where the loss is maximum.

Previous discussions (see Chapter 18 of Volume 1) remind us that phosphate ions (inorganically-bound phosphorus) in soil form insoluble compounds with iron and aluminium under acidic and with calcium under alkaline conditions rendering it unavailable to water. A substantial amount of phosphate ions are also adsorbed in colloidal complexes. Thus, different forms of phosphorus and conditions controlling their release to water are important. Inorganic phosphorus in soil is bound as insoluble calcium phosphate and is adsorbed on colloids. Both these forms are rendered soluble under acidic conditions. Moreover, under acidic conditions, organically-bound phosphorus is mineralized through bacterial decomposition to soluble inorganic phosphate.

Physico-chemical features of pond soils greatly influence the rate of disappearance of added phosphorus from water and the nature as well as magnitude of its transformation. The level of phosphorus in water declines rapidly in acidic, neutral, and alkaline soils.

The ratio of available nitrogen (AN) to available phosphorus. (AP) plays a key role that determine the soil fertility. In slightly acidic and low nutrient soils that have a AN : AP ratio below unity, soil fertility can be enhanced by artificially increasing the available soil phosphorus and the effect is maximum when the ratio is near unity.

3. Potassium: Similar to nitrogen and phosphorus, potassium is also considered as a major nutrient but its importance is shaded by its easy availability in pond soil and water. However, this element does not form any insoluble salts in ponds and its movement from soil to water is favored by hydration of soil colloids and consequently, the pond water is infrequently encountered with the shortage of potassium.

4. Calcium: Calcium helps increase the availability of other ions and in general, ameliorates the chemical condition of an ecosystem. In pond soils, it is present as calcium carbonate. The amount of exchangeable phosphate in pond mud is inversely correlated to the organic matter ratio, so that in organic soil containing calcium the soluble phosphorus remains adsorbed as an exchangeable form and where sediments are low in organic matter, phosphorus becomes fixed in the soluble precipitate.

5. Organic Compounds: Soil organic compounds are varied and complex and the bacterial activity of carbon-nitrogen ratio in the compound. Nutrient dynamics as described by Rajagopalsa and Ramadhas (2002) depends principally on the carbon cycle. Therefore, the accumulation and decomposition of organic matters form an important strategy in pond productivity.

CHEMICAL FERTILIZERS

As noted in volume 1 (chapter 2), a variety of nitrogen and phosphorus fertilizers are used in fish culture. And these should always be used at recommended rates to increase the productivity of pond ecosystems otherwise they have potential to increase the accummulation of toxic forms particularly ammonia and nitrite. Fertilizers that supply nutrients in quantities far in excess of those taken up by plankton and other fish food organisms can result in contamination of both pond water and soil as a whole. Nitrates and phosphates are the chemical most often involved. Hence use of fertilizers far in excess of phytoplankton uptake should be discouraged. Discussions regarding the impacts of chemical fertilizers on cultured ecosystems have been discussed at length elsewhere in volume 1 and in this volume.

ORGANIC MANURES

Most of the tropical and sub–tropical fish farmers use a variety of animal manures at much higher rates than chemical fertilizers since manures have much lower concentrations of nutrients. Bacterial decomposition of organic manures requires more oxygen and the amount of manure that can be added to a pond depends upon the biochemical oxygen demand of the manure. Addition of manure to a pond ecosystem not only increases the potential for low oxygen concentrations with consequent fish mortality but also contamination of water with heavy metals and therefore exist greater risk for food safety.

Whilst there always exist several detrimental impacts on fish such as disease spread, mortality, suppressed reproductive potential of fish in particular and on aquatic ecosystem in general, the use of organic manure technology (See panel 5.1) has been found to be highly encouraging that has helped fish farmers in improving their ponds within limited financial resources. This technology helps control the seepage of water from ponds where percolation of water through the pond mud is very high.

ROLE OF ORGANIC MATTER IN SOIL FOR FISH CULTURE

Portions of the nutrient elements consumed in animal feeds are found in the voided excrement. As a generalization, three-fourths of the nitrogen, four-fifths of the phosphorus, and nine-tenths of the potassium ingested in feeds are voided by the animals, and appear in the manures. For this reason, animal manures are valuable sources of both macro- and micronutrients. Organic manures have a relatively low nutrient content in comparison with commercial fertilizers, and a nutrient ratio that is considerably lower in phosphorus than in nitrogen and potassium. Modern commercial fertilizers generally carry 20-40 times the nutrient content of manure and have higher ratios of phosphorus to nitrogen and potassium than found in manures. However, since rates of manure application are commonly 10,000 kg/ha, the total nutrients supplied under most practical conditions are substantial and sometimes more than are necessary for the current year's fish crop. In India, a significant proportion of nutrients, especially nitrogen and phosphorus added to fish culture ecosystems come from farm manures.

The most practical generalization is the need to balance the nitrogen-phosphorus ratio by supplying phosphorus in addition to that contained in manure. This generalization has great practical significance in relation to fish culture ecosystems because the use of nitrogen and phosphorus fertilizers in most of the tropical countries commonly exceeds the amount of these nutrients removed by harvested fish.

SOIL TYPES IN RELATION TO FISH CULTURE

The primary factor limiting more effective utilization of soils for fish production is their type. It has been noted that laterite, red loam and red yellow soils were comparatively less productive (fish yield varies between 3,000 and 4,000 kg/ha/year) than grey, brown, alluvial, and medium black soils (fish yield varies between 4,000 and 7,350 kg/ha/year). The small reserves of essential elements in former types of soil can, to some extent, be increased if organic manures are added after fish harvest.

MICRONUTRIENTS

The amount of micronutrients in chemical fertilizers must be carefully controlled than the macronutrients. Difference between the toxicity and deficiency levels of a given micronutrient is extremely small. Hence, micronutrients should be added only when their need is certain and the amount required is known. For better condition of ponds, micronutrient fertilizers are used in many developed fish farms to supply specific nutrients other than nitrogen, phosphorus, and potassium.

OXIDIZING AGENTS

Several oxidizing agents (especially potassium permanganate) are used for controlling pathogenic micro-organisms, phytoplankton, and oxidizing bottom soils. These agents reduce the rate of oxygen consumption by chemical and biological processes. In water, potassium permanganate rapidly oxidizes labile organic matter and other reduced substances and is transformed to relatively non-toxic manganese dioxide which is precipitated. Since potassium permanganate is toxic to phytoplankton, it has the effect of reducing the generation of oxygen by photosynthesis. Consequently, precautionary measure should be taken to avoid this situation. Chlorine compounds such as chlorine gas, calcium hypochlorite,and sodium hypochlorite are also considered as powerful oxidizing agents and their concentrated forms are strongly irritant. Calcium hypochlorite is frequently used in ponds to oxidize organic matter and to reduce the biological oxygen demand.

Chlorine gas is hydrolyzed in water and forms hypochlorous acid (HOCl) and hydrochloric acid (HCl). This HCl is completely ionized and depending upon temperature and pH, HOCl ionizes to hydrogen ion (H^+) and hypochlorite (OCl).

$$HOCl = H^+ + OCl^-$$

In slightly alkaline waters (pH 7.4), 50 per cent ionization of HOCl occurs and in highly alkaline waters, there occur more OCl than HOCl. Hypochlorous acid and hypochlorite are termed as *free chlorine residuals.* The relative distribution of these two reactive species is important since disinfecting efficiency of HOCl is 40 to 80 times to that of OCl. Chlorine readily oxidizes Fe^{2+}, Mn^{2+}, and hydrogen sulfide and organic matter and reduced to non-toxic chlorine ion. Chlorine reacts with ammonia to form chloramines and addition of chlorine results in a direct increase in free chlorine residuals. The amount of chlorine that must be added to reach a desired level of free chlorine residuals is termed as *chlorine demand.*

Some organic substances of water and soil react with chlorine residuals to form polychlorinated biphenyls, dioxins, and trihalomethanes. These substances are suspected as carcinogens and they contaminate farmed fish. The first two compounds are, however, thermally stable that resist oxidation and hydrolysis and hence stay in ecosystems for prolonged period. The major threat associated with the presence of these chlorine residuals in finfish and shellfish is the increasing concern both in domestic and international markets.

Chlorine dioxide (ClO_2) and chloramine have advantages over chlorine gas and calcium hypochlorite for use in water containing organic matter and ammonia in substantial amounts. Their disinfecting power are higher than that of chlorine. Residuals of these compounds are degraded very rapidly than chlorine residuals. Since there is no evidence of accumulating these compounds in the tissues of aquatic organisms, there is no possibility of food hazards associated with them. Consequently, use of chlorine compounds in fish culture does not pose a significant food safety risk, though there is possibility of ecosystem contamination by reaction products of chlorine in effluents and/or surface flow-off.

Formaldehyde solution (or so called *formalin*) is a very common and widely used disinfectant because the compound destroys pathogenic fungus and microorganisms. Although it is applied to the total pond water, application is generally limited to small pools of muddy water (or puddles) after harvest of fish. It is also used as a disinfectant in hatcheries. Since the compound is easily degraded in the ecosystem, no food safety hazard has been noted to be associated with its use in fish culture.

Quaternary ammonium compounds (see volume 1, Chapter 10) are sometimes used in hatcheries and farms to disinfect larvae, fingerlings, tanks, and farm equipments. These compounds are frequently used in ponds at the rate of 400 kg/ha that brings about no dangerous effects than other oxidizing agents.

ANTIBIOTICS

Antibiotics are chemical compounds used to treat infections caused by bacteria, fungi, protozoans, and worms. The use of very selective antibiotics is now widely accepted as one of the key substances in the strategy for transporting of fish and fishery products. And extensive use of antibiotics in shrimp and fish culture has raised a number of issues involving product quality, human health, and the ecosystem in general. According to one estimate, antibiotic residues persist in the bottom mud and consequently may lead to the development of pathogenic micro-organisms which are resistant to antibiotics. Use of chloramphenicol, for example, has shown to cause increased bacterial resistance in shrimp hatcheries. Antibiotic residues get deposited in fish fillet that may definitely lead to rejection of products in export markets. Until recently, residue of the antibiotic oxytetracycline, for example, were detected in farm-reared shrimp from India and Thailand and caused rejection of shipment to Japan. However, this circumstance brings into sharp focus that availability of antibiotic residue is critical for exported fish product and in future, attempts will have to be made to their efficient use and monitor of antibiotic residues both in fish flesh and in ecosystems so as to ensure that the product is of superior quality and a large expansion in exports will be made possible if the product is zero defect. This calls for urgent and in-depth investigations of the trends in antibiotic use in fish culture so as to reflect useful insights for quality products.

PLANT EXTRACTS, COAGULANTS, AND PISCICIDES

Extracts of grapefruit seed, garlic, neem, and passion fruit are used in fish culture either directly or by mixing them with lime. These are also used in shrimp/fish feed.

Several coagulants such as gypsum, zeolite, ferric chloride, and aluminium sulfate are widely used in pond waters to flocculate and precipitate suspended clay particles that make the turbid water very clear. The coagulant gypsum (calcium sulfate), for example, is dissolved in water and as a result calcium and sulfate concentrations increased. These ions are extracted by aquatic life from the ecosystem. Aluminium and ferric ions derived from the use of aluminium sulfate and ferric chloride rapidly precipitated as aluminium and iron oxides. These two compounds and those that form ammonia upon reacting in the pond soil can increase soil acidity. Because of their high potential to create acidity these should be handled with utmost care to prevent skin irritation from contact. However, continuous and substantial use of acid-forming coagulants must be accompanied by the application of lime.

Zeolite is an alumino-silicate clay of high cation exchange capacity and is widely used in aquaculture ponds. Farmers use it with the aim of reducing ammonia concentration through ion exchange, fostering physical cover over sediments to prevent leaching of metabolites to overlying water, that helps remove suspended solids and consequently water color is greatly improved.

The most commonly used piscicides in fish culture include tea seed cake, derris root powder, mahua oil cake, and the like. Though these substances have toxic effects on fish as they contain saponins, on degradation in pond bottom, their nutrient contents which are essential to pond productivity are liberated. This undoubtedly represents a considerably better condition of fish culture ecosystems which is common in many farms where farming strategies are highly favorable to the environmental factors.

PESTICIDES, HERBICIDES, AND ALGICIDES

Whilst several kinds of herbicides and algicides are generally used in fish culture at recommended rates to reduce/control the abundance of nuisence aquatic plants including algal blooms, agricultural washes discharge substantial quantities of pesticides and their degraded products into the aquatic systems with adverse impacts on aquaculture industry should not be ignored. Although frequent mention has been made in Volume 1 and in this Volume on acute and long-term effects of agricultural chemicals on fish and aquatic ecosystems in relation to fish culture, attention must be focussed on management protocols and to the maintenance of stable ecosystem structure throughout the fish-growing season. The sentence "ecosystem impact on agricultural chemicals" is central to such a brief discussion that follows.

* When agricultural chemicals are carelessly used, they can result in adverse effects on aquatic life. Even when properly used, chemicals have a number of side and residual effects that cannot be avoided. Their persistence and ubiquitous nature, having a tendency of some compounds to concentrate in organisms via the food chain, may increase their toxicity to fish and man.

* The fate of pesticides in an aquatic environment is presented diagrammatically in Volume 1 (see Figure 12.20). Pesticide accumulation in the ecosystem compartments is

dependent on both chemical nature of the compounds and the manner in which they are applied which make them more effective. This leads inevitably to the persistence of potentially toxic chemicals in fish culture ecosystems and contamination of the ecosystem relates to their non-degradable nature.

* Precipitation and leaching of pesticides may find their way eventually into water bodies. The transfer and fate of pesticides in fish culture ecosystem are generally influenced by their concentration and degradation by biotic and abiotic factors. Indeed, microbes and sunlight are the two important forces that cause degradation of ecosystem chemicals. Chemicals that are used for myriad purposes are transferred to soil where their fate is largely determined not only by soil micro-flora and fauna but by the chemical environment of the soil as well. Very little is known about the interaction between insecticides and chemical fertilizers on the role microbial flora in the accumulation of residues of persistent chemicals in soil. Their interactions are, however, either antagonistic or synergistic or additive depending upon the type of chemicals deposited and their interactions among abiotic factors.

* The risk of ecosystem inherent to chemicals stems from differential vulnerability of different biotic communities to their toxic action and differential response of fish and their food species to their non-lethal but possible long-term effects. Moreover, the rapid development of resistance among vectors, three problems that are now being faced are disruption of natural controls, accumulation of more chemicals and their degraded products, and the pesticide trademill, bringing on as an inevitable consequence of the use of more and more chemicals and their applications, escalating the cost of production of food from aquatic ecosystems.

PROBIOTICS

Probiotic denotes a substance which stimulate the growth of microorganisms, especially beneficial ones, such as those of the intestinal flora. Conventionally, probiotics are applied in fish culture through feed. Until recently, however, they are also being widely used to pond water or sediment. On the basis of their mode of application, different kinds of probiotics are broadly grouped as feed probiotics and pond probiotics. The former type offers several advantages such as enhancement of fish growth, improved food conversion, prevention of intestinal disorder, avoidance of residual effects of antibiotics in the tissue, and pre-digestion of anti-nutritional factors. Fish health deteriorates rapidly along with the increase in environmental stress such as ammonia, nitrate, nitrite, and hydrogen sulfide. These are, in general, very harmful to fish and aquatic ecosystems. Potential benefits of probiotics to ecosystems are manifold (For further detail, see Chapter 19). However, several microbial supplements can be used to mitigate such conditions in ponds. Biocontrol (elimination of the pathogenic microorganisms through antagonistic properties) (See Panel 5.2) and bioremediation (See Chapter 19) are two important strategies of pond probiotics.

5.4 Effects of Aquatic Insects on Fish Culture

Though insects are everywhere, there are some that belong to the order Heteroptera that include bugs which have forewings half homy and half membranous and wings remain folded flat and parallel over the abdomen. The bugs are common in aquatic, semi-aquatic, and terrestrial habitat and are serious pests of crops, fruits, vegetables and trees. The aquatic fonns, however, live in water and exhibit structural adaptations to perform their vital functions suited to their special habitat, especially for locomotion and respiration which involve the process of aerial breathing.

Bugs can be classified into two main groups: (1) Short-horned bug having very small antennae (feelers) generally concealed under head such as water boatman, back swimmers, giant water bugs, water scorpion and (2) long-horned bugs with distinct antennae in front of the head. These include water skaters, assassin bugs, stink bugs, and bed bugs, flat bugs, squash bugs, and lace bugs. In freshwater fish culture ecosystems, a number of aquatic bugs exist (figure 5.1) and they cause severe damage to carp nursery and rearing ponds to a greater extent. These bugs pray on carp hatchlings, fry, and early fingerlings. Some common aquatic bugs are noted below.

1. Water boatmen: They are medium-sized with a short rostrum with shorter front legs that are flattened like longer hind legs and are fringed with combs of bristles used in swimming. Both the nymphs (young) and the adults are aquatic and keep their dorsal

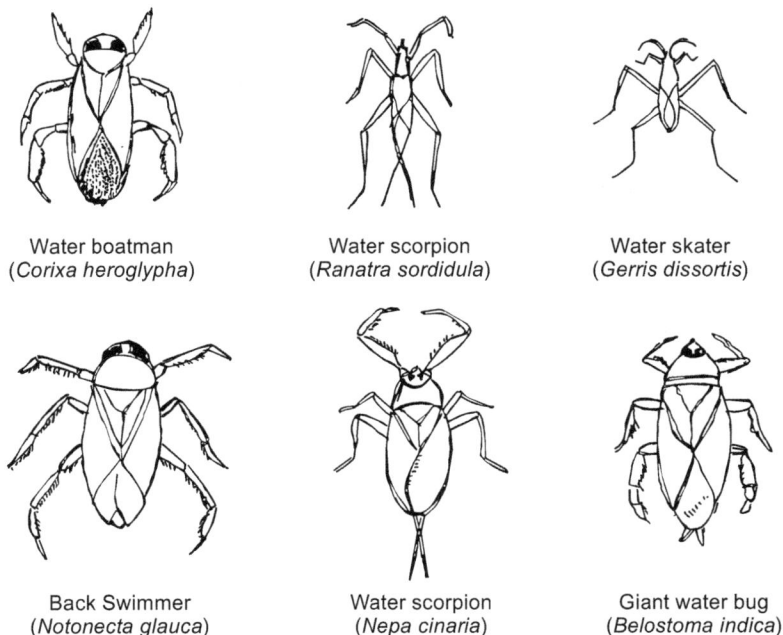

Water boatman
(*Corixa heroglypha*)

Water scorpion
(*Ranatra sordidula*)

Water skater
(*Gerris dissortis*)

Back Swimmer
(*Notonecta glauca*)

Water scorpion
(*Nepa cinaria*)

Giant water bug
(*Belostoma indica*)

Fig. 5.1 : Some common aquatic insects that cause damage to hatchlings and fry.

side upwards during swimming. Adult disperse over the surface of water in large swarms. They feed on the ooze at the bottom of the water and collect diatoms and algae for their nourishment. The common examples are *Corixa heroglypha* and *Micronecta striata*.

2. Back swimmers: These medium-sized bugs swim upside down and their long hind limbs are fringed with bristles for swimming. The dorsal side is pale but the ventral side is brownish, which provides an effective counter-shading mechanism against predators. The rostrum is sharp and used to suck body contents from prey. They are fierce carnivores and feed on young fish. A typical example of a water boatman is *Notonecta glauca*.

3. Giant water bugs: They are very large, oval, flat, leathery, brown bugs with a broad head and short, inconspicuous four-segmented antennae. Eyes are large, rostrum is long, wings are large and strong forelimbs are used for catching insects, snails, and small fish for food. The common examples are *Belostoma indica* and *Sphaeroderma rusticum*.

4. Water scorpions: They are oval-shaped, flattened with a terminal, long and thin respiratory siphon at the end of the abdomen and large grasping fore-legs. Head is small with short antennae. Prothorax is long and forms a short of neck. The bugs live in calm waters and climb on aquatic plants. Because of their body shape and locomotory behavior, they have been misnamed as water scorpions. Scorpions are never aquatic. They are carnivorous and lie in wait for prey (fish), which seize with their front paired legs. The common examples are *Rantra sordidula* and *Nepa cinaria*.

5. Water striders (Skaters): They are predatory, slender with long and thin legs that live on the surface flim of ponds, lakes, edges of rivers and slow-moving streams. Tarsus on legs are fitted with sets of non-wetting hair which allow the bugs to run and stand on the surface of water with an amazing speed. Common examples are *Gerris dissortis* and *G. conformis*.

5.5 Effects of Wild Fish on Fish Culture*

Escaped cultured fish can drastically affect wild populations both genetically and ecologically. Escaped farmed fish can affect the genetic diversity of local populations provided that the escapers are fertile and that their ecological requirements for reproduction are fulfilled. Where cultured fish are genetically distinct from local stocks, the disability of the loss of local adaptation through interbreeding exists. Although these risks are common to a variety of species, they have been most widely investigated in Mahseer and Trout due to their conservation status and their importance in both recreational and commercial fisheries.

Some farmed fish such as salmon are well-known to spawn in many countries. Genetic changes in wild populations could include incorporation of genetic variations previously absent in wild populations, reduction in natural variations through inter- breeding and loss of local adaptations. Loss of local adaptations could lead to local extinction of wild populations. This causes a threat to fish culture which utilize genes from wild populations

---

* For further study, see McKinnell et. al. (1997).

to blend commercially advantageous traits in cultured broodstocks. Loss of wild populations and genetic variability within wild populations would deplete the gene pool available to fish breeding programs.

Escaped fish may also have ecological consequences. If cultured exotic species are escaped from farmed ponds or tanks, adverse effects on native populations may exhibit with serious consequences. The classic example is that of the silver carp which has drastically declined the native fish *Catla catla, Labeo rohita*, and *Cirrhinus mrigala* of Govindasagar reservoir in Himachal Pradesh, India. The Culture of native fish must therefore be preferred on ecological grounds. Native species may also perform better than introduced ones owing to thier local adaptations particularly with regard to disease resistance.

Pathogenic organisms may be transferred from wild to cultured fish and vice versa. Accidental or deliberate introduction of wild fish infected with parasites, fungus, and bacteria may result in the extinction of cultured fish from ponds. Epizootic Ulcerative Syndrome was believed to have been introduced in Indian waters from neighbouring countries in the mid 1990s, causing severe damage to both farmed and wild fish.

5.6 Effects of Amphibians and Reptiles on Fish Culture

Larval forms of amphibians and reptiles are harmful to a greater or less extent to farmed fish as they competed with fish for food. The most harmful amphibians are frogs which swallow hatchlings and early fry. During rainy season, ponds are infested with tadpoles and during harvest, tadpoles and fry are mixed together which make the sorting of fry difficult.

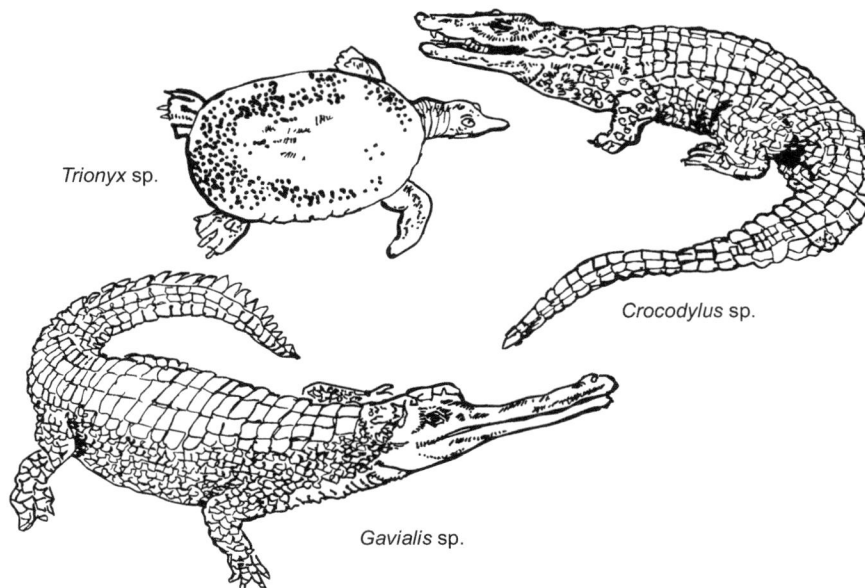

Trionyx sp.

Crocodylus sp.

Gavialis sp.

Fig. 5.2 : Some tropical retiles that cause severe damage to fish.

To remove amphibians from ponds it is necessary to destroy their eggs from the bank of the pond by using a scoop net. Use of quicklime is more effective to destroy tadpole populations. The adults can be removed by wire trap nets.

In most tropical countries, certain reptiles particularly snakes and crocodiles (Figure 5.2) destroy fry and fingerlings. The destruction is, however, severe in open water than in ponds.

5.7 Effects of Birds on Fish Culture

Recurring visit of a variety of birds (Figure 5.3) to many fish culture ecosystems and catching small fishes from them is one of the most dramatic scenes in many regions. The loss of fish from farms caused by birds is more severe in nature as they descend into water in flocks and eat fish. The nursery, rearing, and initial grow-out ponds (upto 2 months) are more susceptible to predation by birds.

A total number of about 8,600 species of birds have been noted across the world and only 1,750 species are resident within Indian sub-continent. Of the total 28 orders of living birds of the world currently recognized, 20 orders are represented in India by 62 families and 24 sub-families. Of them, 26 families and 18 sub-families belong to the order Passeriformes (Perching birds). The remaining orders contain both water and land birds. Although aquatic birds are very few in numbers, they are known to cause devastation in fish culture. Some common inland water birds and their orders are noted in Table 5.2

Interactions between fish culture and birds are very significant to fish production. Fish farms are often constructed near marshy lands and mangroves that are important nesting and over-wintering regions for many species of aquatic birds. Shallow waters provide ideal conditions for birds to prey on cultured fish stocks. Several species such as kingfishers, cormorants, fish eagles, herons, pelicans, and ducks are known to take fish from ponds. A large variety of birds are considered as the most destructive predators.

HERONS

Herons are long-leged birds and live around shallow and standing waters in which they walk. They remain still, standing on one leg with the water upto the knee and the neck is folded between the shoulders. When a fish approaches, neck shoots out rapidly. The bird destroys huge quantities of fish. A bird can able to swallow whole fish that varies between 10 and 15 cm long.

KINGFISHER

The kingfisher is a beautiful blue-green bird with irridescent glints. It has a straight, strong and pointed beak. The fishing method of this bird is spectacular. They dive on their prey (fish) like lighting from tree trunks where they perch. A bird can swallow small fish (4-6 cm long) in one gulp.

Grus sp.

Pelicanus sp.

Milvus sp.

Phabacrocorax sp.

Ansar sp.

Spoon bill

Alcedo sp.

Cygnus sp.

Heron

Fig. 5.3 : Some common birds that cause severe damage to fish. These birds are known to take fish from fish culture ecosystems.

PELICANS

These birds hunt fish by co-operative effort with rounding up and herding the fish into the shallows and scooping them up with the aid of their bill-pouches.

DUCKS

Though ducks are integrated with fish culture for high production, they often cause severe damage to hatchlings, fry, and young fish. Ducks are harmful as they disturb the fish with their ceaseless movement over water surface and they also damage the hatchlings on the spawning beds and in fry ponds. They are expert diver and when alarmed, vanishes below the surface in the twinkling of an eye.

CORMORANTS

Large flocks of cormorants can dive fish into shallow areas of freshwater jheels and tanks by flapping their wings and then prey on them. Cormorants consume large-sized fish (15-20 cm in length) than herons. Cormorants are most dangerous predatory birds.

General description of some common aquatic birds which are harmful to aquaculture are given in Chart 5.1.

5.8 Transmission of Fish Diseases by Birds

Several types of birds such as storks, terns, ducks, swans, cormorants, darters, cranes, and coots are known to transmit a number of parasites. Herons are the final host for fish tapeworms (cestodes) and gulls, herons, grebes, and goosanders are the final hosts for fish flukes (trematodes). These infect both farmed and wild fish in their larval stages. These birds daily travel miles of distances for food and reproduction. Breeding of these migratory birds in particular occur after monsoons when rivers are in spate and fields are flooded. At this time, fish juveniles and fingerlings are available in large numbers that forms the food for new-born offspring and young birds. The parent birds generally gulp the fish and disgorge into the mouths of their offspring after travelling long distances. However, it is quite possible that partly devoured diseased fishes are sometimes dropped off accidently in ponds underneath by these birds. Moreover, physical contamination of water by birds also occurs as a result of constant dips in the pond water while hunting fishes.

5.9 Control Measures of Pest Birds in fish Culture Systems

Several methods of controlling predatory birds are practised by farmers with varying degree of success. While raceways and small farms can be covered with nets of suitable mesh size, it is very difficult to save extensive farm ponds in this way. Damage to harmful birds by traditional devices such as false firing, flash guns, sirens, gill nets, shooting, bamboo rattles, habitat manipulation, beating of drums, fire crackers, placing of kites, and hanging of colored balloons around fish farms have all been tried with initial success but in the course of time these devices are not likely to be significant as the birds become habituated with these devices and neglect them.

Mechanical control of pest birds involves the use of (i) automatic bird scatters, (ii) pyro-techniques, and (iii) bio-acoustics. In the first case, acetylene gas is generated from calcium carbide and the gas is then ignited to produce a loud sudden noise. Pyro-techniques involve the use of fire and light accompanied with noise. In bio-acoustics, warning and embarrass calls of birds are used. Though all these methods are adopted to bring about the removal of birds from fish farming areas, the management cost involved in this regard may sometimes negate the benefit achieved.

Many progressive fish farmers sometimes use certain chemical substances to control pest birds as chemical substances hold an immence potential against them but they must be species-specific, highly effective, economical, and safe to fish culture ecosystems.

   (i) Sticky Substances: Some sticky chemical substances (such as mixture of gun arabic, coal tar, and the milk of jackfruit tree) are coated on perches which induce erratic behavior and consequently birds are dispersed.

  (ii) Frightening Substances: Frightening substances such as methoxychlor, carbaryl, dalapon, and 4 - aminopyridine are used in baits at recommended rates. These substances induce erratic behavior, fright, and disposal of birds.

 (iii) Repellents: Certain chemicals such as tetramethyl thiuram disulfide, lindane, 5 per cent malathion, 5 per cent benzene hexachloride, aldrin, and napthalene dust are used as bird repellents. On ingestion, they cause intestinal disturbance leading to emission of warning calls and removal of birds.

  (iv) Fumigants : The most commonly used fumigants include hydro cyanic acid, methyl bromide, anhydrous ammonia, and aluminium phosphide. These are strongly recommended for killing trapped birds or fumigating the nests.

The above methods have been susscssful in many regions. But unfortunately, most of the farmers overlook or neglect these factors during fish farming which may definitely cause severe loss to the tune of about 45 per cent. Therefore, growing awareness among farmers in this regard will definitely help to solve the problem of pest birds in fish culture.

5.10 Effects of Mammals on Fish Culture

Mammals such as otters, mink, musk rats, brown rats, water rats, and water shrews (Figure 5.4) may also cause problems and proximity to known haunts of these animals is often considered when the construction of new fish farms is contemplated. Otters attack large-sized fish, and consume the best parts. Fencing and use of traps are the best device to catch them. Different types of rodents mentioned above are very dangerous as they prefer large burrows on the banks and dykes. At the same time, they also destroy eggs, fry, and food reserves. Effective control of these rodents includes poisoning, trapping, trap netting, shotting, and by fumigating their burrows with sulfur.

5.11 Strategies for Limitation of Impacts*

High nutrient loading that are dangerous to intensive fish farming, can be reduced by 'no-loss' recirculation system where nutrient-enriched waters are microbiologically and chemically treated before return to the culture system, to merely permitting the water to move through a series of settlement tanks and ponds to eliminate particulates. Discharge of waste substances from fish ponds can be massively reduced simply by sustaining water storage potential and not evacuating between fish harvests.

Composite fish culture with a combination of prawns and crabs is extensively used particularly in developing countries both to maximize the usage of food inputs.

The practice of retention of several species of aquatic macrophytes in composite fish culture ponds allows eutrophicated waters to return more normal situations for several weeks or months before further stocking is becoming common and may become a strategy of keeping the culture pond in good condition. Aquatic plants remove most of the anmonia developed in fish ponds and, at the same time, oxygenate the water. Such strategy should be recommended to farmers.

Short-term (a few weeks or months) uncultivated farm ponds to break disease cycles are generally regarded as having production advantages. This has positive benefits both to the future farm productivity and to fish culture ecosystem.

Otter

Bandicoota
indica

Bandicoota
bengalensis

Fig. 5.4 : Some common mammals that cause damage to fish farms.

---

* For further detail, see Tucker et. al. (1996).

5.12 Sustainability of Fish Culture

Since inputs and wastes in extensive and semi-intensive fish culture practices are very less and the harvested fish are locally consumed, they have received much less criticism. In contrast, intensive fish farming system has been severely criticised by many experts due to an inefficient use of fish meal and for causing degradation of ecosystem due to nutrient-loaded waste accumulation. Whatever may be the criticism, it seems plausible that both scientific and economic aspects of the sustainability of intensive fish farming are definitely complex. Different criticisms are, however, likely to become more evident as fish production levels continue to increase.

5.13 Conclusion

Commercially important cultured fish have environmental requirements, particularly in terms of water and soil qualities and can degrade that ecosystems with toxicants. In general, extensive culture system has lower impact than intensive one, although even extensive culture has some local impacts. Since fish production levels are gradually increasing in many parts of the world and it is likely that new species and hybrid fish will not only be considered for culture to utilize unexploited ecosystems in full potential but to gratify new markets. Research and development on the environmental requirements of potential and farmed fish will also continue to increase.

A number of major geographical factors that influence farmers, their activity, and production potential of the system concerned would have to be kept in mind when fish farming strategies are taken into consideration. There is increasing farmers' awareness of environmental and aquaculture welfare issues which must be apprehended by the farms and by factors. The laws that are made on superior culture practice under varied geographical conditions will foster adequate data so that the local effect of effluents can be severely reduced which is being developed in developing world and is mostly fulfilled in many developed nations.

References

Gavine, F. M., M. J. Phillips and A. Murray. 1995. Influence of improved feed quality and food conversion ratios on phosphorus loadings from cage culture of rainbow trout, *Oncorynchus mykiss* in fresh-water lakes. Aquaculture., 26 : 483-495.

Hennessy, M. M., L. Wilson, w. struthers, and L. A. Kelly. 1996. Waste loadings from two fresh waterAtlantic salmon juvenile farms in scotland wat. Soil and Wat. Pollut. 86 : 235-249.

Jensen, F. B., M. Nikinman and R. E. Weber. 1993. Environmental perturbations of oxygen transport in teleost fishes: causes, consequences and compensations., in Fish Ecophysiology J. C. Rankin and F. B. Jensen (eds), Chapman and Hall, London.

Kelly, L. A., A. Bergheim and J. stellwagen. 1997. Particle size distribution of wastes from freshwater fish farms. Aquacult. Internat., 5 : 65-78.

Kelly, L. A. and A. W. Karpinski. 1994. Monitoring BOD outputs from land-based fish farms. J. Appl. Icthyol. 10 : 368-373.

Konstantinov, A. 5., V. V. Zdanovich and Y. N. Kalasihnikov 1988. Effect of temperature variation on the growth of eurythermous and stenothermous fishes. J. Ichthyol., 28 : 61-67.

Massik, Z and M. J. Costello. 1995. Bioavailability of phosphorus in fish farm effluents to freshwater phytoplankton. Aquaculture Res. 26 : 607-616.

Mc Kimell, 5., A. J. Thomson and others. 1997. Atlantic salmon in the North Pacific. Aquacult. 28: 145-158.

Munday, B. W., A. Eleftheriou and others. 1992. The Interaction of Aquaculture and the Environment — A Bibliographical Review. Commissioners of the European Community. DGXIV/D/3, Brussels.

Rajagopalsa C. B. T. and V. Ramadhas. 2002. Nutrient dynamics in freshwater fish culture. Daya Publishing House, Delhi, India.

Tacon, A. G. J., M. J. Phillips and U. C. Barg. 1995. Aquaculture feeds and the environment — the Asian experience. Water Sci. Tech. 31: 41-59.

Tucker, C. S., S. K. Kingsbury and others 1996. Effects of water management practices on discharge of nutrients and organic matter from channel catfish, (*Ictalurus punctatus*) ponds. Aquaculture, 147: 57-69.

Questions

1. What are the geographical factors that significantly affect fish culture?

2. Discuss how social and economic factors are directly related to fish culture, production, and marketing.

3. What are the environmental factors necessary to cultivate fish in ponds? State why fish should be cultured within the range of environmental tolerance. Discuss their effects on fish production.

4. What factors are more dangerous to fish culture? Discuss.

5. Discuss the role of carbon, nitrogen, phosphorus, and calcium in fish culture.

6. Discuss the role of oxidizing agents in fish culture.

7. What are the edaphic factors inevitable for fish production in ponds? Discuss how these factors influence the production potential in a fish culture industry.

8. Discuss how antibiotics and probiotics are related to fish production.

9. Discuss how pesticides, herbicides and fungicides affect the aquatic ecosystem in general and fish production in particular.

10. Discuss how aquatic insects, wild fish and birds affect fish culture industry.

11. Write notes on the following:

    (a) Aquatic bugs, (b) Interactions between birds and fish culture, (c) piscicides and coagulants, (d) oxidizing agents, (e) micronutrients, (f) alkalinity, (g) algal blooms, (h) ammonia (i) hypoxic condition, (j) biotic factors, (k) topographical factors.

12. What are the geographical factors that largely influence fish culture?

PANEL - 5.1
USE OF ORGANIC MANURES TO CONTROL SEEPAGE OF WATER FROM FISH PONDS

Impenetrable soil is one of the basic requirements that helps construct an ideal fish pond. Such fish culture ponds should have potential to retain water in adequate quantities for prolonged periods. This situation makes the pond ecosystem more productive. Unfortunately, this criterion is very common in most regions of the world especially where the soil is coarse, loose and porous in nature and does not hold water for a long time and at the same time, substantial amounts of nutrients are lost along with the percolating waters. Geologically, it has been classified as *lateritic soil*. Such characteristic feature of this kind of soil has become a serious problem for the construction of fish ponds because all these ponds are dried up in a very short period when monsoon season is over.

A few methods such as (1) construction of ponds in Reinforced Cement Concrete (RCC) and (2) lining of bottom and sides of fish ponds with plastic sheets (such as polyvinyl chloride and polyethylene). These methods were introduced experimentally in the early 1980s with great enthusiasm since they had remarkable ability to retain water considerably. Unfortunately, their high cost in relation to their benefits has made them uneconomical and their use has like-wise been abandoned.

Though it is not possible to retain water in such pond ecosystems completely, a partial retention involves the use of organic manures in huge quantities. The most common management practice to overcome the problem of water retention is the application of organic manures of plant and animal origin in several instalments. Examples of organic manures include mainly compost and cattle manures.

After application, organic manures generally accumulate in the pond bottom. The enzymes and soluble nutrients contained in the manures dissociate in water while the cellulose and other undigested matters settle at the bottom of the pond. For years, the residues of organic manures have sunk into the water, which by reducing oxygen availability has inhibited their oxidation and consequently, has acted as a partial preservative.

As one generation of fish crops follows another, layer after layer of organic manures are deposited in the pond. The constitution of these successive layers alters as time goes on because a sequence of undigested plant materials occurs. The succession is by no means regular or definite, as a slight change in water level may alter the sequence entirely. The profile of an organic deposit is, therefore, characterized by layers that differ in their degree of accumulation and in the nature of the plant tissue consumed by the cattle.

After adding each time to the pond, proper spreading of organic manures and frequent towing of drag net helps deposit in the mud of the pond bottom which ultimately turns into muck 1ayer (also termed as *detritus*). Undigested particles also help in providing reinforcement to the muck layer. Repetition of this process helps improve and strengthen the muck 1ayer. This muck layer in combination with undigested matters forms an impervious barrier between the bottom soil and pond water, thereby reducing the percolation of water. With the formation of detritus of some thickness, percolation rate of water decreases to a minimum.

Inorganic fertlizers may also be applied as a part of the technology in subsequent stages for increasing phyto- and zooplankton populations. Development of plankton in excessive amounts accompanying with their death and decay to the pond bottom also adds to the detritus, thus enhancing its impervious character.

A consecutive three-years study was made on a large scale to evaluate the use of this technology at Dhon Fish Farm in Satara District, Maharashtra. The farm has an area of about 12 hectare, consisting 100, 27, and 5 nursery, rearing and stocking ponds, respectively. These ponds were characterized by massive percolation of water, the entire farm was not suitable for fish culture for about 4 years. In this farm, different culture sequences, manure and fertilizer treatments were initiated on all these ponds in the year 1988. From this study, the following conclusions can be drawn:

1. Application of cattle manure in large quantities (5 tons every time) helped maintain the productivity of farm ponds.

2. Nutrients that are available to plankton are likely to account for the higher level of fish production.

3. This study clearly exhibited significant reduction in the loss of water through seepage. At the same time, this technique of water conservation had helped several fish farmers in improving their fish ponds within their limited financial resources.

4. The pond water and nutrients maintained at high levels are economically viable.

The results on these ponds remind us the importance of high ponds productivity in the production of robust fish in the area concerned. Ponds that are kept highly productive by the application of organic manures, inorganic fertilizers, and lime and by the choice of high yielding fish varities are likely to have water retention capacity than comparable to less productive fish ponds. The quantities of water retained at the pond soil are definitely dependent on the amount of inputs to be added.

Nutrients and freshwater deficiencies in many fish culture ponds have been diagnosed in most fish production centres of the world. In some regions where lateritic soils are spread over extensive areas, the extent of these deficiencies is much less well–known. Very limited research suggests that there may be large areas with deficiencies of one or two of these ingredients. Adoption of cattle manure and/or compost management principles if established for fish producing regions will definitely reduce the risk of seepage of pond water considerably.

PANEL - 5.2
BIOCONTROL OF PATHOGENS

A number of informations have emerged regarding hazardous effects of synthetic pesticides on fish and ecosystem health. On the basis of these informations aquaculturists have been forced to develop alternative methods which should not cause pollution and should be non toxic. And biocontrol method is such a technique which involves control of diseases by biological agent(s) (living macro- or micro-organisms) other than disease-producing organisms.

Simply defined, the term biocontrol refers to "the reduction of disease-producing activity of a pathogen or parasite in its active or dominant state, by one or more organisms, accomplished naturally or through manipulation of environment, host or antagonist, or by mass introduction of one or more antagonists". An antagonist is, however, an organism which has inhibitory relationships with other organism(s). Some antagonists *Bacillus subtilis, Aerobacter cloacae, Agrobacterium radiobacter, Streptomyces* sp., *Aspergillus* sp., *Arcella* sp., *Gephyramoeba* sp., *Vampyrella* sp., *Geococcus* sp., are known to reduce infection of the ecosystem by the pathogen or to reduce the severity of attack by the pathogen.

Introduction of Antagonists: In this method, a potential ant antagonist is isolated from a specialized niche, multiplied on nutrient media and then introduced in the same habitat for microbial interactions and keep the ecosystem healthy. An antagonist can be used for a number of pathogens in different habitats.

Role of organic material in biocontrol : Addition of fresh organic materials such as chitin, oil cake powders, compost and farm yard manure to fish ponds promote the activity and multiplication of the pathogens. The amendment of aquatic ecosystems by such method should be one of the best techniques for producing robust and disease-free fish without polluting the ecosystem to a large extent.

Genetic Engineering of Biocontrol Agents: It is known that several interactions always exist among micro-organisms in any ecosystem that provide an opportunity for genetic engineering of microbial biocontrol agents. The effectiveness of these agents can be substantially increased by gene splicing, gene cloning, and transformations. Such genetically engineered biocontrol agents help to destroy the pathogenic organisms or reduce their virulence. Research findings of yesteryear on the used of probiotics in fish culture have brought to light on the significant role of genetically-engineered biocontrol agents in helping fish culturists to produce healthy fish for the current and future generations.

Table 5.1 Temperature requirements of some cultured fish

Common name	Scientific name	Note
Rohu	*Labeo rohita*	Upper lethal temp37.0–40.0°C
		Optimum temp. 34.5°C
Catla	*Catla catla*	Upper lethal temp. 37.0–38.0°C
		Optimum temp. 34.5°C
Grass carp	*Ctenopharyngodon Idella*	Upper lethal temp. 40.0.–45.0°C
		Optimum temp. 39.5°C
Common carp	*Cyprinus carpio*	Upper lethal temp. 43–46°C
		Optimum temp. 38°C
Europen eel	*Anguilla anguilla*	Larvae, optimum temp. 19–21°C
Gold fish	*Carassius auratus*	Larvae
Nile tilapia	*Oreochromis niloticus*	Temperature related to pond depth
Rainbow trout	*Oncorhynchus mykiss*	Large fish
Silver carp	*Hypophthalmichthys*	Larvae, optimum temperature
	molitrix	32° C; upper lethal temp. 43.5–46.5°C, optimum 39.0°C
African catfish	*Clarias gariepinus*	Larvae and post larvae, optimum temp. 30.0°C
Europen whitefish	*Coregonus lavaretus*	Larvae, optimum temp. 19–21°C
African catfish	*Clarias gariepinus*	Larvae and post–larvae, optimum temp. 30°C

Source : Data collected by the author from different published literature

Table 5.2 Some common inland water birds, their orders, families, common and scientific names

Order	Family	Common Name	Scientific Name	Habitat
Gaviformes		Divers and Loons		Ponds and Ditches
Coraciliformes	Alcedinidae	Kingfisher	*Ceryle rudis* *Halcyon syrnesis* *Alcedo atthis*	Ponds, streams and ditches
Podicipediformes		Grebes	*Podiceps ruficollis*	All inland stagnent waters
Procellariiformes		Petrels and Shear water		Pelagic
Pelecaniformes	Pelecanidae Phalacrocoracidae	Pelicans Cormorants	*Pelecanus philippensis* *Phalacrocorax niger* *P. carbo, P. fuscicollis*	Jheels and Tanks
	Anhingidae	Darter or Snake Bird	*Anhinqa rufa*	
Ciconiformes	Ardeidae	Herons and Egrets	*Andreola grayii*	Ditches, Ponds and Marshy areas
	Ciconiidae	Storks	*Ibis leucocephalus* (painted stork)	Ditches, Ponds and Marshes
			Anastomus oscitans (open bill stork)	
			Ciconia episcopus (white-necked stork)	

Anseriformes		Swans	*Cyqnus* sp.
		Geese	*Branta* sp.
		Ducks	*Ansar* sp.
Gruiformes	Turnicidae	Bustard Quails	
	Gruidae	Cranes	
	Ralidae	Rails	
Charadriiformes	Jacanidae	Jacana	

Chart - 5.1
Brief Description of Some Common Reptiles, Birds and Mammals
which are Harmful to Aquaculture

REPTILES

1. Gharial: A large crocodile, *Gavialis gangeticus,* (exceeding 6m in length) of the Family Gavialidae and found in deep, fast-flowing rivers in parts of the nothern Indian subcontinent. Snout is narrow, stout, and well adapted for seizing swimming fish. The false gharial, *Tomistoma schlegeli,* (Fam. Crocodylidae) occurs in rivers of other parts of Southeast Asia.

2. Turtle: Any of more than 250 species (Order: Chelonia) of reptiles having a bony shell overlaid with horny shields. The body is broad, short, and enclosed in a box comprised of bony plates. Living forms lack teeth, instead they have horny beaks. Most species are semi-aquatic or aquatic; some are terrestrial. Fish-eating turtles such as the common snapping turtle found in North and Central America and the alligator snapping turtle found in the U.S. are found in many slow moving bodies of freshwater. They prey and bite fishes with their powerful jaws. Some turtles are fish-eaters.

3. Crocodile: There are three main categories of living crocodilans: true crocodiles, alligators and the gharial. True crocodiles comprise some thirteen species found in the tropical Americas, Africa, Asia, the East Indies, and Australia. Members of some of the species have been known to develop a taste for fish flesh.

Fish-eating reptiles are widely distributed in marshy areas, rivers, lakes, jheels, and the edge of the sea in warm parts of the world. They consume large quantities of fish from these water bodies than in ponds and ditches.

BIRDS

1. Heron: Herons are stork-like birds, which, together with bitterns, belonging to the Family Ardeidae. They are slender, with long S-shaped necks and legs and a long pointed bill. In flight the neck of most herons is retracted, the broad wings flap slowly, and the feet project beyond the tail. When hunting, a heron will stand motionless in the shallows, head on its sudden lunge of its bill. Consumption of fish by heron as much as 100 kg per year have been reported from many parts of the world. They build stick nests in trees. Common example is *Ardea cinerea* (grey heron) with some species.

2. Kingfigher: Kingfishers are cosmopolitan, small to medium-sized birds with heavy-looking, sharp beak, short legs, rounded wings, and brilliant plumage. They nest in holes in banks or tree trunks. *Alcedo atthis* and other species, Fam. Alcedinidae.

3. Pelican: Pelicans are a family of large-sized, gregarious aquatic birds allied to cormorants, and well known for their very large beaks and extraordinary throat pouches. They appear ponderous creatures, but are majestic and spectacular both in flight and on the water. Most of the species are surface-feeders, dipping their heads under water or scooping prey up in their throat pouches. They are distributed in warmer latitudes. *Pelicanus* with several species.

4. Duck: Ducks are closely related to swans and geese, but they are smaller, with shorter necks, and beat their wings faster in flight. They are much more aquatic than geese. The drakes (males) are generally more brightly coloured than the females. There are over 110 speices and distributed world-wide. Some like sea-ducks, sawbills and pochards dive for their food. Other like teal, mallard and widgeon are surface-feeders that dabble or up-end. Ducks are adapted to any aquatic way of life in having webbed feet and a water-repellent plumage that gives them added buoyancy.

5. Cormorant: Cormorants are slender-billed coastal and riverine birds, usually dark brown or black. The thirty-two species of the family occur world-wide from the tropics to both polar regions. They are foot swimmers and strong divers, with webbed feet and stout legs and often hunt for fish, crabs, shrimps and other aquatic food in small or large groups. Before diving they wet their plumage thoroughly, and after hunting they fly to a rock or branch to dry out, extending their wings in a characteristic pose. They breed colonially in untidy twig, sea weeds and guano nests, laying 2 to 4 eggs.

6. Spoonbill: Spoonbills are large, tall and pinkish wading, chiefly freshwater marsh birds akin to ibises, with unusual flattened spoon-tipped beaks. Mainly tropical or sub-tropical, they breed in colonies, often building stick nests in trees. *Platales* with other species.

7. Crane: Cranes are ground-living birds, looking rather like herons but related to rails and bustards. Cranes make up a family of some 14 speices with representatives in most parts of the world; they include demoiselle and stanley cranes. Cranes are tall, long-legged, short beaks, long-necked birds, typically with white or grey plumage and tail plumes. They perform ceremonial dances and possibly pair for life. Many species migrate in large flocks, flying in lines or wedge formations. *Grus* and many species.

8. Gull: Grebes are a family of some twenty species, related to bay ducks. They differ from other ducks in having lobes toes, pointed bills, slender erectly held necks, and practically no tails. They have world-wide distribution. Little grebes (*Tachybaptus* sp.) are the commonest European species. They nest on ponds, lakes and rivers. Fam. Podicipedidae.

9. Gull: Any of more than 40 species of heavily built gulls are long-winged, web-footed seabirds typically having white plumage with a grey or black mantle. They are gregarious surface-feeders or scavangers, roosting in flocks and often nesting in large colonies. Most species live in the Northern Hemisphere. Fam. Laridae with forty-six species.

It should be added as a note that cormorants, herons, kingfishers, and pelicans are considered the most destructive predators as they consume between 100 and 200 kg of fish per year from freshwater fish farms.

MAMMALS

1. Otter: Otters are twelve speices of semiaquatic carnivores, with world-wide distribution. Long-bodies and streamlined with short limbs and webbed paws, they have a wide, flexible tail and rich brown, waterproof fur. They inhabit rivers, freshwater lakes, and ponds, and are considered as the fastest aquatic mammals, cruising along at 10 km per hour. Otters feed mostly on fish, usually the rather sluggist species. The Eurasian otter (*Lutra lutra*), the Oriental short-clawed otter (*Aonys cinerea*), and Cape clawless otter (*A. capensis*) catch their food with their front limbs. The large sea otter of North America (*Enhydra lutris*) feeds on shellfish. They are inquisitive and playful; a favourite sport is sliding down a mud bank and pluging into water. Fam. Mustelidae.

2. Mink: Mink are nocturnal, semiaquatic carnivores (weighs upto 1.6 kg) related to otters. There are two species, the North American mink (*Mustela vison*) and the rarer European mink (*M. lutreola*). Fam. Mustelidae.

3. Shrew: Shrews are mouse-like (weighs upto 14 kg) insectivores or carnivores found in Africa, Northern Hemisphere and are characterized by a pointed snout and dense fur. They are solitary, burrowing, terrestrial animals, but a few are adapted for an aquatic life. Water shrews such as the American *Sorex palustris*, have hair on the feet and tail which help them to swim and buoyancy. Most species live in ground litter, but some live in burrows or trees and a few are semiaquatic. Water shrews are noted for their continuous and intense activity around fish farms, which necessitates almost continuous feeding. Water shrews feed on hatchlings and early fry. Fam. Soricidae, many species.

Farms with dense population of finfish and shellfish offer suitable grounds for above-mentioned animals. Hence, fish farmers consider them a threat to their stock and must take all possible measures into consideration.

6

Culture of Carnivorous Fishes

The culture of carnivorous fish is becoming an important aspect of environmentally sound and sustainable aquaculture. This involves manipulation of derelict, swanp, and marshy water-spreads and their use for culture of certain species of fish which are suitable in such waters. These fishes are consumed by therapeutic persons. Most of the carnivorous fishes possess white creamy flesh with little or no inter- muscular bones. This feature has special attraction for domestic consumption and export.

The culture practice of carnivorous fishes in such neglected water bodies is essential not only for sustainable off-take from the resource but also for providing nutrients to alleviate malnutrition. Utilization of neglected waters through state-of-the-art technology rather than keeping them or otherwise preventing their being unutilized, is a judicious cultivation practice.

6.1 General Consideration

In times past, the cultivation of carnivorous fish was not very popular and important source of income for rural people. Later, it has become so important that in several Asian countries, catfish raising ponds have been reported to occupy more than 40 per cent of the total water area. According to an estimate, India has about 1.4 million hectare of derelict and swampy water areas that can be well utilized for cultufe of these types of fish. It has also been reported that in India, about 20,000 hectare of water areas (including carp nurseries, marshes, and swamps) are judiciously utilized for the culture of carnivorous fish and fish production from this area is about 40,000 kilogram (data for the year 1980). A case study was carried out in 1995 and noted that the culture practice comprises about 38,500 hectare and about 86,000 kilogram of robust fish can be produced. Though these figures may not be accurate, however, there is beyond doubt that the rearing of carnivorous fish is an important economic activity. Though harsh environment of derelict aquatic ecosystems significantly affects their survival and growth and becomes impossible for their existance, there is every possibility of better growth and higher production to the tune of about 6 tonnes/ha/year.

Non-availability of fish seeds severely restrict their culture operations, since production depends upon individual pond operators utilizing their own ponds for local markets. Mass mortality of hatchlings and fry must have serious disruption of culture systems. From the early ninetees of the 20th century, induced breeding and larval rearing of commercially important species of carnivorous fishes are being carried out on commercial basis. Application of different techniques have triggered the expansion of carnivorous fish culture in several tropical regions. But unfortunately, such expansion could not fulfil the needed demand of farmers in some regions. Of course, the recent interest in expanding the cultivation of carnivorous fish has changed the entire situation. In India, about 70,000 hectare of derelict waters have been brought under culture of these fishes and production of more than 50,000 tonnes of fish are highly encouraging.

6.2 What is Carnivorous Fish?

Carnivorous fish are those which largely depend upon animal tissues for their sustenance. Predatory fishes always feed on populations of other animals, called *prey*. Predatory fishes are, however, essentially carnivorous so far as their diet is concerned and therefore are not suitable or considered for culture in carp ponds.

Most of the carnivorous fishes are called as *catfish*. Catfish is a group of bottom-dwelling fish which typically has several whisker-like barbels around the mouth. About 2,000 species are known, most of them are freshwater but about 50 species in two families live in the sea. They are widely distributed in all tropical and temperate continents. Most of them have scaleless skins but many South American species have a covering of large and hard scales. Indian sub-continent harbours about 142 species of catfish belonging to 13 families and 43 genera.

Carnivorous fishes grow at rapid rate than herbivorous species particularly in their early stages. In India, a few species of carnivorous fish are considered for culture in freshwater ponds. However, the culturable species of carnivorous fish include two main types such as Air-breathing fish and other carnivorous fish.

6.3 Types of Carnivorous Fish

Although a large number of carnivorous fish are found in different freshwater systems of India, the most common and important species which are now being considered as ideal candidates for culture in ponds/tanks are as follow.

AIR BREATHING CARNIVOROUS FISH

Air-breathing organs are generally found in many freshwater fishes of tropical and sub-tropical regions of the world. These organs are rarely developed in marine fishes. Air-breathing fishes are, however, categorized into two groups such as Murrels and Feather- backs. Some common species are noted below.

Murrels or Snake heads (Family : Channidae)

Murrel, *Ophiocephalus punctatus* (Bloch)

Common murrel, O. *striatus* (Bloch)

O. gachua (Hamilton)

O. *leucopunctatus* (Sykes)

O. *stewartii* (Playfair)

O. *orientalis* (Hamilton)

O. *micropeltes* (cuv. and val.)

Large murrel *O. striatus* (Hamilton)

Magur (Family : Clariidae)

Clarias batrachus (Linnaeus)

C. *macrocephalus* (Linnaeus)

C. *dussurnieri* (day)

Singhi (Family : Heteropneustidae)

Heteropneustes fossilis (Bloch)

H. *microps* (Gunther)

Koi (Family : Anabantidae)

Anabas testudineus (Bloch)

A. *oligolepis* (Bleeker)

A. *scandens*

Featherbacks (Family : Notopteridae)

Chital, *Notopterus chitala* (Hamilton)

Folui, *N. notopterus* (Pallas)

OTHER CARNIVOROUS FISH

Freshwater shark (Family : Siluridae)

Wallago attu (Schneider)

Pangas (Family : Pangasiidae)

Pangasius pangasius (Hamilton)

P. *stuchi* (Fowler)

P. *larnaudi* (Fowler)

Mystus (Family : Bagaridae)

Mystus aor (Hamilton)

M. *seenghala* (Sykes)

M. vittatus (Bloch)

M. gulio (Ham–Buchanan),

Butter catfish, *Ompok pabo* (Hamilton)

O. bimaculatus (Bloch)

O. pabda (Hamilton)

6.4 Importance of Carnivorous Fishes

The air-breathing carnivorous fish are well known for their nutritive, invigorating, and therapeutic qualities and therefore, these fishes are recommended by physicians as diet for patients. Besides these, there are other important features for which they are considered as an important group of economic species.

1. These fishes are rich in iron and copper and low fat content.
2. Easily digestable flesh and hence recommended for childrens and patients.
3. They are fast-growing fish and thrive well in swamps and marshes.
4. They can also withstand brackishwater environments to a less degree.
5. Most of the people of India prefer these fishes over major carps and, therefore, are in great demand.
6. Due to their prolonged freshness out of water and high nutritive value with less intermuscular spines, some air-breathing fishes are highly esteemed as food fish and command a high price in the market.

6.5 Advantages of Air-breathing Fish Culture

1. Due to hardy nature of these fishes, their culture and transportation involves comperatively low risk and simple management.
2. They are best suited for monoculture but their culture with carps instead of bottom-dwelling fish (such as *Cyprinus carpio* and *Cirrhinus mrigala*) is also economical.
3. They are cultured in cement cisterns and in almost all types of small and shallow ponds.
4. Fertilization in culture systems is not necessary.
5. These fishes can easily be cultured in pens. Pens are constructed by putting split-bamboo mat or nylon net in cordoned-off portions of derelict or swampy water areas.
6. Swampy or derelict water which cannot be used for carp culture without proper reclamation can undoubtedaly be put to immediate use for air–breathing fish culture.
7. Carp nursery and rearing ponds practically remain unutilized after seed production and harvesting operations and these types of ponds can be used for

air-breathing fish culture for short duration because in order to produce marketable size, they require only 3-4 months for culture period.

8. Since air-breathing fishes can able to tolerate water of low dissolved oxygen content, they are also cultured in sewage-enriched waters.

9. Farmed catfish have generally higher proportions of muscular fat (3 per cent) than wild catfish (0.4 per cent). Their polyunsaturated fatty acids (PUFA) are of n3 series whereas in case of terrestrial animals the polyunsaturated fatty acids are of n6 series. The major PUFA in catfish is 20: 5n3 and 22:6n3 which fish can be obtained from the diet. The ratio of n3/n6 fatty acids is reported to be 0.6 and 4.8 per cent in farmed and wild catfish, respectively. Such reduction of essential fatty acid content in farmed catfish has great therapeutic importance.

6.6 Biology and Culture of *Clarias batrachus* (Fig. 6.1)

This species is widely distributed in most sub-tropical and tropical countries with its prevalence in derelict and swampy waters. The highest recorded size of *Clarias batrachus* in India has been reported to be as 450 mm. The farm fish can grow to a Weight of around 1.5 times more than their wild varieties.

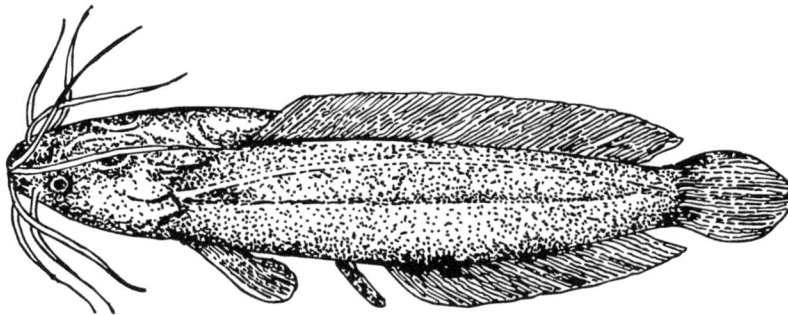

Fig. 6.1 : The most common Indian catfish, *Clarias batrachus*

Characteristic Features

The body is elongated, gradually tapering at the posterior end and devoid of scales. It measures upto 40 cm in length and brown or greenish-black in color. Head is depressed with tip and sides and covered with osseus plates. Dorsal fin is long and without spine which extends from the neck to the caudal fin. Adipose fin is not present. Caudal fin is more or less round, pectoral fins are provided with spines. Air bladder is connected with the internal ear by weberian ossicles. The fish is distributed in Africa, South and West Asia, and India.

Food and Feeding

The early post-larval stage of the fish (less than 15 mm in length) is mainly planktophagus and subsists on zooplankton such as *Diaptomous, Moina, Cyclops, Cypris* and some invertebrate eggs. In the later part of post-larval stage (16-30 mm in length) they consume insect larvae and worms, while in the juvmile stage (31-100 mm in length) the fish principally depends on insects, shrimps, and organic debris found in pond mud.

Though the adult fish exhibits a wide range of food preference, insects constitute the main source of food. The fish has poor vision and hence they cannot locate its prey with accuracy. Using their barbels, they consume food materials whatever food-like ,substances come across during movement. Since the fish has a very poor gustatory sense, they fail to sort out non-edible and edible food materials. In general, with the increase in weight of the fish, food consumption gradually decrease. It takes about 10-12 hours to pass the ingested food from oesophagus to the rectum.

Maturity

This fish attains sexual maturity when it is one-year old. During breeding season, the secondary sexual characters in fish are well developed and therefore, sexes can be distinguished with accuracy by the appearance of two distinct features. *Firstly,* the genital papilla in female is short, oval and slit-like; while in male, it is long and pointed. *Secondly,* the male fish looks slender and more streamlined. On the other hand, the female looks heavier with its distended abdomen.

Fecundity and Spawning

Though the fecundity rate is extremely variable in different regions of the country and also in different length range of the fish, on an average, the fish generates 500 and 50 number of eggs per gram of ovary and body weight, respectively. The color of eggs is pale or greenish-yellow or yellowish-brown. Spawning generally occurs in rainy season and it depends on the geographical conditions. In West Bengal, for example, the fish breeds between June and October. In Assam, it breeds in May-June. During rainy season, however, the fish migrates to the inundated ponds, paddy fields, canals, and other water-logged areas. The fish deposits fertilized eggs in holes (20 cm in diameter and 25 cm in depth) constructed in pond bank below the water surface where young fry develop and then exhibits parental care of the brood. Each hole contains fry that range from 2,000 to 15,000.

Fertilized eggs are demersal, adhesive, and spherical in form. Immediately after fertilization, the eggs (1.5 mm in diameter) begin to swell and attain a size of about 2 mm in length. After 20-24 hours and at temperature between 26° and 30° C, the fertilized eggs hatch into larva, the length of which varies between 4.5 and 6 mm in length. The hatchling has laterally compressed body with its head protruded with a large yolk sac. The larvae start feeding before completion of the yolk absorption. By 10-15th day of development, aerial respiration begins.

Fry Collection from Natural Sources

In many areas of West Bengal, Bihar, Karnatak, and Assam, the seed of this fish is available in large numbers. During post-monsoon and early winter months, fry are collected from water-logged low-lying areas or paddy fields using traditional fishing devices.

Breeding

The success of clarias culture not only depends upon the production potential of seed but also adequate management of brood stocks. For successful spawning, healthy broodfish should be considered. This species attains maturity at the second year with body weight varying between 150 and 200 g. Broodfishes are stocked during June-February in well-prepared cement tanks (3 X 1 X 1 m each) having an inlet and an outlet at desired levels. In each cistern, soil is added to make 2 feet thick layer at the bottom. Broodfishes are fed daily with a mixture of fish meal, trash fish, rice bran, and groundnut cake fortified with minerals and vitamins at the rate of 10 per cent of the body weight. It should be added as a note that specially prepared paddy fields constitute one of the most preferred breeding grounds for this species.

Breeding of clarias is executed either through environmental manipulation in the pond or through hormone injection. In the first case, specially designed ponds of 0.1-1.0 hectare having an average depth of 1 m are recommended. In the pond bottom several pits (15-20 cm deep) are constructed for nesting purpose. Moreover, a perimeter trench of 40-100 cm depth is also constructed along the bank of the pond. Pond is fertilized with cow manure at the 10,000 kg/hectare rate.

During this period the pond bottom is kept exposed where paddy is cultivated. Before the onset of rainy season, brood fishes are stocked into the trench and fishes are fed regularly at 10 per cent of the body weight. During monsoon season, the pond bottom is flooded with water and the depth of the pond water is kept at 25-30 cm. Broodfishes congregate into the pits where spawning occurs. After one week, water level in the pond is reduced and the spent fishes migrate to the trench. Fry are collected with the aid of a small hand net.

For induced breeding of clarias , carp pituitary extracts at different doses (10-300 mg/ kg of fish) are widely used with varying degree of success. In general,however, 25 and 35 mg of carp pituitary extracts per kilogram body weight of male and female, respectively, are injected into the body of fish. Of course, if the male broodfish is robust, there is no need to inject the extract. Since male clarias does not release sperm, the testes are removed, made into pieces, macerated with the aid of a pestle and morter and squeezed into distilled water or physiological saline solution and then milt is collected .

Injection of male and famale fish with ovaprim at a dose of 1 mg and 2.5 mg/kg, respectively also exhibited better results in terms of smooth stripping and percentage of fertilization. Field experiments have shown that 16-17 hours of latent period (time between hormone injection and stripping) is congenial for smooth release of eggs and better fertilization.

In addition to ovaprim, another inducing agent called *ovapel* is also widely used by many fish seed producers for its reliability and easy availability in readymade forms. The preparation contains D-Ala6, Pro9NEt-GnRH, and metoclopramide. The first two ingradients are mammalian gonadotrophic realising hormone analogue and the last one, is the water-soluble dopamine receptor antagonist (DRA). The concentrations are in the form of 18-20 micro-gram/pellet (for mammalian GnRH analogue) and 8-10 mg/pellet (for DRA). However, a dosage of 14.3 and 7.0 pellets per kg of female and male clarias, respectively, are enough for ovulation that gives encouraging results. The pellets are pulverized in a morter and dissolved in distilled water before injection.

After 10-15 hours of injection, females are stripped in plastic or enamel trays. The milt is then thoroughly mixed with the eggs with the help of a feather. The fertilized eggs are then transferred to a flow-through hatchery for incubation. In general, females take active part and provoke male to cooperate during spawning.

The flow-,through hatchery consists of a metallic stand on which a row of small ,plastic tubes (12 cm in diameter and 6 cm in height) are placed. Water is supplied from a overhead tank through a common pipe to all the tubes. . Each tube is provided with a control tap and an outlet at a height of 4 cm. The eggs are distributed in the tube and a very slow current of water is passed to maintain the water of its hardness, dissolved oxygen, pH, and temperature at 150 mg/l, 7.5 mg/ 8.0, and 28°C, respectively.

Rearing of Hatchlings

Hatchlings are transferred to rectangular or circular plastic containers for 15 days. During the entire period, water management is an important aspect. Oxygen must be provided by aerators to the containers. Accumulation of metabolic wastes and uneaten feed are not uncommon in rearing containers. To avoid these situations, daily replenishment of water (about 80 per cent of the total volume) with refiltering it to maintain 15 cm water depth is generally recommended.

The newly hatched larvae retain yolk sac which is absorbed in about 4 days. Therefore, it is wise to provide artificial feed from fifth day onward. Mixed zooplankton, artemia nauplii, molluscan meat and tubifex may be provided as feed. Hatchlings attain to 2-5 g in weight within 15 days in indoor rearing.

Rearing of Fry

Rearing of fry is done in well-prepared nurseries usually of 400 to 3000 m² water area with depth not exceeding 50 cm. In the middle of the pond, a small pit is constructed so that fry can easily be collected after dewatering of nurseries. About 300 fry are reared per square meter area of the pond. Fry are fed twice daily with mixed rice bran and ground trash fish (1:9 ratio) at the rate of 1kg/lakh fry. The average production of fry is about one million/ha/year. Fry attain to 5-7 cm in size within one month of culture with survival rates varying between 30 and 45 per cent.

For better survival of fry, however, rectangular plastic or cement cistern (8 X 1.5 X 10 m) is considered. All cisterns are provided with about 10 cm thick layer of soil at the base of each cistern and water level of 30 cm. Cattle manure and single superphosphate at the rate of 4 kg and 200g respectively are added to each cistern. These cisterns are exposed to sunlight for several days to facilitate plankton growth. Each cistern is then stocked with fry at the rate of 25-300/m^2 Fry attain a size of about 8 cm within a month which are then stocked in grow-out ponds.

1. Construction of Cement Tanks For Fry Rearing: Generally oblong or rectangular tanks are constructed for rearing of fry. Construction of a pond using cementing materials starts by ramming plenty of hard core in the bottom and sides. About three inches thickness of concrete layer are stretched without breaking. Over the concrete, a wire netting is placed and again followed by another three inches thickness of concrete layer. The concrete is allowed to dry in the sun and sprinkled occasionally with water. The same system is also followed for the sides and ends. About 2 inches thickness of concrete layer is made which is followed by wire netting and then concrete layer. For smooth wall and floor of the tank, brushing is necessary.

Tanks can also be constructed with cement slabs or blocks with thickness varied between 4 and 6 inches. Before setting slabs, hard core should be used on the floor rammed. The slabs are placed in suitable position and then water-proofed.

Since most of the carnivorous fish are cultured for more than 8 months in shallow ponds, seepage in such ponds must be taken into consideration. On the other hand, seepage is more intense in natural ponds. To prevent seepage and to construct ponds (if necessary) in porous soils, a material called *bentonite* is widely used. It is a chemically-treated, pulverized colloidal clay, and is the by-product of chemical factories. The material is spread on the bottom of the pond and then mixed with fine earth. Compaction of the mixture of soil and bentonite is made by a roller. In case of large ponds, however, plough and harrows are used. The quantity of bentonite required for one hectare of pond is 250 tonnes.

Fry and fingerling ponds must be provided with inlet and outlet pipes. This system is inevitable because exchange of water will keep fry more active and at the same time helps remove toxic substances along with water currents.

Efficiency of Different Stimulants for Breeding

The percentage of fertilization of eggs largely determines the number of hatchlings to be produced during induced breeding as shown in Table 6.1. Suitable stimulants and their effective doses are essential if admirable number of eggs are obtained. Table shows that the percentage of hatching is higher (60 per cent) than either pituitary extracts (25 per cent) or human chorionic gonadotrophin (10 per cent). This points to an obvious conclusion that the use of ovaprim is the most practical means of encouraging high percentage of hatching that has great practical significance in relation to better production and profit.

Culture Practices

The culture of *clarias* is best suited to perennial or shallow waters with a depth varying between 4 and 6 feet. Besides this, culture in derelict water bodies and weed-infested waters are also practised. The traditional carp nurseries and rearing ponds after harvesting fingerlings could also be used for culture of *clarias* from January to June each year and the entire stock is harvested before monsoon. In cases where culture is undertaken in weed-infested waters,floating fixed cages made of split bamboo mats or synthetic fibre (2 m X 3 m X 0.7 m) may be installed.

1. Three Systems of Culture: Although there is no specific culture system of air-breathing fishes, on the basis of their habitat and habit, some indigenous systems such as extensive, intensive, and semi-intensive have been developed. In case of extensive culture system, the species breed, spawn, and grow normally in either paddy fields or in ponds where fertilization /manuring and feeding programs are not followed. This practice of farming indicates poor survival and growth.

Semi-intensive culture system involves regular checking of water quality and feeding of fish. Of course, this system of culture is followed in some states of India, the only disadvantage of this system includes poor production. For this reason, this type of culture system has likewise been replaced by intensive one which is now being widely practised.

Through intensive culture system, the production potential of *clarias* in some Asian countries has increased considerably during the nineties with an annual production to the tune of 100 tonne/hectare. This production figure has, however, been reported from Thailand ponds. In contrast to this, the potentiality of catfish production in India is not as great as in the case of other Asian countries. This is due to the fact that non-availability of quality fish seed, fish meal, and trash fish at a low price and high cost of water exchange have been the limiting factors for high production of catfish in India. For high production, however, following suggestions have been recommended.

(i)   Culture of *clarias* is undertaken in shallow (0.8-1.0 m water depth) and small (0.05-1.0 ha) ponds with at least 15 cm of muck at the pond bottom.

(ii)   In semi-intensive systems, stocking densities of *clarias* at the rate of 5,000 -10,000 number/ha for polyculture and 50,000 -100,000 number/ha for monoculture are recommended. In intensive systems, stocking of fingerlings at the rate of 200,000 number/ha is recommended, however.

(iii) A cost-effective diet should be prepared using locally available ingredients.

(iv)   The carrying capacity of ponds can be increased through water exchange once in a month to eliminate toxic wastes developed in ponds.

For effective management with recommended techniques, the area of pond water should not be more than 0.1 hectare. In cases where the species is cultured in perennial shallow ponds, utmost care should be taken to eliminate existing predatory fishes. For this purpose, mahua oil cake at the 2,500 kg/ha rate is used. After 15-20 days of mahua oil cake application, the pond bottom is racked by dragging a few bricks fitted with a rope.

Table 6.1: Estimated Doses of Different Types of Stimulants Used in Controlled Breeding of *Clarias batrachus* and their Effectiveness of Mass Production

Stimulant	Male Average (mm)	Male Average (g)	Dose of stimulant (mg/kg)	Female Average (mm)	Female Average (g)	Dose stimulant (mg/kg)	Average number of eggs stripped out	Average number of ferti-lized egg	Per cent of ferti-lization	Average number of hatch-lings obtained	Per cent of hatch-ling
Pituitary gland	241	94	25	292	150	40	4,368	1,966	45	493	25
Human Chorionic Gonadotrophin	233	87	30	284	140	50	2,683	942	35	95	10
Ovaprim	243	100	1	249	115	2	5,040	4,032	80	2,421	60

Source : Basu *et al.* (2000)

Table 6.2: Nutrient Specification for Catfish Feeding

Ingredient	Clarias batrachus	Heteropneustes fossilis
Digestible protein (per cent)	30-35	25-30
Lysine (per cent)	1.50	1.20
Methionine + Cystine (per cent)	0.70	0.55
Lipid (per cent)	5-8	5-8
Calcium	0.5	0.5
Phosphorus	0.7	0.5
Vitamin A (IU/kg)	1,000-2,000	800-1,500
Vitamin C (mg/kg)	50-100	50-75

In a heavily silted pond bottom, limestone at the 300 kg/ha rate is applied which will reduce the toxicity of the pond water.

2. Stocking: Catfish permits high stocking density which triggers in achieving high production with adequate feeding. The rate of stocking density of *clarias* varies depending on the type of culture system. For examples, in case of monoculture in stagnent ponds, the stocking rate of 4-6 fingerlings per square meter water area is recommended. For culture with carps following polyculture method, the species is stocked instead of common carp/mrigal at the rate of 3 fingerlings per square meter of pond bottom in addition to the recommended stocking density of carps.

3. Feeding: Feeding of fish with low grade dried, marine trash fish and rice bran generally stimulates the growth rate. If marine trash fish is not available, fish offal, slaughterhouse offal or dried silk worm pupae mixed with rice bran and oil cake in the ratio of 2 : 2 : 2 should be provided to the culture system as supplementary feed. A mixture of biogas slurry, rice bran, and oil cake (2: 2: 2 ratio) has proved to be highly economical. At present, highly improved feed has been developed (Table 6.2) for fast growth and high production under intensive culture system. For better feed utilization under semi-intensive culture system in stagnant ponds, the species may be daily fed at the rate shown in Table 6.3. Feeding is done either by broadcasting feed in small quantities into the pond or by lowering the feed baskets into the water.

Table 6.3: Feeding Schedule of *Clarias* in 0.1 hectare Pond Stocked with 10,000 Fingerlings

Period (month)	Kg/day	Ratio of rice bran : trash fish
First	2.4	3 : 1
Second	4.8	1 : 1
Third	8.0	1 : 3
Fourth	16.0	1 : 3
Fifth	12.0	3 : 2
Sixth	8.0	3 : 1

Table 6.4: Growth Characteristics of *Clarias* batrachus in a Farmed Pond

Age (Day)	Average weight (g)
60	15.6
88	16.2
105	22
151	35
181	54.3
210	60.0
238	62.5
297	65.0
309	66.0
330	70.0
358	73.0

Source: Jana and Das (1980a)

4. Water Management: In intensive and semi-intensive culture systems where feeding rates are high, there may be excessive accumulation of metabolites such as ammonia and hydrogen sulfide and occurrence of algal blooms in ponds. Treatment of ponds with potassium permanganate (@ 300 mg/l) is recommended. Spreading of floating aquatic plants in the pond control algal blooms and at the same time removes considerable amounts of ammonia from water. Frequent aeration and exchange of water are also important to alleviate these problems.

5. Growth and Production: Under semi-intensive culture systems, fish attains a weight, on an average, upto 145g within 6 months. Following the recommendation of culture techniques, periodical checking of the growth of fish is very important that makes the culture system more profitable. In each month, fish is sampled from the stock by a cast net and their length and weight are calculated. If there is any fluctuation in the growth rate of fish, the management procedure is evaluated accordingly. However, the growth feature of *Clarias* in a well-managed farm pond is shown in Table 6.4.

Production potential of *Clarias* from farmed ponds (Figure 6.2) has been evaluated by a number of experiments. As a generalization, monoculture production of *Clarias* and mixed culture with *Anabas* and *Heteropneustes* are highly encouranging.The production potential of this species, however, depends on the type of culture, duration of culture period, feeding rate, stocking density, and the circulation of water in the system. The circulation of water is highly effective at the rate of 4 hours per day.

Fig. 6.2 : Harvesting of *Clarias batrachus* from an experimental pond. Note that the weight of individual fish is significant and has good market value.

Table 6.5: Mixed Culture of *Clarias batrachus, Heteropneustes fossilis* and *Anabas testudineus*

Area of pond (ha)	Stocking density (Number/ha)	Type of pond	Total production (kg/ha)	Duration (month)	Remark
Mixed Culture					
0.05	75,000	Derelict and Swampy	1,200	7	No feed and fertilizers
0.05	70,000	Derelict and	2,250	8-10	Feed and fertilizers are used
Mono-Culture of *Clarias*					
0.1	60,000	Shallow and perennial	5 tonne/ha	6	Rice bran + marine trash fish
0.1	50,000	Shallow and perennial	3,360	5	Rice bran + marine
	50,000	Shallow and perennial	4,015	6	trash fish
0.18	50,000	Shallow and perennial	3,645	8	Rice bran + marine trash fish
0.18	1,00,000	Shallow and perennial	35,000	7	Rice bran + marine trash fish plus frequent water exchange
0.20	1,00,000	Derelict/Shallow/ perennial + recirculatory filtered pond	21,900	6.5	Feed + fertilizers
0.07	2,00,000	Shallow	7,270	6.5	Monthly water exchange + feed
0.1	5,00,000	Shallow	40 tonne/ha	12	Feed + fertilizers

6. Harvesting: In general, harvesting of *Clarias* is done during summer season when the level of water decreases to a considerable extent. For successful harvesting, ponds and paddy fields should be dewatered and fish are collected using scoop and hand nets. If carp nursery ponds are used for catfish culture, the practice of dewatering could be helpful because it serves two purposes. *Firstly,* harvesting of catfish becomes very easy and *secondly,* carp seed rearing program is initiated.

6.7 Biology and Culture of African Catfish, *Clarias gariepinus* (Burechell)

The fish *Clarias gariepinus* popularly called as *Thai Magur,* is known for its rapid growth and size under tropical climatic conditions. The fish is reported to grow to a maximum weight of about 15 kilogram. Under Indian conditions, however, it attains to a size of about 1.5 kilogram within a year if adequate food is available for the fish with a stocking density of 5 per square meter. In the late 1970s, the fish has entered into India from Africa and Thailand via Bangladesh and has gained wide popularity among Indian farmers.

In some states of India such as West Bengal, Kerala, and Assam, progressive fish culturists beleive that the African catfish has several attributes that would make it suitable for culture in various water areas such as rivers, lakes, ponds, swamps, beels, and cement tanks. Fish farmers have accepted well for stocking and culture of this species because of its rapid growth, high survival rate, reproductive potential, good flavor and taste, moderate price in markets, and high fecundity rate (a female weighing 2 kg of body weight may produce about 1,40,000 eggs). In different countries of the world, the fish has been described as different specific names such as *Clarias lazera, C. Mossanbicus, and C. sengalensis.*

One of the most important features is that the fish exhibits canibalistic behavior and highly predacious that might create a menace to the indigenous species. It has been noted that even 8-day old fry can chase, bite, and severely damage other species of fish causing about 5 per cent daily mortality and consequently, there is drastic decline in the growth of fry.

Typically they are carnivorous and feed on a variety of food items that ranged from zooplankton to fish. The fish is able to tear pieces of cadavers with the help of small teeth on its jaws and to swallow the whole body of prey such as fish and frog. These feeding habits are found to be suitable for culture in small seasonal bodies of water for better production. Bamboo cage culture of this species should be recommended if the culture is adopted in polyculture ponds. The culture technique is similar to that of the Asian catfish. However, since no special culture technique has been widely recommended, production potential of this species is difficult to interpret.

Under natural conditions, the fish generally mature at an age of 2-3 years in Africa and spawn between the onset and the end of the monsoon season. In India, however, the fish sexually mature within 7-8 months when they attain a weight of about 500g. Breeding generally occurs at the onset of monsoon, the stimulus is associated with the increase in water level and inundation of marginal areas where the fish is spawned.

An inducing agent, the ovaprim, has been used at the following dosages with success. The males and females are injected at 1 and 2.5 ml ovaprim/kg of body weight, respectively, and they are released in ponds where aquatic weeds grow. Courtship behavior lasts for 3-4 hours and the female lay eggs in batches. As soon as spawning is over, the spent fish moves into the deeper region of the pond. The fertilized eggs, when attached to aquatic plants, are transferred to a hatching hapa/pool. Hatching time varies between 28 and 32 hours depending upon the temperature of water. The yolk sac larvae are hidden underneath the vegetation. After hatching, the larvae are reared in a specially prepared nursery tank which is teemed with zooplankton. A ovaprim-injected female fish having a weight of 2 kilogram can generate about 2,25,000 eggs with a survival rate of 77 per cent.

Recommendation of National Committee on Culture of *Clarias gariepinus*

A meeting of the National Committee to oversee and regulate the introduction of exotic aquatic species in Indian waters was held in New Delhi on the 9th October 1997. The committee has recommended that the existing stock of the African catfish and their hatcheries must be destroyed. It was also stressed that this species should not be allowed to cross with the endemic species *Clarias batrachus* because the gene pool of this lucrative Asian catfish will be lost and consequently, the fish will be disappeared for ever from our natural environment.

Inspite of such recommendation, the introduction of this African catfish into the Indian waters has already encouraged among fish farmers to cultivate it in their own ponds for profit. Some progressive fish culturists have reported to fishery experts that they are culturing this species for several years in their carp culture ponds and they did not observe any adverse impact on the survival and growth of carps and even freshwater prawns. Most of the farmers claim that the species is not threat to carps and prawns. At present, many European, Southeast Asian, and African countries consider this species as an ideal candidate for culture in ponds; of course, after proper evaluation of its biology and feeding as well as breeding behaviors.

It should be kept in mind that once introduced into the Indian waters, it is very difficult to exterminate the population of this species and hence there is no other alternative except to welcome the fish for culture. But caution must be taken so that detrimental effects on indigenous fish fauna can be avoided. It is, however, hoped that the culture of this fish will definitely open up new vista for use of derelict water areas for future generations.

6.8 Biology and Culture of Giant River Catfish, *Mystus seenghala* (Sykes)

North and North-Western states of India generally prefer giant-sized catfishes for the table due to their excellent flesh quality and flavor. Among different types of giant catfishes, however, only some species such as *Aorichthys (Mystus) seenghala A. aor, Wallago attu, Silondia silondia*, and *Pangasius pangasius* are available in different river systems for commercial culture.

The fish is distributed in India, Mayanmar, Burma and Bangladesh. Commercial catches of this species come from different river systems of India and their tributaries such as Yamuna, Ganga, and Krishna. Substantial quantities of this catfish are generally landed from the Ganga river system. An increase in fishing intensity around the breeding grounds and pollution of water clearly indicates fishing stress of this stock. The stock can well be maintained if suitable culture management strategies are adopted. This involves an in-depth understanding of ecological conditions in relation to their habitat, breeding and parental care, seed resource, and rearing of juveniles.

Characteristic Features

The body is elongated and compressed with rounded abdomen. Head is large snout spatulate, mouth sub-terminal, wide and transverse, lips are thin with subequal jaws and villiform teeth. There are four pairs of barbels, one each of nasal, maxillary and two mandibular. A distinct interneural shield in between basal bone of dorsal fin and occipital process is present. Dorsal fin with 7 rays and a spine. Adipose dorsal fin is long. Pectoral fins with 9 or 10 rays with a serrated spine. Pelvic fins with 6 rays. Caudal fin is forked with 19-21 rays.

Sexual Dimorphism

Sexes of *A. seenghala* can easily be identified by the presence of morphological features. The male attains to 365 mm in length and have a muscular conical reddish pink papilla situated above the genital opening. In female, however, the papilla is absent. As soon as the breeding season commences, a white creamy secretion is observed in males.

Breeding Behaviour

The fish clearly exhibits parental care. The males select soft, sandy, and clear bottom of the river where they construct a pit within which the young ones are nursed and guarded in the nest till they attain a size of about 40 mm in length. For construction of suitable nest, males select the places of the river where shallow depth with feeble current of water and aquatic weeds exist which favor egg laying in the nest.

Culture

Fry of *A. seenghala* (about 20 mm in length) are collected from the nesting sites of the river. Initially fry are reared in one-tenth inch mesh nylon cages (1:1:1 metre) and these cages are kept afloat in running waters of the river for a period of one month. When fry attain a length of about 100 mm (fingerling stage), they are stocked in well-prepared ponds at the stocking density of 10,000 number per hectare. During the entire culture period, fishes are fed with semi-boiled trash fish. For better production of this species, *Amblyopharyngodon mola, Puntius ticto, P. sophore* and *Ambassis ranga* are also stocked in culture ponds as forage fish to serve as food for *Aorichthys seenghala*. It has been observed that the fish attains an average weight and length of 710 g and 510 mm, respectively, in about 20 months of culture period with more than 80 per cent survival. This point clearly indicates the good prospect of pond culture and culture practices can be taken up for commercial production.

The parental care of the male fish for feeding of fry has a direct impact on the survival of the young ones in confined waters. The problem becomes more complicated due to intermittent breeding of this species. Since induced breeding of this fish has failed to produce quality seed, there is no other alternative to collect young fish from natural resources for culture in pond ecosystems unless further elaborate studies on induced breeding are carried out. Therefore, suitable device should be developed to collect seed from the nests so that the problem of producing the stocking material is solved.

6.9 Biology and Culture of *Wallago attu* and *Ompak* spp. (Fig. 6.3)

Wallago Attu

The fish *W. attu* is considered, to some extent, as an important and an additional catch from most of the Indian river systems and their tributaries. This fish is also found in some large and medium-sized reservoirs.

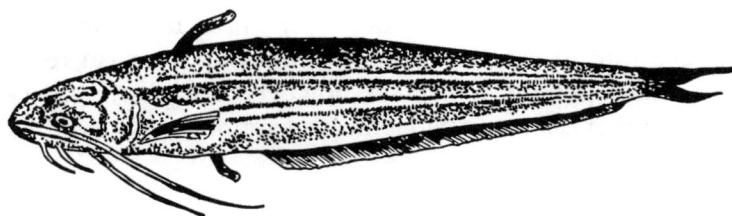

(a)

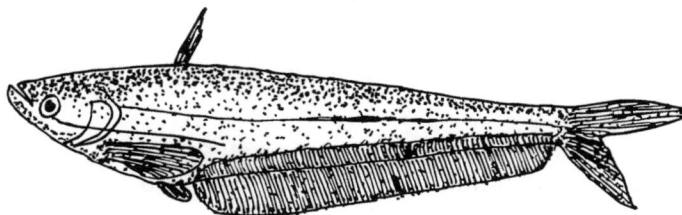

(b)

Fig. 6.3 : The most common Indian catfish, (a) *Wallago attu* **and (b)** *Ompak* **sp.**
Though these species are not cultured in ponds on commercial basis, substantial amounts are harvested from other freshwater resources and the species *Ompak* **has high market value.**

The fish is characterized by a large-sized body and attains to a length and weight of about 1 meter and 5 kilogram, respectively. The body is compressed and elongated. Teeth are present on premolar, mandibular and premaxilla. Nostrils are separated from each other by a short distance and situated at the tip of snout. Barbels are well developed and generally four in numbers. Antero-dorsal fin is short with 4 or 5 rays, and devoid of adipose fin. Paired fins are laterally inserted. Each pectoral fin is provided with a strong spine. The fish is distributed throughout India, Sri-Lanka, Bangladesh, Thailand, Malaya, Indonesia, Vietnam, and Pakistan.

It is a very hardy fish and can thrive well under varied aquatic ecosystems. The fish is highly carnivorous and consume almost any kind of food from dead body of animals and debris to medium-sized fish fry/ fingerlings due to presence of well-developed teeth. This species is never cultured scientifically because accidental intrusion into the carp culture ponds drastically decline the production potential of other species of commercial importance.

Different river systems and their tributaries where the fish survive, inundate low lying and spill-areas during the monsoon. These impounded waters are known as *jheels* and *Beels* in Bihar, Assam, and Uttar Pradesh. In West Bengal, it is called *Bhery*. When these low lands are inundated with rain water, eggs and fry of this species are transported to nursery and rearing ponds and cause menace to carp fry. Since the fish has good demand among middle class families, culture of this species should be made in separate ponds.

Ompak bimaculatus

This fish is also an another important catfish and forms an important catch in inland waters. It has very good taste, has no intermuscular spines, and fetch a high price in the market that varies between -Rs. 225.00 and 275.00 (data for the year 2011).

The body is elongated and compressed with rounded abdomen. Head is small, broad, and depressed. Snout is blunt and rounded. Mouth is superior, moderately wide with its cleft oblique. Jaws are subequal with villiform teeth. There are two pairs of barbels, one pair each of mandibular and maxillary, the former occasionally rudimentary or very small. caudal fin is forked. The anal fin is very long and extends very close to caudal fin but narrowly connected to it. Dorsal fin is very short and adipose fin is absent. This species is distributed throughout India, Nepal, Pakistan, Bangladesh, Malaya, Thailand, Vietnam, and East Indies. Eight species are known and distributed in these regions.

Fishes are caught from Bherries and reservoirs during monsoon and winter seasons. However, there is no organized culture-based fisheries in India and adjacent countries.

Ompak pabo

This fish constitutes an important component of riverine fisheries in Indian sub-continents. It dwells and breeds in rivers and reservoirs and in connected watersheds in flooded condition.

The adult fish exhibits sexual dimorphism. Mature males exhibit roughness of the first ray of pectoral fin and the lower side and possess a narrow and pointed genital papilla and white milt oozes on applying mild pressure on the abdomen. In mature female, the pectoral fins are smooth, and the genital papilla is thick and muscular. The brood fishes attain maturity after one year of culture with the average body weight of about 85 g. Fry can be cultured in a polyculture pond which has already been stocked with the Indian major carps. The method of culture is similar to that of carp culture.

Induced breeding of this species has been found to be more successful (percentage of hatching is about 70 per cent) using synthetic hormone ovatide at the rate of 3.0 ml/kg body weight of male and female. The major water quality variables in the breeding and larval rearing of this fish are as follows: dissolved oxygen, 7-8 mg/l; pH, 7-8; hardness, 110-130 mg/l; alkalinity, 130-150 mg/l, and temperature, $30.0 \pm 2.0°C$.

For successful rearing and better survival of larvae, feeding is the most important criterion. Formulated feed mixed with piscimix has been found to combat with mortality and enhanced growth of fish. Piscimix powder is mixed with the formulated feed at the rate of 10 mg/kg of feed.

It is needless to say that production of these important catfishes from inland waters can never be as high as production in ponds and tanks. This is due to the fact that great fluctuations in water ,levels cannot afford suitable ecological conditions for fish food organisms resulting in poor production. Therefore, suitable planning and development of these catfishes and also other catfishes such as *Pangus* and Featherbacks (to be discussed latter on) should be undertaken on commercial basis to increase their production.

6.10 Biology and Culture of *Heteropneustes fossilis* (Fig. 6.4)

It is an important air-breathing fish and similar to clarias, it is also widely distributed in swamp and derelict waters in some tropical countries. It has been identified as a potential candidate for culture in these waters. It commands good consumer preference due to its taste and high protein, iron, and low fat contents. It is widely distributed in India and Burma.

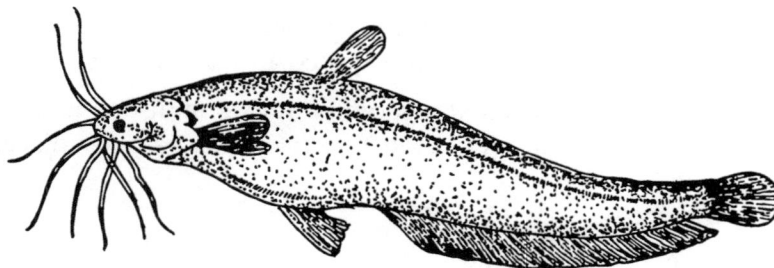

Fig. 6.4 : The most common Indian catfish, *Heteropneustes fossilis*. Similar to *Clarias batrachus*, this species is well known for its nutritive, invigorating and therapeutic qualities and hence they are considered by physicians as diet for patients. Hence, their intensive culture is badly needed.

Characteristic Features

The body is elongated, laterally compressed, and devoid of scales. It attains a length of 30 cm. Head is flattened, eyes with circular margins. There are four pairs of long barbels at the anterior part of the head. Dorsal fin is short and without spine, ventral fin is situated at the level of the dorsal fin, pectoral fins are strong and provided with poisonous spines. Anal fin is elongated which reaches upto the caudal fin separated from it by a notch. The gill opening is wide, and the membranes are not being confluent with the skin of isthmus and separated by a deep notch.

Food and Feeding

It is an omnivorous fish and a bottom-feeder. In addition to artificial feed, they mostly prefer ostracods, algal matter, worms, insects, organic matter, and gastropods as natural food organisms. Post-larvae, however, consume plankton.

Maturity, Fecundity and Breeding

This fish arrives at sexual maturity when both females and males are one year old. Breeding season of the fish extends from June to October. In contrast to *Clarias*, fecundity rate has been found to be very high; depending on the size of the fish and the rate of fecundity varies between 2,500 and 46,000.

Breeding of *Heteropneustes* is conducted in two ways: (i) natural breeding in paddy fields and (2) induced breeding. For natural breeding, a number of male and female fish are stocked in paddy fields where fertilized eggs are hatched into larvae which are collected by means of cast nets. No parental care has been observed.

Induced breeding of this fish has been quite successful. The fish has been successfully bred by homoplastic pituitary extract at the rates of 70-80 mg/kg of female and 60-65 mg/kg of male fish. For large-scale production of fish seed, carp pituitary extract at 160-200 mg/kg of body weight is more effective. This extract permits to release huge quantities of eggs. After 10 hours of injection, testes are dissected out and squeezed out in 0.96 ml saline solution. The females are stripped by applying gentle pressure on the abdomen and eggs are allowed to fertilize with the sperms. Many farmers do not follow stripping method; rather, two injected males and one injected female are kept in a breeding tank. However, the duration of spawning varies between 2 and 6 hours. Injected fish mate in the column or near surface layer of the ambient water and the number of eggs released at each mating varies between 40 and 150. As a generalization, a female fish weighing 100 g, releases about 8,500 eggs. The eggs are slightly adhesive in nature and the color of eggs is either green or brown.

The spawning efficiency of this species using synthetic hormone ovatide is very high, rates of 0.4 -0.6 mg/kg of body weight being common (Table 6.6). Most of the farmers use this hormone which is subject to very high fertilization and spawning rates. Therefore, the spawning success will definitely have significant effects on production potential of this valuable species.

Rearing of Hatchlings

The incubation period of this fish is temperature dependent. At 27-30°C, however, the incubation period varies between 20 and 24 hours. By the end of the 4th day, the yolk material is completely absorbed and the larvae start feeding from 5 th day. Generally larvae consume rotifers and ciliates but they are severely affected by the presence of different types of crustaceans such as *Daphnia, Cyclops* and *Diaptomus*. Their presence lead to mass mortality of hatchlings. This problem can, however, be alleviated by rearing them in hapas made of nylon cloth having 50 mm mesh size. Aerial respiration starts from 8 th day of development and their rearing is carried out in well-prepared nursery tanks, the survival rate of about 75 per cent being very common.

The state-of the-art rearing of early hatchlings involve daily feeding that comprises (i) boiled eggs and rotifers (1-2 days), (ii) rotifers and ciliates (3-12 days), (iii) artemia (4-7 days), and (iv) crushed molluscan meat and zooplankton (13 days). It should be kept in mind that cannibalism of hatchlings is one of the most serious constraints for their survival rate (less than 30 per cent) that limits the development of culture strategies of this species. This constraint can, however, be overcome following the application of balanced diet and other management protocols that can simultaneously enhance the survival rate of fish.

Culture Methods

Similar to *Clarias*, the culture of this fish is receiving more attention after its therapeutic attribute has been realized. Three culture systems such as intensive, extensive, and semi-intensive are adopted. Culture of this species in carp nurseries and in cement cisterns is also employed for high production.

Table 6.6: Estimated Doses of Ovatide Used in Breeding of *Heteropneustes fossilis* Under Controlled Conditions

Fish weight (g) Male	Female	Dose of ovatide (mg/kg body weight)	Latency period (hrs)	Number of eggs spawned	Per cent of fertiliza-tion rate	Per cent of hatching
60						
	75	0.2	14	1,508	65.3	65
55						
70						
	68	0.4	10	4,358	93.3	82
60						
75						
	80	0.6	13	3,220	56.6	70
65						

Source: Marimuthu *et al* (2000)

The production potential of this fish has been evaluated by a number of trials in different states of India especially in West Bengal, Bihar, and Assam. Monoculture of this species and mixed culture with *Anabas* and *Clarias* exhibit better production (Table 6.5 and 6.7) and adoption of the recommended culture system by many farmers is highly encouraging (Figure 6.5).

Table 6.7 : Mono-culture of *Heteropneustes fossilis*

Area of pond (ha)	Stocking density (No./ha)	Type of pond	Total production	Duration (month)	Remark
0.1	2,50,000	Shallow and seasonal	4.6 tonne/ha	4	Rice dust and cattle manure as feed
			4.9 tonne/ha	6	Groundnut cake + rice bran + fish meal + cattle manure
0.04	2,50,000	Seasonal and shallow	15 tonne/ha	12	Oil cake + slurry + rice bran
0.2	2,50,000	Seasonal and shallow	36,000 kg/ha	7	
0.015-0.1	50,000	Shallow and seasonal	4,510 k g/ ha	4-11	

Source: Data compiled from different published literatures

Fig. 6.5 : Harvesting of *Heteropneustes fosilis* from an intensive culture pond. Similar to *Clarias,* this species has also market potential.

1. Major Production Constraints: Tables 6.5 and 6.7 illustrate the production potential of this species in different water bodies of India. The system of culture and feeding schedules are the most significant constraints. Nutrient deficiencies in feeds provide serious limitations to all culture systems. Excess water in many derelict and swamp areas is a more serious problem in several regions. The heavy monsoon rains account for this constraint.

6.12 Biology and Culture of *Pangasias pangasias*

Successful culture of murrels, gourami, trout, and tilapia have serious limitations to temperatures, food, and environment. As a consequence, gangetic carps have continued to monopolize in ponds and have enriched waters where large-growing species are few in numbers. In several places of India, the culture of pangas in freshwater ponds or reservoirs has great practical significance in relation to its production. Easy availability of fingerlings, preference to molluscs in ponds, high resistance to pollution, existence in both fresh and brackish waters, and stocking of this species with the Indian major carps in ponds are some of the important criteria which permit mass production of this fish not only for its high market price and for the table but for excellent taste. These attributes must be taken into account.

General Considerations

Among the three important species of the group Pangas, *Pangasius pangasius* is the latest addition to the list of air-breathing teleosts. It is a hardy fish and can withstand polluted environments without any difficulty. In contrast to other catfishes, it is a facultative air-breather and aerial respiration occurs when necessary. It is a large-sized catfish and maximum as well as minimum weights of this fish have been recorded as 26.0 kg (1,340 mm in length) and 2.5 kg (665 mm in length), respectively.

Three species have been reported and considered for commercial culture such as *Pangasius pangasius* (Hamilton) *P. larnaveli* (Bocourt), and *P. sutchi* (Fowler). These species are very important and are considered as catfish fauna of major freshwater river systems.

At the onset of monsoon season, juveniles and fry are found within the tidal regions of many river systems. The foul feeding habit accounts for its low market value between January and May. Condition factor clearly indicates that the fish is in best condition in late winter and early summer months.

Distribution

'The fish is distributed in different river systems of Thailand, Java, Malaya, Archipelago, Burma, and East Coast rivers of India from the Mahanadi and northwards to the Godavary, Krishna, Cauvery, and Tapti. It shows a long range migration from the estuarine lower delta to the middle of the Jamuna waters, the length of which varies between 1,000 and.1,200 kilometers from the estuaries.

Spawning

Spawning occurs in between June and August. On an average, each gram body weight of the fish generates about 200 ova. During spawning months both males and females aggregate in almost equal proportions at the higher reaches of the river, the larvae and fry drift to the tidal reaches and ultimately entered into the estuaries. Larvae and fry are collected from these regions by ring nets.

At about 8 day, the larva attains 10 mm in lengtht (Fig. 6.6) Head is large, caudal fin is greatly spread, adipose fin is prominent, minute dorsal fin but fairly well developed anal fin. Posterior to the vent, there are seventeen myotomes. When the larva attains 20 mm in length, fin rays are developed which are complement to an adult. Pelvic fins appear as small buds. In the late larval stage (30 mm in length), the body is deeper with compressed head and barbels are shortened.

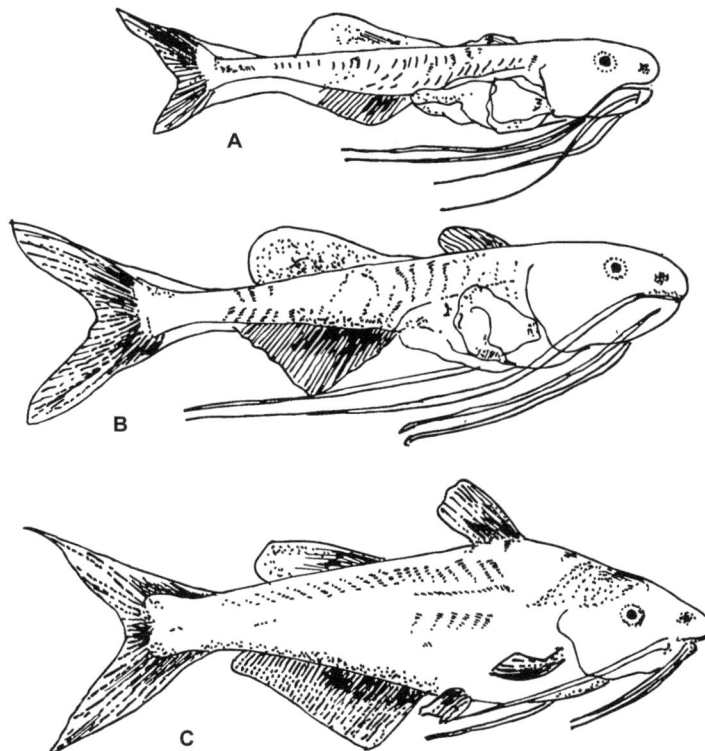

Fig. 6.6 : Different stages in development of *Pangasius pangasius*.
(a) larval stage of 10.2 mm (total length), (b) larval stage of 13 mm (total length) and
(c) fry of 30 mm (total length)

Feeding Habits

In general, crustaceans, insects, molluscs, and plant matter constitute the main food items. However, juvenile fish consume mysids, insects, may flies, ants, beetles, water-bug, odonata nymphs (dragon fly and damsel fly). Adults consume huge quantities of molluscs and plant materials. Large-sized fish feed on other teleosteans such as *Amphipnous cuchia, Cynoglossus* spp. *Setipinna phasa* and *Ambassis* spp.

Culture Methods

The culture of the fish *Pangasias pangasias* in carp ponds is suitable because it does not adversely affect the survival and growth of major carps. However, the fish is cultured in two ways such as cage culture and pond culture. Cage culture of pangasias has been described in Chapter 9. A short description of pond culture is noted below.

Monoculture of pangasius or polyculture with the Indian major carps have been found to be very profitable. They are cultured in well-prepared derelict waters, the average depth of 6 feet being very common. In predominantly catfish farming areas, intensive farming systems are generally preferred. The hardy nature of the fish makes high density culture possible. But nonetheless, excessive stocking densities and very intensive feeding schedules have created several management problems in some areas.

1. **Brood Stock Management:** Generally brood fish of pangasius are about 2 years old. Though brood fish can be obtained from rivers, progressive fish farmers prefer cultured brood fish grown on a diet rich in animal protein in a healthy environment. Smaller brood fish of about 1 kg are stocked at the rate of 300-400 kg/ hectare. The natural food produced in a well-fertilized pond can sustain about 370 kg/ha. Depending on the availability of natural food, daily feeding is generaly recommended. Additional food consisting of fresh or frozen meat or low-priced meat products is given at the rate of 10 per cent of body weight which are believed to meet the additional needs of minerals and vitamins for brood fish during gonadal development.

2. **Breeding and Fry Production:** Though breeding of this fish occurs in rivers and their tributeries, induced breeding has been carried out by hypophysation, although it is not practised to any appreciable extent in production farms. For induced breeding of pangasius, hypophysation with carp and catfish pituitaries at doses ranging from 30 to 80 mg gland/kg female in two injections have been found to be adequate under controlled conditions. However, the use of several synthetic hormones has given satisfactory results. When the eggs hatch, fry are collected from the breeding grounds and are transferred to fry ponds for rearing. Fry can be raised to fingerling stage in ponds and moved to fingerling ponds after they have grown to 20-40 mm in length.

3. **Stocking Ponds:** Stocking of *Pangasius* to market size takes a little less than one year after hatching. The ponds are prepared for stocking by controlling weed fish using mahua oil cake or tea seed cake or derris root powder. New ponds are

fertilized with mustard oil cake, limestone, and single superphosphate at the rates of 375, 100, and 25 kg/ha/ month, respectively. The stocking density depends mainly on the quality and quantity of the water supplied and desired size of the market fish. In ponds with dependable water supply, the stocking density of about 13,000/hectare is common. At this stocking density, the fish would weigh 600-700 g at the end of the growing season.

Polyculture involves the stocking of fingerlings of pangus and carps at the rates of 2,300 and 1,700 (*Catla catla* 500 : *Labeo rohita,* 700; *Cyprinus carpio* 250, and *Cirrhinus mrigala* 250), respectively with a total stocking density of 4,000/ha. Such culture system requires heavy feeding (@ 6% of the body weight of fish) and usually twice a day with high animal protein content (35 per cent). A well-managed pond is reported to give very satisfactory results with production rates varying between 2.0 and 3.0 tonnes/hectare/year.

4. **Harvesting:** Seining and draining are the two important methods of harvesting catfish from ponds. Draining is an effective method since it permits better management of pond soils, but in cases where pond refilling requires pumping, it involves an additional cost. On the other hand, seining is also permits partial harvesting. The use of mechanized seining equipment is very effective as it reduces the labor requirements.

6.12 Biology and Culture of Featherbacks (Figure 6.7)

Two species of fish such as *Notopterus notopterus* (Hump-backed razor fish) and *N. chitala* (Razor fish) are called as *featherback* and are considered as high-priced commercial fish. Of the two species, *N. chitala* is a very important food fish and is found in different freshwater ecosystems. The fish is widely distributed in marshes, lakes, reservoirs, and rivers of India, West Africa, Burma, Malaya, Pakistan, Bangladesh, Thailand, and Indonesia.

Characteristic Features

The body is strongly compressed, has a humped back and covered with minute scales. The color is coppery brown or greyish along the back with 15 or 16 silvery transverse bars. The head is small in comparison to body size and with large mouth. Snout is obtuse and convex. Eyes are moderate, superior and not visible from below ventral surface. Lips are thin with equal jaws. Dorsal, pectoral, and pelvic fins are very small. Anal fin is ribbon-like with 100-110 rays and confluent with the caudal fin. Its flesh is said to be uncommonly rich and well-flavored.

Habitat

The *N. chitala* is a rheophilic fish, but it has established itself in lakes and confined waters. They prefer the weedy reaches of surroundings. They also prefer the flood plains and stagnant backwater habitats. They generally move in shoals during day time for seeking shelter in the midst of vegetation. In ponds and puddles, they live in low-oxygenated waters by the help of their air-breathing habit; but in rivers, they live in fairly well-oxygenated waters.

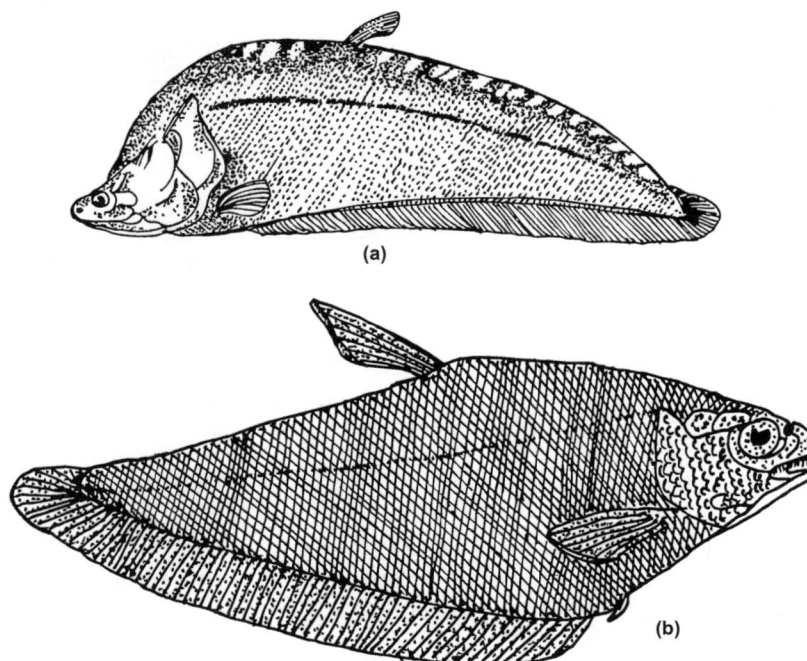

Fig. 6.7 : A high-priced commercial fish, (a) *Notopterus chitala*. It is widely distributed in different freshwater ecosystems of several tropical countries. It is unfortunate that the fish is now on the verge of extinction. Hence, there is urgent need to prevent them from destruction. (b) Another important featherback fish *N. notopterus* of commercial importance is also shown in this figure.

Food and Feeding Habits

The larvae consume zooplankton; but at the fry stage, they exclusively consume carp fry and aquatic insect larvae. Early fingerlings actively feed on mosquito larvae and carp larvae. The adult fish is purely carnivorous and feed on insect larvae, crustacea, and molluscs.

Reproduction

1. **Maturity and Mating :** The fish attains to first maturity in the third year of life, male being at the later part of the second year. A fully mature male gonad produces very small quantity of milt when gentle pressure is given on the abdomen. Mature female fish also does not discharge ova freely since the ova first fall into the body cavity before expulsion.

    In tanks and ponds, spawning pairs are observed near the spawning grounds where they are in search of suitable nests. Any masonry structure or stony objects near the banks are preferred. In these places, spawning is accomplished.

2. **Fecundity :** As a generalization, the average fecundity rate of this fish is 4,000/ kilogram body weight. However, the number of mature ova for a 4 kilogram fish has been estimated as 16,000. The fish exhibits parental care to their progeny.

3. **Spawning** : The fish is a seasonal spawner and is known to breed during summer (April-May) and monsoon months (June-August). The fish breeds both in lotic and lentic environments. In lentic environments, however, it breeds in shallow or moderately deep waters near the banks where submerged objects such as stones, brick-walls, tin containers, wooden box, and other hard objects are available to deposit glutinous eggs which permit firm attachment to these objects. The male fish then emits milt over the deposited eggs. About 500 eggs are laid by the female at a time and then the nest is very carefully protected by the parent fishes. The rate of fertilization of eggs generally varies between 60 and 70 per cent.

The fish has unique spawning habits. Only one ovary develops during breeding season with eggs of several different stages of development. Male fish exhibits persevering attendance on the fertilized eggs. By continuous fanning movement of the tail, the male fish removes sediment and other debris which are deposited over the eggs. The incubation period with an average temperature of 33° C varies between 5 and 7 days, however.

Culture Methods

The culture method of this species generally involves (i) the collection of seed either from natural breeding grounds or through hypophysation following refined methods, (ii) stocking of fingerlings in ponds for several months, (iii) the rearing of larvae in rearing ponds, and (iv) brood stock rearing and breeding by careful management. However, featherbacks are purely carnivorous and consequently, they are not at all cultured along with carps since they cause considerable damage to carps and other freshwater farmed fish. Therefore, monoculture of this fish may be initiated to enhance production for at least domestic consumption.

The fish forms an important fishery of all the major rivers of India and in some reservoirs (such as Hirakund). The fish can be found in deep and confined waters where carp culture is not feasible.

Since the fish is piscivorous in nature and feed on young fish, it is customery to stock carp fingerlings in polyculture ponds under judicious management and therefore, stocking of juvenile featherbacks as a minor population in polyculture systems must be reckoned with as a wise practice for culture of this species. And monoculture may, however, be possible if adequate seed is available to fish growers.

The species makes a considerable contribution to the featherback fisheries and aquaculture- in several states of India such as Assam, Bihar, West Bengal, and Orissa. But unfortunately, there is no organized fisheries for this fish in India.

Appraisal to Featherbacks

Since current technology is now available among experts, there is potential for the area concerned to produce and conserve them both for the table and future generations.

Whether this potential will be reached will depend on the wisdom that is used in manipulating and utilizing potential water resources for these species.

Due to pullution of most of the river systems, the population of these important species is declining to a considerable extent and their breeding grounds have been severely damaged. Therefore, farm–growing fingerlings should be stocked in unpolluted water bodies so as to improve fish stocks and to prevent them from extinction. At the same time, *in situ* conservation of these endagered fish is a comprehensive system of protected areas which are managed with different objectives. Of them, fish sanctuary is important which forms a solid basis for conservation of this species. It should be noted that fish landing records clearly indicate a sharp decline in many freshwater resources, the landing rate of the country's total inland fish production is about 5 per cent. This figure is highly alarming from extinction point of view.

The main requisite for increased production to transplant them from polluted regions to other ones or from rivers to lakes, ponds, and tanks to ascertain the guarantee of their survival. The range needed goes ones or from rivers from experts to the farmers. Researchers whose interest relates to the solution of water pollution problem are needed. Likewise, technicians and progressive farmers who are actively engaged in breeding and culturing programs are inevitable.

The fish has good market especially in West Bengal, Assam, and Bihar. While the back region of the fish contains numerous intramuscular spines, the abdominal part is highly flavored, good taste, and rich in nutrients. These attributes have made the abdominal part more lucrative to consumers. The abdominal part is so tasty that it is sold in the market at Rs. 190.00 per kilogram but the cost price of the back portion is very low, it is less than Rs. 50.00 per kilogram.

It is a matter of grave concern that the fish is now on the verge of extinction. This perilous situation has been forced fishery scientists to preserve them in their natural ecosystems. But in many regions, human needs have become so important that fish sanctuaries have already become impossible to protect the stock of fish. In order to survive, protected areas will either have to be economically valuable or ecologically inevitable. However, the best way of conserving this species in sanctuaries is to place an economic value on them. Sanctuaries are similar to savings accounts – the fish resources held in these areas are only precious if they are not harvested. Once harvested, the account will become empty and it can take several decades to fabricate again. However, if the stock of this species is not properly maintained right now, it would never be possible to observe them in the market and on the table and the fish will definitely be appeared in the Red Data Book of India.

6.13 Biology and Culture of *Anabas* (Figure 6.8)

Popularly known as the '*climbing perch*', the celebrated walking fish belongs to the genus *Anabas* found in India, Bangladesh and in some other Asian countries. It is well-known for its walking behavior (see Panel 6.1) and hardiness which permit easy handling in cultivation. In India, only three species have been reported: *Anabas testudineus, A. oligolepis* and *A. scandens.*

Characteristic Features

The body is laterally compressed and covered with ctenoid scales. It attains the length of 20 cm. While dorsal and anal fins are long and spinous, the pectoral and pelvic fins are small. Caudal fin is well developed. The operculum bears backwardly-directed spines. This species is distributed in freshwaters of India, Ceylone, Burma, Malaya Peninsula, and Archipelago.

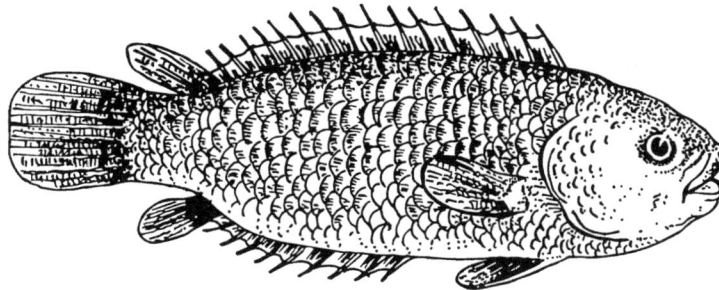

Fig. 6.8 : The Indian climbing perch, *Anabas testudineus*

Food and Feeding Habits

The climbing perch is a voracious feeder. It accepts almost any food from 'idli' to cockroaches. Though it is an omnivorous fish, it shows preference to insects and their larvae.

PANEL -6.1
Walking Behavior of *Anabas*

The most peculiar characteristic feature of *Anabas* is that the fish often goes on ramble over land and from pond to pond. Its eyes are capable of aerial vision. During walking, the fish uses its fins and gill cover. Although it can move on land in an upright position,the general movement is effected with the fish lying on lateral side. In this position, the tail fin and the gill covers are used. The gill covers directing the ground opens out and gains hold with the aid of its spinous margin. In the intervening time, the tail fin strikes against the ground, the fish jerks forward and the gill cover comes back to normal position. This process is repeated. It has been observed that both gill covers open out and close in at the same time. When the fish moves in an upright manner, the paired fins are well spread, gill covers open and close persistently and the tail fin lashes sidewise violently. The fish cannot move in a straight line.

PANEL -6.2
Structure of Grinding Apparatus of *Anabas*

The grinding apparatus of the fish consists of two sets of teeth, one on the roof of pharynx and the other on the floor. The dorsal set of teeth roofing the pharynx is in these patches. An anterior 'T'-shaped patch consists of a number of blunt and elongated conical teeth. Two patches of teeth are arranged side by side in close connection with the 'T'-shaped patch. In these two patches, a number of teeth with blunt surfaces are crowded in such a way that a hard, uneven surface is formed very much similar to the face of grinding stone. The ventral set of teeth which forms the floor of the pharynx consists of two triangular-shaped closely set teeth. Each patch is composed of a number of blunt and hard teeth. Paddy grains and insects are well ground between these two sets of pharyngeal teeth, one dorsal and the other ventral in position.

In general, food varies from filamentous algae to that of pure carnivorous nature. It should be pointed out that the presence of pharyngeal teeth are better used in crushing the food items. A well-developed grinding apparatus (see panel 6.2) has been developed in the pharyngeal region (Fig. 6.9) for efficient crushing and feeding. Margins of jaws are provided with rows of small conical teeth. They are helpful in killing and crushing insects. The fish is found in paddy fields where it feeds on paddy grains. In this way, it destroys paddy crops. The phenomenon of destruction of paddy crops is interesting too. Just before the paddy crop is ready, the plant bends. The fish is then jumps out to bite the grains. The fish swallows the paddy grains, then it makes grinding sounds which are clearly audible from a distance . As a result of grinding, paddy is dehusked, husk is vomited and the paddy grain is swallowed.

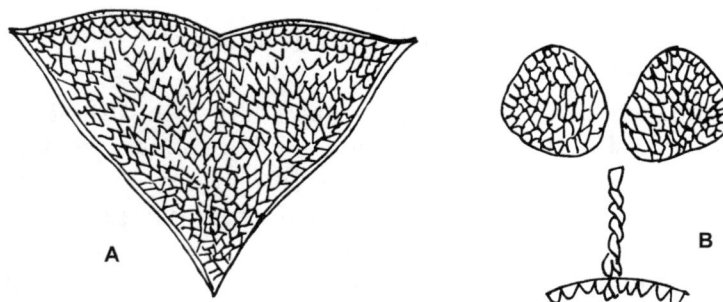

Fig. 6.9 : Grinding apparatus of *A. testudineus*. (a) Dorsal and (b) ventral sets of teeth

Breeding and Larval Development

The fish grows upto 30 cm in length. Sexual maturity starts with early rains and from the month of September, an abrupt decline in excellence is observed. Breeding occurs only during the monsoon season. The fecundity rate of this fish varies between 5,000 and 35,000. It is interesting to note that during rainy season, the fish migrates into carp nurseries where breeding takes place. This behavior is very important for seed production through natural phenomenon.

The induced breeding of *Anabas* under controlled conditions using carp pituitary extracts have been found to be successful in most of the states of India. At the time of monsoon, male and female fishes can be identified externally by the appearance of some features such as (1) the ventral region of the male fish becomes smooth, (2) the urino-genital region of the female is enlarged, and (3) while male shows distance between the base of pectoral fins smaller than the length of isthmus, the female shows this distance greater. Live and sexually matured fish can be collected from the market but if not available, spawners are generally collected from swamp and derelict waters. These are reared in 10 square meter cement cisterns and during rearing, fishes are daily fed with locally available balanced diet.

A fully-grown female weighing about 55g generates about 19,000 eggs. Males and females are injected carp pituitary extracts at the rate of 20 mg/kg of body weight. The fertilized eggs are small, yellowish in color, more or less circular (0.7 mm in diameter) and freely float on the surface at random. Eggs look like crystal clear tiny glass spheres with no adhesive filament or any adhesive surface. Eggs are transparent which obviously develops into embryos. The incubation period varies between 18 and 22 hours and at temperatures varying between 27° and 30°C.

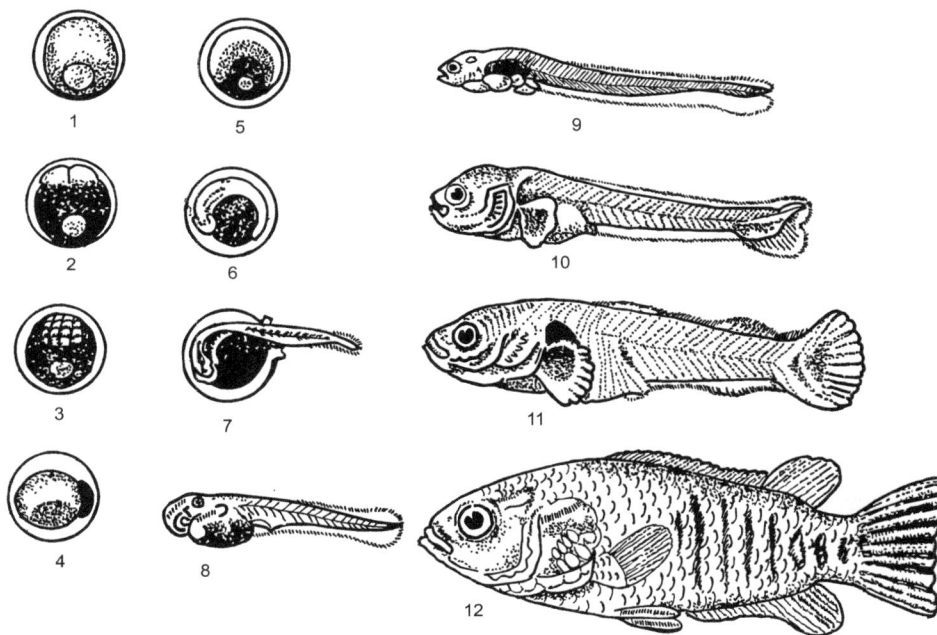

Fig. 6.10 : Life history of *Anabas testudineus*. 1-3, cleavage of egg; 4-7, formation of embryo within the egg; 8-12, larvae at different stages of development.

The larvae are small, 1.9 mm in length and float upside-down position at the surface of water with the yolk sac directed upwards at the water surface (Fig. 6.10) The larva is devoid of mouth. By the end of 5th day, the post-larvae attain to about 6 mm in length, heads are enlarged with prominent eyes. On 10th day, the length of post-larvae has been measured as 9 mm; caudal fin shows eleven distinct rays and dorsal fin with 9 soft rays. By 15 th day of incubation, it attains to 15 mm in number of length and scales are fully formed. The number of vertebral band increases from 7 to 10. By 20th day, the length of post-larvae becomes 24 mm, fins show their usual number of spines and rays. At this stage, it looks like a miniature adult.

Rearing of Larvae

Larvae require precise management over the entire development period so that high survival rate is achieved. At the end of 5 th day, larvae readily accept minute zooplankton, ciliates, and flagellates. Different species of zooplankton such as *Daphnia* and *Cyclops* are harmful to the young larvae of fish. Therefore, damage to the larvae can be considerably reduced by rearing them in a mosquito net basket which is suspended in water in such a way that the roof of the curtain remains within the water and the corners of which are fitted with bamboo poles immersed in pond water. This system will minimize the rate of mortality of early fry. Utmost care must be taken to reduce the effect of cannibalism among developing larvae.

Culture Methods

The fish can be cultured either in shallow derelict waters or in cement cisterns or in fixed/floating cages made of bamboo mats. Sometimes it is advisable to culture in barrel-shaped bamboo cages (3 m in length and 2 m in diameter).

1. **Cage Culture:** The size of the rectangular bamboo cage should be 2m × 1m × 1m where fingerlings are stocked at varying rates. Cages are generally used for grow-out and necessary fry or fingerlings are produced in ponds, lakes, and cisterns. Fixed and floating cages are used for grow-out of this fish. The stocking density varies with the size of the cage, but in fixed cages about 25 fingerlings/m^3 are generally stocked and the duration of culture varies from 4 to 10 months. The growth rate of fish is largely based on the productivity of water and management practices, which include the density of cages in the culture site and the distance between cages. However, the average growth characteristic of Anabas in a pond ecosystem is shown in Table 6.8. The average production of 3.0-7.0 kg/m^3 has been recorded from several farm ponds.

2. **Pond and Tank Culture:** The most Common and widely practised system of culture of Anabas is in earthen ponds and in tanks. In pond culture, attempts have been made to produce fish by stocking several species of catfish such as *Clarias* and *Heteropneustes* and the production is highly encouraging. In case of mixed culture, progressive fish farmers generally adopt moderate to high stocking density.

Depending on the management options, the stocking rates vary greatly. Monoculture is, however, most common in *Anabas* culture, and polyculture systems are considered necessary in some situations to make the operation economically attractive. In areas where climatic conditions are conducive to yield fish, high stocking density (at the rate varying between 50,000 and 1, 25,000/hectare) for monoculture has been recommended. The practice of stocking *Anabas* in ponds using traditional feed is a common form of monoculture in several regions of India. The stocking density, feed, and production potential of mono-and mixed culture systems of air-breathing fish are shown in Table 6.9. Table shows that the monoculture of *Anabas* generally gives an encouraging result and the production -varies between 125 and 1,800 kg/hectare while for mixed culture of this species with other catfishes and murrels, the production range is higher than that of monoculture – it varies between 525 and 2,250 kg/hectare.

Table 6.8 : Growth Characteristics of *Anabas testudineus* in a Farmed Pond

Age (Day)	Average growth (g)
60	5.7
89	6.4
116	6.7
146	7.0
179	9.6
207	10.5
231	11.8
267	12.5
290	21.5
329	26.0
360	29.0
387	31.0

Source: Jana and Das (1980b)

The culture of this species in cement tanks is most common in some Asian countries such as India, China, and Bangladesh. It is essentially a best culture method where culture systems can easily be manipulated with satisfactory results. The size of each tank measures as 10 feet × 8 feet × 8 feet and duration of culture period varies between 4 and 8 months, the duration period of 6 months is highly economical for the fish and hence considered by farmers. A thick soil bed of about 6 inches is prepared in each tank. Cattle manure at the rate of 10,000 kg/ha/ 6 month is added to each tank and is thoroughly mixed with the soil. Tank is then filled with, water and the level of water is kept upto 4 feet. Each tank is stocked with 600 fingerlings. A frequent water flow (15 day interval) is maintained in the tank which is sufficient to flush waste substances. Around 80 per cent of the stock attains a weight of about 50 g and can be marketed. Each tank can produce two crops per year with survival and production varies between 75-85 per cent and 8-14 kg/m, respectively.

Table 6.9: Mono-Culture and Mixed Culture of *Anabas* in Ponds

Area of pond (ha)	Stocking rate (Number/ha)	Feed used used	Duration of culture	Production (kg/ha)
0.03	80,000	-	3	125
0.03	60,000	Rice bran + mustard cake	6	1,800
0.02	1,25,000	Rice bran + mustard cake	12	700
0.04	40,000 + Hf*	-	10	525
0.03	70,000 + Hf* + Cb** (2 : 5 : 3)	Fish meal + rice bran	7	1,200 (11.5, 51.4, and 37.1 per cent respectively)
0.04	100,000 + Cb** + Hf*	Rice bran + fish meal	10	2,250
0.07	43,000 + Murrels	Mustard cake + rice bran + silk worm pupae		915
0.10	80,000 + Hf* + Cb**	Mustard cake + rice bran + silk worm pupae		1,550

Source: Data compiled from different published literatures
* *Heteropneustes fosilis*, ** *Clarias batrachus*

6.14 Biology and Culture of American Catfish (Figure 6.11)

A variety of catfishes such as the channel catfish, the blue catfish, the white catfish, the flathead catfish, the bullhead catfish are commercially important which are cultured in USA. Among these species cultured, the most important is the channel catfish. The channel catfish, *Ictalurus punctatus* is highly significant from culture standpoint because of its efficient food conversion, high production, and acceptance by consumers. As food, this fish stands the head of the class. Its firm, flaky, and white meat has an excellent flavor. However, some characteristic features of these catfishes are summarized below.

The Channel Catfish

The channel catfish, *Ictalurus punctatus* (Rafinesque) is distributed all over the USA and is abundant in the large streams and lakes. It ranges from Montana to the Ohio Valley and Southward through the Mississippi Valley to the Gulf of Mexico and to Mexico and Florida.

The fish is bluish or olive above, fading to silver on the sides and white on the belly. As the fish grows larger and reaches maturity, it becomes darker. The smaller fish generally have scattered dark circular spots on their sides but these are lost by the time the fish weighs 8 kilogram. Though the fish may reach a weight of 38 kilogram, eight kilogram is generally considered large. Three or four kilograms are the most desirable for the table.

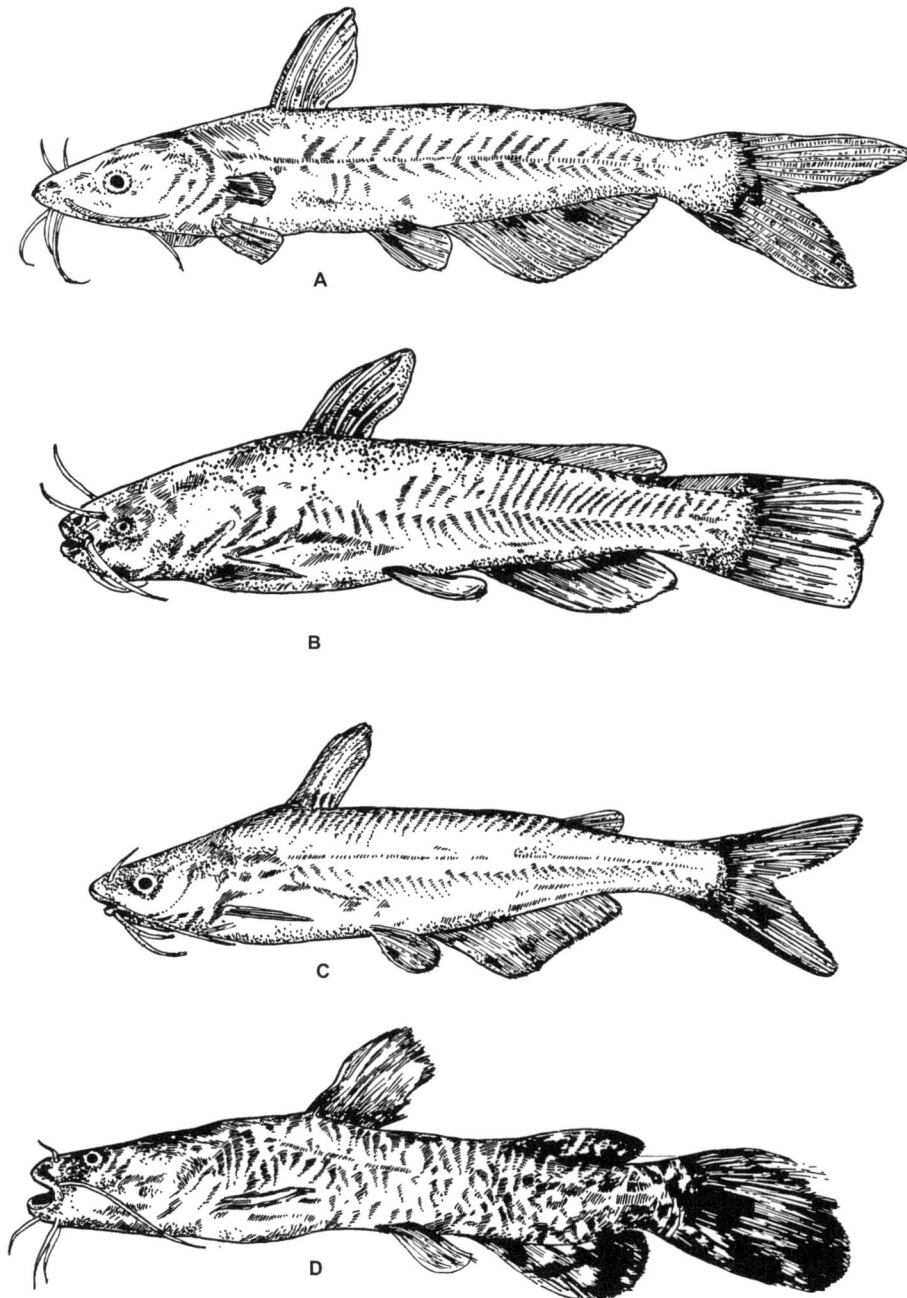

Fig. 6.11 : **Some common American catfish. These species are highly imporant because of high production and acceptance by consumers. In the USA, intensive systems of catfish farming are very common. A.** *Ictalurus punctatus*, **B.** *I. melas;* **C.** *I. furcatus;* **D.** *Pylodictis olivaris*

The fish generally makes its nest in a hole in the bank, beneath a submerged stump or in a similar schedule spot. Spawning takes place in the spring, usually in June, depending on the water temperature. The male performs the bulk of the parental duties such as cleaning the nest and a guarding the eggs and young. The eggs are about two-thirds the size of a garden pea and a three kilogram female will lay from 5,000 to 7,000 number of eggs. Eggs are adhered to each other and to the bottom of the nest, forming a firm gelatinous mass, and a rich golden yellow in color. During the 8 day incubation period, the male hovers over the eggs, fanning them gently and driving away intruders. The fry remain in the nest about a week under the watchful eye of the male and some observers have reported that the male continues to guard the young for a few days after they leave the nest.

The fry feed largely on insect larvae and gradually change to larger insects and other foods. The fish also consumes mussels and snails. During the day, they rest in deep holes and brush piles. At night, they move into fairly shallow water.

The channel catfish is very healthy, being free of parasites. The hardiness of the fish coupled with his carnivorous appetite enables the fish to thrive in water that is not suited to other game fish.

The Blue Catfish

The blue catfish, *Ictalurus furcatus,* occurs in the Mississippi River and Gulf Coast drainages westward into Mexico. It has spread into the Mississippi, Ohio, and Missouri drainages as far as Ohio and West Verginia, Minnesota, and the Dakotas. In Missouri, it is most abundant in the larger streams and in the impoundments built on them. The fish resembles to the channel catfish in having a deeply-forked tail, and in its bluish olive back, silver sides and white belly. The tail is long, with 31-33 fin rays, the outer edge is straight and it looks like a barber's comb.

The fish breeds in June when the water warms to 70 degrees and to make use of nesting places much like the channel catfish. The young reach a length of 4 inches by the end of their first year.

The fishes are omnivorous, eating just about anything available. Crayfish, frogs, small freshwater mussels, insects and their larvae, and small fish are included in the range of their food as well as a great variety of other materials, living and dead. They depend more on smell and touch than on sight to locate their food. Sense organs are located in the barbels which are helpful to them in food finding. This method of feeding is advantageous to a species which inhabits turbid waters where visibility is very low.

Though blue catfishes up to 45 or 55 kilograms are caught often enough to keep it interesting, smaller fish in the 4 to 8 kilogram range are more common and are better for consumption. The meat of the fish is firm and white, very similar to that of the channel catfish in flavor and texture.

The White Catfish

It is a race of semi-albino blue cats inhabits the lake and river systems of USA. They are pale silvery white in color with beady black eyes, but in other respects are typical blue catfish. They are locally calles as white catfish (*Ictalurus catus*), but the use of this name tends to create confusion since it is properly applied to another species of catfish native to the Atlantic Coast drainages.

The Flathead Catfish

The flathead is one of at least 11 known species of Missouri catfish. There are only a few of a family that has many members and is widely distributed. The family is not well-known because many of the species live exclusively in salt water and many others are exotic to North America.

The flathead catfish, *Pylodictis olivaris* (Rafinessque), has a number of common name such as mud cat, Morgan cat, flathead, and yellow cat. The head is always large, much flattened and the lower jaw protrudes like a bulldog. The adipose fin on the back is as long as the anal fin and the caudal fin is large and almost rounded in contrast to the forked tail of the channel catfish.

Fish weighing under 8 kilogram are usually a much-mottled slate color with some brown or yellow spots. Large-sized fish are generally yellow and may or may not be mottled.

The fish favors sluggish, large rivers and big lakes. It ranges throughout most of the Mississippi drainage from Minnesota south to the Gulf and in other large waters from Alabama to the Rio Grande and Mexico. The fish is more abundant in the southern states of USA. The bluehead catfish is considered the largest of catfishes because occasionally enormous individuals are reported. This species is, however, very important in commercial fisheries of the Missouri and the Mississippi rivers and that part of the St. Francis River which borders Arkansas.

The flathead's habits of nesting and loafing in cavities make this fish especially vulnerable to hand-fishing. The fish has very little chance to escape the expert noodler and taking a spawning flathead from its nest is as destructive and unsporting as shooting a hen quail brooding her eggs. A submerged stump, or a hollow log, and a pile of boulders is an excellent resting site for an old fish. Flatheads weighing 50 or 60 kilogram are not uncommon.

The Bullhead Catfish

Brown, yellow, and black bullheads are native to Missouri but the black is widely outnumbered than other species. Most black bullheads (*Ictalurus melas*) are from 6 to 10 inches long when they are caught. They are well distributed but are found most frequently in sluggish and muddy waters.

It is difficult to distinguish black, brown, and yellow bullheads. The brown bullhead, *Ictalurus nebulosus*, is a more slender fish and the base of its anal fin is about half the length of the body; the barbels are grey or black and the spines on the front fins have strong barbs. The back is dark yellowish-brown and is mottled with dark–green. It seldom grows more than a foot long. The yellow bullhead, *I. natalis*, grows somewhat larger than other members of the family and it is a chunky fish. The belly is bright yellow, the back is yellowish mottled with a darker color and the barbels are white. The black bullhead, *I. melas*, has grey or black barbels and its back is black to greenish brown. The sides are green to golden and underparts are greenish yellow or bright yellow but never white.

Bullheads are hardy and remain alive out of water for a long time. Although they may be found in the same area as other fish, they live and flourish where game fishes cannot survive and are found in waters in which oxygen content is very low. They have been known to survive in lakes that freeze almost solid.

These fishes consume insects, some crayfish, and other crustaceans, molluscs, and frog. Their vegetable diet consists of plant remains picked while searching about the bottom. They feed mostly at night.

Bullheads spawn in spring and early summer. They make nests on mud or sand in shallow waters, often using some natural depressions. The nest is guarded by a parent until the young are hatched. They are then cared for by the male, who herds them in schools until they attain a length of about one inch. Black bullheads are prolific breeders and under favorable conditions, they build up large populations. Female bullheads lay from 6,000 to 14,000 eggs.

Bullheads are harvested by pole and line method. The greater part of the catch comes from streams in the north and west and from small tributaries of the Missouri and Mississippi rivers. Commercial fish farmers catch substantial amounts of fish. The flesh is firm and tasty when the fish is taken from unpolluted waters but when collected from polluted waters may have a disagreeable and muddy flavor; but if the catch is kept alive in clean water for a few days, the flavor is improved.

Culture Methods

In the USA, pond and raceway culture systems are generally employed and the intensive systems of channel catfish farming are very common. Raceways are prepared by concrete or earth. The degree of production intensity depends largely on the availability of water supply. Small-sized raceways with water supply of high volume and high velocity are employed for highly intensive production, but in case of larger raceways, a relatively lower rate of water flow is used for semi-intensive culture systems. Circular and linear tanks are also used for channel catfish culture.

The culture of channel catfish in ponds to table size takes a little less than two years after hatching or at least one year from the fingerling stage. The usual market size generally varies between 500 g and 1.5 kilogram. Fingerlings are generally stocked in grow-out ponds in the spring and are harvested in about 8 months in October or November.

Cage culture of channel catfish is a greater desirability when the cages are established in open waters rather than in ponds. Fish are stocked in cages at the rate of 65 fingerlings per cubic meter. Complete floating feeds are given at the rate of 4 per cent of body weight per day.

Tank culture of channel catfish is reported to give very encouraging results. A circular tank of 6 metre diameter and 0.6 metre depth affords the same production as a 0.5 hectare pond. High stocking rates are possible if tanks are provided with aerators. Complete feeds are fed twice a day at the daily rate of 3.5 per cent of the body weight when the water temperature varies between 23 and 27° C. Compulsory daily feeding makes the catfish farming more attractive and profitable to farmers. Protection of fish stocks from bright light seems to give better growth rates and hence the tanks should be covered.

1. **Broodstock Management:** Generally brood fish of channel catfish are about 3 years old and weigh at least 1.5 kilogram. A female of this size can be expected to spawn 6,000 -9,000 eggs. Although brood fish are collected from natural waters, farmers prefer cultured brood fish. The brood fish are reared in brood ponds on a diet rich in animal protein in a healthy environment. Smaller fish of about 1.5 kg are stocked at the rate of 350-450 kg/ha. A well fertilized pond can sustain about 340 kg/ha but supplementary feeding is essential when fish are stocked at higher rates. Daily feeding at about 4 per cent of body weight is recommended. Additional food consisting of fresh or frozen meat, beef liver or other meat products is given at the rate of 10-15 per cent of the body weight.

2. **Spawning and Production:** The channel catfish can be spawned in ponds, especially designed pens, and in aquaria. Condition for a suitable nest is essential for pond spawning since natural spawning sites such as holes in banks or submerged stumps in natural habitats may not be available in ponds. Nests may be made of wooden boxes, hollow logs, milk cans, metal drums, and concrete tiles. The number of nests required to construct is dependent on whether fertilized eggs are allowed to hatch in the spawning pond itself or to be removed and hatched indoors in hatching troughs. The nests are placed around the edge of the pond at depth varying between 20 and 140 cm with the open end facing the centre of the pond. The number of brood fish to be stocked depends on whether the eggs will be hatched in the pond or will be removed from the hatching indoors. If hatching occurs in the pond, a stocking density 50 female brood fish is per hectare is generally recommended. In some situations, however, the stocking density can be raised to 120 females/hectare.

To ensure mating between selected pairs of brood fish, pen spawning is generally recommended to farmers. Different sizes of pens are used, but pens are not larger than 2 m X 3 m X 1 m. Pens are made of wood and welded steel wire mesh (2.5 X 5 cm mesh) and are embedded in the pond bottom with about 0.3-0.6 m of the pen above the water. A spawning nest is kept in each pen where a selected pair of brood fish is stocked. After spawning, the female fish is removed from the pen and returned to the brood pond. Either the mass of eggs is collected and hatched indoors or the male fish is left to hatch the eggs. If hatched in the pen, fry swim out of the nest through the wire mesh into the pond. If the eggs are removed for hatching, each pen is stocked with a new male and female pair or a second female after removing the spent female.

After spawning, fry are collected, kept in containers and are transferred to fry ponds for rearing. Artificial hatching methods in pen spawning are desirable because spawning nests can be further utilized rapidly and the farmer will predict the exact number of eggs to be hatched. At the same time, it will drastically reduce the loss of predacious fish and insects.

Hatching troughs, made of sheets of stainless steel or aluminium, are widely used in commercial farms. Each trough is segmented by partitions across the middle and the each segment has an inlet and outlet pipes. These segments permit to segregate eggs of different stages of development. The eggs are placed in a deep basket, made of 0.6 cm meshed cloth and suspended by wire from the side of the trough. Paddles are attached to a revolving shaft that cause the movement of water on the eggs.

Though fry are reared in ponds, trough rearing is strongly recommended because it permits greater control over the entire rearing period and fry are less exposed to predators. Troughs are 4 X 40 X 40 cm in size, made of wood, metal, plastic, or fibre glass. A water flow of about 0.06-0.3 l/minute and a temperature range of 24-30°C are maintained. About 8 days after hatching, floating and non-floating types of commercial feeds having 30 per cent protein and other nutrients are fed six times a day. Many farmers use diets that contain about 50 per cent protein. As fry grow, the protein percentage is gradually reduced. In a few weeks fry attain a length of 3.5 cm (fingerling stage).

3. **Culture of Fingerlings in Stocking Ponds:** The channel catfish fingerlings weighing an average of 20 g each are stocked in well-prepared stocking ponds at rates ranging from 1,000 to 35,000 numbers per hectare; of course, the suitable stocking density varies between 1,000 and 10,000. Fish are stocked in the month of January-February. Fish are fed daily at 6 per cent body weight, adjusted every two weeks for growth. The feed generally consists of floating pellets containing more than 30 per cent protein. The culture period usually varies from 6 to 10 months depending on the water retention capacity of pond soils because the water retention time in a culture pond has a marked influence on the pond fertilization and the residual effect of lime applications. This is due to the fact that if water is rapidly flushed through ponds, nutrients may be released before they can affect phytoplankton populations and the role of liming will be decreased to a significant extent.

(i) **Pond Preparation:** Before stocking with catfish fingerlings, wild fishes are eliminated from ponds by rotenone treatment. Ponds are then stocked with requisite number of fish. New ponds are fertilized with liquid or granular fertilizers. Fertilization of many catfish ponds in the USA usually consists of 10-12 periodic application of chemical fertilizers, with the first application being made in early spring or late winter. Fertilization is generally initiated between 15th February and lst March at lower latitudes ($30°$ to $35°N$), but the application of fertilizers is delayed until end March or later at higher latitudes. Fertilization beginning in mid-February does not soon bring to light plankton in shallow (average depth 1 metre) and small (less than 0.2 ha) ponds and in deeper ponds, phytoplankton blooms also do not develop until May or June even after several application of fertilizers.

It should be kept in mind that liquid fertilizers are more effective than granular ones for catfish ponds. Because of differences in soil and environmental conditions among localities, the benefits of a fertilization technique may vary greatly. In cases where liquid fertilizers rare not available, granular fertilizers are used in many places. Generally diammonium phosphate ($N:P_2O_5$ K_2O ratio of 18 : 46 : 0) is applied by broadcasting at 20 kg/ha/month.

(ii) **Feeding:** The channel catfish are fed expanded feeds that are manufactured by extrusion processing, which allows them to float. This type of feed is advantageous because the farmer can see how much the fish are consuming. In most commercial fish farms the feed is blown onto the surface of the water using pneumatic dispenser mounted on or pulled by vehicles. Feed is scattered over a large area to provide feeding opportunities for as many fish as possible. It is desirable to feed on all sides of the pond, but this is not practical because prevailing winds dictate that feed must be distributed along the upwind side to prevent it from washing ashore.

Feeds are given once or twice a day when water temperature is $25°C$. The additional time and management required for multiple daily feeding make the practice unattractive to most farmers. During disease episodes or during extremely hot weather when feeding activity is poor, it may be beneficial to feed every other day or every third day.

Feed allowance is affected by several factors such as fish standing crop, fish size, water temperature, and water quality. Water temperature and fish size have an impact on feed consumption by the catfish. Feed consumption increases as the water tanperature increases until a temperature of about $32°C$ is reached and subsequently begins to decrease. As the fish size increases, the feed consumption as a precentage of the body weight decreases and the feed conversion efficiency is reduced (Table 6.10)

Fish farmers limit or restrict feeding of catfish in ponds during the winter. Since the metabolism of fish is a function of temperature, catfish grown in temperate regions do not feed as much or as consistently in the winter months as during the warm season. Catfish farmers, in the United States generally follow the

**Table 6.10 : Feed Consumption (FC) and Feed Conversion Ratio (FCR)
for Different Sizes of Channel Catfish**

Fish size (g)	FC	FCR*
27	4.0-4.5	1.1-1.2
45	3.5-4.0	1.3-1.4
136	2.5-3.0	1.4-1.6
272	2.0-2.5	1.6-1.8
340	1.5-2.0	1.8-1.9
454	1.3-1.5	1.9-2.0
908	1.1-1.2	2.0-2.2
1362	1.0-1.1	2.2-2.4

* Total feed applied divided by net fish yield

following feeding regimes for fish held in ponds during the entire culture period: (i) regular feeding but at a reduced daily allowance, (ii) feeding is discontinued when water temperature decreases to a designated level in the fall and feeding is resumed following spring and (iii) feeding is performed only on warm days during the cool weather season. The feed is a three-sixteenths inch diameter pelleted formulation containing high kilo calory (1,200-1,500 kcal) and at least 35 per cent protein with one-sixth of the protein supplied by fish meal. Maximum feeding rate of 112 kg/ha/day is generally recommended for channel catfish ponds stocked with fingerlings at the rate of 17,000/hectare.

(iii) Aeration : (See Panel 6.3). Fish consume more feed and grow best at dissolved oxygen concentrations near air saturation; of course, it is not possible to maintain dissolved oxygen concentrations near saturation at night. Catfish production increases with increase in the average concentrations of dissolved oxygen at dawn. Since high feeding rate is executed in catfish ponds, emergency aeration should frequently be applied particularly during daily morning hours to prevent dissolved oxygen depletion and fish mortality. For this purpose, two types of aeration devices are employed in channel catfish ponds such as emergency aeration and supplemental aeration. For emergency aeration, large paddle-wheel aerators and pumping sprayer devices are the most effective types which will definitely increase the production of channel catfish.

(iv) Production: Culture of catfish accounts for about two-thirds of the commercial aquaculture production in the United States and the production of farm-raised catfish reached 2,55,000 tonnes in 2009. At present, production ranges from 4,000 to 7,000 kg/ha. The increased yields can be attributed to higher stocking densities and improvemtnt in feed formulations, feeding practices, water quality management, and disease control. In addition, a multiple-batch cropping system is used in which fish of different sizes and age groups are present in the pond simultaneously.

Catfish production has been found to increase in direct proportion to feeding rate up to a rate of 112 kg/ha/day. At this feeding rate, the net production in ponds has been estimated as 6,500 kg/hectare. Large (several hundred hectares) and medium (1-2 thousand hectares) sized farms with the development of specialized feed mills and catfish process in plants have made catfish farming a profitable and stable production-related industry in the United States. It should be kept in mind that about 95 per cent of the catfish are produced in Mississippi, Arakansas, Alabama, and Lausiana in approximately 1,70,000 hectare of catfish ponds.

PANEL-6.3
Aeration Device in Fish Culture

Aeration is the term applied to a device that cause mixing of air in the farmed ponds. Many cultivated fish species in tropical regions are gill-breathing but some catfish and murrels are air-breathers. On the contrary, many farmed fish can thrive upon the available dissolved oxygen in pond water, otherwise fish tend to suffocate.

In intensive and semi-intensive fish farming systems where the stocking densities are high enough, per capita availability of dissolved oxygen to fish is very low. Massive inputs of feed also deplete the content of its oxygen. To enhance the dissolved oxygen in water, artificial methods are adopted to cater the pond water with much dissolved oxygen enough for the growth of fish biomass.

1. Time of Aeration Required:

(a) When the dissolved oxygen levels in water fall short of requirement, the pond urgently requires aeration. This is due to the fact that low dissolved oxygen level can cause stress and mortality of fish. It should be kept in mind that the dissolved oxygen levels below 1 mg/l and/or above 7 mg/l lead to physiological stress which could prove to be fatal. As a generalization, the hypo-and hyper-oxygen conditions lead to asphyxiation and gas-bubble diseases of fish, respectively. Therefore, in order to maintain the critical balance of dissolved oxygen level, aeration of pond water is inevitable.

(b) The early morning is the most critical time of aeration when the dissolved oxygen level decreases over-night by aquatic plants and animals.

(c) In cloudy days when the intensity of sunlight is very low, the dissolved oxygen in water also becomes very less due to decrease in the rate of photosynthesis.

(d) In eutrophicated ponds, fish have a tendency to move vertically to catch breath with an open mouth. As the sun rises, this condition cease temporarily; but the situation will again recur in the next morning. In this moment, emergency aeration is badly needed.

2. Equipments Used for Pond Aeration:

A number of equipments are available in the market for pond aeration and can be categorized as follows:

(a) **Paddle Wheel Aerator :** Paddle wheel aerator (Fig. 6.12) is one of the most common types which is widely used. A floating, electric paddle wheel aerator consists of floats, a platform, motor, speed reduction device, coupling, paddle wheel, and bearings. The wheels disturb water surface briskly that make the air to contact water. Adoption of proper aeration device and its optimal use would definitely help to reduce the cost of fish production. However, there is consideratable variation in the design of the paddle wheel and in the speed reduction mechanism of the motor.

Cost-effective aeration device in fish culture involves (i) the selection of multiple wheel aerator for intensive fish farming system, (ii) rigid impellers of aerator's wheel for proper aplash of water to make the air-water contact more effectively, (iii) the setting of the aerator in balanced condition for good splash, and (iv) operation of the aerator when necessary.

(b) **Vertical Pump Aerator:** This aerator consists of a sub-mersible electric motor (3 kilowatt for fish culture) with an impeller fixed to its shaft. The motor is suspended by floats and the high-speed impeller (1,730-3,450 revolution per minute) jets water into the air for effective aeration.

(c) **Pump Sprayer Aerator :** It is composed of a high pressure pump (10 kilo-watt) that discharges water at high velosity through one or more orifices for effective aeration. The impeller speed varies from 500 to 1,000 revolution per minute.

(d) **Propeller-Aspirator Pump Aerator:** The most important components of this aerator include (i) an electric motor, (ii) a hollow shaft that rotates at 3,450 revolution per minute, (iii) a hollow housing inside which the rotating shaft fits, and (iv) a diffuser and an impeller are connected to the end of rotating shaft. During operation, the high velocity of impeller accelerates water to cause a drop in pressure within the rotating shaft. As air is forced down the hollow rotating shaft by atmospheric pressure, air bubbles exit through the diffuser which diffuses and enter the turbulent water around the impeller

(e) **Diffused Air Aerator:** It is composed of an air blower to afford air to diffusers suspended in the water or placed in the pond bottom. Various types of diffusers such as porous ceramic tubing, porous paper tubing, perforated rubber tubing, perforated plastic pipes, ceramic dome diffusers and the like have been used. These aerators generate a large volume of air to effect aeration. Diffusers are frequently clogged with dust and drift particles that are carried with air supply. Slime of bacteria may plug the outlet of diffusers. Often, calcium carbonate forms a deposit over the diffuser which clogs the pore outlet in hard and saline waters. Therefore, washing of the diffuser is badly needed.

(f) **Propeller-Type Aerator:** This type of aerator is most efficient in aeration capability and it sprays water into the atmosphere. A cost-effective diesel/electrically operated propeller-type portable aerator has been developed in India.

3. Mechanism of Aeration:

A motor-operating propeller of an aerator moves large volume of water into the atmosphere. The water is propelled through air, broken down into small droplets and then returns to the surface of water that cause wave action. This increases the surface area of the pond and permits more water exposure to atmospheric oxygen. The aeration process involves destratification of thermocline in ponds resulting in thermoregulation of the entire pond system. This makes enough freedom for the fish stocks to move in any direction or manner. In general, aeration to a level of about one horse-power per surface acre (three-fourths of an hectare) should be enough for the fish.

Although paddle-wheel aerator is one of the most popular alternatives for shallow and small ponds but progressive fish farmers generally prefer propeller-type aerator due to its high efficiency. The propeller breaks the thermocline in the pond ecosystem diffusing air into the bottom and covers the entire area of the pond. A paddle-wheel aerator, in contrast, oxygenates only a small layer of the water column. A good aerator indicates that it will work most efficiently, consume less power and inject air directly to pond water. It should be kept in mind that aeration not only provides more than enough oxygen to the pond water but also removes excretory products of fishes and very unpleasant gases developed in ponds.

To sum up, efficient aeration is cardinal for any system of fish cultivation including hatcheries for the maintenance of a healthy pond, providing a source of oxygen not only for the fish but also for the beneficial bacteria in the pond mud. Using an air pump to pump air or an aerator can help to aerate the fanned pond. The bubbles produced cause surface ripples that increase the area exposed to the air where exchange of oxygen can takes place.

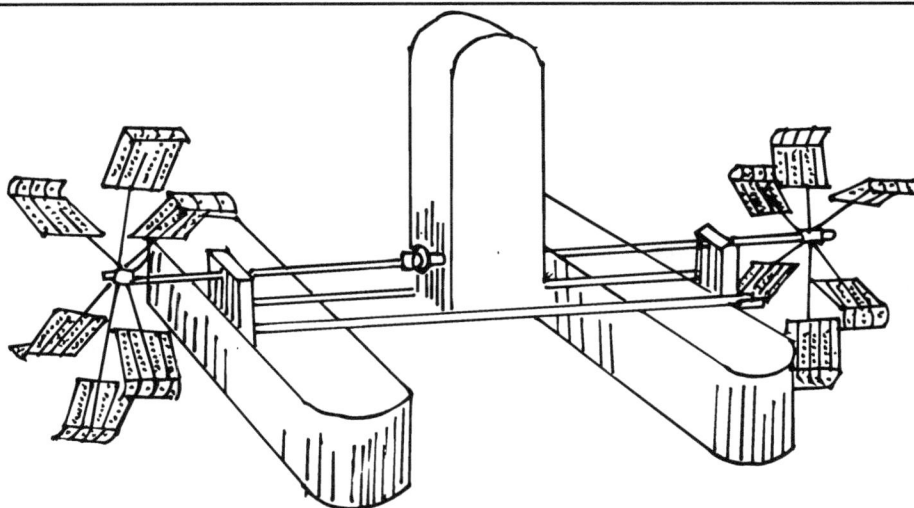

Fig. 6.12 : A paddle-wheel aerator

6.15 Economics of Catfish Farming

The cost of investment for the establishment of catfish farms is comparatively high and since the need for protein-rich feeds, supplies of quality water and the employment of skilled labor, production costs also tend to be high. In most catfish farms, these are counter-balanced by the extreme volume of production and reasonably high prices of the variegated products that can be marketed. Therefore, profitability is likely to be dependent on several factors and consequently, it is not so easy to generalize on the levels of return that can be expected. A well-managed and well-run catfish unit with good marketing should return between 25 and 30 per cent profit on turnover and pay off the invested capital in less than three years.

Compared to 6 months cycle of production for portion-size fish, the culture of large-sized fish furnishes higher return per tonne of fish produced in regions where there is demand for them and for market. As a generalization, farms which attempt series of culture operations starting from hatchling production to market fish are far more profitable than those which execute only one part of the operation. Likewise, larger enterprises exhibit higher returns. The returns on investment are not lucrative in case of smaller units.

The economics of catfish farming show considerable variations between farms, depending on the location and culture systems adopted. In general, culture in large shallow ponds is definitely the most profitable system. Production has to be at least 1,000-2,000 kg/hectare to make a judicious profit. Some farmers adopt integrated farming systems to yield better returns.

6.16 Biology and Culture of Murrels (Figure 6.13)

Murrels belonging to the family Channidae (= Ophiocephalidae) are highly regarded as food fish in the South and Southeast Asian countries. These fishes are also called as *snakeheads* because their heads shows similarity with the head of a snake. They live in ponds, rivers, ditches, pools, swamp, and derelict waters. During rainy season, they migrate from one pond to the other. Their ability to breathe atmospheric oxygen makes it possible to keep them in live condition for long periods out of water and to sell them alive at high prices in the market.

Characteristic Features

The body is elongated and covered with two types of scales such as cycloid scale and ctenoid scale. Head is depressed and covered with large scales. It may grow to the length varying between 15 and 45 cm. Dorsal and anal fins are single, long, and without spines. The mouth is terminal in position. The tail is homocercal. Murrels are distributed in different freshwaters of India, Burma, Ceylone, Malaya, Thailand, China, Pakistan, Philippines, Bangladesh, and Islands of East Archipelago.

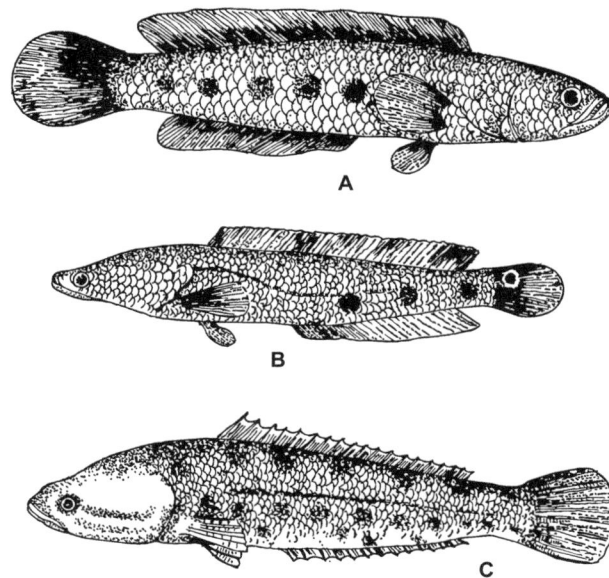

Fig. 6.13 : Some common murrels. (a) *Channa straitus;* **(b)** *Channa marulius;* **(c)** *Channa punctatus*

Varieties

Though several species of murrels are found in different freshwater bodies, only three species that follows are very important from culture point of view.

1. *Channa punctatus*: The fish commonly called *spotted murrel*, is small and grows up to 25 cm in length. The fish is abundantly found in pools, ditches, and ponds. Small fishes, insects, shrimps, and small gastropods (such as *Lymnaea, Viviparus,* and *Planorbis*) are consumed by this fish. Breeding commences in April and continues till September. Fecundity rate varies between 2,500 and 25,000.

2. *Channa marulius* : Commonly known as *giant murrel*, and is found in lakes, reservoirs, rivers, and swamps. It . grows up to 125 cm in length. They consume small to medium-sized living' animals as food and exhibits cannibalistic behavior. Breeding starts from March and continues up to October. Fecundity rate of this fish varies from 2,500 to 40,000.

3. *Channa striatus:* Commonly known as *stripped murrel* and is highly economical than the other two types. It grows up to 85 cm in length. The fish consumes tadpole, small fishes, frogs, and insects as food. The fish breeds in tanks and ponds almost throughout the year particularly during monsoon season. Fecundity rate generally varies between 2,500 and 30,000.

In all the cases, however, sexual maturity is attained at the end of first year. Fertilized eggs are float for a while and then laid in the marginal areas of water bodies that are infested with floating and emergent plants. Fertilized eggs and fry are guarded by the parents till they attain a length of about 30 cm.

Induced Breeding and Larval Rearing

Farmers concern over induced breeding of murrels has resulted in mass production of eggs and hatchlings. Different inducing agents such as carp pituitary extracts, human chorionic gonadotropin (HCG) and ovaprim at different rates ranging from 30 to 130 mg gland /kilogram female in two injections have been found to be adequate. In general, however, carp pituitary extracts are injected to mature male and female fish at the rates of 20 and 100 mg/kilogram body weight, respectively. The efficiency of fertilization of *Channa striatus*, for example, is higher when ovaprim and HCG + fish pituitary extracts are used. The fertilization rates are 95 per cent in case of ovaprim and that of 85 per cent in case of HCG + pituitary extracts. However, the carp pituitary extracts can no longer be taken into consideration because ovaprim and other synthetic inducing agents are now available in markets.

Hatching takes place within 20 hours of injection. After absorption of yolk sac by about 5 th day, intensive care is taken for hatchlings. Since larvae with their heavy yolk sac are not able to move fast, they are very much susceptible to attack by copepods (particularly *Cylops spp*). Therefore, caution must be taken so that this situation can be avoided. For better survival of larvae, suitable 1arval feed amd manipulation of stocking density per unit water area must be taken into consideration. Though feeding of murrel hatchlings with boiled chicken egg-yolk has been found to be suitable under laboratory conditions, the costs involved in large-scale production of murrel under farm conditions may negate the benefit achieved. Hatchlings are, however, ready to consume zooplankton when their air-breathing habit is developed.

Culture Practice

The success of murrel culture largely depends upon the availability of several species of forage fish such as *Oreochromis mossambicus*, *Oxygaster bacaila*, *Puntius* spp., and minnows. The availability of such forage fish will definitely results in high production of murrels. Being highly predaceous and cannibalistic, murrels are generally cultured under monoculture system using, as far as feasible, stock of the same-size group. In some Southeast Asian countries, such as Taiwan, murrels are cultured in carp and tilapia ponds with the presence of forage fish. In several regions of Tropical Asia, farmers have started integrated murrel culture with poultry and pig.

1. Culture in Ponds: Murrel culture in ponds involves several steps that follow.

(i) Broodstock Management: Before the commencement of monsoon season, broodfishes are collected from natural habitats and are stocked in specially-prepared cement tanks (3m X 1 m X 1m) . Good and unpolluted soil of 4-5 inches thickness is provided in each cistern. To facilitate continuous flow and exchange of water, an inlet at the top and an outlet at the bottom of the tank are provided at a desired level. Fishes are daily fed with boiled slaughterhouse offal, trash fish, rice bran, and oil cake at the rate of 6 per cent of the body weight of fish. The preferred size of murrels varies between 600 and 1000 g. These sizes are considered for induced breeding programs.

(ii) Seed Collection: Though seeds are collected from natural breeding grounds, seeds can also be procured through induced breeding under controlled conditions. In the breeding environment, fry and fingerlings move in shoals for search of food in marginal areas of the pond which can easily be seen from a distance. The entire shoal is collected by a fine meshed net. Fry of *Channa punctatus* can be identified by the appearance of a golden yellow longitudinal band running from the posterior extremity of the orbit to the caudal base and a yellow line extending from the snout to the base of the dorsal fin. On the other hand, fry of *C. marulius* can be distinguished by a distinct orange-yellow longitudinal band running laterally from the posterior part of the orbit to the tip of the caudal fin. The body color of *C. striatus* fry is vermillion red. On each side of the body, there is a reddish golden-yellow longitudinal band and a dark band situated ventral to the body.

(iii) Stocking of Fry and Fingerlings in Ponds: Fry (25-30 mm in length) are stocked in well-prepared nurseries where they are stocked for 60-90 days. When they attain fingerling size, they are transferred to the stocking ponds. Before stocking the pond with murrel fingerlings, following precautions would have to be to kept in mind. *Firstly*, stocking ponds must have to be infested with marginal weeds where aquatic insects can thrive well. *Secondly*, adult forage fish (such as *Orechromis mossambica* should be stocked at the rate of 1,400 number/hectare. *Thirdly*, for the production of plankton populattions, ponds must be manured with compost or farmyard manure or cow manure at the rate of 25,000 kg/hectare/year which is dividid into several lots. *Fourthly*, fingerlings (10 cm in iength) are released in July-August at the rate of 3,000/ha. *Fifthly*, artificial feed should be provided with the mixture of rice bran, oil cake and chopped trash fish in equal proportions (1 : 1 : 1) and *sixthly*, ponds should be fenced with wire net or bamboo slits up to a height of 60 cm to prevent fish populations from migration elsewhere.

Under Indian conditions, the recommended stocking rates of *Channa punctatus*, *C. striatus* and *C. marulius* are 40,000, 30,000, and 20,000/hectare, respectively. These stocking densities are adopted in areas where semi-intensive farming systems are adopted. In case of intensive culture systems, the stocking density is higher it is 80,000, 60,000, and 40,000, respectively. High stocking densities require a healthy ecosystem and a highly balanced but nutritive feeds to enhance the growth and survival rates. Therefore, growth-promoting substances such as cobalt chloride should be thoroughly mixed with the feed and than broadcast over the ponds twice a day.

Mixed culture of one or two species of murrels and *Heteropneustes fossilis* in ponds using supplementary feeds containing fresh silkworm pupae and trash fish have been found to be effective in producing appreciable quantities of murrels. Culture of four species of carnivorous fish, for example, such as *H. fossilis, Channa punctatus*

C. *marulius*, and C. *orientalis* in the ratio of 2:2.5:2.5:3 with a stocking density of 34,500 fingerlings/hectare have produced murrels to the tune of 2,300 kg/ha/year.

(iv) **Growth and Production:** The growth of murrels depends very much on the climatic conditions in the farming area. As a generalization, however, the growth of C. *punctatus*, C. *striatus*, and C. *marulius* under Indian conditions varies between 150 and 200, 250 and 300, and 350 and 400g in 6 to 8 months, respectively. The production range of murrels varies from 6,000 to 8,000 and 2,500 to 4,500 kg/ha in intensive and semi-intensive culture systems, respectively.

2. **Culture in Banana Fields (Figure 6.14) :** Banana plant is not only considered the world's largest herb but also is one of the most important and well-known fruits in warm regions and occasionally found on the grocers' shelves. Since the cultivation of banana plants in most of the tropical regins is on the increase, fish culture can easily be taken up with the banana plants for either simultaneous or alternate production of fish and banana from the same place. In Asia and Southeast Asia, the culture of murrels and other wild fish of commercial importance in different agricultural fields is an old-age practice. Culture of carnivorous fish in different ecosystems and their production trends depict that the production potential would be large enough to support nutrition in rural areas. It has, however, been observed that murrel production can be increased several folds if agricultural fields are well utilized.

Fig. 6.14 : Culture of murrels in Banana fields. Banana plants are planted in rows and a trench is constructed between the rows. All trenches are filled with water in adequate depth. Selected fish species are stocked in each trench. Such farming system permits the production of fish and banana from the same place.

The cultivation of murrels in banana fields is very simple and is highly economical. A trench of 2.5m wide and 0.5m depth is constructed in the field for stocking fish. At the same time, water supply must be ascertained to banana fields. Single species of fish particularly *C. straitus* or *C. punctatus* generally stocked at the rate varying between 10 and 20 per square metre (7,000 number/hectare). Fishes are daily fed with chopped slaughterhouse offal. After six months, fishes are harvested.

(i) **Production of Fish and Banana :** Alternate production of fish and banana permits better care of fish and the use of mechanical means for the banana. It also permits an increase in depth of the water during fish production which is highly favourable to growth and reproduction of murrels. After tthe banana has been harvested, the field is transformed into a temporary pond and at the same time, dykes are constructed. These temporary ponds may be used as breeding grounds for murrels.

Production of murrels in banana fields varies considerably and depends on the methods of exploitation, species cultured, soil fertility, and the management protocol adopted by farmers. However, where the method is more widespread, it gives low to moderate production. Simultaneous production of murrels and banana varies from 50 to 270 and 17,500 to 19,000 kg/hectare, respectively. Though production volume of fish is very low in some regions, it is highly economical nonetheless.

(ii) **Loss Ratio of Murrel Culture in Banana Fields :** Murrel cultivation in banana fields has a higher loss ratio than when ponds are used. It is, however, generally between 40 and 50 per cent for fingerlings and 30 and 40 per cent for the table fish. The principal causes of this are the presence of predatory animals and decomposition of plant organic matters. The accumulation of organic matter far in excess than the tolerance limit may not only leads to stunted growth of fish but crop failure as well. The loss can to some extent be reduced by using sufficiently large-sized fields to 10 hectares at least for murrel culture.

6.17 Culture of *Channa micropeltes* (Figure 6.15)

This large-sized murrel is widely distributed in the North-Eastern Regions of India. It lives in numerous tributeries of the river Barak and Brahmaputra. The fish has good demand for the table since the flesh contains fatty substances and has excellent flavor.

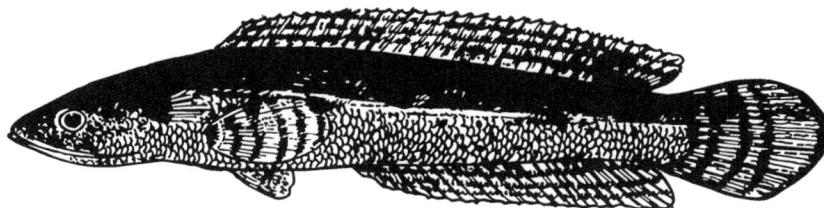

Fig. 6.15 : **A large-sized endangered murral, *Channa micropeltes*. It is widely distributed in the North-eastern regions of India. Though it has good demand for the table, the importance of its culture has not yet been realized.**

CHARACTERSTIC FEATURES

The fish is beautifully colored; the head and back are emerald green with golden yellow on sides and belly. There are four rows of dark spots on either side below the lateral line. Dorsal and anal fins are yellow in color with dark bands. Caudal and pectoral fins are also yellow with dark vertical bands. There is a dark spot at the base of the pectoral fin. It is interesting to note that the color of the body is changed according to their age and locality. The fish is known to attain a length and weight that varies from 6 to 10 inches and 50 to 200 g respectively.

Habitat

The fish lives mostly in holes that are located along the periphery of ponds, canals, and beels. Unlike other murrels, the fish does not normally occur in open ponds or mershy areas. During the autumn, as the level of water in beels and ponds recedes, the mouth of the holes where fishes take shelter, remains above the water surface. Breeding occurs during the southeast monsoon when the beels and ponds are filled with water and submerging the holes.

Culture

Since this fish is at risk of being endengered as they are easily located and caught during the pre-winter months, adequate management must be adopted to prevent then from extinction and their culture in ponds has great practical significance. The culture technique of this fish is more or less similar to that of other murrels. Since the fish is on the verge of extinction, the indigenous system of its culture requires precise control over the entire culture period that must have to be evolved mostly by state-of-the-art technology by skilled farmers thereby permitting not only the yield of fish biomass, but to rejuvenate fish stocks as well. Although intensive culture of this species has great production potential but unfortunately, the importance of its culture strategy is yet to be evaluated and realized.

6.18 Culture of Giant Sea Perch or Sea -bass

The species *Lates calcarifer* commonly called as *Giant sea Perch* or *Sea bass* (Figure 6.16) Sea-bass is a euryhaline and catadromous fish. Sea-bass is regarded not only as a very popular fish for the market in Southeast Asian countries, South Asia and Australia but also as a highly priced food fish. High consumer demand and market prices have led fish culturists to devote considerable attention to the farming of sea-bass. It is only the basses that presently accounts for significant production through culture in freshwater ponds. Some small- to large-scale production of sea-bass has also been reported from tropical countries and from Australia. Recently, fish culture experts are actively involved in investigating the possibility of farming sea-bass under mono-or polyculture systems.

The fish is widely distributed in the tropical and sub-tropical areas of the Westem Pacific and Indian Oceans. Its range of distribution includes the areas such as Australia, Southeast Asia, the Philippines and countries bodering the Arabian Sea. The fish attains a size of about 35 cm in one year.

Characteristic Features and Distribution

Body is oblong, deep and compressed; head moderate, compressed and wide mouth. Eyes are moderate, superio-lateral and not visible from below ventral surface. Lower jaw is longer than upper one. Lips are thin; jawls and palate with villiform teeth. Shoulder bone is serrated. Pre-opercle with a strong spine at its angle and denticulated along the horizontal limb with 3 or 4 spines; Spiny opercle. Two dorsal fins, the first with 7 or 8 spines and the second with 11-13 rays. Both fins are continuous and united at base. Anal fin with 3 spines and 8 or 9 rays. Caudal fin is truncate or rounded. Lateral line is curved and complete with 52-60 scales. Scales are ctenoid.

The fish is distributed throughout India, Pakistan, Bangladesh, Burma, Thailand, Malaya, Africa, China, and Australia.

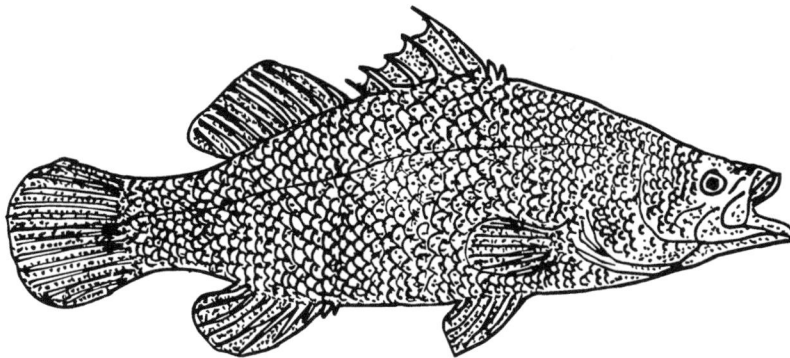

Fig. 6.16 : Any of about 400 species (Class : Osteichthyes, Order: Perciformes, Family: Serranidae) of carnivorous fishes, most of which inhabit in shallow regions of warm and tropical seas. They have a slender body, small scales, large mouth and straight-edged or rounded tail. The spiny frontal section and the soft-rayed rear section of the dorsal fin are generally jointed but may be separated by a notch. Species range from about 3 cm to 1.8m long and may weight 220 kilogram. Several species of bass are commercially important. *Lates calcarifer* is, however, considered as a delicacy and valuable food fish in most countries of the world commending good domestic and export markets. In recent years, it has become the main target species for large-scale farming in freshwater impoundments due to its rapid growth rate, attractive size, better meat quality and carnivorous feeding habit.

General Biology

The adult fish are voracious carnivores while juveniles are omnivorous. Fecundity rate of the fish varies between 2 and 7 million in specimens of 5.5 to 11.0 kg weight range. Under natural environment, spawning occurs round the year with peak during April to October. Spawning generally occurs from full moon to 6 days after full moon during the low tide period in the evening hours.

Newly-hatched larvae of sea-bass measure 1.5 mm in length, with a large yolk sac containing a oil-globule which permits the larvae to float at the surface. The body is slender and pale in color. The eyes, alimentary canal, anal opening, and caudal fins can be seen vividly. The yolk gets almost fully absorbed by 72 hours.

By 18 to 20th day of development, the larvae assume a pale brownish coloration and several vertical stripes are clearly observed. In about 45 days, developing embryos are changed into fry (1.5 to 2.0 mm in length).

Culture Techniques

Though the fish is widely cultured in Southeast Asian countries and in Australia in freshwater ponds and cages, its importance has gained momentum due to high market potential. Development of induced spawning technology and hatchery seed production techniques have solved much of its culture problem. The culture of sea-bass is carried out either in ponds or in floating cages. Cage culture has, however, been found to be more suitable and profitable than pond culture. The culture of sea-bass in freshwater ponds involves several steps that follow.

1. **Criteria For Site Selection:** Like other freshwater fish culture farms, it is suggested to construct the existing survey maps and the chart. This information can be used to delineate the tidal creeks, to locate the areas covered by lagoons, swamps as well as to find out whether there is flooding of the land from the adjacent river system. Other factors include (i) location of the site with reference to sea, brackish or freshwater source, (ii) topography of the area with reference to its general gradient, (iii) nature of soil conditions and the highest water level at the site, (iv) soil characteristics, topography, and freshwater supply, and (v) availability of fish seed and feed, labor, technical assistance, larval rearing, disease control and the availability of marketing outlets and general price levels, culture of live fish for sea-bass larvae, broodstock management and graders for grading fish seed.

(i) *Soil Property*: Soil texture can be adopted as a criterion to determine the suitability of the site for sea-bass pond construction. The soil should be clay-loam with adequate clay content to retain water. Clay soils are preferred for dyke construction because they have good water retention property and excellent compaction features. Acid sulfate soil should be avoided, however.

(ii) *Tidal Amplitude:* In case of gravety-fed fish ponds, the depth of water to be maintained is determined by the height of the incoming tide and the elevation of the pond bed with respect to zero tidal level as it is cardinal that the existing tidal range will permit to fill up the pond. The design of pond depth must allow the most economical construction with regard to ground elevation with respect to tide. The optimum pond bottom elevation should be the one which permits the pond to be drained at any day of the year and filled with water to the desired depth within 7 days. Low areas that may require wider and higher dykes should be avoided to reduce costs.

A site that receives moderate tide fluctuation ranging from 2 to 3m is, however, desirable for sea-bass culture. This tidal fluctuation helps complete drainage at a very low cost during the low tide even when the pond is 1.5 m deep. At the same time, the pond should also receive water during the critical spring tides.

(iii) *Water Quality* : Similar to other fish culture systems, huge quantities of fresh-or brackish water is cardinal for sea-bass culture and hence only those areas where adequate quantities of water are available round the year should be selected. The flow of water must be sufficient to meet the demand of fish farms.

In principle, surface or ground water can be used mainly for sea-bass hatcheries. Surface water from rivers, lakes, and open sea may be relatively less expensive to use. Under special circumstances, it becomes necessary to filter the water and where there is a high content of silt, it is also necessary to have a settling tank. As a generalization, sand or gravel filters with back-flushing will make the water suitable for the hatchery. This will definitely eliminate the risk of water contamination considerably.

2. **Pond Design and Construction:** Sea-bass culture can be undertaken in two types of ponds such as (i) Nursery ponds and (ii) Grow-out ponds. Larger the farm area, cheaper the cost per hectare of farm would be because of saving on the construction of periphery bund. Thus large farms should be planned to cover an area over 20 or 30 hectares. The size of a pond, which is considered good for managanent, should be between 0.5 and 2.0 hectares in area. In the case of small-sized ponds, bunds occupy considerable space in the farm enclosed by the periphery bund. Consequently, the pond area considerably reduced resulting in an increase in cost per hectare of productive water area.

On the other hand, the shape of the total farm area should be more rectangular than oblong to reduce the cost of periphery bund. In a small-or medium-sized farm, a series of nursery and grow-out ponds are constructed.

(i) *Nursery Ponds*: The surface area of each nursery pond generally varies between 550 and 2,000 sq. m (0.05 and 0.2 ha) with a water depth ranging from 50 to 80 cm and salinity range from 10 to 30 ppt. Ponds should be rectangular to make them less economical and also provide better management from all directions. Each pond should be provided with inlet and outlet facilities to facilitate easy water exchange. Both inlet and outlet pipes should be covered with nylon net to prevent the escape of fry from the pond and entry of predators.

* *Preparation of Nursery Ponds*: The preparation of nursery ponds essentially involve complete drying during summer season. In case the pond bed cannot be dried, quicklime and ammonium sulfate at the rate of 500 and 100 kq/ha, respectively, should be applied for 0.5 m water depth. About 7 days after the application of lime and ammonium sulfate, poultry droppings and diammonium phosphate at the rate of 250 and 50 kg/ha, respectively, are necessary for better results. After a gap of 10 days, water is allowed to enter the pond to increase the level up to 0.8 m.

Since sea-bass is a carnivorous fish, a variety of natural food organisms such as *Artemia, Nauplii, Brachionus, Moina,* larvae of *Acetes*, shrimp, small freshwater fishes etc. are necessary. These organisms are cultured in separate concrete/PVC

(Polyvinyl chloride) tanks and are introduced in nursery ponds at the rate of 50 numbers/fry for the first 10 days. During rearing in nursery ponds, fry are sorted out size-wise. This sorting helps prevent fry from canibalism. Consequently, survival rate increased considerably. Sea-bass fry of 1.0 -2.0 cm in size are stocked at the rate of 1 lakh/ha until they grow to 8-10 cm in length.

* *Collection of Fry and Fingerlings From Natural Sources*: Sea weed beds and mangrove areas with salinity ranging from 1.0 to 20 ppt. and tide-fed low-lying areas are the principal nursery grounds for sea-bass. These nursery grounds are teemed with sea-bass fry and fingerlings. Fry and fingerling collection essentially involves the use of different types of nets in areas where they are concentrated. A seine net of 1 cm mesh size is operated by a team of four farmers each holding one of the four wings of the rectangular net which measures 1 m x 5 m. The lift net consists of a wooden circular ring and a net with mesh size of 1 cm is fixed in such a way that it forms a cone-shaped structure. Dewatering of small tidal pools that contain huge quantities of sea-bass fry is also practised where hand nets are used. In some areas, however, drag nets are used for collection of fingerlings over extensive water areas. While fry are stocked in nurseries for several days, fingerlings are cultured in grow-out ponds, cages and paddy fields.

* *Induced Breeding of Sea-bass*: Induced breeding is carried out either by allowing the broodfish to spawn in the tank by itself with or without hormone injection or by stripping. In the former case, several environmental factors such as changing of temperature and salinity come into play and spawning occurs rather easily and intermittently for about 3-7 days. Stripping of broodfish is one of the best methods of obtaining fertilized eggs for mass production of sea-bass fry. Generally, human chorionic gonadotrophin (HCG) is generally used to induce spawning of sea-bass. Injection is given either behind the pectoral fin or between first and second dorsal fins. The dosage of HCG varies from 50 to 1,000 IU/kg body weight of fish.

As soon as fish is allowed to release mature eggs through vent by gentle pressure on the abdomen, the male fish is immediately squeezed out to exude milt over the eggs. The milt and eggs are gently stirred by adding some amount of water until these two components are well mixed. After repeated washing with saline water, fertilized eggs are kept in plastic bags containing water of adequate salinity (28-32 ppt) and are transferred to the hatchery. Each nursery tank is stocked with fertilized eggs at the rate varying between 8,00,000 and 10,00,000. Hatching takes place in about 12-17 hours at temperature and salinity level of 27-30°C and 20-30 ppt. respectively.

(ii) *Grow-out Ponds*: For farming sea-bass, rectangular pond with an area varying between 2,000 and 20,000 sq. m. (0.2 and 2.0 hectare) are considered desirable, both for harvesting by netting and for good swimming area for fish. Both inlet and outlet pipes are guarded with nylon screen of 2-3 cm mesh size.

Sea-bass culture in grow-out ponds essentially involves rearing of fry till they attain a marketable size of about 400 g. In such ponds, however, four types of culture systems such as monoculture, polyculture, cage culture, and culture in paddy fields are generally recommended.

* *Monoculture* : Under this culture system, juveniles ranging from 6 to 10 cm in length are stocked at the rate of 10,000-20,000 number/ha. Artificial feeds such as trash fish, fish meal etc. are given 5 times daily at the rate of 10 per cent of the total fish biomass during the first 60 days following decrease in the feeding rate by 5 per cent. Since the feed conversion rate of this species varies from 7-10 : 1, the cost of trash fish is very important from commercial point of view involving sea-bass culture. Hence, utmost care must be taken so that over-feeding can be avoided. Water quality variables should always be kept in safe limit.

* *Polyculture* : Polyculture involves the culture of sea-bass in combination with one [(such as Tilapia (*Oreochrormis mossambicus*)] or more species of fish (such as eel and Indian major carps). Of these, tilapia is considered as an ideal candidate as a forage fish in polyculture systems. Since tilapia is a prolific breeder, production of hatchlings and fry in large numbers in grow-out ponds at regular intervals will definitely sustain the survival and growth of sea-bass throughout the culture period (Table 6.11) Hatchlings and fry serve as natural feed for sea-bass. For this purpose, tilapia is stocked at the rate of 15,000 numbers/ha with female to male ratio of 3 : 1.

Table 6.11: Production of Seabass in Polyculture Ponds

Total pond area (ha)	No. of fish stocked per hectare		Average initial wt. (g)		Period of culture (day)	Final average wt.(g)/survival (%)		Production (kg/ha)	
	Seabass	Tilapia	Seabass	Tilapia		Seabass	Tilapia	Seabass	Tilapia
22	3,000	12,000	15	50	176	682/71	123/42	1,450	136

Source : Mondal and Mondal (2000)

As a generalization, seabass juveniles of 6-10 cm in length are stocked in one hectare pond at the rate of about 5,000 numbers. When the sea-bass attains a size of about 50 g in weight, mesh size of inlet and outlet pipes is changed to 4 cm to permit ingress of forage fish into the pond. To maintain the normal growth of seabass, constant supply of forage fish into the grow-out ponds is necessary.

* *Cage Culture*: Cage culture of seabass in freshwater ponds in Taiwan and in some areas of India (especially Andhra Pradesh) has been very successful and hence their culture is highly encouraging. Most of the tanks and ponds can be brought under seabass cage culture. Generally floating net cages are installed in suitable areas. Cage culture has been found to be more appropriate and profitable than pond culture.

The convenient size of the cage is 5.5 m x 6.0 m x 2.0 m with mesh size of 2 to 4 cm depending upon the size of the fish to be kept. Each cage is stocked with fingerlings at the rate of 1,000 numbers. Floating cages are stocked at higher rates than fixed ones.To avoid cannibalism, grade-sized fish are stocked. Fish are fed with ground trash fish at the rate of 10 per cent of body weight with degree of regularity. The rate of survival varies between 50 and 80 pre cent. For raising table-sized fish, net cages are generally used.

It has been estimated that about 150 kg of fish can be produced from each cage. About 80 cages of the above-mentioned diamention can be installed in one hectare water area from which 5 -8 tonnes of fish can be obtained.

Culture in Paddy Fields: Sea-bass is farmed in paddy fields. For growing sea-bass, the level of paddy fields is raised to about 50 cm. The recommended size of ponds is about 4 hectare for easy management, but some consider a size of 8-10 hectare optimal.

Depending on the environmental conditions, it takes 2-6 months to attain an average size of about 600 g. About 100-150 g of sea-bass on an average are collected at the time of paddy harvest. These are then stocked from March to June in small ponds built in paddy fields at the rate of 5,000 number/ha where they are cultured. During spawning season a massive number of sea-bass fry along with small weed fishes and prawns also enter into the paddy fields adjacent to creeks. These weed fishes and prawns are utilized by sea-bass fry as food.

Paddy fields can be fertilized with organic manures such as compost, cattle manure, and vegetable waste all of which help increase indirectly the productivity of paddy fields. Fish are regularly fed with trash fish or fish meal.

Production of sea-bass in paddy fields generally ranges from 200-800 kg/ha. At the same time, the production of paddy in 'Rabi' and 'Kharif' seasons has been reported as 1,200 and 3,500 kg/ha, respectively.

To sum up, despite high potentiality of extensive water areas, attention has yet to be given towards sea-bass farming in India. Such water areas might be used for profitable sea-bass farming provided the availability of seed must be ascertained from different seed potential areas.

It must be remembered that Taiwan, Thailand, Japan, and China are the four leading countries of the world in sea-bass farming. Some other countries such as Hong Kong, Australia and Singapore are currently developing sea-bass farming industry. These countries, however, produce huge quantities of sea-bass for consumption and export. And due to absence of intra-muscular bones along with the excellent taste of flesh, sea-bass has enormous market potential.

6.19 Culture of Eel*

While eels are regarded as a delicacy in some countries of the world, most of the countries are not consumed at all or have limited demand. Japan and Western Europe have been the principal regions where there is high demand for eels. The earliest form of eel

* For further detail, see Chen (1972), Tesch (1977) and Usni (1974).

culture in ponds is the extensive system of lagoon farming along the Mediterranean coast. And in lagoons, eels are extensively cultured along with grey mullets (*Mugil cephalus*), seabass (*Lates calcarifer*), and Sea-bream *(Chrysophrys major* –Red sea-bream and *Sparus aurata* – Gill-head sea-bream).

Among different species of eels distributed all over the world, the most important ones from large-scale aquaculture view point are *Anguilla anguilla* (= *vulgaris*) in Europe, *Anguilla japonica* in Japan and Taiwan, A. *rostratus* in USA, and A. *bengalensis bengalensis*, A. *bicolor bicolor* and, A. *nebulosa nebulos* in India. Another species of eel such as *Amphipnous* is also cultured in the Indian sub-continent. A brief description of two species of eel is given in Figures 6.17 and 6.18.

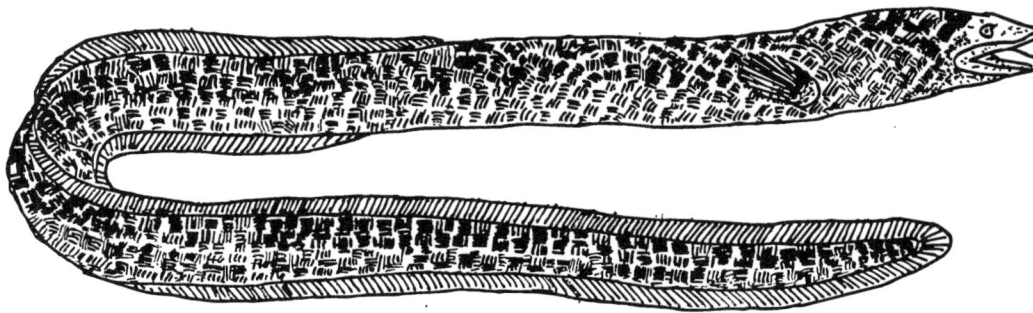

Fig. 6.17 : Any of more than 500 fish species (Class: Osteichthyes, Order: Anguilliformes, Family: Miraenidae) – commonly caled as *eel* that are slender, elongated, and generally scaleless, with long dorsal and anal fins that are continuous around the tail tip. Eels are found in almost all fresh, brackish and marine waters. Freshwater eels are, however, active, predaceous fish with small embedded scales. They grow to maturity in freshwater and return to the sea in autumn where they spawn and die. Delicate transparent and pelagic larvae, called *leptocephali* or *glass eel* or *elvers*. The duration of the larval life in different species varies from 3 months to 2-3 years. They drift to the coast and make their way to upstream. They later grow in fresh- or brackish waters for some years and at this phase, they are called *yellow eel*. On nearing sexual maturity, the adult yellow eel cease to feed, acquire a silver coloration, called *silver eel;* leave the fresh- and brackish waters and migrate to their spawning ground in the open sea where they breed. Freshwater eels, considered as valuable food fish, including species ranging from 4 inches (10cm) to about 11.5 ft (3.5m) long. Eels are widely distributed acros the world. There are 17 species in the genus Anguilla (Shaw) of which a few species such as Indian eels (**A. bengalensis bengalensis, A. bicolor bicolor,** and **A. nebulosa nebulosa**), Japanese eel (**A. japonica**), European eel (**A. anguilla** and **A. vulgare**), and American eel (**A. rostratus**) are highly important from commercial point of view.

Characteristic Features of Eel

1. *Anguilla* **spp. :** Body is elongated, cylindrical, band-shaped, and broader than body. Abdomen is rounded. Head is long and compressed. Gill openings are situated in the pharynx in the form of moderate slits near the base of pectoral fins. Snout is pointed with terminal mouth, cleft of mouth is wide and extending to posterior margin of orbit. Eyes are very small, superior, in middle of the head and not visible from below ventral surface. Lips are thick and well developed. Villiform teeth on jaws and palate.

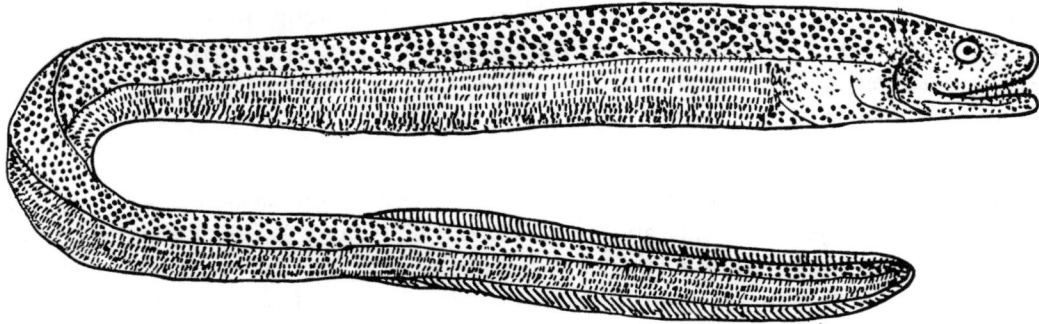

Fig. 6.18 : A very few species of *Amphipnous* (especially *A. cuchia, A. albus, A. fossorius, Ophisternon bengalense* and *Synbranchus* sp.) of the Class: Osteichthyes, Order: Synbranchiformes, Family: Amphipnidae which are snake- or eel-like in structure live in marshes near coasts. They are also considered as valuable food fish and they range about one meter long. Amphipnous is widely distributed in India, Mayanmar, Africa, Indo-Australian Archipelago, Central and South America, Nepal, Bangladesh, Pakistan and Burma.

Dorsal fin is inserted near the tip of the snout than caudal base with 220-305 rays and devoid of spines. Anal fin equally long with 200-250 rays. Caudal fin is continued round the end of the tail. Rudimentary scales are embedded in the skin.

2. *Amphipnous* **spp. :** Body is much elongated, eel-like and compressed. Abdomen is rounded. Distinct head, short and with a dome-shaped muscular occipital region. Snout is bluntly rounded. Mouth is wide, terminal, cleft reaching anterior margin of the orbit. Eyes are superior, upper lip is thin which is the overlapping part of the lower lip. Jaws and palate with villiform teeth. Gill openings are triangular. Scales are minute and not uniformaly distributed. Caudal tip is bluntly rounded, lateral line is well-developed. Dorsal and anal fins are in the form of folds or ridges. Dorsal fin commences from behind the vent or behind the anal fin. Scattered black spots and short yellow lines on the body which is darty pale red below and dark-green above.

Biology of Eel

Adult yellow eels are known to be catadromous species and migrate from different inland water areas into the sea for breeding and the glass eel or elvers or leptocephali return to inshore waters and eventually move up the rivers. While the Indian short-or long-finned and Japanese eels spawn in the sea very near to the coast, the European form performs stupendous journey (about 4,000 kilometres) from their native place to the Sargasso sea area of the Atlantic ocean to spawn. The leptocephali of *A. anguilla* and, *A. japonica* reach the continent 3 and 1 years after hatching, respectively, and enter the river after they have metamorphozed into elvers. The migration from the sea to the river starts when the temperature of water increases considerably – particularly in April-May and continues until the end of August. Once undertake the upriver journey, they exhibit conspicuous stamina and capability to overcome obstacles and barriers and form shoal(s).

Culture Methods

Eels are cultured in five different methods such as (1) Standing water method, (2) Running water method, (3) Recycled running water method, (4) Net enclosure method, and (5) Tunnel method. Of all these methods, the first three ones are generally followed.

1. **Standing Water Method:** In this method, occasional /daily exchange of water (5–10 per cent of the total volume of pond water) is cardinal. Phytoplankton populations rapidly developed in water which, in turn, increase the content of its oxygen through photosynthesis and thus provide an ideal condition for the growth of eel.

2. **Running Water Method:** This method essentially involves continuous flow of freshwater to the pond and simultaneously an equal amount of water is drained to keep the pond water at a constant level. This method permits intensive stocking and feeding for high production.

3. **Recycled Running Water Method:** In this method, eel culture ponds are generally resorted to reconditioning and recirculation. In certain cases, it may also be considered necessary to reduce the risk of disease and parasite infections considerably through continuous use of water from external sources. Oxygen can be enriched through aeration and most of the carbon di-oxide is dissipated, but elimination of metabolic waste products (especially ammonia) involves more complex systems, which besides mechanical filtration and re-aeration definitely involve biological treatment. Latest designs generally exercise one or more bypass treatment units such as for denitrification, oxygenation, and the like. Of course, such recirculation should make it economically viable to culture eel under different climatic conditions.

   For intensive culture of eel (and also sea-bass), the most commonly used and relatively more economical treatment would appear to be biofiltration, which may consolidate down-flow filters (such as trickling filters), up-flow filters or horizontal flow filters. Different types of filter media such as plastic, oyster shells, anthracite, activated carbon, diatomaceous earth and their combinations, and sand gravels are widely used. Besides serving as strainers, they provide surface area for the growth of micro-organisms. Through microorganisms and oxidation, ammonia is converted into nitrite and then nitrate. The nitrate may further be combined with ions in water to form salts or reduced to nitrogen gas through denitrification. With a retention time of about 30 minutes and a hydraulic loading rate of 1.0 -3.7 litre/second/m^2, about 48 per cent of the initial ammonia can be eliminated.

4. **Net Enclosure Method:** The culture of eels in net enclosures has been carried out over extensive areas to reduce the cost of pond construction and to avoid problems of water supply. Since eels grow better in warm-water environments, purified heated water from thermal stations or industrial sources or from hot springs, if feasible, may be used.

5. **Tunnel Method:** Tunnel method involves the rearing of eel in cement tanks (surface area and depth are 1 m^2 and 1 m, respectively) with inlet and outlet tanks. Water is allowed to enter the main tank through a 25 cm diameter pipe and then drained through the outlet tank. This system of farming is suitable for intensive production. The recirculation system has also been adopted on a limited scale for culture to allow intensive tank culture in regions where the supply of water is extremely limited.

Culture Systems

In general, two types of pond culture systems are carried out such as (1) mono-culture and (2) polyculture. The former system is adopted in Japan and the latter is in Italy and Taiwan. In polyculture systems, however, eels are the main species stocked in frestwater ponds where silver carp (*Hypophthalmichthys molitrix*), big head carp (*Aristrichthys nobili*), common carp *Cyprinus carpio*), mud carp (*Cirrhinus melitorella,*) and stripped mullet (*Mugil cephalus*) are also stocked.

Culture Tenchiques

Eel culture essentially involves the collection of seed eel(termed as *elvers*) from the wild, selection of site and design of eel farm, stocking of elvers and eels in ponds, feeding, maintenance of eel farm, disease control, and other management protocols.

1. **Collection of Elvers:** Elvers are generally collected in different times of the year at the lower reaches of the rivers when they ascend from the sea. In India, however, elvers of *Anguilla bicolor bicolor* and *A. nebulosa nebulosa* are collected from close to the banks of the rivers at night during October-March. In Europe, elvers are collected during winter and spring or in the beginning of summer in June and July when they ascend the rivers. Seed elvers measure 50-100 mm in length and about 1-2 g in weight. The elvers for restocking or rearing are collected during December. The best catches are usually obtained from February to April.

   Elvers are collected with large wire-meshed sieves or large nets (such as Scoop net, Bag net, Dip net, and Plankton net) dragged from a motorized boat. Drag net is the best gear used to collect elvers from the pools, small streams, rivers, and from the surroundings near the sluice gates.

2. **Transport Elvers:** The catches are kept in temporary holdings such as hapas or floating boxes till they are transported to shore facilities. While circular-shaped with flat bottom galvanized metal cans (20 to 40 litre capacity) are extensively used for short distance transport, special motorized vehicles having a series of tanks are used for long distance transport of elvers in large numbers. These tanks are insulated and have an inbuilt aeration system. Nylon nets are suspended inside the tanks to protect elvers from injuries.

   In some cases, specially-designed wooden boxes are also used to transport elvers over long/short distanes by road and rail. The wooden box contains 10 frames (each measuring 8 cm x 40 cm x 5 cm) and each frame can hold about 1,000-5,000 elvers depending on their size, duration of transport, and temperature of water. Frames are placed one above the other inside the box which is closed with a lid. The top and bottom frames are provided with crushed ice and saw dust, respectively, and the remaining frames contain elvers. Melting ice will trickle down the frames which helps moisten the elvers and the ice water is finally absorbed by saw dust.

3. **Selection of site and Design of Eel Farm:** Like sea-bass culture, similar steps should be taken into consideration for eel culture. This involves the availability of fresh or brackish waters in huge quantities in the area concerned, topographical features of the site, construction of dykes, arrangements for water supply and drainage system.

   Eel farms should be constructed in such a way so as to meet various requirements in different phases of culture. Initially, the elvers are stocked in nursery ponds (150-600 sq. m and 70 cm deep) where they are reared for about 3 months. They are then transferred to fattening ponds (size vary from 0.3 to 4.0 ha and 1.0-1.5 m deep). Pond bottoms are muddy in nature and the banks are reinforced with stones and brick or lined with concrete slabs.

   An ideal eel farm should have a series of nursery and fattening ponds each having its own independent water supply and drainage facility.

4. **Stocking of Elvers and Eel in Ponds:** In running water ponds, the stocking rate of elvers is higher than those of stagnent ponds. Also, the stocking density varies considerably from country to country depending upon their culture methods. In Japan, for example, elvers are stocked in ponds at the rate of 3 lakhs/ha or 30m^2 (average weight of elver is 0.2 g). In Taiwan, ponds are stocked at a very high rate (3 million/ha). In fattening ponds, however, the stocking rate of young eel is about 3,000 kg/ha (at the rate of 20 eel/m^2 each having 15 g weight).

5. **Feeding:** In the initial phase of culture, young elvers are fed at the rate of 20 per cent of their body weight with crushed earthworms or crab flesh or minced clam meat. After 15 days, crushed earthworms are mixed with fish meat. When they attain a size of 20 g within a month, they exclusively feed on fish meat. Specially formulated artificial or concentrated dry feeds are also widely used for eel culture. The feed is mixed with water and a paste is made which is kept on a mesh tray on the feeding areas. Mesh trays are suspended from above the water surface. Since growth rate of eels is very rapid during March-October, it is advisable to provide as much food as feasible to make the culture system more remunerative.

   Intensive eel culture system essentially involves feeding with artificial dry feeds (either pelleted or granular form) and in contrast to trash fish/chopped fish, the conversion ratio of various forms of artificial dry feeds has been noted to be very high – it is 3.4-4.4 : 1. Artificial dry feeds contain fish meal (65 per cent), fish concentrates (5 per cent), multi-vitamins (1 per cent), lysine (0.2 per cent), antioxidant (0.2 per cent), and binding substances (0.2 per cent).

6. **Management of Eel Farm:** Management of eel ponds includes the supply of adequate quality and quantity of water, optimum pH of water (7.8-9.0), maintenance of the flow of water at a constant level in case of running water ponds, and the pond should have a minimum depth of about 65 cm. Most of the eel ponds are provided with motorized paddle wheel aerators to aerate the water for oxygenation, particularly at night and in the early morning. Some farms have a corner of the pond which is partitioned off with wooden boards into a pool at the inlet with an entrance for the eels. Water wheels are installed in these pools. When depletes the content of its oxygen, eels congregate in these pools. Some farms install compressed air blower at the pond bottom to increase the content of its oxygen. Sorting the stock and restocking them in separate ponds are also a part of the management strategy. This is extremely important to have eels of marketable size. Japanese and Indian consumers prefer 100-250 g eels and this size is reached in one year after stocking of young eels. In Taiwan and Europe, still large-sized eels (0.5 -3.0 kg in weight) are preferred and this requires a second or third year of growth.

7. **Disease and Mortality of Eel:** Eels appear to be more susceptible to various diseases such as cotton cap disease (fungal infection), red disease (putrifective lesions in the liver or kidney), white spot disease (sporozoa-like species of *Myxidium* and *Myxobolus*), gas embolism, gill erosion (caused by bacteria), and the crustacean parasite (caused by *Argulus*). The cause severe mortality to eels than many other freshwater fish species. They also cause severe damage to elvers and to adult eels. Fluctuations in water temperature and accumulation of uneaten feed are direct or indirect causes of mortality of adult eels and elvers. Eels are very susceptible to low dissolved oxygen levels in stagnant ponds. Susceptibility to diseases and the resultant mortality seems to be accelerated by over-wintering practices. These factors essentially contribute to the occurrence of high mortality in eel ponds.

8. **Harvesting and Production:** Eels are harvested with different kinds of nets. While partial harvesting of eel ponds does not require draining, ponds should be completely drained at the time of final harvesting.

(i) *Partial Harvesting*: Eel farming in most areas of the world involves partial harvesting in summer months and stocking at regular intervals. For thinning of stock or capture of marketable eels, a scoop net is used in the feeding region where the fish muster at the normal feeding time. At the time of fishing, freshwater is allowed to enter into the ponds so that the content of its oxygen is not reduced. At times when eels fail to concentrate on a particular area in large numbers, a seine net having a fine-meshed bag at the centre is used. The net is dragged from the deeper region of the pond to the inlet and finally the stock is removed by small dip nets. Seining is repeated several times which helps aerate the pond.

(ii) *Complete Harvesting*: Complete harvesting of eel ponds in winter months is carried out by draining the ponds in warm days. Draining in the morning hours is most effective because eels will not take shelter in the pond mud and will move with the flow of water. If eels still remain, the ponds are again refilled with water partially and drained again at night. The remaining eels will move with the drained water. When the harvest is completed, the pond bottom is treated with calcium oxide and stirred several times and then allowed it to sun drying.

(iii) *Sorting*: Soon after harvest, eels are sorted out size-wise. Smaller eels (below 120 g weight) are used for further rearing if harvested in summer or if harvested in winter, they are kept for over-wintering and rearing during the next year. Before shipment of live eels, they are starved for a few days and kept in a limited space. This helps reduce the accumulation of fat and destroy off-flavor. The starvation period determines better sanitary conditions in the containers in which harvested eels are transported.

For better conditioning the elvers and adult eels, several buskets are stacked one after another each containing 3 kg eels and are provided with a shower of water from above the stack for several days.

The size of farmed eels at harvest generally varies according to species found in different countries. In Japan, for example, the Japanese eel is harvested when they attain a size that may vary between 100 and 200 g within two years from elvers stage. Partial (30-50 per cent of the total stock) and complete harvesting is done at the end of first and second year, respectively. In Taiwan, harvesting is done when eels attain a weight of over 200g. Ponds are harvested only depending on the demand in markets. In European countries, eels are harvested when female and male attains a size of 500 and 150 g, respectively.

(iv) *Production*: Per hectare production data of eels are extremely variable depending on the type of culture techniques and topography of the region concerned and the data obtained so far from different countries of the world are highly disturbing. Eel production in running water impoundments is, however, several times higher than that of still water farming (Table 6.12). And most of the Asian and European countries produce huge quantities of eels each year for large-scale consumption and export.

Average Table 6.12 : Average Production of Eel in Some Countries

Country	Running water (kg/ha)	Still water (kg/ha)
Japan	26,350	6,125
Taiwan	10,000	4,000
Germany	65,000	15,000
India	25,000	6,000

9. **Marketing:** While eels are shipped to markets in live condition in vinyl bags each containing about 10 kilogram fish with 1 kilogram ice for short distance, trucks or rails are used for long distance journey where eels can be packed in polythene

boxes filled and charged with ice and oxygen, respectively. Aerated tanker trucks are also used for transport of live eels for several days journey. Each truck can carry 15 tonnes of live eels in adequate quantities of water with continuous aeration by a compressor.

Dead eels are quickly frozen for market. Eels are frozen either whole or after gutting and cleaning, in blocks in an air-blast. Each eel or block of frozen eels is glazed to prevent oxidation. They are then wrapped in polyethylene and sealed at -20°C. In many countries, hot-smoked eels have good market potential. Since eels are cooked at the time of smoking process, they are ready to eat. The smoked eels can be better canned in vegetable oils. Most of the Asian farmers export large quantities of fresh and processed eels to Europe.

To sum up, in contrast to tropical regions, very low to moderate temperature of pond water in Europen countries is one of the most striking problems in the way of rapid develppment of eel culture industry. Several methods that follow have been developed to eradicate the problem. *Firstly*, use of polyethylene multi-span covers over the pond; *Secondly*, artificial waming of the water with insulated sides and bottom under a green-house and *thirdly*, use of warm water effluents from thermal power stations. However, commercial attempts to farming eel, use of continuously circulated and purified heated waters, insulation of tanks, and simultaneous expansion of air-freight import from different eel-growing countries of Asia are some of the major developmental attributes to the profile of eel culture industry which must be kept in mind. And the costs involved in eel culture in European countries may not negate the profit derived. Since refined technology is a major plus point of any aquaculture industry, the consequence of staying with older technology may be severe on the cost and quality fronts.

6.20 Harvesting and Marketing of Carnivorous Fish

Harvesting of catfish and murrels is extremely difficult and painstaking. Harvesting schedules in catfish culture depend very much on the stock of fry used and the climatic conditions in the region concerned. However; draining and seining are the two most effective methods of harvesting catfish and murrels from ponds. Draining is an effective method that helps remove the pond water so that the entire stock can be exposed to air. Draining involves the construction of a short but narrow canal to serve complete removal of water from ponds. Of course, draining of water by using diesel engine involves relatively higher costs. Draining permits easy harvests of almost all fishes from ponds. At the same time, pond bottom soil is exposed to the sun to make the pond ecosystem more effective for the cultivation of next crop. Though draining allows better management of pond soils, it involves an additional cost where pond refilling is done by the pump set. Harvesting from cages does not create any problem. Partial harvesting of carnivorous fish culture ponds is done by seines, but differences in catchability have been noted between ponds. This is due to the fact that these fishes avoid seines by lying on their side on the pond bottom and therefore, repeated seining is inevitable to catch a bulk amount of the stock. Though the use of mechanized seining equipments may significantly reduce the labor cost, larger investments for equipment may negate the benefit achieved. In many situations, however, complete harvesting of catfish and murrels culture ponds is not feasible.

In small-scale farms, the catches are sold fresh and in live conditions at the harvesting site or in the nearby markets. Larger farms generally transport the entire stock that is sent to and receive by a fish trader for selling in urban markets to make a reasonable profit. Catfish and murrels are marketed as whole and dressed fish or steaks and fillets. Fish of 500-600g are well-suited for dress-out product. Steaks and fillets are made from larger fish. The processed products are packaged for quick-freezing or for marketing fresh and wrapped in polyethylene bags.

6.21 Factors Affecting the Production of Carnivorous Fish

Energy level in the diet is one of the most important factors for the survival and growth of carnivorous fish. Very low energy level drastically affects the utilization of protein. On the other hand, excess energy creates nutrient imbalance and gets deposited in the adipose tissues of fishes that reduce the market price. The optimum protein for energy ratio in catfish diet per 1 g of protein digestible energy (DE) requirement is 10 kcal (90-100 mg protein/kcal DE). Catfish easily digest the energy present in animal feeds (such as fish meal, meat, and bone meal) than that of cereal grains. At the same time, balanced diet of catfish and murrels must contain protein (30 -50 per cent of the diet on dry matter basis), fat (15-20 per cent), minerals, and vitamins in trace amounts for better production.

In general, the growth of catfish in swampy and derelict waters principally depends on the availability of natural food organisms. In farmed ponds, however, stocking density, initial size of fish stocked, and the feeding schedules largely influence the metabolism and growth of catfish. Regular feeding at the rate of 6 per cent of the body weight has been found to be highly encouraging and considered as principal factor for the growth of catfish in ponds.

Wide growth patterns of different types of carnivorous fish have been noted in different types of water bodies. This is attributed primarily to toxic effects, high turbidity of water, and wide fluctuations of abiotic and biotic factors of water and soil. By removing the turbidity and toxic pressure and manipulating feeding schedules fish production can be increased. It is certain that not all factors are equally harmful to catfish growth, but with edible fish there will be obvious concern for both the growth and hygenic condition of fish stocks. The need for management of water bodies where carnivorous fishes are cultured should be an inevitable part of any planning when their commercial culture is considered.

Since catfish production is highly dependent upon the use of supplementary feed, concentrations of total ammonia nitrogen (combination of un-ionized ammonia nitrogen and ammonium nitrogen) becomes higher and varied between 3 and 6 mg/l in ponds receiving large amounts of feed. It has been noted that pH of water markedly increased in catfish ponds during daylight hours and the potential for high ammonia nitrogen concentrations exists when total ammonia nitrogen (TAN) values are high and during afternoons, pH values vary between 8.5 and 9.5. As a generalization, at high temperature (30°C) and pH (9.0), about 45 per cent of the TAN is in un-ionized form. In some situations, high concentrations of TAN in ponds in winter time are common when water temperatures decline followed by decline in fish growth. Therefore, 20 -30 per cent of pond water should be replaced by unpolluted water before the critical period for TAN.

Deterioration of water quality definetely limits the amount of feed that can be applied and the quantity of fish that can be produced within a given aeration regime; of course, the amount of aeration must be limited by cost factors. If enough aeration is, however, applied to prevent dissolved oxygen depletion at high feeding rate, fish production will likely be limited by high concentrations of ammonia nitrogen in culture systems.

Development of various carnivorous fish culture systems have been forced to access the availability of the seed. In contrast to carp seed, the carnivorous fish seeds occur in a highly restricted niche in nature. Swamp and shallow water areas as well as paddy fields are known to be preferred habitats especially for air-breathing catfishes and snakeheads and the availability of seeds is highly restricted to a particular season after the monsoon when the levels of wild water resources significantly reduced to a minimum and the collection of seeds becomes comfortable. In most of the cases the water level recedes enough to make the water areas very suitable for collection operations. Fish seeds grow to a considerable size, making the stocking materials more expensive. They have fixed breeding periods and distribution of their young ones exhibit definite seasonal abundance. The present limited demands of carnivorous fish seed may be met from natural resources as more new areas are explored for their potential. But the establishment of carnivorous fish culture system as an industry for high yields, state, of-the-art infrastructure for controlled seed production is cardinal not only to effectively support the industry but also to eradicate the problems of carnivorous fish culture considerably.

6.22 Probable Future Expansion of Carnivorous Fish Cultivation

In yesteryear between 2004 and 2007, carnivorous fish culture practices have spread rapidly in some Asian countries. Whereas in 1970, these practices were used for only 5-8 per cent of the total waterspreads, by 1980 this figure has risen to about 15 per cent and in 2009 to 35 per cent.

Future projections suggest that this trend will continue, and that by the year 2025, about 54 per cent of the country's wetland will be brought under cultivation of high-priced carnivorous fishes. This expansion is highly encouraging because of its probable constraint on yield. To utilize wetlands for fish culture, however, engineering features should be considered. If such features are extensive, costs may exceed the benefit derived.

6.23 Conclusion

Most of the Asian countries consume considerable quantities of murrels and catfishes since they serve as an excellent bulk-filling material particularly in rural populations. The importance of carnivorous fish is evident from the fact that they contain valuable amino acids and minerals especially iron and hence recommended by physicians as diet during convalescence.

Despite high nutritive value of these fishes, there is no great demand for them owing to the conservative food habits and religious considerations of the people; of course, most of the carnivorous fishes are consumed by the middle and upper class families particularly in Indian sub-continent. In this context, the culture practice of carnivorous fishes must be emphasized. In India and in some other Asian countries, none of the carnivorous fish is cultivated on commercial basis at this moment. They are collected at random for the limited demand in local markets, though fishery experts have, to a limited extent, tried to improve the culture practices in ponds and cages.

Murrels and catfishes offer great promise for farming and certainly deserves more scientific attention than they are receiving at present in developing nations. Growing awareness among the people and at the same time, the state-of-the-art technology should be adopted in such a way that it is not impossible to produce more and better carnivorous fish from different freshwater resources. These fishes, however, will certainly continue to play an important role in different regions for consumption and market.

Future developnents will rely on maximizing the off-take from large number of small/medium-sized water bodies found in many countries of Asia. It is likely that carnivorous fisheries will be based on culture aspects supported by technically efficient hatcheries. Although it is unlikely that the catfish fishery will return a profit in the near future, there is scope for increasing the market value of catfish products following the adoption of better methods of processing and preservation.

References

Banerjee, S. R. 1974. Hypophysation and life history of *Channa punctatus. J. Inland Fish Soc.* India, 6 : 62-73.

Banerjee, S. R. and D. Prasad. 1974. Observations on reproduction and survival of *Anabas testudineus. J.Inland Fish Soc.India*, 6 6-17.

Boyd, C. E. 1979. Water quality in warmwater fish ponds. Alabama Experimental Station, Auburn University, Alabama.

Basu, D., G. C. Ran a, B. K. Mondal, K. K. Sengupta, and P. K.Dhar. 2000. Studies on the comparative efficacy of ovaprim, HCG and piscine pituitary gland in induced breeding of *Clarias batrachus* (Linn.). Fishing Chimes, 19 (10 & 11) : 103-104.

Chen, T.P 1972 Eel culture in Taiwan. FAO Aquaculture Bulletin Number 5.

Deomempo, N. R. 1995. Farming systems, marketing and trade for sustainable aquaculture. *In* Report of the ABD/NACA Regional Study and workshop on Aquaculture Sustainability and Environment. Network of Aquaculture Centres in Asia-Pacific, Bangkok., Thailand.

David, A. 1963. Fishery biology of the Schibeid catfish *Pangasius pangasius* and its utility and propagation in culture ponds.*Ind. J. Fish* 10 (2): 521-600.

Jana, B. B. and R. N. Das. 1980a. Relationship between environmental factors and fish yield in a pond with air-breathing fish, *Clarias batrachus*. *Aquacultura Hungarica (Szarvas)*, 11 : 139-146.

Jana, B. B. and R. N. Das. 1980b. Changes of chemical composition of water in the fish farming ponds with air-breathing teleosts. *Proc. Ind. Acad. Sci. Section B,* 89 (1): 1-8.

Marimuthu, K., M. Muruganandam *et. al.* 2000. Induced spawning of the Indian catfish, *Heteropneustes fossilis*, using a synthetic hormone, Ovatide. *Fishing Chimes*, 19 (10 & 11) : 105-106.

Mondal, B. K. and A. K. Mondal. 2000. Culture of Sea-bass, *Lates calcarifer* (Bloch) in West Bengal-A case study. *Fishing Chimes*, 19(12),: 13-15.

Tesch, F. W. 1977. The Eel-Biology and Management of Anguillid Eels. Chapman and Hall Ltd.,London, 434 pp.

Thakur, N. K. and V. K. Murugesan. 1979. Breeding of air-breathing fishes under controlled conditions. Souvenir in Commemoration of the ICAR Golden Jubilee Year 1979, CIFRI, Barrackpur, West Bengal, Part -II,129-134.

Thakur, N. K. and P. Das. 1986 a,b, c. Synopsis of biological data on *Anabas,Clarias*, and *Heteropneustes*. Bulletin Number 39, 40, and 41,CIFRI, Barrackpur, West Bengal.

Usui, A. 1974. Eel Culture. Fishing News (Books), West Byfloet and London, 186 pp.

Wee, K. L. 1982. The biology and culture of snakeheads. *In*: J.E. Muir and R. J. Roberts (Eds.). Recent Advances in Aquaculture. West View Press. Boulder Co. pp. 180-211.

Questions

1. Define the term carnivorous fish. Name some common carnivorous fish generally cultured in Asian countries. What are the constraints of their culture? Distinguish between catfish and murrels. Why carnivorous fish is very easy to handle during culture operation?

2. State the importance of carnivorous fish culture. What are the advantages of their culture?

3. Discuss the food, feeding, fecundity, and breeding of *Clarias*.

4. Discuss how fry of *Anabas* are reared in ponds/tanks.

5. Discuss the steps involved in the culture of *Anabas* in ponds/tanks.

6. Do you think that the African catfish should be recommended for culture in Indian waters? Discuss in support of your statement. Is there any possibility to culture in carp ponds? Discuss.

7. Discuss the culture methods adopted for *Aorichthys seenghala* and *Wallago attu*. Why the fish *Ompak* spp is yet to be considered for culture inspite of their high market price? Is there any opportunity to culture under Indian conditions?

8. Discuss the maturity, fecundity, and breeding of *Heteropneustes fossilis*.

9. Discuss the culture methods of *H. fossilis*.

10. Why pangas culture in ponds is highly profitable? Discuss its culture methods generally adopted in ponds.

11. What steps should be taken to conserve featherback fishes for future generations? Is there any scope to produce more featherbacks from freshwater resources?

12. What are the main characteristic features of the fish *Anabas*? Discuss the food, feeding habit, breeding, and larval rearing of this fish.

13. Discuss the culture methods of *Anabas*.

14. Mention some important features of the following fish:

    (a) Channel catfish, (b) bullhead catfish, (c) flathead catfish,

    (d) blue catfish, (e) *Clarias*, (f) *Heteropneustes*,

    (g) *Anabas*, (h) *Pangas*, (i) Murrels, (j) *Eel*

15. Discuss the culture methods of channel catfish adopted in USA.

16. What are the steps involved in the culture of murrels ?Discuss.

17. Discuss the process of harvesting and marketing of carnivorous fish.

18. What are the features which affect the production of carnivorous fish in ponds/tanks.

19. How can derelict and swamp water bodies be utilized for the culture of catfish.

20. Discuss the culture methods followed for sea-bass in freshwater impoundments.

21. Discuss how sea-bass is cultured in grow-out ponds.

22. Discuss the culture techniques of eel in freshwater ponds.

7

Culture of Exotic Fishes

The culture of commercially important introduced fish in different freshwater ecosystems has serious limitations with regard to the devastating consequences of native fish stocks, but the expansion of their culture in several Asian countries to increase food fishes are highly significant nonetheless.

Though freshwater fisheries and culture of commercially important indigenous fishes in India was not uncommon as far as aquacultural records can be traced, introduction of foreign species either accidentally or deliberately into the Indian waters since the beginning of the 20th century and their culture either alone or in combination with the Indian major carps has been extensively employed for little more than 30 years and has fulfilled the primary objective of increasing fish production. The culture of exotic species is not only becoming an important issue for the developnent of rural economics but their introduction has become a serious threat to the existance of native species. Any freshwater fish brought from neighbouring countries used to promote production by applying suitable methods is considered as an *exotic fish*. Exotic fish production in several Asian countries through polyculture has increased several folds. Due to disagreeability of these fishes to consumers, their culture in selected areas is declining.

7.1 Expansion of Exotic Fish Culture in India

Technology is in part responsible for the significant expansion in the production of fish particularly which has occurred in the developing countries. Since the beginning of the 20th century, a number of foreign species were introduced in India mainly for the development of culture fisheries and game fishing (Table 7.1). High population growth, unemployment generation, and malnutrition have encouraged the culture of exotic food fishes in Indian fish culture systems.

During the 1970s advances in polyculture system have enormously changed the entire aquaculture scenario. Exotic food fish production has been dramatically increased and game fishes have helped, to some extent, foreign currency earnings. Fishery scientists have demonstrated the culture techniques of exotic food fishes in rural areas. The consequence is the exceptional production of exotic fish. Among different exotic fish, some species that follow are important so far as their culture aspect is taken into consideration: (1) *Puntius javanicus* (Tawes), (2) *Osphronemus gourami*, (3) *Oreochromis mossambicus* (Tilapia), (4) *Ctenopharyngodon idella* (Grass carp), (5) *Hypophthalmicthys molitrix* (Silver carp), and (6) *Cyprinus carpio* (Common carp) Figure 7.1.

Table 7.1: Some Important Foreign Species Introduced into Indian Waters

Species introduced	Year of introduction	Source	Purpose
I. Food Fish			
Ctenopharyngodon idella	1959	Japan	Culture and weed control program
Hypophthalmichthys molitrix	1959	Hongkong	Culture
Cyprinus carpio (Chinese strain)	1957	Bangkok	Culture
C. carpio (European strain)	1939	Ceylon	Cul ture
Oreochromis mossambicus	1916	Java	Culture
Osphronemus goramy	1916	Mauritius	Culture
Tilapia mossambicus	1952	Bangkok	Culture
Puntius javanicus	1972	Indonesia	Culture
Tinca tinca	1870	England	Culture
Carassius carassius	1870	England	Culture
II. Game Fish			
Salmo gairdneri	1907	Ceylon, Germany and England	For planting lakes, streams and
S. trutta fario	1901	England	reservoirs
S. salar	1968	USA	
Oncorhynchus nerka	1968	Japan	
Salvelinus fontinalis	1959	Canada	
III. Larvivorous Fish			
Gambusia affinis	1928	Italy	Mosquito
Lebistes reticulatus	1908	South America	control

In the middle of the 20th century when the usefulness of inland fish cultivation was realized principally for food and economic importance, Indian fishery scientists observed a lot of problems in inland fishery resources. One of the principal problems was to inability of breeding Indian major carps in confined waters. At the same time though there are cretain disadvantages of Indian carp cultivation in many ponds where algal blooms and aquatic weeds are most predominant, some foreign carps such as common carp, grass carp, and silver carp can be conveniently cultured in such ponds. It has also been observed that some introduced fish are considered as food and aquarium species which are used for pest control. For the last 100 years or so, however, about 300 species of fish have been introduced from different countries of the world for various purposes.

Though the main objective to introduce foreign species in different countries is to culture them with selective native fish species, such attempts have created serious disastrous effects on different countries. In many cases such introduction has resulted in complete or partial extermination of many native species. According to one estimate, about 15 per cent of the total fish (160 fish species) introduced in 120 countries of the world has been found to be beneficial. Hence enough precaution must be taken before their introduction for the purpose concerned is recommended.

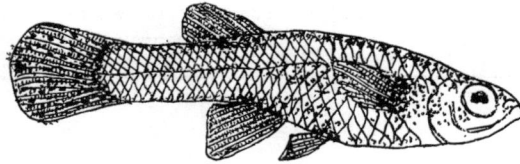

1. *Gambusia affinis*

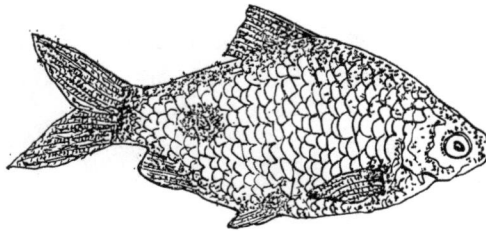

2. *Puntius javanicus*

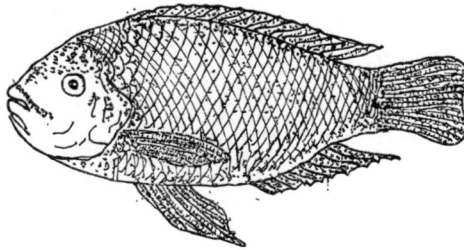

3. *Oreochromis mosambicus*

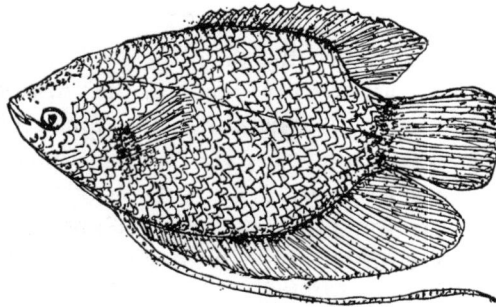

4. *Osphronemus goramy*

5. *Cyprinus carpio*

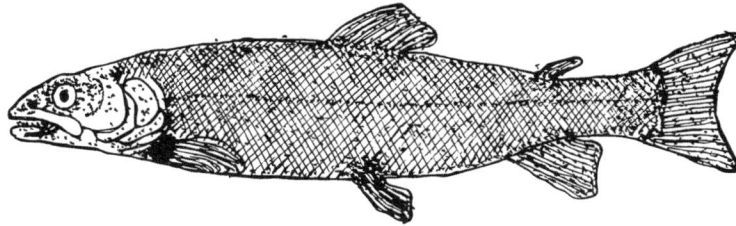

6. *Salmo salar*

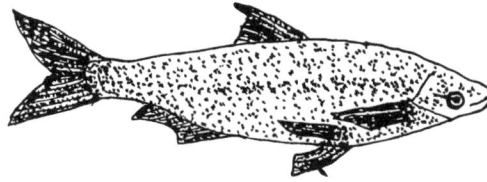

7. *Hypophthalmichthys molitrix*

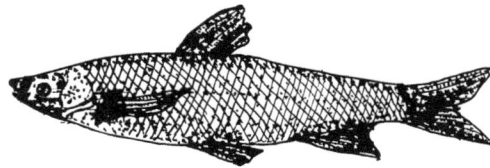

8. *Ctenopharyngodon idella*

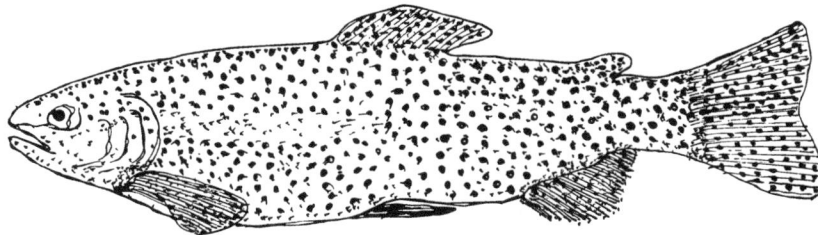

9. *Salmo gairdneri*

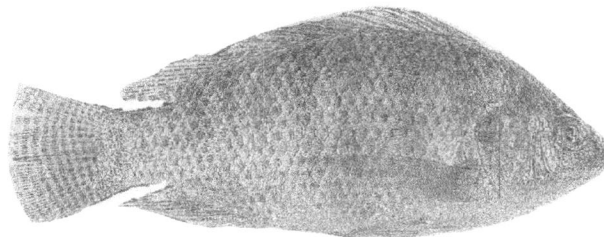

10. *Oreochromis nilotica*

Fig. 7.1 : Several species of fish were introduced in India from different countries of the world for the development of food fish and game fishing. These species have been well acclimatized under Indian conditions. Though most of the introduced fish are used as food, their culture has proved disastrous in many inland fishery resources. Some representatives are shown in this figure. For further detail, see sections 7.14 and 7.15

A brief description of breeding and culture methods of some exotic species and their impact on native fish fauna have been added insight into several sections that follow.

7.2 *Puntius javanicus* (Blfr.),Order: Cypriniformes,Family: Cyprinidae

The tawes, *P. javanicus* was introduced in India in 1972 from Indonesia for experimental culture. It is one of the high priced and fast-growing compatible minor carps for culture in confined ecosystems along with other major and minor carps. The fish is well-suited in North-eastern and Eastern regions of the country.

Characteristic Features

The body is compressed and moderately elevated with rounded abdomen. Head is short and blunt snout. Mouth is small, terminal and oblique. Eyes are moderate and not visible from below ventral surface. A supra-branchial organ is present. Scales are moderate and ctenoid type. Caudal fin is rounded in shape. Anal fin with 9-12 spines and 19-21 rays. A single dorsal fin is inserted above the tip of pectoral fin, with 11-14 spines and 11 or 12 rays. It is widely distributed throughout India, Sri Lanka, Pakistan, Mauritius, Malaya, Malay Archipelago, China, and Thailand.

Breeding

The fish breeds in ponds of Indonesia where riverine conditions are stimulated as in the case of dry bundhs in India. Hatchlings and fry are collected during monsoon season from different river systems of Indonesia. Induced breeding of this fish is, however, carried out by administration of fish pituitary extracts or other inducing agents.The mature female fish is injected twice at an interval of six hours at the conducive weather conditions. The first and second doses of pituitary extracts have been recommended as 5 and 8 mg/kg body weight, respectively. The male is administered at the rate of 6 mg/kg body weight at the time of second injection to the female; of course, even higher doses (7-8 mg/kg) may be given if the production of milt is not quite free at the time of testing brood fishes. A female fish weighing about 800 g can generate about 1.3 lakh eggs. Hatching takes place within 16 hours of fertilization. Spawning and hatching occurs at pH 7.5 and at water temperature that varies from 25° to 27.5° and 28° to 30°C, respectively.

The induced breeding of any carp in Indian sub-continent is mostly carried out in stagnant ponds which have received adequate fresh rain water, do not have algal blooms, and any weed as well as predatory fish. In such environments, the breeding hapas are fixed. Within these hapas, the injected brood fishes of both the sexes are kept for spawning. A breeding hapa is a rectangular box-shaped cloth container off fine mesh which is provided with a cover. It has an opening on one side (width-wise) afforded with cloth loops and button holes to enable proper closing.Through the loops a bamboo stick is passed to close the opening after keeping the injected male and female fishes inside the breeding hapa. The breeding hapa is fixed with the aid of bamboo poles of suitable length. 'The hapa is immersed partially in gently flowing water.The inner hapa is made of nylon cloth of large mesh size. The eggs are released into the inner hapa where they undergo development and hatch out from the egg shell. The hatchlings exhibit wriggling movement within the outer hapa while unfertilized eggs, dead embryos, and cast off egg shells remain within the inner hapa. After fertilization, the parent fish are removed.

In cases where induced breeding is done on a large scale, different types of hatcheries such as Chinese Circular Hatchery (see Volume 1, Chapter 11), low-cost modified chinese hatchery, aluminium bin hatchery, and glass polythene jar hatchery have been established (See Panel 7.1) Supply of clean, unpolluted, and oxygenated running water permits successful hatching of eggs, increased survival of hatchlings, and young fry.

Hatchlings are left undistributed till the third day after hatching when they are released in nursery ponds. After collection, fry are stocked in nursery ponds. Real success in production of quality seed of this exotic species is principally based on the adequate preparation of nursery ponds and scientific way of nursery management,thereby raising maximum number of advanced fry.

PANEL 7.1
FISH HATCHERIES FOR MORE FISH SEED

In case of the inland water fisheries, fish seed is produced in fish farms and they are stocked into the water bodies in order that the population of fishes does not decline over years of harvesting. The fish seed is obtained from three sources: (1) seeds of indigenous species are collected from the rivers, (2) seeds are obtained by carrying out dry and wet bund breeding, and (3) by inducing the fishes to breed in fish farms. Generally, commercially important species of fishes used for breeding at fish farms are the Indian major carps, Rohu (*Labeo rohita*), Catla (*Catla catla*), Mrigal (*Cirrhina mrigala*) and the exotic carps such as. Common carp (*Cyprinus carpio*) and its three varieties: scale carp, mirror carp and leather carp, Chinese grass carp (*Ctenopharyngodon idella*) and the silver carp (*Hypophthalmicthys molitrix*). These fishes are reared in ponds till they reach maturity and achieve 'condition' to breed. Except the common carp and its three varieties, all these fishes are induced to breed in confined waters by a technique termed as *Hypophysation*. This is the technique in which the pituitary extracts from its own species or allied species or synthetic chemical substances are injected into the breeders (Fig. 7.2) in order to induce them to breed in confined waters.

The eggs obtained in this way undergo fertilization soon after spawning. The fertilized eggs are transferred to the hatchery made of inner and outer hapas. The hapa is rectangular box-shaped (or mosquito net-like) container made of cloth immersed partially in gently flowing water (Figure7.3) The inner hapa is made of nylon cloth of large mesh size and the outer one is of fine mesh. The eggs are stocked into the inner hapa where they undergo development and hatch out of the egg shell. The hatchlings wriggle out into the outer bapa while the dead embryos, unfertilized eggs, and cast off egg shells remain in the inner hapa. The hatchlings are then transferred to the nursery ponds where they are fed and well maintained till they are transported to great distances. In the hapa hatchery, the mortality of fish seed is very often as much as 60 per cent. Sometimes, the loss may reach upto 90 per cent. In this context, however, the problem of fish seed mortality has been evaluated and five principal types of hatchery such as (i) Low-cost Chinese hatchery, (2) Aluminium bin hatchery, (3) Low Density Polyethylene (LDPE) jar hatchery, (4) Glass jar hatchery, and (5) Chinese circular hatchery, have been developed. All these hatcheries aim at providing oxygen-rich water to the embryos. To achieve this, hatchlings are maintained in cool and circulating water. The flow of water not only provides adequate oxygen to fish eggs and embryos but also prevents infection of embryos and rotting of dead materials. It removes fecal matter and excretory product of fishes which otherwise undergo decomposition and aggravate dissolved oxygen concentration. Slow current of water maintains even distribution of eggs, rolls them and frees them from gravitational and other pressures. Also, the cost of fabrication of hatcheries, recurring expenses involved in operating them, their efficiency, handling cost, and simplicity of the design are looked upon while designing fish hatcheries.

1. **Low-Cost Modified Chinese Hatchery:** This hatchery is the modification of Chinese hatchery to meet the requirement of farmers under Indian conditions (Figure 7.4). It consists of a circular cistern made of cement concrete having an inlet pipe fitted obliquely at an angle of 40° to the tangent of cistern wall and an outlet pipe from the centre of the bottom leading to another rectangular cistern constructed below the bottom level of the circular cistern. Gentle flow of water (0.3m to 0.6m/second) is maintained by controlling the inflow and the outflow at a constant level of 90 cm water in the circular cistern. Broodfishes are stocked into the cistern at the rate of 10 kg per 1.5 cubic metre of water where they are induced to breed. Immediately after spawning, broodfishes are removed from the cistern and a cylindrical strainer is fixed in the middle of the cistern to prevent eggs from escaping through the bottom outlet pipe. The strainer is a cylindrical bamboo frame wrapped with cloth and fixed to the bottom of the circular cistern around the opening of the outlet pipe. The eggs are held outside the strainer till they hatch. The hatchlings are either kept in the circular cistern or transferred to the rectangular cistern till they are marketed.

   There are several advantages of this type of hatchery. *Firstly*, the non-recurring expenditure towards fabrication of this type of hatchery is very low. *Secondly*, the mortality of hatchlings is generally not more that 10 per cent and *thirdly*, its efficiency is claimed to be almost hundred per cent.

2. **Aluminium Bin Hatchery:** Several models of this type, depending upon the magnitude of fish seed production, have been designed. The shape of this hatchery may be circular or rectangular. In all these cases, there is a perforated egg container kept inside the bin. Gentle flow of water is maintained and there is an arrangement to prevent hatchlings from escaping.

A single unit circular bin hatchery essentially consists of outer and inner cylindrical-shaped aluminium bin (Figure 7.5a). The outer bin (diameter 51 cm and height 55 cm) has the capacity of 81 litres and is provided with two inlet pipes at the top. Additionally, there is an extra outlet pipe in the bottom to remove the spawn. Two outlet pipes are fitted with wire mesh screen at either end that helps prevent the hatchlings from escaping. The inner bin is cylindrical in shape, having perforated wall and a sliding plunger lid that can be slid like a piston and can be adjusted to any desired height. The inner bin contains adequate quantities of eggs. The lid holds down the floating eggs from rising towards the surface. Three- (Figure 7.5b) and six-unit rectangular hatcheries are the modified version of the circular type.

The operational procedure involves loading of fertilized eggs in the hatcheries and the maintenance of water circulation. Eggs from the breeding hapa are loaded into the egg container and the plunger lid is slid down. Water circulation in bin hatcheries is maintained at the speed of 10-14 per cent volume of the hatchery container per minute. The egg container is removed after the hatchlings have wriggled out into the outer bin. The removal of egg container after hatching eliminates the decaying material that is likely to reduce the dissolved oxygen level of water.

These models are portable and can be installed at convenient field stations. They are highly useful in wet and dry bund breeding operations. About 2 lakhs of eggs can be loaded into one unit of these models. In addition, these models are of low cost, rust proof, and easy in handling

3. **LDPE Jar Hatchery:** This type of hatchery is modelled more or less on the same line as aluminium bin hatcheries. There is an improvement over the bin hatcheries by employing an array of showers and aerating devices. This makes the hatchery more riverine flood conditions. At the same time, bore hole water is used for circulating the hatchery complex which can be recycled.

The LDPE jar hatchery complex consists of 24 jars arrayed in 4 units, having 6 jars in each unit. Each jar hatchery (Figure 7.6) is made of low density polyethylene and is shaped like a bucket (lower diameter 32 cm, upper diameter 44 cm, height 44 cm, and 40 litre capacity). It is fitted with an inlet pipe near the bottom and an outlet pipe near the top. For effective hatching, an egg container is used. The container is an aluminium frame (lower diameter 22.5 cm, upper diameter 33 cm-and height 37.5 cm) wrapped around with mosquito net cloth (2 mm mesh). About 2.5 lakhs of eggs can be loaded in an egg container.

The bore hole water is pumped into the overhead tanks where it is aerated to raise the dissolved oxygen content from 3 to 10 mg/l before it is led to the hatching jars. The flow of water in the egg containers, after loading the containers with eggs, is maintained at the rate of 2 to 6 litre/minute/jar depending on the percentage of fertilization taking place and the species of carp from which the eggs have been obtained. Immediately after hatching, the egg container containing egg shells is removed. The water from the jar is conducted to the nearby plastic/cement tank by means of an open conduit. After emerging from the jar, the hatchlings flow along the conduit and are received in the hatching hapas which are partially submersed in the container. Overhead showers above the hapa keep the water cool and well-aerated to increase the content of its oxygen.

The construction and fabrication costs of this model are relatively high. The operation cost is, however, low and its efficiency is more than 94 per cent. Also, the time required for hatching is more or less half than that of hapa hatchery.

4. **Glass Jar Hatchery:** In this type of hatchery, the egg container is not used and an upward flow of water is maintained. Depending upon the magnitude of fish seed production, several jars are used The upward flow of water in jars having an inlet at the bottom and the spout acts as outlet at the top is the main feature of this hatchery. The vertical flow of water not only provides well-oxygenated water but also facilitates speedy escape of hatchlings. The hatchlings exhibit an upward movement and the upward movement of water helps them to rise to the surface and escape along the overflowing water. In fact, the vertical flow of water cushions the eggs and makes them weightless so that proper development of the embryos can takes place. The flow of water is so gentle and well-controlled that the water-hardened fertilized eggs of different ages assort themselves vertically inside the jar. While the younger eggs, being heavy, remain at the bottom, the older ones lose their specific gravity and rise towards the surface. Hatchlings at the surface of jars passively flow out through the spout into the conduit and then into the hapas kept inside the nearby cisterns.

The hatchery complex consists of (1) overhead tank (length 2.66 m, breadth 2.66 m, height 2.4 m) with a capacity of 32,000 litre of water, (2) 12 cement cisterns (length 2.05 m, breadth 2.05 m, height 1.0 m), (3) 12 overhead showers above the spawning cisterns besides the usual breeding cisterns and nursery ponds.

The hatchery complex is situated on an elevated ground so that water can be drained out from the hatchery complex. Water from the irrigation canal is made to flow in an open channel into the main storage tank that serves the fish farm. The water from the main storage tank is pumped to the overhead tank from which pipe lines with control valves are originaed to supply water to the jars, cistems, showers, etc.

A total of 16 jars are fixed vertically in the circular holes made in 2 rows (8 holes in each row) in the wooden top of each table. One of the jars in a row is made of glass and the rest are of fibre reinforced with polythene. The glass jar helps visual checking of the eggs in each row of jars. The jars are fixed in such a way that their spouts are directed outwards over the conduits placed below each row of jars.

A hatchery jar (Figure 7.7) is cylindrical in shape with a funnel-like bottom to which an inlet pipe is fitted. The jar is provided with a spout at its top end to permit the overflow of water from the open conduit fixed below each row of 8 jars. Water from the conduits flows into the hapa of cisterns via plastic hose pipes fixed below the bottom of the conduits which are closed at both ends.

About 0.5 lakh eggs can be loaded in each jar. The operation of jar hatchery is very simple and its maintenance is easy involving less man-power and expertise. The broodfishes are induced to breed inside the breeding hapas and water-hardened eggs are carried in plastic buckets and immediately released into the hatchery jars. A gentle upward flow of water is maintained at the rate of 800-1,000 ml/minute/jar for eggs of grass and silver carps and at the rate of 600-800 ml/minute/jar for eggs of major carps. After hatching, the flow of water is marginally increased to facilitate the rapid escape of hatchlings into the hapas inside the cement cisterns. The flow of water is maintained in the cisterns till hatchlings become 3-day old before they are stocked in nursery ponds or marketed.

5. **Chinese Circular Hatchery:** The most world-wide large-scale carp breeding practice to increase fish seed production is the construction of Chinese circular hatchery. Of course, such hatchery is generally used in large fams, although at present it is also installed in medium-sized farms. The greatest benefits of this type of hatchery, however, relate to high survival rate of hatchlings and production of quality fry. Better management systems involved in this type of hatchery helps promote quick supply of hatchlings to farmers. The production largely depends on the reduction of hatching time by 50 per cent than hapa hatchery. The design and construction of Chinese circular hatchery have been described in Volume 1 (See Section 11.29).

Culture Methods

Similar to Indian major carps, the culture of this species involves the rearing of fry and fingerlings in rearing and stocking ponds, respectively, harvesting and marketing. However, three day-old hatchlings are stocked in a well-prepared nursery ponds at the rate varying between 40 and 50 lakhs/hectare. The preparation of nursery, rearing, and stocking ponds are similar to that of carp culture ponds and have been described in Volume (see Chapter 20). Before stocking with fry, the pond is treated with agricultural limestone at the rate of 500 kg/ha/1 m depth of water to raise the pH for release of nutrients into the water during grow-out period. Subsequent fertilization programs involve the application of cattle manure, urea, and single superphosphate at the rates of 2,500, 170, and 250 kg/hectare, respectively at 15 to 20 days interval and is continued till harvest. Fingerlings (average length 100 mm, and weight 15 g) are then stocked at the rate varying between 4,000 and 5,000 number/hectare.

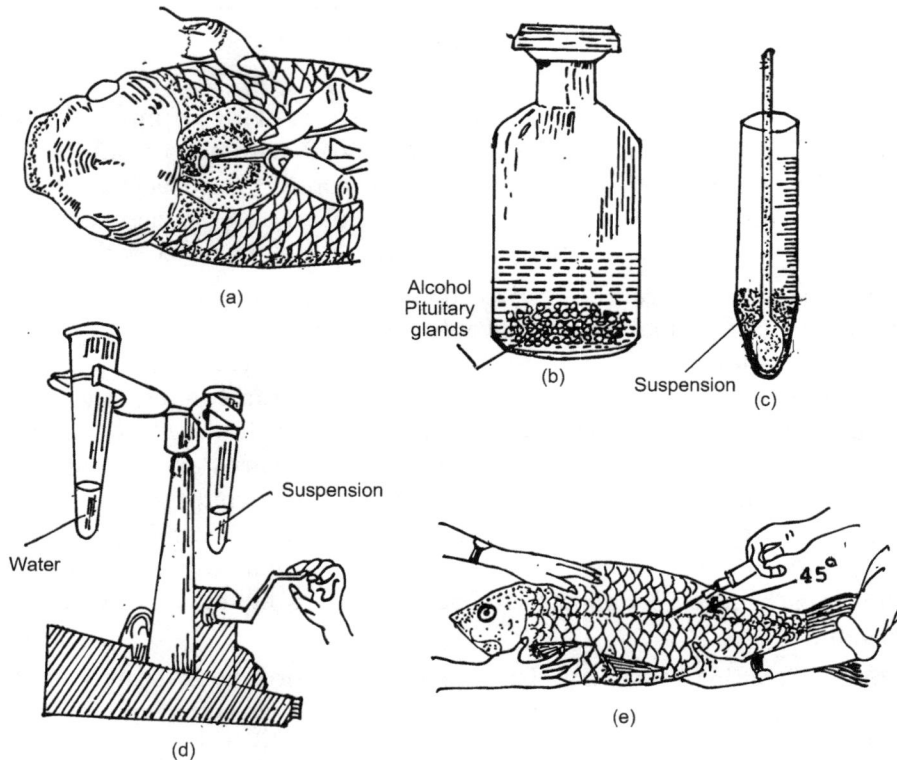

**Fig. 7.2 : Diagrams showing the sequential steps involved in induced breeding of fish.
(a) collection of the pituitary gland from a fish, (b) preservation of pituitary glands in absolute alcohol, (c) a tissue homogenizer, (d) a hand centrifuge machine, and (e) method of injection**

Food Habit

While fry thrive well on phyto- and zooplankton populations, the bigger fishes feed exclusively on aquatic macrophytes such as *Azolla pinnata, Lemna* spp. *Ottelia allismoides*, and *Vallisneria spiralis*. These plants are kept daily in a fixed place of the pond within a bamboo quadrant (2 X 2.5 m) tied with a nylon rope netting.

1. **Production Potential:** If the weight of fish stock and the number of fish per hectare is well-balanced, maximum return is possible provided that the application of supplementary feeds to the pond must be ascertained. And, an average harvest to the level of at least 1,150 kg/hectare/year is possible if monoculture system is adopted. When favorable conditions prevail such as optimum water temperature, use of supplementary feeds, and nutrient concentrations, production to the tune of 1,500 kg/ha/year may be obtained. It should be noted that paddy fields that are infested with soft and succulant aquatic plants may be utilized for the culture of this species.

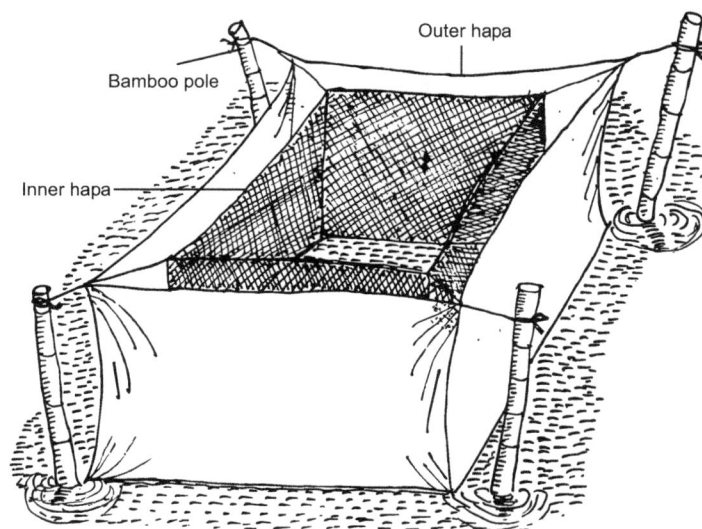

Fig. 7.3 : A hatching hapa is composed of an outer hapa (2m x 1 m x 1m) and an inner hapa (1.5m x 0.75m x 0.5m) and made up of marking and moquito net respectively. Eight ends of the inner hapa are tied with the corresponding 8 ends of the outer hapa. Hapas are partly immersed in pond water. A hatching hapa contains about 1 lakh fertilized eggs. After 15-16 hours, hatchlings emerge and they travel through the pores of the inner hapa. When all hatchlings are stocked in the outer hapa, the inner one containing dead and unfertilized eggs as well as egg shells are immediately removed. After 3 days, hatchlings are stocked in nursery ponds. In a breeding farm, series of hatching hapas are placed for large-scale hatchling production.

7.3 *Osphronemus gourami* (Lacepede), Order: Perciformes, Family: Osphronemidae

The *O. gourami* is a naturally occurring freshwater fish inhabited in Malayasia, Cambodia, Vietnam, Java, Mauritious, Indonesia, and Thailand. The fish was first introduced in Calcutta (India) during the first half of the nineteenth century from Java and Mauritious. The second and third consignments of this species was brought from Mauritious in 1865 and 1916, respectively, which has established itself fairly well in Chennai including the Cauveri river system. From Chennai it was introduced into other states of India where the fish has been successfully bred.

The fish thrives both in brackish and freshwaters, and being an air-breather, is tolerant to faul waters where the concentration of oxygen is very low. As a generalization, gourami is a freshwater fish which is sensitive to hot and cold and therefore suitable for culture in tropical waters. The fish can grow well and reproduce at 15°C and at an altitude of 720 m.

Gourami is a slow-growing fish and consequently cannot be cultured along with the major carps. The fish usually matures when it attains about 12 cm in length at the end of third year. However, in waters with ample forage fish, both in Mumbai and Cauvery river system, the fish grows to 10-15 cm and 15-20 cm in length respectively in the first year. In

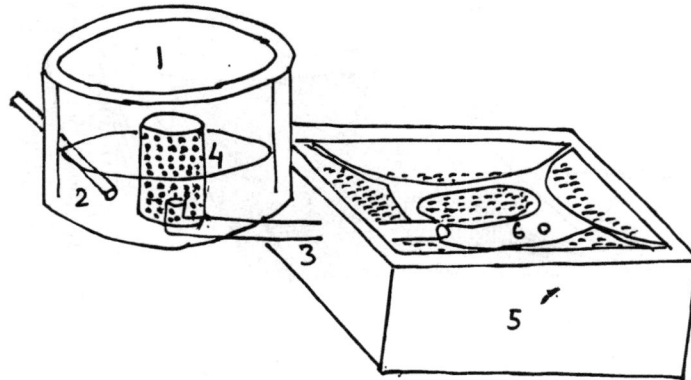

Fig. 7.4 : Low-cost modified Chinese hatchery. (1) Circular cistern, (2) Obliquely-fixed inlet pipe, (3) Outlet pipe, (4) Bamboo frame, (5) Cistern and (6) Hapa for maintaining hatchlings.

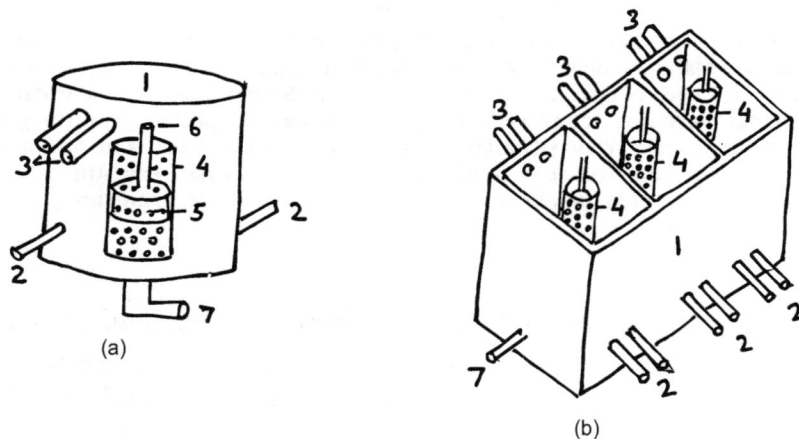

(a)

(b)

Fig. 7.5 : Aluminium bin hatcheries. (a) One-unit circular type and (b) Three-unit rectangular type. 1. Outer bin, 2. Inlet pipes, 3. Outlet pipes, 4. Inner egg vessel, 5. Plunger lid, 6. Rod adjusting plunger lid and 7. Drainage pipe

Cauvery river system, however, it reaches the table size of about 40 cm and 1.2 kg in length and weight, respectively.

Food

The adult fish subsists on aquatic plants but it also prefers to small frogs, insects, and small fish. Fry consume zooplankton, insect larvae, and soft aquatic plants. In general, however, the fish is purely vegetarian and prefers soft leaves of the plants especially *Nymphaea*.

Characteristic Features

The body is short and more or less elongated with rounded abdomen. Head is compressed with concave upper profile. Snout is rounded. Eyes are large, lateral and situated in middle of the head. Dorsal fin is inserted above the base of pectoral, with 15 or 16 spines and 10 or 11 rays. Spinous portion is longer than soft part. The soft part may be prolonged with filamentous tip. Anal fin with three spines. Caudal fin is rounded. Scales are cycloid.

Breeding

In open waters, the fish breeds in rivers and ponds during dry season (generally March-May) and in confined waters, it breeds throughout the year. Similar to *Clarias*, this fish is a nest builder and do not show any parental care. At the onset of breeding, the parents search for a suitable place where nests are constructed. Generally, nests are constructed within a week which is very much similar to that of bird's nest. Nests are camouflaged with green vegetation. Each nest is an oval-shaped having 40 cm in length and 30 cm in width, with a circular mouth of about 10 cm in diameter. When nest is constructed, a free space is kept below the nest. The space permits the fish to lay eggs and to produce water currents by means of fins. Generally submerged regions of ponds lined with stones/crevices provide adequate support to the nest.

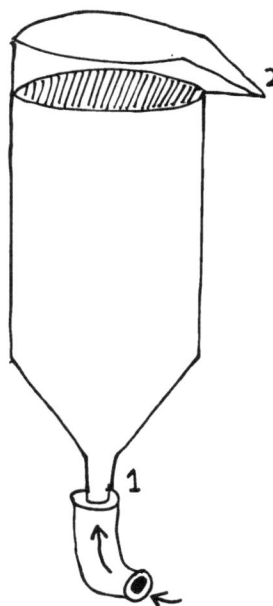

Fig. 7.6 : A single glass jar. 1. Inlet pipe, 2. Spout

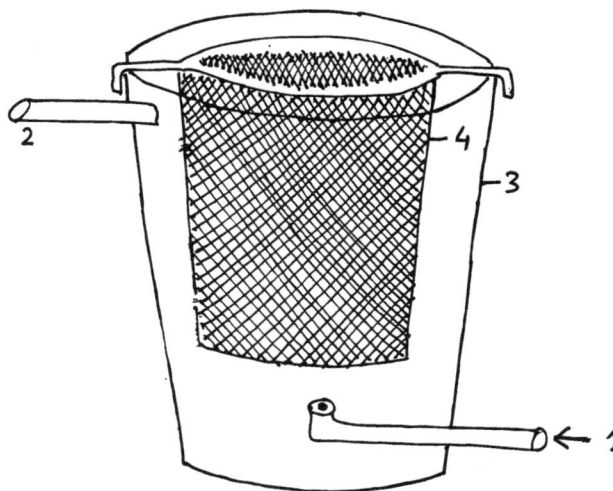

Fig. 7.7 : Low-density polyethylene with egg vesel. 1. Inlet pipe, 2. Outlet pipe, 3. Outer vessel and 4. Inner egg vessel

Each nest contains larvae varying between 600 and 3,000 depending upon the spawning capacity of the fish and environmental factors such as oxygen and temperature.

At the time of spawning, females lay eggs in batches within the bottom of the nest. Males collect a few nest materials, cover the eggs to protect then from intruders. It is assumed that either males or both the sexes guard the nest. A five year-old female fish may generate about 4,000 fry from each nest.

Fertilized eggs are yellowish in color, semi-transparent, pelagic, and non-adhesive. The diameter of the egg varies between 1.0 and 2.6 cm. At 36 hours of incubation, fertilized eggs are hatched into larvae. Larvae are transparent, 5-6 mm in length and devoid of mouth, fins, and gills. They either float upside down or attach themselves to aquatic plants with the help of the gland present dorsal to the head. Five day-old larvae feed on micro-organisms such as rotifers, ciliates etc. which are available in abundance after the nest materials have been decomposed.

1. **Breeding Methods:** Suitable aquatic plants and other nesting materials are the main prerequisites for breeding of this species. Though the breeding method varies in different countries, the method has been briefly described in a general way that follow.

   Breeding of gourami is carried out either in ponds where aquatic plants are grown in the marginal areas where nests are constructed or in ponds where the depth and area vary from 50 to 60 cm and 50 to 100 square metre, respectively having muddy bottom and steep bank. In these ponds, however, about 60 broodfish are kept.

   Male broodfish are characterized by the presence of black or darkish color at the base of the pectoral fin, whitish or light-brown color chin, flat caudal fin (when laid down), and devoid of lump in the inter-opercular space. On the other hand, female broodfish are characterised by the presence of whitish and yellowish colored pectoral fins and chin, respectively, and with hump in the inter-opercular space.

   The pond is provided with funnel-shaped nest frames made of bamboo sticks (1 metre long). Coconut fibres and tufts of plain leaves are also provided to encourage the fish for nest construction. Eggs are laid in nests which are attached to aquatic vegetation with adequate space below to allow the parents to fan the eggs during incubation. However, eggs when float to the surface of water, are collected and transferred to hatcheries. Hatcheries are either round glass troughs or small cement aquaria. Clean and unpolluted freshwater is circulated through hatcheries for oxygenation and temperature is maintained between 30 and 32°C. A few aquatic plants such as *Hydrilla* and *Vallisneria* are placed in the hatchery so that the larvae can attach to them for rest. In cases where circulation of water is not possible, water is changed at least twice a day. Whilst a week-old larva consumes copepods and cladocerans, a two weeks-old fry consume polished rice bran along with their diet.

Culture Methods

The fish gourami can be cultured either individually or in combination with other fish such as murrels, carps and air-breathing fish. In view of its tendency to consume small fish, it is suitable for culture gourami with carps only in the fatenning ponds. In derelict ponds, the fish can be stocked as a culture partner with murrels. It should be added as a note that due to slow-growth rate, the fish gourami cannot be culture with major carps for short intervals.

Fry can be reared in well-prepared rearing ponds. They are stocked at the rate of 10,000/hectare for about three months. Fingering ponds are generally 3 hectares in size and when early fingerlings are stocked at the rate of 5,000 - 6,000/hectare, fish of 120-600 g can be produced in about 12 months.

Grow-out of gourami to market size takes at least two years after hatching or one year from the fingerling stage. The usual market size varies between 120 and 600 g, though most of the fish are harvested when they attain a size varying between 800 g and 1.0 kg. However, the growth rate of gourami varies in different localities (Table 7.2). Table indicates that the fish takes a lot of time (at least 3 to 4 years) to attain more than 3 kilogram.

Table 7.2: Growth Rate of Gourami in Ponds

Locality	Growth rate (g)			
	1st year	2nd year	3rd year	4th year
Coimbatore (India)	680	–	–	–
Madras (India)	110	550	25,000	3,000
Maharashtra (India)	112	454	2,300-2,700	3,200-3,600
Indonesia and Philippines	120-450	450-680	2,400	3,800

Source : Menon et al (1959), Kulkarni (1946), Hora and Pillay (1962)

In India, fishes are kept in aquatic-infested perennial ponds where fishes are bred, allowed fry to grow under natural conditions and are harvested when they attain a market size that varies from 200 to 400 g/year.

Owing to its large-sized fleshy body, delicate flavor, herbivorous nature, and hardiness, it is suitable for culture in weed-infested ponds, shady cisterns, and in ornamental ponds. However, if it is cultured individually, production potential of gourami will definitely increase to a less extent for the table especially for the people of North-eastern regions and West Bengal. Due to its slow-growth rate, it is not at all recommended for culture with major carps and consequently, integrated culture of gourami with other fish has likewise been abandoned.

Though the fish is on the verge of extinction, it is very popular till today. It should be kept in mind that it fetches a high price in the market as in the case of air-breathing fishes and are sold as much as Rs. 180.00 per kilogram (Personal communication with fish farmers). However, this endangered fish must be revived through state-of-the-art culture, technology and if necessary, by the introduction of fresh and genetically-engineered stocks.

7.4 *Oreochromis mossambicus* (Peters), Order: Perciformes, Family: Cichlidae

Tilapia is the name used for many of the cichlid fishes (a very large family of freshwater fishes) native to Africa. Scientifically, it is used as a general name for edible species of the African genus *Tilapia*. They have been widely introduced into a large number of tropical and sub-tropical countries across the world in the last four decades either accidentally or deliberately. They breed in almost all types of freshwater ecosystems. Being herbivorous or omnivorous, it is necessary to feed the fish accordingly. They can be reared in fresh, brackish and even sea water. At present, however, the status of tilapia as a culture candidate has risen as a result of determined efforts made by farmers and technicians. The species has become so popular that farmers have begun to describe tilapia as the 'aquatic chicken'. It is clear that state-of-the-art technologies are now available for raising some of the species or hybrids of tilapia on a profitable basis; of course, with several problems are yet to be elucidated.

Characteristic Features

The body is short, more or less elongated with compressed head; abdomen is rounded. Mouth is terminal, large, cleft extending to below anterior border of eyes. Snout is rounded, eyes are large and lateral; thin lips with equal jaws. Teeth are present in two or more series. Dorsal fin is inserted above the back of pectoral with 15 or 16 spines and 10 or 11 rays. Spinous portion is longer than soft part. Anal fin with 3 spines, caudal fin is rounded. Scales are cycloid, air-bladder is large and simple.

Cultivated Species

At least 80 species of *Tilapia* have been described; of course, with considerable confusion on the taxonomic status of many of them. For commercial aquaculture, however,the most important species of tilapia includes : *Tilapia zillii, T. mossambica, T. hornorum, T. rendalli, T. nilotica, T. aurea, T. melanotheron, and T. Spilurus.* It is assumed that very few pure strains of these species are used in farms and natural cross-breeding has occurred in many regions.

Distribution

Tilapia includes fresh- and brackish-water species common in Africa. Its abundance in many of the Great lakes of Africa gives considerable economic importance in that continent. The only genus belonging to the family *cichlidae* represented in Indian waters is *Etroplus* which is also found in the Middle East.

There are 60 species of tilapia distributed over a large part of Africa. The size varies with the species. For example, *T. nilotica* and *T. galilaca* attain a size of about 21 and 17 inches, respectively. *T. mossambica* is known to grow to a maximum length of 16 inches.

Rivers of the east coast of Africa from Algoa Bay in South Africa to the West Shabeli river in Somaliland are the main natural habitats of the fish *T. mossambica*. It has been transplanted to many countries of the world such as Philippines, Thailand, Pakistan, Ceylon,

Malaysia, Java, India, Vietnam, Mossambique, Madagasker, Zambia, Uganda, Bangladesh, Indonesia, and Tanzania.

Though essentially a tropical fish, tilapias have been successfully introduced for commercial culture in sub-tropical regions and even in temperate regions for indoor culture under controlled temperature conditions.

Food of Tilapia

Tilapia is a herbivorous fish, the fry feed on zooplankton and epiphytic algae. 'The adult fish subsists on Myxophyceae, Bacillariophyceae and Chlorophyceae. When the concentration of phytoplankton decreases, fish consume insects, worms, fish larvae, crustaceans, and detritus. These organisms can easily be produced by fertilizing fish ponds.

The fish has a good preference to artificial feeds such as rice bran, mustard oil cake, groundnut cake, fish meal, kitchen refuse, and flour when cultured in freshwater ponds. For better growth, feed must contain about 30 per cent protein.

Breeding and Spawning

In tropical and sub-tropical countries, the fish breeds throughout the year but in temperate regions, reproduction is interrupted during winter season. The fish is reported to breed every 2 months. The maximum interval between successive spawnings is 45 days and that of minimum is 15 days. The frequency of spawning also varies depending upon the temperature of water, stocking density, climatic conditions and availability of food. The fish has water temperature preference of 26°C for successful breeding circumscribed by the lower and upper water temperatures of 15 and 40° C, respectively. However, if the temperature of water is maintained in aquarium above 21°C, tilapia breeds throughout the year.

Depending on the spawning type, the tilapias have been categorized under three main groups. *Firstly,* the substrate spawners which construct nests on the pond bottom and spawn in them, have retained the name *Tilapia. Secondly,* the mouth brooders which incubate fertilized eggs in the mouth of the male or female are under the genus *Sarotherodon,* and the *thirdly,* genus *Oreochromis* has been constructed which spawn in nests on the pond bottom and incubate the eggs in the mouth of the female fish. Since the fish readily breeds in almost all types of waters, no special breeding method has been adopted for mass production of hatchlings and fry. A simple method has, however, been described in a general way to produce a large number of *Oreochromis mossambics* fry using 800 m² earthen ponds of about 1 m depth. Female and male fish are stocked in one pond at an average density of one per two square metere in the sex ratio of 1 : 1. They are fed on a protein-rich diet for about a month. The male fish constructs a shallow pit of 6 cm deep and 25 cm in diameter. Sand and other debris are removed from the pit by mouth. Courtship behavior occurs within the pit where the female lays eggs in one lot and instantaneously the male fertilizes the eggs. The male fish leaves the nest and fertilized eggs are taken up within the mouth cavity by the female.

As soon as the spawning has occurred, the broodfish are then transferred to a second pond where they are fed in the same manner as in the case of first pond. Feeding is continued in the first pond till the fry reach a size of about 4 cm and an average 4000 fry are available at the time of harvesting. By this time, spawning occurs in the second pond and the broodfish are transferred back to the first pond for further spawning. Thus the spawning cycle is repeated till substantial quantities of fry are obtained for grow-out ponds.

1. **Rearing of Fry:** Removal of hatchlings and rearing them in special enclosures under controlled conditions not only helps increase spawning frequency and thereby overall fry production but also maintain the genetic purity. This is highly significant in hybrid production. Nursing of normal fry or mono-sex hybrids is carried out in nursery or rearing ponds where they are stocked at the rate varying between 50,000 and 100,000/ha depending on the size of fingerlings to be raised. If sex-wise segregation of fry is done manually, it is necessary to grow them to a size of at least 50 g or even a larger size of 100g so that secondary sexual characters can easily be distinguished.

Secondary Sexual Characters

Sex of tilapia can be identified when they attain a length of 35 mm. The male and female tilapia can easily be identified by the appearance of body color and genital papilla. The genital papilla of male fish is long and pointed. They have two openings in front of the anal fin. The genital papilla of female is round, blunt and have three openings. The body color of female is greyish without red margins of median fins, while the male is black in color with red margins on the anal, dorsal, and caudal fins.

Culture Methods

As noted earlier, tilapia exhibits parental care and the care bestowed on the developing young, thus ensure a high survival rate and rapid increase in numbers. These features account for the suitability of these species for culture in different types of waters. The fish has considerable popularity in many Asian countries for culture and has firmly established itself in various types of water bodies such as lakes, ponds, marshes, reservoirs, pools, and swamps.

Under favorable conditions, tilapia may attain a weight of about 800g at the end of the first year. The fish is cultured either individually or in combination with carps, murrels, and prawns. While fingerlings of tilapia are stocked either in (i) ponds/cages (ii) irrigated paddy fields (at the 2,500/ha rate), (iii) sewage-fed ponds (at the 6,000/ha rate) under mono-culture system, poly-culture generally involves stocking of carps along with tilapia (grass carp: silver carp: common carp: catla : rohu : tilapia = 250 : 1,000 : 2,500 : 500 : 625 : 625). Tilapias constitute an ideal candidate of real significance to fish culture in many Asian countries. Due to its hardiness, wide range of salinity tolerance, and frequent spawning habit, it has paved the way for culturing them in a variety of waters. Tilapia is cultured in a variety of ways that follow.

1. **Culture in Grow out Ponds:** The culture of tilapia is generally oriented to produce more fish of marketable size of at least 350g. When the culture system is based on unsorted stock, this can be accomplished by low stocking densities, intensive fertilization and feeding so that most of the stocks will have reached an agreeable size before they become sexually mature and initiate breeding animation. For this reason, a low stocking density of about 5,000/hectare is recommended to farmers. Culture ponds are regularly fertilized with cattle excreta, urea, single superphosphate, and lime at the rate similar to that of carp culture ponds.

2. **Culture in Cages:** Cages are sometimes used for grow-out of tilapia and fry or fingerlings are produced in land-based facilities such as ponds, lakes, and cisterns. Special double-walled cages are used for spawning of tilapia in open waters. Generally floating and fixed cages are used for grow-out of tilapia. While the floating cages are used in deep lakes, the fixed cages in shallow eutrophic lakes. The stocking density of tilapia in floating cages with the size of the cage but as a generalization, up to 25 fingerlings/m^2 of 3-4 cm length are generally considered. Artificial feeding is generally not carried out except in waters with low productivity. In 6 and 9 months, each fingerling grows to an average size of 230 and 330 g, respectively. The growth rate of tilapia is largely based on the productivity of the area concerned and the management protocols including the density of cages in the water body and the distance between cages.

   The stocking density of *O. mossambica* in fixed cages ranges from 25 to 50 fingerlings/m^2 and the duration of culture generally varies between 4 and 12 months. Without supplementary feeding, however, fingerlings (5 cm in length) stocked at 25 per square metre have been found to attain about 135 g in four months. The average production to the tune of 6.5 kg/m^3 with an average weight of 120 g per fish has been recorded in several Asian countries.

3. **Culture in Sewage Fed Ponds :** As in the case of carp culture in sewage-fed ponds, the culture of tilapia in similar ecological habitats has been widely accepted. A record production of tilapia to the tune of 7,000 kg/ha/year has been noted in Indonesia. Under Indian conditions, however, production of tilapia in domestic sewage-treated ponds is impressive that may vary from 4,850 to 9,534 kg/ha (Table 7.3).

   Prior to application of substantial quantities of sewage and sludge in fish culture, their dilution should be so effected that a positive dissolved oxygen balance is maintained and the concentration of unwholesome substances such as ammonia and hydrogen sulfide are kept below lethal limits. The oxygen necessary for biochemical reactions is obtained from freshwater used for dilution and through photosynthesis by green algae and aquatic vegetation.

   A sewage irrigated pond can be utilized for massive production of tilapia. This is due to the fact that tilapia populations have been found not to be affected even at high level of ammonical nitrogen (5.40 mg/l). In such ponds the net fish crop attributed to sewage ranged from 8,000 to 9,350 kg/ha/year. Mixed culture in such ponds with *Clarias batrachus* and tilapia is highly encouraging and the production has been estimated to be as 1,320 and 2,000 kg 8/months, respectively.

While tilapia has been cultured successfully, extreme care is necessary not only to keep the pond aerobic throughout the day and night by regulating the sewage inflow to prevent fish from mortality due to oxygen depletion in water but also to avoid the accumulation of sludge in the bottom mud. Accumulation of bottom sludge creates and anaerobic conditon of the pond bottom that adversely affect the nest building habit of tilapia.

Table 7.3 : Production of Tilapia in Sewage-fed Ponds

Experi-ment number	Fertilization using effluent (in litre)			Stocking density (kg/ha)	Dilution ratio (Sewage: water)	Yield (kg/ha)
	Pre-stocking	Post-stocking	Total			
1.	22×10^5	358×10^5	380×10^5	17,000	1 : 3	9,350/year
2.	30×10^5	189×10^5	219×10^5	35,000	1 : 2	4,850/7 month
3.	234×10^5	238×10^5	472×10^5	20,000	1 : 2	9,534/14 month

Source: Ghosh *et al* (1980)

4. **Culture in paddy Fields:** Fish adapted for rearing in paddy fields must fulfil the following conditions: (i) they must be able to withstand shallow water, (ii) they must be able to tolerate high temperatures and low dissolved oxygen level, (iii) they have rapid growth, (iv) they can resist cloudy water, and (v) they will not exhibit a tendency to escape. The production of farmed fish in paddy fields includes carp, murrels, catfish, and tilapia. Fish production in paddy fields can be carried out with either captured or farmed fish and production can be either alternate or simultaneous. In the first case, fish and paddy are harvested alternately. When production is simultaneous, then the fish and paddy are grown together. Generally paddy and fish are harvested once per year each. In some cases, a triple harvest in one year is adopted: one for fish and two for paddy. Other methods, though more complicated, can afford five harvest of fish or rice over a period of two years. It should be kept in mind that different techniques adopted in the cultivation of fish in paddy fields differ considerably from region to region.

    Alternate production admits better care of fish and paddy as well as mechanical means, pesticides, and insecticides for the paddy. After the paddy has been harvested, the field is converted into a temporary pond. At the same time, embankments and outside trenches are also constructed. The pond is then fertilized with organic manures such as rice straw, compost, vegetable wastes, cattle manure, and liquid manure all of which help to increase the production of the pond. In this method, about 10,000 of 1 cm fry are stocked per hectare for the production of fingerlings. Simulataneous production of tilapia and paddy has advantage. *Firstly*, it allows perfect use of the available ground. *Secondly*, it is good for paddy because it creates hygienic medium through the control of algal blooms. *Thirdly,* it produces animal proteins cheaply because the fish is a complementary harvest and the cost of production is virtually, nothing.

Though the simultaneous production system is the real paddy-fish cultivation, it presents certain disadvantages. First, in deep water all varieties of paddy cannot tolerate. Second, certain soils cannot be kept under water for prolonged periods. Third, the flow of water must be greater than would be necessary for paddy cultivation and this limits to promulgate the paddy fish cultivation. And fourth, Paddy-fish farming drastically limits the adoption of modem agriculture techniques particularly mechanization, fertilization, manuring, and the use of pesticides as well as herbicides. For these reasons, paddy-fish cultivation is being abandoned in many developing areas of the world.

When the simultaneous method of farming is adopted about 7,000 fry of 2-3 cm in size are stocked along with a few hundred adult tilapia just one week after the planting of padi. The first harvest is done about 40 days later and the large-sized fish are sold. The under-sized fish are returned to the field for several weeks.

Production of tilapia in paddy fields is significant. Methods which are based on cultured fish generally give low to moderate production. The production of tilapia in Vietnam, for example, varies between 250 and 380 kg/ha/4-months. In Taiwan where paddy-tilapia farming is more widespread and covers thousands of hectares, harvests vary between 200 and 450 kg/ha/4-months. The yield of paddy has not been seriously affected by the integration of this species. In India, the culture of tilapia in irrigated paddy fields has, however, shown very low production – it generally varies between 30 and 50 kg/ha/4-months.

The production of tilapia in paddy fields may increase to a great extent if ecological conditions permit for the fish to grow. But even if the production of tilapia is much higher, other selected exotic fish species will continue to be the bulk of the fish crop.

5. **Culture in Carp and Murrel Ponds:** Tilapia and carp (*Labeo fimbriatus*) have been growing together in ponds. In areas where tilapia is not the principal object of culture, the possibility of tilapia culture exists for its being used as a forage fish. In this case, the fish has equal value because it will be a useful species of developing 'balanced' fish populations in ponds. A community of tilapia, common carp, and gourami has been found to work extremely well and the total production is highly encouraging.

   In ponds where murrels are stocked, tilapia is used as forage fish. Production of murrels is highly related to the stocking density of tilapia. According to one estimate, the production of tilapia when cultured individually, was doubled than the total production of tilapia and murrels. On the other hand, the total production of murrels was very low when tilapia was not stocked in the same pond.

   A well-managed composite culture pond may yield about 900 and 5,000 kg/ha/year of carp (grass carp, common carp, silver carp, catla, and rohu) and tilapia, respectively.

* *Food and Feeding of Tilapia:* A variety of food-stuffs have been used in tilapia ponds. These includes plant leaves, rice bran, oil seeds and oil cakes, manoic and brewery wastes. Progressive fish culturists sometimes use chicken diets or the more expensive trout foods. In most of the cases, foods are manufactured using locally-available cheap ingradients. A variety of food formulations are considered for mass production of tilapia such as (i) 65 per cent rice bran, 25 per cent fish meal and 10 per cent copra meal; (ii) cotton seed cake (80 per cent), wheat flour (10 per cent), cattle blood meal (8 per cent), and calcium phosphate (2 percent) and (iii) 20 per cent protein, 40 per cent polished rice, 12 per cent wheat bran, 10 per cent peanut oil cake, 9 per cent fish meal, and 1 per cent oyster shell.

* *Harvesting of Tilapia*: Harvesting schedules depend very much on the stocking density of fry and climatic conditions in the region. If unsorted stocks are used or if the sorted stocks contain some females, harvesting is done before wild spawning has occurred.

Similar to catfish, seining and draining are the two important methods of tilapia harvesting from ponds and rice fields. Partial harvesting is generally done by seines, but differences in collecting fish have been observed between species and hybrids. All male hybrids of *Tilapia hornorum* x *T. nilotica* are reported to be collected more easily from ponds. While *T. hornorum* can be collected easily, *T. nilotica T. aurea* and *T. mossambica* avoid seines and therefore, repeated seining are necessary to collect a substantial amount of the stock.

7.5 Disadvantages of Tilapia Culture in Carp Ponds

One of the possible dangers from the culture of tilapia in India is its incompatibity with the existing species such as carps. Their prolific breeding habit and the elaborate parental care leads to the over-population in carp culture ponds, resulting in severe competition for food, space, and oxygen,thus yielding a population of stunted individuals. These stunted individuals have no good market value and consumers do not use for the table. At the same time, the habit of tilapia in constructing pits at the pond bottom during their breeding season and the method of searching food are very harmful to the growth and development of bottom-dwelling organisms which constitute the food for many species of fish. For this reason, the culture of tilapia in many Asian countries has become unpopular among fish farmers.

Carp nurseries are severely affected by the presence of tilapia. Carp hatchlings and fry are not only used by tilapia as food but compete with the food of carp fry as well.The presence of tilapia in nursery and rearing ponds drastically reduce the survival of carp hatchlings and fry by at least 70 and 55 per cent, respectively. Also, the growth of *Cyprinus carpio, Catla catla, Labeo rohita* fry/fingerlings in rearing/stocking ponds has been found to reduce by at least 40 per cent. The percent reduction in growth and yield of carp is highly variable and depends upon the stocking density, species combination, and the interaction of abiotic-biotic factors of water and fish growth. However, production of Indian major carps in cement tanks (10 m^2 each) fertilized with urea, superphosphate, cattle manure with supplemental feeding has been found to be affected by the presence of tilapia (Table 7.4)

Table 7.4 : Production Potential of Indian Major Carp Fry in Cement Tanks in Combination With Tilapia. Duration of Experiment was 90 Days

Species	Stocking		Harvesting		Average weight increase (g)	Calculated production (kg/ha/yr)
	Total wt (g)	Average wt (g)	Total wt (g)	Average wt (g)		
Labeo rohita	1.4	10.5	2.5	30.0	19.5	78
Catla catla	2.5	15.0	3.8	34.5	19.5	78
Cirrhinus mrigala	1.0	10.0	2.5	32.0	22.0	88
Oreochromis mossambicus	2.5	5.5	6.0	53.5	48.0	192

Source : Sarkar (1983)

7.6 Mono-sex Culture of Tilapia in Ponds and Production Potential

High reproductive potential of tilapia in culture ponds is the main constraint that prevents its successful farming on commercial basis. Unwanted propagation of different species of tilapia results in over-crowding and dwarf populations. Among different species of tilapia, *Oreochromis mossambica* (Peters) and, *O. niloticus* (Linn.) have great aquaculture potential in India and neighbouring countries. The problem of high reproductive rate can, however, be overcome by adopting a mono-sex culture technique. Mono-sex culture of tilapia in well fertilized ponds is more advantageous than mixed sex culture. Production of mono-sex tilapia in semi-intensive culture system has been found to vary between 2,500 and 9,000 kg/ha/year. Although mono-sex culture produces large-sized tilapia (weight ranged from 300 to 550 g per year) for the table, accidental introduction of female tilapia into the mono-sex culture ponds will definitely reduce the gross production of mono-sex fish through breeding and over-crowding by the females.

Role of Androgen in the Production of All-male Tilapia

It is needless to mention that triploid all-male and all-female populations of tilapia are suitable for aquaculture systems. All-male or a relatively high proportion of male tilapia can be produced by administering androgens orally to the sexually undifferentiated fry or by immersion into the steroid solution. Androgens (male sex hormones) such as 17-alpha methyl- testosterone (17α MT), 17- alpha ethyl testosterone, mibolerone, 19-norethisterone, mesterolone etc. are used for production of all-male tilapia fry. On the other hand, oestrogens (female sex hormones) such as oestrone, diethylstilbestrol, 17 -β - oestradiol etc. are also used to yield all-female tilapia fry. Also, all sterile triploid tilapia can be obtained by chromosomal manipulation. Generally, androgen sex-reversal tilapia is preferred than oestrogen sex-reversal one because all male tilapias increase in weight up to 70 per cent and thus desired marketable size of tilapia can be attained in a short period of time. However, the dosages of male sex hormone required for producing hundred per cent male *Oreochromis mossambica* varies from 1 to 100 mg/kg diet fed over one week. A dose of mesterolone of 200 mg/kg feed may be recommended for the production of all-male *O. niloticus* populations. Genetic engineering coupled with hormonal sex-reversal and selective breeding and adoption of super-intensive farming techniques may yield tilapia up to 350 metric tonne/ha/year.

Mono-Sex Tilapia Culture in Wastewater Fed Ponds

A wastewater treatment technology has been developed by the Vorion Chemicals and Distilleries Limited (VCDL), Chennai, Tamil Nadu, India, where effluents are converted into medium which permits successful culture of catfish and hybrid tilapia (grey tilapia X red tilapia). The culture of hybrid tilapia in such waters has great practical significance. It grows to a size of about 500 g in four months. The production potential of all-male tilapia in super-intensive, intensive, and semi- intensive culture systems in distillary effluent-treated waterspreads has been estimated to be as 350, 120, and 50 metric tonnes/ha/year, respectively. The stocking density of all-male tilapia in these culture systems varies between 20,000 and 1,25,000/ha. The VCDL's tilapia culture station is regarded as one of the largest and well-managed farms in the world and it is hoped that this typical tilapia farm will definitely encourage the large-scale farming practices in other regions in the near future where tilapia constitutes an important fish both for domestic consumption and overseas market.

7.7 Interactions of Tilapia and Carp in Ponds

There are two prime reasons for concern over the detrimental effects of hatchlings and fry when tilapia is stocked in carp culture ponds. Section 7.5 reminds us that the production of major carps has been seriously affected by the presence of tilapia. A second and equally important reason is that there is severe competition between tilapia and the carp food in culture ponds. Excessive population growth of wild tilapia takes place in ponds, ultimately create a harsh ecosystem with disastrous effects on carp populations. This situation can be overcome if mono-sex (all-male) or hybrid tilapia is stocked in carp culture ponds and the effects of interaction between tilapia and carp growth have been found to be synergistic. According to an estimate, however, when mixed sex population of *O. niloticus* (15 g each) and *Cyprinus carpio* (10 g each) were stocked in Egypt ponds at the rates of 110 and 1,420 kg/ha, respectively, net production of tilapia increased with increase in nitrogen fertilization (nitrogen was applied at the rate of 1 kg N/ha/day and 3 kg N/ha/day till harvest). Total net production of tilapia ranged from 755 to 1,100 kg/ha/6-months and that of tilapia and common carp ranged from 988 to 1,393 kg/ha/6-months. In well-fertilized and feed pellet application ponds stocked with common carp, hybrid tilapia (male *O. aureus* X female *O. niloticus*), *Hypophthalmicthys molitrix* (silver carp), and *Ctenopharyngodon idella* (grass carp), the yield of hybrid tilapia was obtained to the tune of 547 to 905 kg/ha and a total yield of 1,880 to 2,655 kg/ha. A carrying capacity of 2,000 - 3,000 kg/ha has been found in fertilized ponds stocked with *O. niloticus*.

Marked increase in tilapia yield in combination with carps does not make a hassle where suitable techniques are being followed. In composite fish farms, tilapias are sometimes cultured in limited areas and low to moderate amounts of fish are harvested. If the strategy is not adequate for culture, carp grown in tilapia-infested ponds is apt to exhibit less production. This fact emphasizes the need for adoption of state-of-the-art technology in producing tilapia stocks and for serious attention to meet the need.

7.8 Blue Revolution of Tilapia

The international centre for living Aquatic Resources Management has declared that a new type of freshwater fish *Oreochromis niloticus* (Nile tilapia) is sparking a Blue Revolution in many Asian countries and accordingly, the FAO (Food and Agricultural Organizations) has claimed that the blue revolution of tilapia will provide the global food fish supply as the capture fishery potential gradually declining. Several Asian countries such as China, Philippines, Thailand, Malayasia, Vietnam, and Indonesia have developed genetically engineered farmed tilapia or so-called *super tilapia* the growth of which is more fast, reach a marketable size in four and six months when reared in cages and ponds, respectively without feeding. The fish also survive well than the wild variety. Intensive culture of super tilaipa reduces the cost of production thus increasing the income of farmers. The income potential from a super tilapia culture pond is highly encouraging. A farmer having one hectare pond can earn about Rs. 95,000.00 (1,950 US dollars) per year by intensive farming of super tilapia. This scenario will undoubtedly encourage the fish farmers to cultivate this genetically-improved fish in future. The production has reached at such a high level that the tilapia is being considered as *Aquatic Chicken*. China, Indonesia, Taiwan, and Thailand are considered as leading exporters of tilapia to the European and US markets.

7.10 *Ctenopharyngodon idella* (Valenciennes),Order: Cypriniformes, Family: Cyprinidae

The grass carp, *Ctenopharyngodon idella*, also called as *White Amur*, is a large chinese freshwater fish and farmed for food in southeast Asian countries. The fish is a natural inhabitant of the rivers of China and in the river *Amur* in Russian Federation. This species has been introduced in many countries of the world and is now being extensively cultured in India, Thailand, Russian Federation, Ceylon, Burma, Taiwan, HongKong, China, Japan, Hungary, Poland, Australia, Singapore, Israel, and Philippines. This species was, however, introduced in India in 1959 from China and stocked in the ponds of Central Inland Fisheries Research Institute, Cuttack, from where its fry were distributed in different regions of the country for cultivation.

Characteristic Features

Body is moderately elongated, sub-cylindrical anteriorly and compressed posteriorly with rounded abdomen. The head is depressed and flattened with obtusely rounded snout and terminal mouth. Lips are thin and devoid of any lobes. Upper jaw is slightly longer than lower and protractile. Barbels are absent. Dorsal fin is inserted slightly ahead of pelvics with rays, anal fin is short with 10 rays and well-forked caudal fin. Scales are cycloid and large. Lateral line is continuous.

Tolerance to Ecosystem

Although grass carp is a freshwater fish, it can tolerate in slightly brackish water ecosystems. The fish can able to tolerate wide range of temperature (16-40°C), pH (5-9.), turbidity (125-215 cm), dissolved oxygen (1-25 mg/l), total alkalinity (80-620 mg/l), salinity (7-8 ppt), and ammonia-nitrogen (0-3.5 mg/l). These features clearly indicate that this species can be used either in mono-culture or in mixed culture system with the Indian major carps under wide range of environmental factors. Due to this tolerance capacity, the culture practice of this species has spread out rapidly in many tropical countries.

Importance of Gras Carp

Grass carp, though most common in India and other adjacent countries, is considered for controlling aquatic vegetation. The fish is considered as an aquatic weed cleaner and therefore is preferred by progressive fish farmers. Almost all kinds of vegetable matters are consumed. Because of its fast growth rate, it is considered as an ideal candidate for culture successfully along with Indian major carps. Some fishery scientists have, however, advocated that there is no report as to whether the fish is highly efficient for weed control program and for culture along with the major carps.

Though grass carp consumes aquatic vegetation more than its own body weight, incomplete digestion of food stuffs generally occurs as evidence by the fact that the excreta contains semi-digested food particles. These undigested food matter serves either as pond fertilizer or as food for other fishes such as silver carps. It has been estimated that for three grass carps, one silver carp can be reared without any investment.

Food of Grass Carp

Fry of grass carp consume crustaceans, rotifers, unicellular algae, and chironomid larvae. Early fingerling (15-20 mm in length) consumes small-sized macrophytes. When the fish attains to a length of 35 mm, macrophyte forms the bulk of its diet and thereafter the fish becomes a vegetarian. Among different macrophytes, *Wolffia, Lemna, Vallisneria, Azolla,* and *Hydrilla* are the most important genera and consume them well. The quantity of macrophytes consumed by grass carp has been given in Volume 1 (see Chapter 4).

In regions where aquatic plants and other vegetable food-stuffs are not available in adequate amounts, the culture of this species is very difficult due to its high food requirement. Since they consume a lot of vegetable matters a disagreeable smell always exists in its flesh that makes the fish less attractive to consumers.

Maturity, Fecundity and Spawning

The maturity of grass carp is highly variable and depends on the climatic conditions. Under Indian conditions, for example, the male and female fish mature in the second and third years, respectively. But in China, the fish is reported to attain maturity at the age of 5 years when they weigh 7 kg. The fish, however, attains to maturity in about 4, 5, 7, 8, and 10 years in Turkmenia, Krasnodon, Europe, Ukraine, and Russian Federation, respectively.

The female grass carp weighing 4.5 and 7.0 kg may generate about 3.8 and 6.2 lakh ova, respectively. Under Indian conditions, the fecundity of this species in relation to body weight is shown in Table 7.5. The fecundity is very less (average number of eggs is 600 per g ovary weight)and for this reason, adequate quantity of seed is not available. Also, due to low survival rates of fry and fingerlings, the fish is not available all the time in the market.

Table 7.5 : Body Weight and Fecundity of Grass Carp, *Ctenopharyngodon idella*

Fish weight (g)	Weight of ovary (g)	Weight of fish: weight of ovary	Total number of eggs produced
4,766	540	6.7 : 1	3,72,600
4,880	744	6.5 : 1	5,63,900
5,476	656	8.3 : 1	3,96,200
5,830	880	6.6 : 1	4,41,700
7,036	553	12.7 : 1	3,08,800

Source: Selected data from Alikunhi *et al* (1963)

Similar to Indian major carps, the fecundity of exotic ones particularly grass carp and silver carp is fairly high. In general, fishes which do not exhibit parental care generate large number of eggs possibly to compensate for the loss of eggs and fry as a result of unfavorable environmental conditions.

Grass carp though mature in ponds, yet they do not ordinarily breed there and are considered difficult to spawn in confined waters. Similar to other carps, the maturity of grass carp is extremely governed by the age, the type and quantity of food available in the ecosystem involved, and certain environmental conditions. On attaining the proper age, the gonads begin to mature gradually with the advent of breeding season. Both temperature and light govern the sexual developnent and spawning activities in this fish. At the same time, rainfall and monsoon floods are, to a great extent, also equally important for proper maturity and spawning of grass carp in natural habitats. Sudden dilution of pond or river water has a very favorable influence on the full maturity and spawning of this fish. If monsoon is delayed, there is relative delay in attaining the maturity and spawning of grass carp; but continuous or adequate rainfall expedites maturity. However, this Chinese carp breeds naturally in flooded river conditions during monsoon and definitely this season is very suitable for taking up induced breeding programs.

Breeding

Similar to major carps, the sequential steps of induced breeding of Chinese carp, grass carp and silver carp involve (1) collection, stocking and management of broodfish, (2) selection of broodfish and formation of a breeding set, (3) collection and preservation of pituitary glands, (4) preparation of pituitary extracts, (5) determination of dosage, (6) injection protocol, and (7) creation of breeding environment. However, in rivers, flood and rain, water current, increase in dissolved oxygen content of water, optimum water temperature, spawning grounds (shallow innundated areas), and spawning congregations are some of the most important environmental factors necessary for natural spawning.

1. **Collection, Stocking and Management of Broodfish:** The broodfish may be produced in composite fish-seed farms or collected from outside sources and stocked them in well-prepared broodfish ponds. The size of a broodfish pond varies from 0.2 to 0.6 ha with an average depth of about 1.5 m. The Chinese carps may be stocked at the rate of 1,500-2,000 kg/hectare.

    Adequate quantities of supplementary feed (ground nut cake + rice polish/wheat bran in 1 : 1 ratio) are given daily to the stocked broodfish. Suitable aquatic plants or fodder grass must be provided for grass carp.Under optimum feeding conditions, the fish attains to maturity and generate large number of quality eggs. As a generalization, both natural food and supplementary feeding are necessary to raise a good stock of broodfish. Periodical examination of selected broodfish such as their condition, progress in maturity, and for any serious parasitic infections is badly needed. Proper management of broodfish ponds includes periodical aeration, re-circulation of water, and treatment of diseased broodfish.

2. **Selection of Broodfish and Formation of Breeding Set:** Both age and size of the broodfish are important considerations in the selection of robust broodfish for hypophysation. In general, 2 to 4 years old carps and weighing 2 to 4 kilogram are the best for induced breeding. To avoid injury, broodfish are transported either by a hand net or by using special containers charged with oxygen. The transported broodfish should be immediately disinfected upon arrival with the solution of potassium permanganate (@ 10 mg/l) or acroflavin (@ 1 mg/l).

    The success of induced breeding operation entirely depends upon the selection of male and female brodfishes. At the onset of breeding season, some   characteristic features of carps (both Indian major carps and Chinese carps) appear. In case of males when the abdomen is gently pressed, milt will ooze out. The females with soft, bulging abdomen, pinkish, and slightly protruding vent are considered to be good for breeding. In grass and silver carps, the eggs are copper-red and blue in color, respectively, indicating that the eggs will produce robust fry after fertilization.

    It should be added as a note that, it is very difficult to select a fully-mature female grass carp because the swelling manifestation of the abdomen is also appeared immediately after feeding. To avoid this problem, approaching females are attenuated in breeding hapas for 8 to 10 hours at least by digesting the food materials.

A breeding set generally consists of one female and two males. Such a set is kept inside the breeding hapa after the injection of pituitary extracts.

3. **Collection and Preservation of Pituitary Glands:** Generally carps themselves serve as suitable donor fish. The pituitary glands must be collected at the onset of breeding season from ripe and gravid fish which are mature or approaching to maturity. The hormonal potency.of pituitary glands in these fishes increases with the advancement of maturity, reaches the maximum at the onset of breeding and lost as soon as spawning is over. However, pituitary glands from immature fishes and those where regression of gonads has just commenced or the spent fish should be avoided.

The pituitary gland is collected after removing the head from the body. A portion of cranium roof is first chopped off to expose the brain. The pituitary gland is generally located below the brain just posterior to the crossing of the optic nerve. In Chinese and Indian Major Carps, the gland is situated behind the floor of the brain box. A membrane termed as *duramater*, that covers the gland is carefully detached, the gland is exposed fully and taken out without causing any injury (Figure 7.2a). The gland is also easily collected by making an aperture in the region of the foramen magnum on the posterior side of the brain without any deformation of the head. Through this aperture the gland is taken out by a bent needle.

After collection, the pituitary glands are stored safely for several months without any loss of potency. Some common methods adopted for the preservation of pituitary glands include (a) preservation in alcohol, (b) preservation in acetone, and (c) freezing of glands.

(a) *Preservation in Alcohol*: Since pituitary hormones are readily solube in water and not in absolute alcohol, the latter acts as dehydrating and defattening agents. The glands are kept in air-tight glass phials containing fresh absolute alcohol (Figure 7.2b). These phials are stored in a dark place at room temperature inside a desiccator containing anhydrous calcium chloride. During storage, it is advisable to change the alcohol once in 90 days.

(b) *Preservation in Acetone*: The pituitary glands are preserved in acetone which is also considered as a dehydrating and defattening agent and are kept inside the phials. Acetone is kept in phials and are placed in a refrigerator. After 40 hours of storage, glands are removed from the phials and are dried with filter paper at room temperature. These dried glands are preserved in air-tight sterile phials and are kept either in a desiccator or a refrigerator.

(c) *Freezing of Glands*: In regions where electric power supply is available in large farms and fish farmers are able to bear the cost of power, the glands, after collection, are preserved in refrigerator so that the glands are easily available during hypophysation.

4. **Preparation and Preservation of Pituitary Extract:** For preparing the pituitary extract, required quantity of glands are taken out of the absolute alcohol, dried with a filter paper, kept inside a tissue homogenizer (5-20 ml capacity), and then macerated with a little physiological saline solution or distilled water (Figure 7.2c).

The macerated gland suspensions are injected either without centrifuging (if intraperitoneal injection is given) or after centrifuging (if intramuscular injection is given) the suspension. For intramuscular injection, the gland suspension is centrifuged either electrically- or hand-operated machine and the supernatant liquid is used for injection (Figure 7.2d)

After maceration of the glands, one-third of the total volume is made by adding distilled water to make the hormone extract at desired concentration in which the hormone extract is obtained. After refrigeration for 24 hours, two-thirds of glycerine is added. The hormone suspension is again stored for another 24 hours and thereafter it is centrifuged. The supernatant solution is stored in air-tight bottles. As a generalization, a concentration of 40 mg of pituitary/ml of extract is suitable for use; of course, the extract can be diluted to obtain the precise dose for injection. Though this method of preservation has been found to be effective for one year at least, still it is recommended to preserve the extract ready about a month in advance of the breeding season.

5. **Determination of Dosage :** Successful breeding of fish requires precise determination of hormone extract which is a very important consideration. The problems involved in dosage standardization are, however, very complex because the dose of hormones principally depends upon several factors such as the size and stage of sexual maturity of the recipient fish, potency of the glands, freshness, stage of maturity of the donor fish, preservation of pituitary glands, and biological factors of the breeding environment. In general,the convenient dose of the hormones is expressed as weight of the recipient fish (mg/kilogram body weight).

For weighing the pituitary glands, dehydrated and defatted glands are placed on a clean filter paper and gently wiped off the glands with the filter paper. The gland is then placed inside a clean glass phial of known weight with cork and the gland is weighed on a chemical balance. Though this method is very accurate for research and development-related large-scale fish farms it makes more difficult to fish farmers in remote areas. Most of the farmers may not be able to afford such costly balance to weigh the glands. For this purpose, a very simple guideline has been proposed. In general, 1-2 whole glands from the donor fish of about the same weight as that of the recipient fish would be highly effective.

6. **Injection Protocol:** The dose of hormone injected is generally higher for grass and silver carps than that of Indian major carps.* The female fish is given the first

---

* Mature Indian major carps are injected twice at an interval of 6 hours. The first and second doses have been estimated respectively as 2-3 mg and 5-8 mg per kg of body weight. Males are injected at the rate of 2-3 mg per kg of body weight at the time of second injection to the female, and even higher doses (4-5 mg per kg) may be given if the production of milt is not quite free.

and second doses of 4 and 7-10 mg/kilogram body weight of the recipient fish, respectively. The males are injected at the rate of 3-6 mg/kg body weight of the recipient fish at the time of second injection to the female.

In some situations when the recipient fish are found to be in lose condition especially in the late season, a single unconscious dose* is injected to the female fish. During favorable weather conditions, the Chinese carp females are often given three injections at an interval of three hours, the total dose being 10-14 mg/kg body weight of the recipient fish. However, 2-4 mg in 0.1 ml of the extract and a dilution of 0.2 ml/kg body weight of the recipient fish are most effective and widely recommended. It should be kept in mind that 0.2 to 0.6 ml of extract is ideal for broodfishes weighing between 1 and 4 kg. For still large-sized broodfishes (weighing up to 10 kg), though not considered ideal for injection, the dose could be increased up to 1 ml. For a broodfish of grass carp, the dose of injection, quantity of glands, and volume of extracts are given below.

Sex	Weight of recipient (kg)	Dose (mg/kg)	Quantity of gland (mg)	Volume of extract (ml)
Female	3 First injection	4	7	0.7
	Second injection	6	15	0.7
Male	3	5	9	0.7
	3	5	9	0.7

Though the above discussion regarding injection protocol provides some idea about determining the dose and dilution of the pituitary extract for one breeding set in usual practice, hormone extracts of known strength are prepared at a time when large number of sets in a commercial farm are taken into consideration.

Generally two types of injections such as intraperitoneal and intramuscular are recommended. In the former case, the injection needle is inserted under a scale at the base of the pelvic fin and is pushed through the abdominal wall into the body cavity and then directed parallel to the ventral surface. On the other hand, intra-muscular injection is very effective and convenient than that of the former method. At the time of injection, however, the needle is first inserted under a scale parallel to the body of the fish and then the muscle is pierced at an angle of 45° (Figure 7.2e). In general, a 2 ml hypodermic syringe with 0.1 ml graduations with a locking arrangement is suitable for use. BDH No.1 needle is used for fishes weighing between 1 and 3 kg and BDH No.10 for large-sized fish.

The recipient fish are put on a soft cushion. At the time of injection, one person places his hands on the head and the other person holds the caudal peduncle with one hand and with the other hand the hormone suspension is injected in the region of caudal peduncle.

---

* Unconscious doses generally vary from 10 to 14 mg/kg for female Indian major carps and 15 to 20 mg/kg for grass carp and silver carp.

(Figure 7.2e). The injected fishes are at once put in a particular position inside the breeding hapas and the mouth of the hapa is closed to prevent the injected fishes from escaping the pond.

7. **Creation of Breeding Environment:** A favorable environment for induced breeding of carps generally involves the setting of breeding hapas in the breeding pond and breeding-related environmental factors such as dissolved oxygen, pH, temperature, hardness of water and rainfall. Since efficient hatching of fertilized eggs is the key step to fish culture strategy, the breeding environment must be created to ascertain successful hatchling and increased survival of hatchlings as well as fry. While fertilized eggs are placed in hapas for small farms, the large-sized commercial farms generally use either modem glass jar hatchery or Chinese circular hatchery where breeding environment is highly favorable to fertilize eggs in large numbers for hatching.

At present, a number of synthetic hormones (particularly ovaprim) are widely used in induced breeding programs and the results are highly encouraging. The recommended doses of this hormone have, however, been noted in Volume 1 (see Section 11.28)

(a) *Stripping Methods*: The injected female and male fish are kept in separate breeding hapas. The time required by the fish to become ready for stripping principally depends up on the temperature of water. Stripping generally involves squeezing the abdomen of both the sexes to release eggs and milt. At first, the female is stripped by gently squeezing the abdomen. The eggs are collected in a plastic tray, bowl, trough, or basin. A good female may generate eggs to the extent of 50 per cent of the body weight. A 3 kg female, for example, generates about 0.7 kg eggs. Following the stripping of the female, males are also stripped and the oozing milt is poured on to the eggs. A 3-4 kg male fish may produce an average of 15 ml of milt.

Eggs and milt are gently mixed at once for about 5-10 minutes using a feather or a plastic spoon. The eggs are then washed with fresh and clean water for about 30 seconds. Washing is repeated up to 5 times. The water-hardened eggs are ready for incubation either in hapas or in hatchery units (See Panel 7.1).

Culture

Grass carp hatchlings of unifom size and age are stocked in a well-prepared nursery ponds at the rate of 1.5 million/ha with a survival rate of 75 per cent. Stocking densities of 3-5 million/ha have been found to be suitable for high production of fish by using supplementary feeds.

Fry grow well and the survival as well as growth are high if aquatic plants particularly *Hydrilla* and *Najas* are present in adequate quantities (Table 7.6). The conversion ratios of several species of aquatic plants have been reported to be as 1 : 10-22 (mean 20.0 g), 1 :

11-26 (mean 17.8), and 1 : 12-20 (mean 16.0 g) for *Hydrilla*, *Najas*, and *Potomogeton*, respectively.

Table 7.6: Growth of Grass Carp in Ponds Fed With *Hydrilla* sp.

Stocking rate (number/ha)	Initial measurement		Final measurement		Duration of culture (day)
	Length (mm)	Weight (g)	Length (mm)	Weight (g)	
125	26.5	215	40.3	837	21
375	24.4	210	37.5	770	32

The growth rate of grass carp in weed-infested ponds is extremely variable in different countries. For example, grass carp attains to an average daily growth rates of 3, 1.7, 5.0, 12.0, 5.0, 4.7, 8.0, 10.0, and 2.8 g in China, Hongkong, Malayasia, Isreal, India, Thailand, Malacca, Siberia, and Singapore, respectively. In Europe, the fish has been reported to attain 6-8 kg in 6 years in fattening ponds.

Since monoculture of this species is not recommended in many Asian countries, it is advisable to culture them along with other carps if the fish is adequately fed with water-loving plants. For growing table- size grass carp, composite culture is the most commonly practised system. Under Indian conditions, the fish is capable of attaining to a weight of even 4 kilogram within a year. In addition to grass carp, a number of other species such as common carp, silver carp, and Indian major carps are stocked as secondary species because this system of farming permits higher production of fish biomass.

Previous discussions regarding the composite fish culture (see Volume 1, Chapter 21) remind us that different species, according to their habitat (such as upper, middle, and lower layers of ecosystems) are stocked together in suitable ratio so as to utilize natural food available at different niches of the pond. The mutual benefit between different species are fully utilized. The following mutual benefit will illustrate this point. Addition of grass carp in the pond stocked with silver carp and mud carp (*Cirrhinus molitorella*) have been found to be beneficial. Grass carp has a great hunger but unfortunately they are not able to digest the fibrous tissues. The fish is able to digest only the broken plant cells. Therefore, the excreta contains a lot of undigested plant materials that helps fertilize the pond ecosystem. In absence of grass carp, the production potential of the pond has been found to be very low and the culture pond will turn into fertile condition which is conductive to the growth of grass carp. In China, the principal species for pond culture, includes grass carp, bighead carp (*Aristichthys nobilis*), and mud carp; whereas others such as common carp, tilapia, white Arnur Bream (*Parabranis pekinensis*), Mandarin fish (*Siniperca chautsi*), Red Eye (*Squaliobarbus curriculus*), and Wuchang fish (*Megalobrama amblycephala*) are used as secondary species. In India, however, grass carp and silver carp are mostly cultured in combination with the Indian major carps and common carp (*Cyprinus carpio*) for significant production.

Most of the farms follow the system of multiple stocking and harvesting in rotation. In cases where stocking densities are high enough, the large-sized fish are harvested and replaced with adding more fingerlings. Thus, mud, black, and common carps are harvested two times a year and grass carp up to 6 times a year. Production of grass carp and silver carp varies considerably from region to region and evem from farm to farm. The average yield varies from 300 kg/ha is feasible in extensive to 3,000 kg/ha in semi-intensive culture systems and fish production at the rate of about 8,500 kg/ha in intensive system of farming. The culture of grass and silver carps has been well adapted in many Asian countries and is also environmentally sound. Developing countries whether it is tropical, temperate, or sub-tropical, must take the opportunity of this system of fish farming.

7.11 *Hypopthalmichthys molitrix* (Valenciennes), Order: Cypriniformes, Family: Cyprinidae

The fish is naturally found in the river systems of West River, Kwangsi, Kwangtung in South and Central China, and the Amur basin of Russia. The species has been introduced in many countries of the world and is scientifically cultured with other carps in India, Nepal, Japan, Ceylon, Taiwan, China, Thailand, Philippines, Pakistan, Singapore, Israel, Russia, and other countries. In 1959, however, a few hundreds of silver carp fingerlings was brought from Japan to the Central Inland Fisheries Research Institute, Orissa and after hypophysation, fry were distributed to various states of India.

Characteristic Features

The body is stout and compressed. Abdomen is strongly compressed with a sharp keel from breast to vent. Head is moderate; mouth anterior, large, wide, and snout is bluntly round. Lips are thin. Eyes are small, anterior and sub-inferior. Upper jaw is slightly protruded upward and longer than the lower jaw. Barbels are absent. Dorsal fin is inserted behind pelvic fins, or above tip of pectoral fins with 10 rays. Anal fin has 14-17 rays. Caudal fin is forked. Scales are minute and cycloid type .

Tolerance to Ecosystem

Though silver carp is a freshwater fish, it is also able to tolerate in slightly brackish water. Similar to grass carp, the fish can tolerate a wide range of temperature (15-36°C), pH (5.0-8.5), turbidity (125-170 cm), dissolved oxygen (2.5-28 mg/l), total alkalinity (80 620 mg/l), free ammonia (0.3.5 mg/l), and salinity (5.0-8.5 ppt). The fish requires high oxygen for better survival and transport. Reddish patches over the body may appear during handling and have been recognized as haemorrhagic septicemia.

Hybridization

The cross breeding between silver carp and other carps such as Indian major carps (*Labeo rohita*,*Catla catla*, and *Cirrhinus mrigala*),common carp, grass carp, and bighead carp gives rise to combined characters of male and female parents in general, and the hybrids exhibited moderate to fast growth. Hybrids of silver carp and common carp, for example, exhibit poor growth in winter, but the cross-breed of silver carp and bighead carp and its reciprocal cross produced thousands of offspring which grow very well in several Asian countries and the cross exhibited rapid growth.

Food and Feeding Habitat

The fish is a surface feeder with stenophagic feeding habit and is restricted to micro-algae. While larval stages consume micro-algae as principal food, fry and fingerlings depend entirely on dinoflagellates, myxophyceae, flagellata, bacillariophyceae, rotifers,and protozoa supplemented with detritus and decaying macrophytes. Post-larvae feed on zooplankton and when they reach to a size of about 3 cm in length, they subsist on phytoplankton.

The fish is well-adapted to such ecological niches where the water abounds with planktonic micro-algae. It has peculiar structural and anatomical modifications which are correlated with micro-planktophagus feeding habits. The fish has a complex network and numerous gill rackers which can efficiently sieve out micro-algae. Indian major carps are not able to utilize surface phytoplankton most efficiently for nourishment because of the lack of interwoven gill rackers. Though the fish feed voraciously on phytoplankton, experience has shown that the fish is not able to digest some species of algae such as *Microcystis, Anabaena,Oscillatoria* and others and may exhibit extremely poor growth in ponds teemed with these algal species. The quality of flesh is very poor and hence the market price is low in comparison to Indian major carps.

Silver carp has specialized structure of the gill rackers to micro-plankton feeding. Pharyngeal teeth are provided with sharp spines within the buccal epithelium. Though certain structural modifications in the gill structure exist with the changed planktophagus feeding habit, the fully developed gill rackers of the adult fish are longer than the gill filaments. Minute bony bridges between the gill rackers link them up which exhibit like a bamboo curtain form. Its exterior part is covered with spongy sieve membrane forming a filtering net.

Sexual Maturity and Spawning

The adult fish starts sexual maturity when it attains one-year old (2kg in weight). In breeding farms, three years old fish are considered for induce breeding programs. The mature female fish weighing 770 g, has been found to respond pituitary extract.

During the breeding season, the fish exhibits secondary sexual characters. In male, some fine bony ctenoid teeth appear on the pectoral fins. The abdomen is small with slit-like genital aperture. White milt oozes out when the abdomen is gently pressed. Similarly in female, the ctenoid teeth grow only on the terminal parts of the pectoral fin rays. The abdomen is soft with anus being reddish and distinct.

Under Indian conditions, the fecundity of this species weighing 3 and 8 kilogram has been reported as 1,45,000 and 20,44,000, respectively. The number of eggs per gram ovary weight range between 285 and 297, and the body weight between 164 and 175 g. In Japan, however, the fecundity of fish weighing 11 kilogram varies from 13,40,000 to 13,90,000 with an average fecundity of 13,25,000. A comperative statement of the fecundity of silver carp at different size and age groups is shown in Table 7.7.

Table 7.7: Fecundity of *Hypophthalmichthys molitrix* at Different Size and Age Groups. Values are Ranges and Means in Parenthesis

Year of the fish	Weight of the fish (g)	Number of ova per g	
		Body weight (g)	Ovary weight (g)
1st year	590-2,430 (1,735)	75 - 250 (165)	790 - 1,770 (1,165)
2nd year	56-75 (66)	45 - 313 (194)	318 - 950 (730)
3rd year	64-83 (75)	100 - 230 (160)	518 - 1,030 (800)
4th year	80-83 (82)	160 - 180 (170)	667 - 822 (745)

The time of sexual maturity depends on the geographical conditions of the breeding pools. In China, for example, the fish breeds during April-July. In Japan, spawning occurs during June-July. In India, maturity and spawning occur in February-April. Pond-reared mature males are available during March-April and that of females during May-July.

In a fully mature farm-reared silver carp, eggs are pale-bluish in color. Fertilized eggs are demersal and round. Immediately after fertilization, eggs start swelling (4.0-4.8 mm in diameter) and after half an hour, water hardening of eggs is terminated. The incubation period of eggs in stagnant waters varies considerably. At temperature ranges from 20 to 26 and 28 to 30°C, the incubation periods vary from 36 and 20 hours, respectively.

The length of a newly-hatched larva generally varies between 4.9 and 5.5 mm. Each larva has the following characteristic features: length of yolk sac, 3.3 mm; body length, 1.4 mm; number of myotome, 44; and non-pigmented eyes. The pectoral rudiments of the larva do not appear. After 48 hours of hatching, the yolk sac is fully absorbed and post-larva is developed which measures about 7.5 mm in length and consume natural food organisms.

Breeding Techniques

Though natural breeding of silver carp occurs in the dry-bundh, 'Eco- hatchery'and 'Bangla Bundh' is reported under the influence of artificially-created water currents and hormone injection, their breeding by crude Human Chorionic Gonadotropin (HCG) alone and in combination with fish pituitary gland (FPG) in the ratio of 70 per cent HCG and 30 per cent FPG ensures better survival and yield of hatchlings. The average dose of HCG at the rate of 540 IU/kg of female in two intra-muscular injections at water temperature ranging from 27 to 33°C has been found to be effective. The second dose is given 6 hours after the first. Single dose of HCG at the rate of 120-150 IU/kg is used to the males when female receives determinative dose. When mixed solution is used, the first and second injections are received by the females at an interval of 6 hours at the rate of 6 and 10 mg/kg body weight, respectively. The dose of male fish is 6 mg/kg body weight. The ponderous breeding of this fish in eco-hatchery by HCG injection has now become a common practice among fish farmers in South and Southeast Asian countries.

Similar to grass carp, breeding of silver carp requires precise control over the entire breeding operation in the breeding hapa starting from management of broodfish to spawning. However, breeding requires hormone injection followed by stripping. The optimum spawning temperature in China, far East Asian countries, and India varies between 23 and 30.5°C; and in Russian Federation, it ranges from 20 to 23°C. Adequate rainfall and cool weather with water temperature varying between 27 and 30°C are favorable in producing large number of fry in confined waters by induced breeding.

1. **Breeding Technique In China :** In China, the brood fishes are reared under intensive care in ponds in the ratio of 2 male and 1 female for natural spawning and 1 : 1 ratio in case of stripping. Mostly common carp is used as donor fish for pituitary glands. Three to four glands are injected in one kilogram of the recipient fish. Ten to 15 glands are triturated in 1.5 to 2.0 ml normal saline. While the ripe female fish is injected twice at an interval of 6 hours, the male fish is injected only at the time of second injection to the female. Injected brooders are then released in ponds with arrangement of running water, and the temperature of water ranging from 23 to 29°C is highly effective for breeding. After fertilization, the eggs are water-hardened within 30-40 minutes and are collected.

2. **Breeding Technique in Russia:** In this technique, the mature female fish are injected thrice on the back muscles of the fish.The first dose at the rate of 2.4 mg (for 6-12 kg body weight), the second dose at the rate of 20-35 mg (for 10-12 kg body weight), and the third dose is applied at the rate of 25-35 mg. Male is administered 2.5-30.0 mg/kg body weight. Suitable dissolved oxygen concentration and water temperature (20-23°C) are maintained. Fertilization of eggs is made by 'dry method'. The mixture of milt and ova is first allowed to stand for at least 15 minutes with occasional shaking which is followed by washing with clean water. Fertilized eggs are put in Weiss apparatus through which a water flow at the rate of 0.6-0.7 l/minute is maintained. About 30,000 to 75,000 fertilized eggs can be accomodated in this apparatus.

3. **Breeding Techniques in India and Taiwan:** The breeding technique of silver carp in these regions is similar to those followed by other countries. However, the technique followed for silver carp is almost similar to that of grass carp.

Culture Methods

The method of preparation of nursery, rearing, and stocking ponds is similar to those adopted for Indian major carps. In a well-prepared nursery pond, silver carp hatchlings (6 mm in length) are stocked for 10-15 days at the rate of 5 million/ha and the average survival and growth of fry are extremely variable that ranged from 43 to 83 per cent and 90 to 104 mg, respectively. Fry are then transferred to several rearing ponds where they are stocked for at least 60 days each at the rate of 30,000/ha. If the survival rate of hatchlings is very high and rearing ponds are not available in sufficient numbers, a substantial amount of fry is transported elsewhere by containers charged with oxygen. Silver carp fry and fingerlings are generally cultured in rearing and stocking ponds, respectively, along with other carps and has been described in Chapters 20 and 21 of Volume 1. Monoculture of this fish in ponds at different stocking rates, however, did not show encouraging results.

1. Monoculture

Monoculture of silver carp in ponds with better management protocols has shown encouraging results than that of culture in combination with the Indian major carps (particularly *Catla catla*). If silver carp fingerlings (50 ± 10 g) are stocked for 120 days in a well-fertilized warm water pond at the rate of 4,000/ha, the fish may attain to an average weight of about 550 ± 15 g. Silver carp subsists on planktonic micro-algae and has been observed to struggle for food with the fish *Catla*. In order to evaluate the comparison of the yield of silver carp alone and in combination with catla, experiments were carried out for 100 days under identical management protocols in well-manured ponds where catla and silver carp were stocked at more or less comparable weight. The rate of increase in weight of fish in different experimental ponds and the gross production in shown in Table 7.8. While catla and silver carp exhibited more or less similar weight increase when

Table 7.8: Growth and Production of Silver carp and Catla Fingerlings in Three Grow-out Ponds at South 24 Parganas, West Bengal During 1999

Fish	Duration of expe- riment (day)	Date of sampling	Stocking rate (kg/ha)	Total number of fish stocked	Average weight of fish (g)	Gross production (kg/ha)
Catla	100	2.2.99	37	500	75	
		23.2.99			82	
		15.3.99			90	
		5.4.99			101	
		26.4.99			116	217.5
Silver carp	100	2.2.99	38	500	72	
		23.2.99			81	
		16.3.99			93	
		4.4.99			104	
		26.4.99			125	244.0
Catla + Silver carp	100	2.2.99	38	250	60.0	
		23.2.99		+	66.5	
		16.3.99		250	74.0	
		4.4.99			80.0	
		26.4.99			90.5	
					+	
					70.5	
					80.0	
					96.5	
					107.0	
					125.0	173.5

stocked in different ponds, the growth of catla was less than silver carp when they are stocked in combination. Per hectare production from ponds where both the species were stocked together, was very low. The gross production of silver carp and catla when stocked in different combinations (Table 7.9), exhibited significant results. Among different combinations of fish, the ratio of 1 : 2 between catla and silver carp has shown yield in a high degree followed by the ratio of 2 : 1, 3 :1, and 1 : 3.

Table 7.9: Mean Gross Yield of Catla and Silver Carp in Ponds in Different Combinations

Combination of fish (Catla : Silver carp)	Average production (kg/ha/3-month)
2 : 1	285.5
1 : 2	317.8
3 : 1	227.0
1 : 3	192.4

2. Polyculture

The system of polyculture adopted in China along with other species are divided into (1) the Hongkong system, (2) the West River Region system and (3) the Kiangsu and Chekiang systems. The first system includes mullets along with the carp while the second and third systems involve the culture of different species of carps. These culture systems principally depend on the availability of the stocking materials and the nature of local ponds.

In some Asian countries such as Singapore, Thailand, Taiwan, and Malayasia, the silver carp is reared along with grass carp, common carp, bighead carp, tilapia, milk fish, and sea pearch. In India, the polyculture of three Chinese carps (silver carp, grass carp, and common carp) and major carps (rohu, catla, and mrigal) are intensively cultured on commercial basis for the last 35 years, with a record production to the tune of about 10 tonne/ha/year.

In China and Russia, the silver carp is highly significant in commercial fisheries and forms a very important as well as potential fishery in reservoirs and in different river systems of China, Japan,Israel, India, Thailand, Malaysia, and other Southeast Asian countries. Through extensive/intensive culture and commercial production of seed by induced breeding technique, however, the silver carp has now become an important fishery in many countries of the world such as India, Hungary, Sudan, Israel, Bangladesh, Sri Lanka, Pakistan, and Rumania for the last 50 years or so.

Since production of robust fry and fingerlings of silver carp require careful control over the entire breeding and spawning periods, intensive management of broodfishes appears to be the best approach to a sustainable culture for this fish. Hatchery-produced fry must take a number of environmental factors into account during breeding operations.

7.12 *Cyprinus carpio (Linnaeus), Order : Cypriniformes,* **Family : Cyprinidae**

The species *Cyprinus carpio* popularly called as Common carp, German carp or European carp is a native of Russia and China from where the fish has been transplanted to Asia, Europe, and in other countries. At present, the fish finds pleasure in global distribution especially in tropical and temperate regions where it has well adapted itself to a variety of habitats such as lakes, rivers, ponds, streams, ans reservoirs and extremes of environment. The common carp is not only an ideal fish for introduction in high altitude lakes (at an average elevation of 1,500 m) where it thrives very well and breeds but also in plains. This is due to the fact that the fish has a field temperature preference of 25°C circumscribed by upper and lower temperature preference of 40 and 12°C, respectively. Though the fish has been introduced in Asian countries, it is widely cultivated either alone or in combination with other carps. Due to its rapid growth rate, natural breeding, and tolerance to various ecological factors, it constitutes a bulk fishing in different freshwater ecosystems particularly in lakes and reservoirs.

Strains

On the basis of scaling, three phenotypes of common carp have been recognized:

1. **Mirror carp :** *Cyprinus carpio* var.*specularis* (Lacepede) : The body is covered with a few large and bright scales. A large part of the body remains exposed. The species is extremely variable and looks orange in color.

2. **Leather carp :** *C. carpio* var. *nudus* (Bloch) : The body is almost completely devoid of scales except some degenerated scales along the base of the dorsal fin. Sometimes, small scales are also present at the bases of other fins.

3. **Scale carp:** *C. carpio* var *communis*: The body is fully covered with regularly arranged rows of scales. This is the original form of common carp with normal coloration and is now widely cultured in the Far East.

Characteristic Feature

The body is robust anteriorly and more or less compressed with rounded abdomen. Head is moderate and snout obtusely rounded. Mouth is oblique, terminal and cleft not extending to anterior margin of eyes. Eyes are superio-lateral, moderate, situated anterior to head, and not seen from below the ventral surface. Upper jaw is more or less projecting and maxillary bone is protractible. Lips are fleshy. There are two pairs of barbels. Dorsal fin long and inserted above the tip of pectoral fins with three spines and 17 rays. Anal fin is short with 3 spines and 5 rays. Caudal fin is deeply emerginate. Scales are pentagonal, large, and cycloid type.

Food and Feeding Habit

Common carp is an omnivorous fish. The fish feeds on a wide variety of plant and animal matters. When fish are in the fry stage (less than 10 cm in length), they consume zooplankton, but large-sized fishes consume benthic organisms such as chironomid larvae, molluscs, worms, and larvae of Odonata along with heavy intake of vegetable matters and epiphytic organisms. Several genera of phytoplankton such as *Microcystis, Oscillatoria, Closterium,* and Volvocales are also palatable to the fish. However, its habit of sucking food substances from the mud of the pond bottom and margins not only makes the water more turbid but debilitates the base of pond dykes as well.

For the growth of common carp, vitamins and proteins are essential nutritive elements which must be supplied through supplementary feeds, but these nutrient elements are not always essential if adequate amounts of plankton and bottom organisms are available in ponds.

Maturity and Spawning

The strain of scale carp is a prolific breeder in the Indian sub-continent. The weight gain at the first maturity of this strain varies considerably in different countries and its maturity is temperature-dependent (Table 7.10).

Table 7.10: Size at First Maturity of the Fish *Cyprinus carpio* at Different Water Temperatures

Country	Water tempe-rature (°C)	Weight (g) at first maturity	Age at first maturity (year)	Spawning season
India	18-35	80-170	0.5	Year round
Japan	12-30	908-1,360	1.0	Year round
Europe	15-18	500-900	0.2-0.4	April-June
China	15-25	40-45	3-4	May-June
Indonesia	19-30	1,000-2,000	1.0-1.5	Year round

The common carp naturally breeds in stagnant freshwaters. Spawning act takes place in shallow, marginal, and weed-infested waters. The most suitable aquatic weeds require to adhere the eggs are *Najas, Hydrilla, Eichornia* and *Pistia*. These plants form adequate spawning grounds for the fish. Fertilized eggs are small, yellow to light-brown in color, demersal, and cling to the roots of plants. The diameter of the egg varies between 1.0 and 2.0 mm that depends on the size of the brood-fish. Under Indian conditions where the temperature of water varies between 24 and 32°C, fertilized eggs hatch out within 96 hours.

Spawning of common carp occurs during monsoon and winter seasons. Under suitable temperature of water, female and male fish assemble in swarms where water-loving plants grow. They expel eggs and milt while rubbing against each other. At this juncture, the eggs are fertilized and then firmly attached to the surface of aquatic weeds or to any other spawning substrates. The number of eggs that a carp lodges depend on the age of the fish, but the number varies between 7,000 and 20,45,500. Depending on the temperature of water, the length of time necessary for hatching of fertilized eggs varies widely. At the water temperature of 12, 20, and 32°C, fry appear after 7, 5, and 2 days of fertilization, respectively. Each fry is about 5 mm in length with a yolk sac in the abdominal region. A four day-old larva consumes small zooplankton. When the larva attains to a length of about 10 mm (7 day-old), they start normal movement and obtain nourishment from surroundings.

Breeding Technique

In general, fish at the age of 2$^+$ weighing 2-3 kg are considered for breeding. About 1.0-1.5 million eggs are generated from one kilogram body weight of the fish. A fully mature female fish is characterized by the smoothness of pectoral fin, swollen abdomen, and projected vent, while males appear slender, roughness of pectoral fins, and vent is not projected. Eggs and milt are come out while gently pressed on the abdomen of female and male fish, respectively. To avoid unwanted spawning it is desirable to segregate the female and male fish and maintain them in separate ponds. Broodfish are well-maintained on nutrient-rich diet. Artificial feed generally consists of rice bran and mustard/groundnut cake (1 : 1 ratio) fortified with some nutrient ingradients to proceed the development of gonads.

The breeding of common carp is carried out in two ways. *Firstly,* by inducing the fish to breed naturally under the influence of stimulated natural spawning habitat without any administration of pituitary extracts and *secondly,* by hypophysation as in the case of other Chinese carps. In the first case, mature brood fish are selected and stocked in well-prepared ponds/tanks or in breeding hapas. As the fish lays sticky eggs, it is necessary to take adequate care by catering proper spawning substrates for the eggs to adhere on them. Different types of substrates are used for the purpose. In India, for example, fresh and clean submerged aquatic weeds of the genus *Najas* and *Hydrilla* are most commonly used as egg collectors. These plants are kept inside the breeding hapas. For one kilogram weight of female fish, about 2 kilogram weeds are placed. Synthetic fibres or water hyacinth are also used as spawning substrates. In Israel, farmers use branches of cypress, casuarina and pines as egg collectors. In Indonesia, a specially-designed spawning substrate, also called *Spawning mat* made of fibres of the coconut plant or *Arenga* sp. called *Kakabans* are widely used. These mats are placed on long bamboo poles held in place between two pairs of shorter poles driven into the pond mud. In the breeding pond, however, Kakabans are kept in floating conditions just below the water surface.

Fully mature females (2-4 kg in weight) are used at the rate of 10 numbers/ha and the ratio of males to females is 3 : 2. Fishes generally spawn in the early morning hours after their release in spawning tanks/ponds/hapas. Spawning occurs within 10-12 hours after stocking. Eggs can be vividly observed sticking to the spawning substrates. The incubation period of eggs in higher latitudes varies between 7 and 8 days at 16-17°C. In tropical climate where the temperature varies between 30 and 33° C during the breeding season, the eggs hatch out within 40 hours of incubation. After spawning, broodfishes are removed from the hapa and the spawning substrates that contain fertilized eggs are at once transferred to a series of hatching hapas. One kilogram aquatic plant contains thousands of eggs that varies between 40,000 and 80,000. After 3-4 days, hatchlings are collected and then stocked in well-prepared nursery ponds.

In cases where breeding operations are carried out in ponds, well-maintained male and female broodfishes are kept either in separate ponds or in the same pond separated by a screen made of bamboo slits or mosquito nets. Spawning substrates are kept in breeding ponds. A gentle flow of clean water is maintained in the pond. Eggs are deposited by the female on the lower surface of the substrate. The egg-laden substrates are then transferred to hatching ponds. Hatching ponds have a number of controlled outlets. After 8 to 20 days of spawning, young fry are collected.

The technique of hypophysation is adopted when favorable conditions for successful natural spawning in ponds does not occur. Selected broodfishes are directly allowed to involve in hypophysation. A robust female fish having one kilogram body weight can generate about 1.5 million eggs. The eggs measure 0.9 to 1.6 mm in diameter. Though the dosage and intervals of pituitary extract injections widely vary from place to place, the sequential steps involved in hypophysation of common carp are similar to that of grass carp (see Section 7.10).

Similar to grass carp/silver carp, the injected female fish is stripped and the eggs are collected in a tray. A 5 kg female scale carp generates about 1 kg egg mass that contains 7,00,000 - 10,00,000 eggs. Following the stripping of females, males are also stripped in a similar manner and the oozing milt is poured on to the eggs. A 5 kg male may produce about 20 ml of milt.

Milt and eggs are gently mixed using a feather. The adhesive nature of eggs makes them clump together and embarrass fertilization phenomenon.This problem can be alleviated by treating the eggs with a solution of carbamide (30 g) and sodium chloride (40 g). The solution is first poured over the mixture of milt and eggs and then stirred well with a feather for about 10 minutes. As soon as the fertilized eggs are swelled out, small quantities of the solution are added. Fertilized eggs are then cleansed with tannic acid solution (0.05 -0.07 per cent) or about 30 seconds and then cleansed again with freshwater for about 5 minutes. The water-hardened eggs which measure about 2 mm in diameter, are incubated in hatching jars. Depending on the temperature of water, the eggs hatch out within 2-7 days after fertilization. The hatchlings are removed from nursery tanks and then reared in controlled conditions up to the fry stage. About 1 million fry can be reared in a 20 m^2 tank with a water exchange rate of 1 litre/minute/m^2. Fry are raised to fingerling size within a period of about one month.

Culture Methods

The fish common carp is generally cultured either alone or in combination with the Indian major carps. The species can, however, be cultured in various ways such as pond culture, paddy field culture, culture in lakes and reservoirs. As a generalization, polyculture is the most commonly practised system in many tropical countries. Besides common carp, a number of other species are also stocked in polyculture. The common carp has been one of the most versatile fish cultured across the world and its seed production and culture methods are very easy than those of either Chinese or Indian major carps. It is most promising for farmers in developing countries which have no background on fish culture, can easily adopt the culture of common carp in their ponds.

The preparation of nursery, rearing, and fattening ponds are similar to that adopted for Indian major carps. Predatory fishes are first removed from rearing and fatenning ponds and then fertilized with a heavy amount of cattle manure at the rate varying between 2,000 and 5,000 kg/ha. The common carp being an omnivore, utilizes more efficiently the detritus and bacteria-laden organic matter to achieve satisfactory growth. For this reason, this fish is now being successfully cultured in sewage-fed ponds and in ponds having organic matter in high amounts. In cases where treated effluents and organic manures are not available, ponds are generally treated with mustard oil cake/groundnut cake at the rate of 1,000 kg/ha in combination with urea and lime at the rates of 170 and 400 kg/ha, respectively. Under Indian conditions, however, an average weight of about 700 g (without artificial feeding) and 1,000 g (with artificial feeding) is obtained.

In most countries where the culture of common carp is practised, fish an average weight of about 700 g has good market demand, but in some Southeast Asian countries, small-sized fish (about 300 g in weight) are accepted. The marketable size is, however, attained in one year. In temperate regions, it generally takes 2 to 3 years to grow fish to selected weight of 1.2 to 1.6 kilogram.

The various feeds such as rice bran, wheat flour, soyabean cake, potato, fresh and dry pupae of silk worm, and powdered dry fish are employed for better production. It is better if cooked wheat is recommended because it is readily digested and fish is accoustomed with this type of feed.

1. **Culture of Common Carp in Wastewater :** The culture strategy of this species in dairy and domestic wastewater-fed ponds has, at present, attracted to progressive fish farmers and the results are highly encouraging. It is considered as an effective and a challenging option to use polluted waters in fish culture. However, when culture of this fish is carried out in dairy wastewaters, both the growth rate and food conversion efficiency have been found to be low when reared in selected concentrations of the wastewater (such as 2.5 and 5 per cent) inspite of the significant increase in consumption of natural food organisms (Table 7.11). Though such high consumption of food is not uncommon, drastic reduction of natural food consumption in polluted waters is very common in a majority of fish species. Reduced growth may be ascribed to one or more of the following points: (i) accumulation of metabolites in fish flesh, (ii) high energy demand for body maintenance, and (iii) general metabolic stress. Though the amount of food requi red to yield one kilogram of carp flesh from an initial stock of one kilogram in 50 days, for example, has been calculated to be double or more than double (Table 7.12), the culture scenario seems to be glorious. This outlook has definitely encouraged by the fact that the culture of this fish (and also other exotic and indigenous varieties) can be accomplished by simple dilution of concentrated

wastewaters. This does not require any trained personnel and minor changes in the factor of dilution may not produce excess physiological stress on fish due to the moderate strength of the wastewater. This fish has, however, been found to be more resistant to pollutants as compared to minor Cyprinids such as *Barbus stigma* and *Puntius* spp. The physiological - chemical compatibility between the fish species and the wastewater is highly significant which helps develop an adequate strategy for the cultivation of this species in moderately strong wastewaters.

Table 7.11: Feeding Energies of *Cyprinus carpio* Exposed to Dairy Wastewaters. Each Value Represents Average of 80 Individuals ± SD

Concentration of waste-water (per cent)	Initial weight of fish (mg)	Final weight of fish (mg)	Food intake (mg/g live fish/ day)	Growth (mg/live fish/day)	Growth rate (mg/live fish/day)	Coversion efficiency (per cent)
Control	587.26	1,456.81	305.72	869.54	49.74	16.30
	± 82.36	± 152.79	± 18.03	± 72.58	± 3.44	± 1.33
2.5	780.45	1,606.15	404.59	832.45	35.66	8.8
	± 48.74	± 81.99	± 8.36	± 46.63	± 1.91	± 0.55
5.0	712.10	1,270.11	395.82	558.00	26.23	6.63
	± 78.12	± 126.43	± 11.55	± 68.10	± 2.90	± 0.78

Source : Nagendran (1996)

Table 7.12: Balance Sheet of *Cyprinus carpio* Culture

Concentration of waste-water (per cent)	Initial biomass (g)	Food intake (per cent of initial biomass)	Increment of flesh (per cent of total food intake)	Total food required by an initial stock of 1kg of new flesh in 40 days (kg)
Control	47.00	30.55	16.08	6.21
2.5	62.44	40.45	8.70	11.55
5.0	56.96	39.57	6.56	15.33

Source: Nagendran (1996)

2. **Culture of Common Carp in Paddy Fields:** The culture of common carp in paddy fields is highly encouraging in rural areas. While selecting the site for paddy cum-common carp integration, the availability of a flow of diluted wastewater at the rates varying between 60-120 litres/minute is desirable for the proper growth of common carp. Along the perimeter of a paddy field, a canal of 50 cm is constructed. The paddy field is stocked with fingerlings at a stocking density of 6,000/hectare. Fingerlings grow to an average weight of 200 g. With the application of cattle manure at the rate of 2-3 tonnes/ha and supplementary feeding with rice bran and oil cake mixture, the production is lucrative and ranges from 400 to 700 kg/ha during the paddy growing period.

The culture of common carp is highly promising to develop as a relatively cheap protein food in rural areas. Development agencies should cater funds and the necessary technical know-how to the farmers to generate rearing facilities. This would definitely assist in the expansion of common carp culture. Many private organizations produce huge quantities of pelleted feeds for carps which could also be used for common carp.

7.13 Culture of Trout*

The trout species of greatest importance in aquaculture is the rainbow trout (*Salmo gairdnerii*). Native of North America, the fish has been introduced to waters in almost all countries of the world including India. In India, however, trout farming is a growing strategy and is increasingly being cultured in high altitude regions. Its range extends into low latitudes and at higher elevations. Trout culture ponds are located in the upland areas of many tropical and sub-tropical countries of Asia. Though flowing mountain brooks are the native habitat for the trout, it requires deeper ponds (1.5-2.5 m) with cold water (in summer not above 20°C) which must constantly contain oxygen in adequate amounts. Several local strains have been developed through mass selection and cross-breeding for improved culture qualities. Among different species of trout such as brown trout (*Salmo trutta*) and brook trout (*Salvelinus fontinalis*) which are native of Europe and North America respectively, have been introduced to several countries of the world for developing sport fisheries and stocking in natural water bodies; but the rainbow trout has now become the mainstay of large-scale aquaculture on world-wide basis. The two main varieties of rainbow trout which are considered as commercially important are the sea moving form termed as *steel-head* and a land-locked freshwater form. The former grows in saltwater, attaining 6-10 kilogram in about 3 years. The latter form attains to a weight of 4.5 kilogram or more under favorable conditions and is extensively used in commercial fish culture.

The main system of trout culture essentially involves hatchery propagation, rearing of young for stocking, broodstock management, stripping, feeding, and other management schedules. The trout culture ponds are rectangular in shape, each measuring about 30 m X 10 m, with the bottom slopping towards the outlet and a depth of about 1 and 1.6 m at the upper and lower ends, respectively.

Tank and Raceway Culture of Trout

The most widely used system of trout culture is in tanks and raceways. Tanks are 4-10 m in diameter and 1.6 m deep and submerged in the ground, leaving about 30 cm above the ground. The drain pipes are connected to outlet sumps. Many trout tank farms have a central fish grading arrangement. Each tank is provided with a separate outlet pipe which is connected to a separate main outlet leading to a sump where fishes are graded.

Raceways consist of long continuous narrow channels or series of channels divided by cross-walls. They are made of earth, brick or cement concrete, submerged in the ground

---

* For further detail, see Chapter 10 of this volume.

or built above the ground level. The channels have a width of 2-4 m and a depth of less than 1 m. Raceway culture requires a plentiful supply of clean and oxygenated spring water with a constant temperature and flow velocity. Fish are stocked at the rate of 5 kg/m with a water exchange of 2.5 l/minute/m^3.

Though monoculture is the common practice for trout culture, intensive systems are important in most situations to make the culture system economically viable. Where good climatic conditions prevail,double-cropping systems have been introduced. This system generally involves the culture of trout along with channel catfish where the former species is grown in ponds or raceways for several months (November to March) and the Channel catfish from April to October. This system of farming is highly encouraging and is reported to bring down the cost of production followed by increase in returns.

Cage Culture of Trout

Cage culture of rainbow trout is a rapidly growing system and is increasingly being used in both salt- and freshwater environments. Due to the developnent of the design and construction of cage farm units, it has paved the way for culturing trout in cages in more exposed areas that can withstand rough weather conditions. Rainbow trout fingerlings are stocked at the rate of 10-20 kg/m^3 in spring and are harvested in the autumn after a culture of 18 months or they are stocked in the autumn and harvested after one year of culture. When they grow larger, they are graded out and transferred to other cages.

It should be kept in mind that the investment costs for the establishment of trout farms are very high and since high protein-rich artificial feeds, supply of quality water and skilled labor are desirable to run the trout farms economically viable, the production costs also tend to be high. As a generalization, high production costs are compensated for by the excessive production of trout and high prices of the products which will definitely make the trout farming more attractive to farmers. It is beyond doubt that the trout farming is a highly profitable business in many countries of the world.

7.14 Practical Considerations of Exotic Fish Cultute in Ponds

Since introduction of exotic fishes in India, the cultivation of tilapia, common carp, grass carp, and silver carp began to receive due importance as a source of food from inland water resources. As early as the late 1960s, mixed culture of these species with the Indian major carps also called as *Composite Fish Culture* (see Volume, Chapter 21) has been developed as an economical means of intensifying production. But it has now been recognized that introduction of these fishes (and also other exotic fishes) have proved disastrous in many inland fishery resources. In many cases, such introduction has resulted in the disappearance of many native fish fauna. At the same time, their tasteless flesh with numerous intra-muscular spines have no special appeal to consumers. Their flesh is poor in quality and the cost per kilogram of fish in the market is very low. Inspite of these characteristic features, they are strongly recommended for culture either individually or in combination with other indigenous carps. In composite fish culture systems, interactions between exotic and indigenous carps always exist that oscillate the total production potential depending on the manipulation of local advantages for water management. For example, enhanced turbidity of pond water due to browsing behavior of the common carp in the sediment might affect algal growth which, in turn, suppress the growth of silver carp. Moreover interaction between silver carp and catla also worth remembering. The former consumes algal populations 3-4 times more than that of catla, thereby reducing the growth and yield of the latter. Though introduction of exotic carps into Indian waters has dramatically changed the composite culture scenario by increasing unprecedented yield of fish biomass, recent analyses of fish catch from different reservoirs and lakes have shown that the exotic fishes have drastically changed the ecological conditions and upset the balance of introduction of the indigenous fishes particularly mahseers, schizothoracids, catla, mrigal, and *Osphronemus gorami*.

The culture and production of exotic carps in rearing and fattening ponds involve the manuring and fertilization to promote the production of phytoplankton, zooplankton, and bottom organisms. For significant production of exotic fishes, artificial feeding is the most important and great attention should be given to the effect on the growth and health of fish. Consideration should be given to costs and naturally favoring less expensive feeds. It has been reported that production of carps from ponds using only one type of feed often tends to create nutritional imbalance and hence it is necessary to cater mixed feeds.

Pond fertilization with inorganic fertilizers (nitrogen and phosphorus) is inevitable for phytoplankton production which is essential for high production of silver carp. On the other hand, tilapia and common carp are omnivorous and consequently such ponds must be fertilized with organic manures. Grass carp and *Puntius japonicus* may be cultured if quality fish seed is available to fish farmers. At the same time, the availability of aquatic plants must be ascertained.

7.15 Ecological Impact of Exotic Fishes on Indigenous Species

When a species is introduced to a new environment, the trespasser always tries to adjust in the new environment and consequent increase in resistance such as parasitism, competition, and predation. The introduction of fishes into the native waters has become an important issue with their performance in Indian freshwater systens where commercial fish farming is highly impressive. A number of fishes were brought to India from other countries for various purposes such as game fishing, weed and mosquito controls, experimental culture, and for stocking in lakes, streams as well as reservoirs. Whatever the original purpose, such introduction has been proved to be fruitful sources of food production through mixed farming systems and have also been provided the needed stimulus to develop the practices of this type of farming technique. And new technologies for the farming of freshwater systems using exotic and native carps with certain modifications of techniques used in culture systems must be developed.

Gradual increase in the population of exotic fishes in natural waters may occasionally cause ichthyo-pollution such as excessive growth of snakeheads and tilapia, in addition to killing indigenous species and destroying habitats, may exhibit reduced production potential of inland waters. In an interesting report of commercial catches taken from Govindasagar reservoir, for example, some authorities have been forced to describe the reservoir as a drastic change in the composition of Indian carps and mahseer (*Tor putitora*) and reduced yields since three exotic carps (such as common carp, grass carp, and silver carp) are more predominant in this reservoir with slight oscillations in their composition that comprise about 90 per cent of the total fish catch.

The ecological impact of exotic fishes on indigenous species may be assessed from the viewpoint of the human intervention. Through composite fish culture systems, a number of exotic species have been accidiently or deliberately introduced in many lakes, reservoirs, and rivers which has completely or partially reduced the number of native fish. Exotics that disturb the habitats of indigenous fishes, can be of great importance in determining the survival of eggs, spawn, and fry of Indian carps. Some important exotic food fishes, although not often directly related to the death of native fishes, can materially reduce their yield potential and are of great economic significance in relation to fish.

Reduced catches of indigenous fishes from different natural waters where exotic fishes have accidiently been introduced, are not uncommon, even though many such events exhibited distraction of ichthyologists from their work on this line to a large extent. Personal communications with fishermen and other informations from local agencies would definitely provide the best criteria for assessing the extent of damage to indigenous fishes.

Mass extinction of indigenous fishes appears to be accidental intrusion of exotic species when viewed from close proximity. Long-term presence of exotic fishes in many ecosystems may have severe effects on native fauna, even resulting in extinction in part of the indigenous species, or the effect may be insignificant, causing minor depressions in survival and growth. If reduction is severe enough, the exotic species will reproduce with similar ecological requirements. The indigenous species may then persist in low abundance for short/long periods. Possibly this is one of the extreme consequences of inland fisheries. Due to high reproductive potential of several exotic fishes in natural waters (such as tilapia, silver carp, brown trout, and common carp), their population size increases with permanent disturbances of inland waters. Some of the effects of exotic fishes on native species are briefly described below.

Silver Carp

This fish was introduced in Indian waters based on the fact that it is a surface feeder, stenophagic with feeding range restricted to planktonic algae. It has now been ascertained that the fish play a complementary role to *Catla catla* because the latter species is also a surface feeder. Therefore, question may arise why this species was introduced in India and recommended for cultivation with indigenous carp farming systems. Accidental intrusion of silver carp in several natural waters such as Gavindasagar reservoir in Himachal Pradesh has substantially reduced the population of Indian carps (particularly *Labeo rohita, Catla catla,* and *Cirrhinus mrigala*) which has reduced the commercial value of indigenous carps. Silver carp has also appeared in catches in Gomti river of Sultanpur district. This situation has undoubtedly great ecological threat to our inland fishery resources in the near future.

Common Carp

The second important and well-known introduced fish is the Common carp and is widely distributed all over the world. Their presence in freshwater systems has significantly demolished the indigenous snow trout of the genus *Schizothorax niger, S. esocinus,* and *S. curviforms* in Dal Lake of Kashmir. On the other hand, common carp has also affected the fishery of *Cirrhinus mrigala* and *C. reba* in Govindasagar reservoir. The population of *C. mrigala* declined from 9 per cent during 1974-75 to 0.05 per cent during 1997-98. 'The genus *C. reba* has been completely exterminated from this reservoir. Statistical analyses of fish catch records from other inland water resources such as Kumaon lake of Uttar Pradesh, Loktok lake of Manipur, and Pong reservoir of Punjab have shown that the population of Common carp is very high (more than 70 per cent of the total catch). And this condition has led to a sharp decline in native populations of *Tor* spp., snow trout,and *Osteobrama belangeri.*

Severe effects of common carp on indigenous species popultions relate to significant reduction and fluctuation in the population dynamics of the Indian major carps and are more conspicuous each year as evidenced by reduced catches of indigenous species from different freshwater systems. It seems plausible, however, that periodic fluctuations of fish stocks in many freshwater impoundments occur due to the presence of broodfish of common carp and their high fecundity rate as well as potential breeding habit would present obvious threat to indigenous species.

Tilapia

Tilapias are said to be delicious, with no intramuscular bones and with little carcass wastage. As noted earlier, they are prolific breeders, cheap to feed and tolerant to wide range of salinity, temperature, and comparatively free from diseases and parasites. Inspite of these advantages, tilapias have drastically reduced several native fish species in many regions of India. High population density of tilapia has significantly reduced several species of fishes such as *Chanos chanos, Labeo kontius, Puntius dubius, Eutroplus suratensis* and other indigenous carps from different reservoirs such as Tindivanam (State : Tamilnadu, Dist.: Viluppuram), Vallaikulam (State: Kerala), Threethankulum (State : Kerala, Dist.: Tribandram), Vaigai (State : Tamilnadu, Dist.: Theni), Amaravathy (State : Tamilnadu, Dist.: Coimbatore) etc. and lakes such as Powai (State : Madhya Pradesh) and Jaisamond (State : Rajasthan).

Tilapias have received scientific attention to pond culturists primarily because population explosion of tilapia in ponds has drastically suppressed the survival and growth of carp fry with serious consequences to the production potential of major carps. The presence of tilapia has caused extensive damage to fish culture environment. Infestation of freshwater resources by tilapia is, therefore, a matter of serious concern to fish culturists and fish culture ecosystems have changed from bad to worse condition.

Grass Carp

The main aim to introduce this species was to control water-loving plants. Due to its rapid growth rate, it is considered as an integral component of polyculture system. Though this species has not been reported any adverse impact on composite fish culture, its presence in different river systems not only have serious concern to several indigenous fishes but alter the entire ecosystem scenario as well.

Since this fish does not naturally breed even after hormone injection, stripping method is followed. The survival rate of fry and fecundity rate are very less and unless excessive aquatic plants are available/supplied as feed, their growth will be drastically reduced. Moreover, due to unavailability of stocking materials, its culture in composite farming systems is very difficult. However, though this fish is now considered as an uneconomical, it is an excellent fish for control of aquatic vegetation.

Salmonoids

The salmonoids belong to arctic and temperate waters of the Northern Hemisphere. Salmonoids include different species of the genus *Salmo trutta fario* (brown trout), *Salmo gairdnerii* (rainbow trout),*Onchorhycus nerka* (Kokane salmon), *Salmo salar* (land-locked salmon), and *Salvelinus fontinalis* (eastem brook trout). These fishes inhabit both the sea and freshwaters. During the nineteenth century, they have been introduced in many parts of the world. In India also, salmonoids were introduced at different times of the 18th and the early part of the 19th centuries. The above-mentioned species have been successfully introduced in different hills and mountains of India. Since they are stenothermal species, their range of distribution is restricted to the snow-fed rivers (such as Ramganga, Alaknanda, and Bhagirathi), lakes and reservoirs. These fishes have attracted the attention to biologists, fishing experts, and anglers due to their tasty flesh, beauty and gamesomeness.

Though these fishes were introduced fairy well to Indian waters during British regime, reports on their culture are surprisingly infrequent; and in fact, relatively very few studies on their breeding and culture have been recorded from natural populations of salmonoids. This is possible due to their long incubation period, low survival rate, slow growth rate, high managerial effort, and high cost of feed. For these reasons, trouts are not economical to culture for the table and as a food fish for common people, its contribution is virtually nothing.

Economic effects of trout culture in high altitude water bodies may be categorized as follow: (i) rejection by consumers and subsequent loss of interest in culture for the table, (ii) very slow growth of fish due to low temperature of water, and (iii) drastic reduction in trout populations in hill streams due to pollution. However, high mortality results in significant loss of trout populations. But keeping foreign currency earning capability of trout farms in mind, a number of experts have drawn scientific attention to possible losses of thousands of kilograms of trout annually because of the lack of adequate food requirements and other management protocols.

The culture of trout in low land areas must include consideration of the broodstock management and dynamic relationships of the environment, parasites, and pollutants. Some of the consequences of these interactions are reflected on variations in the occurrence or prevalence of pollutants or parasites with age of the fish and with geographical locations. Thus a parasite may be characteristic only to juveniles or adults. Parasitic infections may occur in one region or may be absent in another.

7.16 Processing and Marketing of Exotic Fish

Although common carp, silver carp, grass carp, and some other exotic species of less commercial importance are available in the market in many regions of the country, they are not widely accepted by most of the people. They generally command a low market price whereas the cost of trout and tilapia are very high and the higher profitability of trout and tilapia is due to high price of the product. Consumers' complain due to presence of intramuscular bones of small- and medium-sized Chinese carps has made the culture operation less attractive to farmers.

In many markets, trout and tilapia are preferred in fresh and live conditions. Both small- and large-sized fish are also marketed in frozen, smoked (hot- or cold-smoked), and packaged forms. The product generally depends on the consumers' preference, but the demand for ready-to-cook product for trout or fresh/live tilapia is increasing in most urban areas. Hot-smoking is carried out for small-sized fish. Large-sized fish (2.5 kg) are splitt and then smoked at different temperatures such as 30°C (for 30 minutes), followed by 50°C (for 30minutes), and 80 °C (for 1 hour). The smoked fish are cooled down to 5-8°C before being packed. However, different processing methods have been discussed in Chapter 23 of this Volume.

A low market appeal must stems from poor standards of processing and preservation. Improved standards of processing and preservation would definitely increase the market value of fishery products especially for trout and tilapia. The success of trout and tilapia farming and subsequent processing and packaging through state-of-the-art technology appears to have been due to quality and zero-defect products that are exported to super-markets.

7.17 Conclusion

Exotic fishes have certain characteristic features which provide a lot of utility both in terms of economy and resource creation. In India, exotic fishes play an important role for the marked expansion of polyculture to gratify the demand of both rural and urban people. Some exotic fishes such as tilapia, African catfish, *Puntius*, and common carp are considered by farmers as ideal candidates for commercial culture and hence these species have become very popular in some regions of the country because of low production costs, rapid growth rate, and consumers demand.

The treacherous effects of introduced species on native fish biomass following destruction of the aquatic biotic community as a whole must involve extinction of native fish species. Though extinction is a natural process, introduced species help to expedite the natural process at an alarming rate and this event must be kept in mind. However, recent investigations on the ecological impact of exotic fishes on native ones reveal the need for reassessment of potentiality of exotic fishes selected for culture in India. Today, marked expansion of introduced fish population over extensive water areas are their culture in freshwater ecosystems clearly indicates that the conservation of native fish stocks is badly needed. These events ensure that the farmers must have a production interest of value-added indigenous fish to protect them from extinction.

Tasteless flesh of some culturable exotic fishes and repugnance to consumers have been considered as a problem, yet the rural populations are accoustomed to these fishes. Inspite of this situation, farmers' awareness is necessary and information campaigns are being implemented, stating their harmful effects on native fish fauna.

It should be pointed out, however, that in many countries the selection of exotic fishes for culture is strictly prohibited, either legally or by implementing and enforcing Fisheries Act. In reality, inspite of the existing prohibition, exotic fish culture systems still exist and in several countries the culture is spreading.

References

Alikunhi, K. H., K. K. Sukumaran and S. Parameswaran. 1963. Induced spawning of the grass carp, *Ctenopharyngodon idella* and the silver carp *Hypopthalmicthys molitrix* in ponds at Cuttack, India. *Proc. Indo-Pacific Fish. Coun.* 10(2) : 181-204.

CIFRI (ICAR). 1976. Glass jar hatchery for carps, Central Inland Fisheries Research Institute, Barrackpur, India.

Ghosh, S., L. H. Rao and S. K. Saha. 1980. Culture prospect of *Sarotherodon mossambicus* in small ponds fertilized with domestic sewage. *J. Inland Fish. Soc. India.* 12 (2) : 74-80.

Hora, S. L. and T. V. R. Pillay. 1962. Handbook on fish culture in the Indo-Pacific region. *FAO Fish Biol. Tech. Paper* 14 : 204 p.

Kulkarni, C. V. 1946. Gouramy culture. *Indian Fmg.,* 7: 565-572.

Menon, M.D., S. Srinivasan and B. K. Murthy. 1959. Reports on Madras Rural Pisciculture Scheme, July 1942 - March 1952. 171 p. Madras Govt. Press.

Nagendran, R. 1996. Culture of common carp *Cyprinus carpio* in dairy wastewaters: A feasibility study *In*. Recycling the Resource- Ecological Engineering for Wastewater Treatment. (Eds. J. Staudenmann, A. Schonborn and C. Etnier). Transtec Publications, Switzerland.

Sarkar, S. K. 1983. Influence of cowdung and mustard oil cake on the effectiveness of fertilizers in fish production. *Environ. Ecol* 1: 31-40.

Dwivedi, S. N. 1980. Fish hatchery for seed production, technology transfer in rural areas, Central Institute of Fishries Education (ICAR), Versova, Bombay, India.

Dwivedi, S. N.*et.al* 1980. Fish hatchery and breeding unit. Central Institute of Fisheries education (ICAR), Versova, Bombay, India.

Questions

1. Why the problems of exotic fish production are severe in India?

2. What are the methods of increasing exotic fish production under Indian conditions?

3. Introduction of exotic fish for culture may sometimes create hazardous effects on native fish species. Why ?

4. What is the role of exotic fish culture in India? Why their culture has been encouraged in India?

5. Discuss the culture methods of the following fish species: (a) *Puntius javanicus*, (b) *Oreochromis mossambicus*, (c) *Ctenopharyngodon idella*, (d) *Cyprinus carpio*, (e) *Hypopthalmicthys molitrix.*

6. Why mono-sex culture of tilapia is more advantageous than mixed sex culture?

7. How tilapia population is controlled in carp fish ponds?

8. What are the ecological impact of exotic fishes on native fish species?

9. Among different exotic fish species mentioned in this chapter which species you would like to select best for culture? Give reasons in support of your statement.

10. Is it necessary to culture exotic fish in India? Explain.

8

Fish Culture in Paddy Fields

The quantities of fish produced in paddy fields in the last 20 years or so markedly increased from 50 to 1,800 kg/ha/year in several countries of the world, the average rate of 650 kg/ha/year being very common. High quality fish production generally depends upon the mechanization of paddy cultivation. The cultivation of paddy was already well established in early civilizations. At that times, only paddy cultivation was undertaken only for human consumption, but with the concentration of human populations, the demand for animal protein began to develop gradually and the paddy-fish industry grew very rapidly. The paddy, which is the staple food in almost all countries of the world and since fish are integrated with the paddy farming, will be discussed first.

8.1 Practical Considarations of Paddy-Fish Cultivation

Fish culture in paddy fields has formed the basis of fish culture industry for rural people. The system of fish farming, which supplies rural people with nutrients and food for alleviating malnutrition to some extent is not only the oldest, but also one of the most widespread culture systems in world trade and agriculture. Many countries have based their growth on cereals. Paddy cultivation is now relatively simple and requires considerable labor supplies. Under-developed countries always supply cheap labors and consequently, cultivation is not difficult. At the same time, animal and plant food materials are necessary for life and adoption of paddy-fish farming technology in paddy-growing regions has assured of local markets. The raise of fishes to prominence, however, again have given most of the Asian countries as an advantage at first. The products from paddy-fish farming can be of good quality and large-scale operations and a self-acting mechanism make cheapness of labor a less important factor. But in case of small-scale operations and small as well as marginal farmers, employment generations have great economic importance. In 1970s, the concentration of paddy-fish farming in many countries, particularly in Asia, have drawn serious attention to both fish and paddy growers to a considerable extent.

In suitable regions, the raising of fish in paddy fields is regarded as one of the best and most rational means of using agricultural lands. There are many advantages of raising fish in paddy fields and is considered as most important to the rural economy of the regions where it is practised. Paddy fish farming is highly economical where extensive areas are covered with paddy fields, and the potential for fish culture must be considered and implemented. At the same time, simultaneous and alternate production of paddy and fish (see section 7.4) also help to control weeds, molluscs, and mosquitoes from paddy fields.

Cereal is a collective term for all types of grass-like plants (such as paddy, wheat, rye, maize, etc.) and play an important role from food production standpoint. Although many varieties of paddy exist across the world and many of them are cultivated in many countries, a few varieties of paddy have greatest technical importance in relation to their production potential. Paddy is, however, one of the most common and widespread cereals. Paddy cultivation is done in flooded fields where a temporary aquatic fauna is found. The paddy field is generally considered as a successor of shallow marshes or a low land area which can be supplied with water. In addition to this, deep water paddy is grown in permanent marshes and paddy is also grown in hillsides. Since paddy fields are treated with chemical fertilizers and organic manures, this type of ecosystem creates a unique, productive, temporary and rapidly changing habitat which is often very ideal for raising a variety of fish species particularly tilapia, catfishes, murrels, carps, and prawns on an artisanal and extensive basis.

Fish culture in paddy fields has had a chackered history during the past 160 years when records are available. Long-term records of fish culture activities are not available from any part of the world although apparently this enterprise seems to have existed in Japan, Italy, Russian Federation, Thailand, and Indonesia. Before going to discuss about the paddy-fish farming, however, it is inevitable for the farmers to have a succinct idea about the origin of paddy, their variety and cultivation that follows.

8.2 Origin of Paddy and its Variety

Paddy (i.e. the growing plant or the yellowish unhusked grain) from which rice (i.e. the whitish husked grain) is obtained has been cultivated since time immemorial. The exact origin of paddy is not traceable, but it is believed that the present cultivated species were developed from the wild species found in swampy areas. There are. hundreds of sub-species of paddy but these fall into two groups such as wet paddy and dry paddy, which differ in their growing requirements. In recent years, as the result of genetic research and plant breeding, new and better strains of paddy have been devised such as IR 8, IR 20, and IR 22 'Miracle Rice' of the Philippines.

Paddy must have originated somewhere in China or India. Chinese literary works mentioned paddy as long ago as 3,000 B.C. It has been the staple food of China for a long time and is consumed in many ways. However, from India and China, paddy cultivation spread to neighbouring Asian countries. The Japanese have consumed rice since the historic beginning of the country. Records of paddy cultivation in Sri Lanka date back as far as 500 B.C. in the region where paddy was grown under tank irrigation. Chinese immigrants to Luzon in the second millennium B.C. must have introduced paddy cultivation to the Philippines. Paddy cultivation also spread to Java, Malayasia, and the rest of South-East Asia.

According to some authorities, paddy is believed to have been domesticated in two regions. The first and extensive domestication seems to have occurred in the Indo-Chinese region, at the foot-hills of the Himalayas about 6,000 years ago. In this region, the principal species *Oryza sativa* is cultivated today. In the upper Niger valley, *Oryza glaberrima* was domesticated about 5, 500 years ago. At present, thousands of paddy varieties are grown and Dr. M. S. Swaminathan estimated the number of cultivars at 1,20,00.

In Europe, the Moors first introduced paddy farming to Spain, then to Italy and Turkey. Paddy was first grown in the USA in 1685 in South Carolina. Commercial paddy cultivation in Australia first started in 1920 in tropical Queensland.

8.3 Paddy Cultivation

It is presumed that rice is consumed by most of the people than any other cereals. It is the staple food not only in monsoon Asia, but also in other tropical and sub-tropical areas where climatic and physical conditions permit its cultivation. It is, therefore, a dominant crop in tropical and monsoon lands and flourished in a variety of areas such as plains, swampy deltas and irrigated low-lands and hence it has been possible to integrate paddy with fish culture. And paddy is also successfully grown in the warm temperate lands of both the Southern USA and Europe where some regions have been selected for the cultivation of fish in paddy fields.

Paddy is an annual grass growing to a length of 1 to 2 metres. Different varieties are raised in different regions of the world, and the methods of farming also vary between that grown on dry hillsides (hill or dry paddy)and flooded fields (wet paddy). Farming methods also differ between the different oriental countries and between Western and Eastern growers. Consequently, rice yields vary greatly. The highest yields come from wet paddy which contributes about 95 per cent of the world's output of rice.

Field Preparation

Since paddy farming is very conditioned by the rainfall rhythm, farmers have to get their fields ready before the commencement of the rainy season. Ridges or bunds that separate one field from the other have to be prepared with compact mud, and any damaged ridges have to be repaired. The innundated ground is ploughed to a depth of a few inches. A wooden-ribbed roller is then used to break up any soil clumps and a wooden rake with rows of teeth is used to bury any remaining weeds under the mud. Manures and fertiltzers are added to the soil. The excess water is drawn away, leaving the silty surface covered with about 1 inch of water. The field is then ready for receiving young seedlings from nursery.

Field Management

After transplanting, seedlings are allowed to grow for about 4 to 5 months till harvesting time. The farmers must take note on the level of the water. When seedlings are about 6 inches tall, the water level is about 2 inches. With the growth of the paddy plant, the water level is also raised and remains at about 6 inches for 2 to 3 months. In areas of heavy rainfall, this growing period coincides with the rainy season. If the amount of water in the paddy fields falls short of requirements, water is allowed to enter the field from irrigation canal to obtain the desired level. Selected species of fish are stocked in the paddy field. Adequate care must be taken to prevent birds and other animals from damaging the paddy and fish. It should be pointed out that if fish are integrated with the paddy, only selected variety of paddy that are resistant to diseases and parasites, should be recommended so that the use of insecticides and pesticides can be prevented altogether.

Harvesting

Traditional paddy harvesting in the oriental region is very simple and a curved knife (sickle) is used. Harvesting of paddy is done in the dry season when the weather is relatively free from rain. After harvesting the paddy, the fields are then filled with water. These fields are then converted into temporary ponds in which fry/fingerlings of cultivable fish are stocked for several months (alternate farming). If simultaneous farming system is adopted, fish are harvested first by removing water from paddy fields by regulating the sluice of the irrigation canal. While portions of fish stocks are harvested at the time of draining the water from the fields, the remaining fish are collected simply by hand-picking. When the total fish stock is harvested, paddy harvesting is accomplished. Therefore, both fish and paddy harvesting are labor-intensive.

Since paddy is cultivated mainly by peasants in small farms, and statistics are incomplete, it is by no means easy to assess the annual rice yield of the world. The Food and Agricultural Organization, puts the figure at around 390 million tonnes of which about 95 per cent is grown and consumed within Asia.

8.4 Objectives of Paddy–Fish Integration

The main objective of paddy-fish farming is to improve the income of farmers and to make available an important item in the diet of rural community. Also, this integration provides off-season employment to farmers and at the same time it is mutually beneficial. *Firstly,* when herbivorous fishes are stocked in paddy fields, the weeds can be controlled to a considerable degree. *Secondly,* when. management of water is carried out, it is feasible to control the growth of molluscs and breeding of mosquitoes, thus reducing public health hazards. *Thirdly,* aquatic insects which are harmful to paddy, are consumed by the fish, thus promoting paddy yield. *Fourthly,* the movement of fish in paddy fields causes increased tillering, resulting in higher paddy production. And *Fifthly,* the greater depth of water in paddy fields not only prevents rats from excavating holes in the bunds but also the existing holes are flooded. Inspite of several benefits mentioned above, paddy-fish integration involves additional costs for the farmer which will have to be counter-balanced by income from fish production and marketing.

8.5 Types of Paddy-Fish Farming System

Different techniques are followed in the cultivation of paddy and fish in paddy fields and the techniques differ considerably from region to region. Techniques generally depend on the type of fish to be cultured, varieties of paddy, and different practices followed in paddy farming such as manuring, fertilization, and other management protocols. Paddy-fish integration can, however, be grouped that follows.

Alternate Farming

This system of farming consists of using flooded paddy fields after harvest where one or more fish/shrimp crops are cultured for short duration.

Simultaneous Farming

In this type of farming, both paddy and fish are grown together and they are harvested at the end of rice-growing season. This is the most veritable paddy-fish cultivation.

Complicated Type of Farming

This type of farming system assures an extended period of fish culture that involves transferring the fish stock to specially prepared channels or ditches at the time of paddy harvesting and restocking the fish in the field for further growth. Though the growth of fish is very high, the costs involved in this system of farming may negate the benefit achieved.

Capturing and Rearing Methods

In some cases, the paddy fields are not stocked with fish rather wild fish are allowed to enter the fields where they propulate when in flood and after several weeks or months, fish are harvested. In this method, generally murrels and tilapias are cultured. The production potential is very less.

It should be kept in mind that, whatever the system is adopted, paddy is considered as the main crop and fish is secondary one.

8.6 Infrastructure and Production Techniques of Paddy-Fish Systems

The term infrastructure refers to the basic structures/needs such as availability of quality water, fish seeds which are suited for cultivation in paddy fields, disease-resistant paddy varieties, site selection, field design and construction, farming technology, and arrangement of harvesting as well as marketing needed for the operation of a paddy-fish farm. In this section, we will limit ourselves to an analysis that follows.

Infrastructure

1. **Site Selection :** Paddy fields having clay soil with sufficient water-holding capacity for long duration and less prone to occasional flooding are considered as best for paddy-field cultivation. Uniform contour preferably with one lower end and drainage facilities, higher ground water table, and adequate soil texture generally make the cost of infrastructure relatively less.

2. **Size of The Farm:** Though the size of the farm vary considerably depending on the type of fish culture to be adopted and geographical conditions of the region/ area concerned, the overall size of the farm should be as square or rectangular as possible. The paddy plot may vary from 0.5 to 1.5 or 2 hectare. For one hectare plot, the size (in terms of square metre) required for dykes, trenches, and pond refuge and field is 2,000 (20 per cent), 1,300 (13 per cent), and 6,700 (67 per cent), respectively. However, certain modifications in size may be made as per availability of land.

3. **Field Design and Construction :**

(i) *Dyke:* Since paddy cultivation is very much conditioned by the rainfall rhythm, farmers have to get their fields ready before the commencement of the rainy season. Bunds or dykes that separate one field from the other have to be constructed, usually with compact mud, and any broken dykes have to be repaired. Dykes should be strengthened and raised higher (at least 15 cm) than the maximum flood levels. The excavated soil should be well compacted. This compaction will undoubtedly make the dyke more solid. Side slopes of the dykes should be steep. Crest width of dykes may vary from 50 cm to 2 m in cases where fishes are integrated with horticulture and vegetable (Figure 8.1).

(ii) *Refuge* : Simply defined, a refuge is a pond or trench in the paddy field where fishes are sheltered wnen the field is dried up. When a refuge is constructed, the excavated soil is utilized for making the dykes. No refuge is constructed in traditional paddy-fish systems. The morphometry of refuge varies considerably with the type of lands, hydrology of the field, and the culture system. As a generalization, one-tenth of the field is used for the construction of refuge. The following four designs are followed (Figure 8.2).

Perimeter Canal Type: In this type, paddy growing area is circumscribed by a canal that varies from 1 to 3 m width and 1 to 1.5 m deep. This type is better for large fields on very flat land but is expensive since extra construction is badly needed for the purpose.

Central Pond Type : In this case, paddy growing area lies in the fringes and connected by narrow but shallow trenches (50 cm to 1 m width and 20-70 cm deep) to a central pond, the depth of which is about 1 m. Though the construction of dykes is costly, this type of refuge is better for large low lands and irrigated paddy fields.

Peripheral Trench Type : In this type, trenches (1 m width and 1 m deep) are constructed on one side, preferably in the lower region of the field. This design is very ideal for small farmers in water-logged or irrigated low lands.

Pond Refuge Type : In this case, one or two ponds (1-2 m depth) depending on the size of the field are constructed at the lower region of the field. The pond is locked with the paddy growing area either by two wide trenches (1-2 m wide in case of

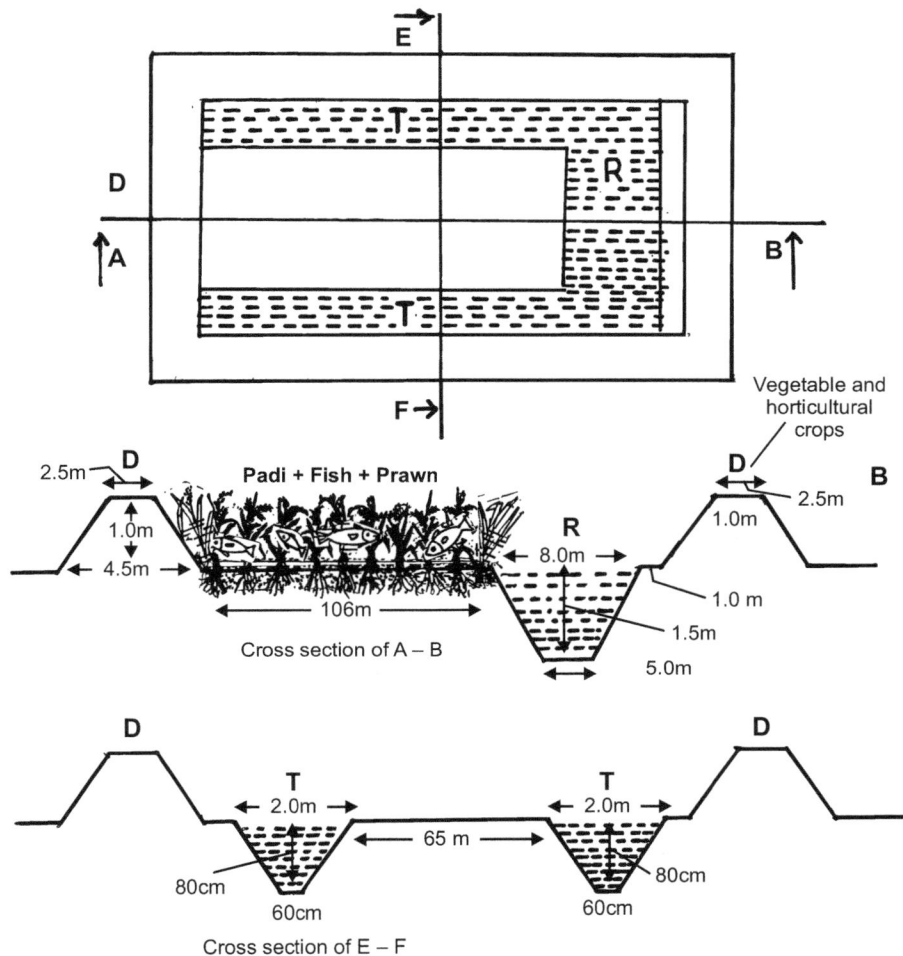

Fig. 8.1 : A rice-rish crop system model for rainfed low lands. The model is based on 1 hectare area and pond refuge with common connecting trenches design. T - Trench, R- Refuge D - Dike.

relatively large fields) or by a number of narrow trenches (in case of small fields). This design is recommended for flat land since it has water holding capacity, less risk, and better management. This system is not desirable in sandy/porous soils where seepage is more intense.

4. **Drainage :** The paddy field requires a good outlet for rapid draining of flood water. At the same time, the field also must have proper inlet facilities when regular supply of water is necessary as in the case of irrigated paddy fields. Both inlet and outlet should be regulated by screens that prevents the fish from escaping. The diameters of outlet and inlet generally depend on the volume of water flow.

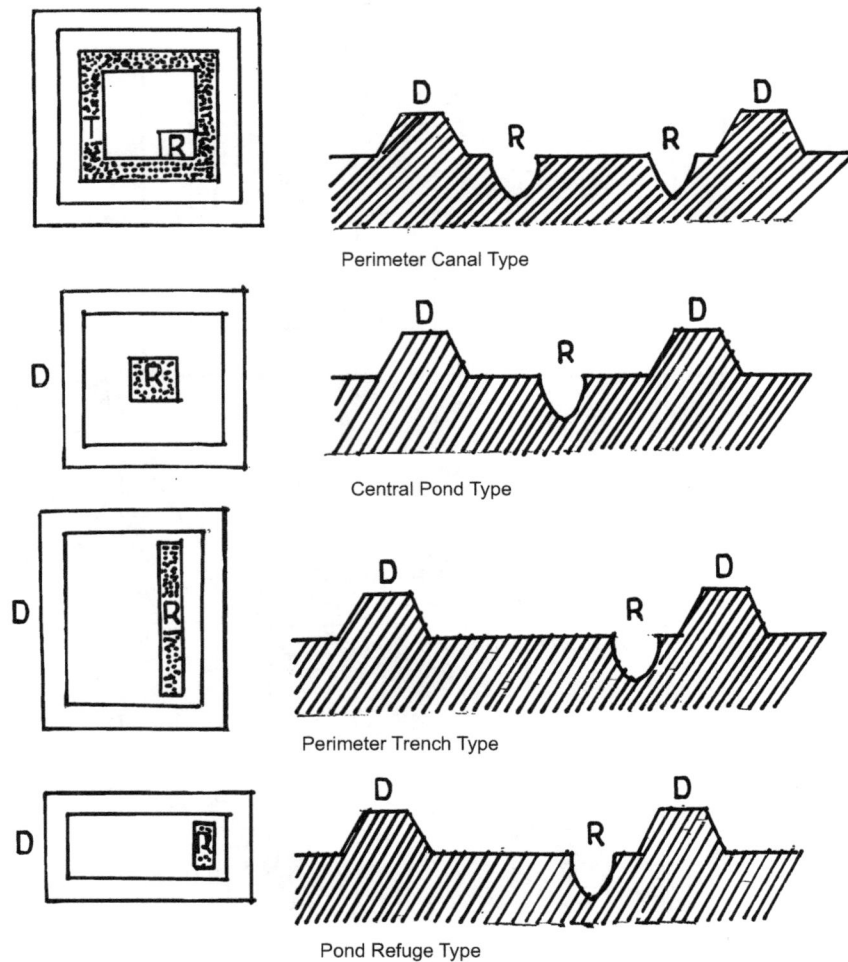

Fig. 8.2 : Various types of paddy-fish field designs.

Production Techniques

Production techniques involve searching of suitable combination of fish species, judicious use of paddy fields as nursery and fingerling ponds, and different management protocols. This makes the enterprise more viable in the long run. However, there is sufficient informations by which fish culturists may use paddy fields for fish culture by adopting methods suitable for local conditions. Fish farmers, in general, use traditional methods or methods taught by extension workers. Published leaflets are also available for fish farmers.

Fish that colonizes and live in paddy fields for long or short periods include mostly marsh-dwelling and air-breathing species found so widely in tropical Asia. Air-breathing species have accessory respiratory organs and can aestivate in wet mud habitats generally found at the edge of paddy fields. Some riverine fish move into the paddy fields where they colonize only when the level of field water remains high and when the flow of water through paddy fields decreases considerably, they again migrate into the irrigated ditches and channels. Four paddy-fish integration systems that follows are generally adopted for better economic returns.

1. **Paddy Fish Seed Production:** In most Asian countries, raising of fry (2-3 cm in length) and fingerlings (10-20 cm in length) with paddy is widely practised in irrigated and favorable rainfed low land ecosystems. In traditional system of paddy-fish seed farming, the water is drawn in the field by regulating the sluice of the irrigation canal that brings wild fish seeds to these fields and as they grow to fingerling sizes, farmers harvest them using trap nets. In improved culture system, one of the four refuge types is used for fish seed production.

(i) *Crop Management*: Improved varieties which are tolerant to pests and diseases are suitable for cultivation in irrigated paddy fields. Selection of paddy cultivars depends very much on the water regime and culture systems adopted. Under shallow water conditions (8 to 10 cm depth), for example, a number of paddy varieties (such as Udaya, Sarasa etc.) of medium duration (120-140 days) during *Kharif* followed by cultivars of 120-130 days duration such as IR 36 during winter season (*Rabi*) are integrated with fish seed production. During monsoon season, a still longer duration semi-tall cultivar (such as Savitri) can be grown under high water depth. If irrigation facilities round the year are assured, a third crop of paddy like 'Ratna' can be considered in between early *Kharif* and *Rabi* crops.

There are many ways by which paddy seedlings can be propagated. However, the traditional method widely practised in most of the oriental countries involves raising of young seedlings in a well-prepared nursery bed first and then transplanted them into the field. In regions where the advent of the monsoon season is not certain, a dry nursery bed is used. In this case, the young seedlings can remain in dry nursery until the monsoon appears.

About 25 day-old seedlings should be planted in the prepared field with proper spacing (20 and 10 cm between rows and hills, respectively). Paddy requires three major essential nutrients such as nitrogen, phosphorus and potassium. Most of the paddy lands have a moderate amount of these mineral nutrients, but if nutrients are deficient in soil, chemical fertilizers or organic manures have to be used. However, fertilization with 60-80 kg of nitrogen, 40 kg of phosphorus, and 40 kg of potash during *Kharif* and 80-100 kg of nitrogen, 50 kg of phosphorus and 50 kg of potash during *Rabi* generally permits better results. Three-fifths of the total nitrogen is added in three equal instalments along with total phosphorus and potash as basal doses at the time of field preparation. The rest two-fifths are added at 30 and 70 days after transplantation of seedlings. Farm manure at the 5-10 tonne/ha rate during land preparation may be used for better results.

Weed control by weedicides should be avoided as far as possible. If the use of pesticides is necessary, however, comperatively less toxic and degraded chemicals may be used. Prior to application of chemicals, the field is drained to the fish stock into the refuge. After 4 to 7 days, the field is again flooded with water at desired level.

During plantation of young seedlings, the depth of water in the field is around 3-5 cm. This level gradually increased upto 10 to 15 cm which is maintained till one week before the harvesting of paddy. At the time of harvesting, the water is drawn away from the field to facilitate easy movement of fish to the refuge.

(ii) *Fish Species:* In tropical Asia, paddy fields harbor a wide range of indigenous fish varieties specially Indian major carps, common carp, catfish (*Clarias spp, Mystus* spp.), snakeheads, and tilapia. In more northern latitudes which experience winters, fish do not occur in abundance and variety Table 8.1) in paddy fields, although they do colonize in fields and die if trapped them in winter months. The diverse fish fauna has enabled an artisanal fishery in paddy fjelds based on indigenous fish varieties in many tropical regions.

Table 8.1 : Numbers of Indigenous Fish Species Recorded in Rice Fields in Different Parts of the World. Except in a Very Few Cases, Their Numbers Must be Considered as Lower Than the Actual Number, Because Many Small Fishes Have no Apparent Economic Value and Hence Have not Been Listed

	Country	Number of species	Remark
I.	**Tropical and sub-tropical**		
	1. Sri Lanka	35	Pesticides are not used
	2. Malaysia	12	Possibly incomplete
	3. Thailand	18	Possibly incomplete
	4. South East Asia	12	Economic species are cultured
	5. Bangladesh	20	
	6. Indonesia	23	Possibly incomplete
	7. China	5	Economic species are cultured
	8. Venezuela	0	Heavy pesticides are used
II	**Outside tropics and sub-tropics**		
	9. France	0	
	10. Hungary	5	Before heavy pesticide use
	11. Russian Federation	6	
	12. Italy	4	

Source : Fernando (1993)

About two weeks old fry are stocked one week after the plantation of seedlings at the rate of 10,000 - 50,000 numbers per hectare depending on the level of management. Supplementary feed that consists of oil cake and rice bran (1 : 1 ratio) is applied at the rate equivalent to total fish biomass. This feeding is continued up to 30 days and then the feeding rate is doubled the initial rate during the rest of the culture period. Farm manure, if necessary, is applied at the 300 kg/ha/week rate. Fingerlings are harvested from the refuge after or during paddy harvest. Production of common carp seed as second crop can be possible with paddy during winter season.

Irrigated paddy fields if properly prapared can be used for breeding and seed production of air-breathing fish particularly *Clarias batrachus, Heteropneustes fossilis* murrels, and *Mystus* sp. Culture of these carnivorous species has several benefits (See panel 8.1). Brooders are injected with carp pituitary extracts (for further detail, see Chapter 6) and are released at the rate of one set (two males and one female) per three metre area one week after transplantation. The field is provided with a series of deep horizontal burrows of 5 inches diameter all around at the water

---

PANEL - 8.1

BENEFITS OF FISH CULTURE IN PADDY FIELDS

Retention of water for at least 4-6 months in paddy fields promotes several conditions favorable for the fish concerned and fish food organisms that permits higher production. The low-lying paddy field with stagnant water is considered as potential area for the culture of *Clarias* sp, *Heteropneustes* sp. Murrels, and *Mystus* sp. in particular. Draining of rain water into the paddy plots during monsoon seasons palys a critical role that permits the stocking of these species. At the time of paddy harvesting, these fishes are collected. The rain water is stored in the paddy fields for several months which are best suited for stocking desirable fish. These water bodies have low dissolved oxygen concentration, high temperature and high turbidity and contains sufficient nutrients to encourage the building up of a productive ecosystem which is suited for good fish production. The production of Paddy has heen reported to have increased by 7, 15 and 17 per cent in Russia, China, and Indo-Pacific countries, respectively when catfishes, for example, were stocked into the paddy fields. The use of supplementary feeds, addition of excretory products by fishes into the paddy fields, and greater tilling effects of soil by the activity of fish generally provide greater yields. Tilling effects of the fish permit the availability of nutrients such as nitrogen and phosphorus which are essential for the growth of fish food organisms.

The greatest benefit of fish cutlure (particularly carnivorous fish) in paddy fields relates to control of insects and their larvae. Their occurrence can be controlled by the introduction of several species of fish into the paddy fields. At the same time, paddy fields offer an ideal environment for fishes to breed successfully.

Special care must be taken to protect the outlets of paddy plots. If the outlets become weak, there is every possibility to escape fish from the plots. The outlets may be covered by a gate in such a way as to retain the water inside the field. During harvesting of fish, the gate is replaced by a wire mesh, water is allowed to drain outside the plot and the fish is then collected.

level height to serve as resting place of fish. The field is gradually flooded up to a level of 8 inches. About 20 days after spawning, fry measuring 1.5 - 2.0 cm in length are collected.

(iii) *Production:* The production potential of paddy and fish is highly encouraging. The average yield of paddy has been recorded to the tune of about 4.5 and 6.0 tonnes/ ha during *Kharif* and *Rabi* seasons, respectively. Similarly, per hectare yield of fingerlings in irrigated paddy fields generally ranges from 100 to 150 kg/6-10 weeks. The fish production figure may be doubled if adequate feeding schedules are followed.

2. **Paddy–Table Fish Farming:** The field design which is adopted for paddy-table fish farming system is similar to that followed in the case of Paddy-fish seed rearing system. For large paddy fields with adequate supply of water round the year, however, perimeter canal type is more advantageous.

(i) *Fish Species* : For shallow trenches and low water regimes, air-breathing fish such as *Clarias batrachus, Heteropneustes fossilis,* and Murrels are suitable for farming. On the other hand, common carp, Indian major carps, and freshwater prawns (*Macrobrachium rosenbergii* and *M. malcomsonii), Labeo bata, Puntius javanicus, Hypophthalmicthys molitrix,* and *Ctenopharyngodon idella* can be stocked in wide and deep trenches where sufficient quantities of water are stored.

Though usual stocking density of fish is generally one fingerling/ m², the under-stocking of fish (about 50,000 number/ha) is made if no supplementary feed is added. In cases where the stocking density is high (about one lakh/ha), daily feeding with rice bran and fish meal (1 : 2 ratio) for carnivorous fish at the rate of 5 per cent of fish biomass and with rice bran and mustard oil cake (1 : 1 ratio) for carp and prawn is followed.

(iii) *Production:* Under shallow water ecosystems having three high-yielding paddy varieties and air-breathing fish, per hectare total annual yield of paddy and fish to the tune of 13.5 and 2.3 tonnes, respectively can be achieved. Fish production, however, may be reduced by around 50 per cent if no supplementary feed is used. In wide and deep trenches, per hectare production of paddy and table fish has been reported to be about 5 tonnes and 480 kg, respectively.

3. **Fish Culture in High Altitude Terrace and Valley Paddy Fields :** The importance of paddy-fish systems in rainfed high altitude regions should be taken into consideration. Most of the oriental countries have extensive potential area in high altitude regions where paddy-fish farming provides additional income to hill dwellers. This system of farming is confined to regions at an altitude that varies between 1,000 and 5,000 feet above the mean sea level. Paddy farming in these regions is characterized by growing tall or semi-tall and long duration cultivars.

Uplands may be terraced but each of the terrace fields is laboriously and skilfully levelled to create low land flooded conditions. The edge of each terrace is bounded by a man-made ridge or stone wall that contains the water which gradually flows down the terraced slope. In this way, the silt-laden waters that are eroded from the upper slopes are caught by each terrace to irrigate and enrich the fields. And the population perssure is so great that hill-terracing for paddy-fish cultivation is practised more and more widely all over the monsoon Asia.

Slopes may be cut into a series of terraces with sufficient level ground on each terrace for cultivation, and an outer wall at the edge to retain the soil and to slow down the flow of rain water down the slope. Terracing is widely used in monsoon Asia for wet paddy cultivation, as the excess water and silt can be retained at each terrace to form flooded paddy fields. Many tree crops are planted on terraces to prevent them from erosion. Terraces are also used in semi-arid and temperate regions where slopes are steep. Terracing enables farmers in mountainous regions to utilize the steep ground for paddy cultivation that is integrated with fish. Among fish, common carp, *Clarias* sp. , and some cold water fish species are mostly grown as they grow best in such climatic conditions. At low altitude regions, Indian major carps can also be recommended and integrated with Paddy cultivation.

In deep water valley areas, a production level to the tune of 1 - 3 tonne/hectare of Paddy and 300 to 700 kg/ha/year of Indian major carps has been recorded.

4. **Rain-fed lowland Paddy Ecosystem and Fish Culture :** Deep water (1 m depth) and semi-deep water (60-100 cm depth) areas provide better opportunity for fish farming along with paddy. Intermediate lowlands with high water regimes can serve the purpose with proper refuge systems.

In improved system of paddy-fish farming, pond refuge type is suitable where Indian major carps, common carp, Chinese carp, grass carp, mirror carp (*Puntius spp.* and *Labeo bata*) and freshwater prawns are widely used for better production. Prawn juveniles and fish fingerlings in equal proportions (1 : 1 ratio per square metre) are stocked in well-established paddy fields immediately after the accumulation of water in the field and refuge. Organic manuring and regular supplementary feeding to fish and prawn permits better production. In open deep water ecosystems, prawn and fish can be raised in pens made of bamboo strips or synthetic materials. Harvesting of fish and prawn is carried out after paddy harvest or the culture period can be extended up to 5 months in refuge with periodical harvesting of table-sized fish.

In semi-deep low land and deep water ecosystems, the yield of paddy generally varies from 2 to 5 tonne/ha and 1 to 2 tonne/ha, respectively. At the same time, the yield of fish and prawn under these ecological conditions also varies between 0.3 and 1.7 tonne/ha depending On the level of management. Pen culture system gives satisfactory yields to the tune of about 630 kg/ha/season.

8.7 Paddy Field as an Aquatic Habitat

The aquatic phase of a paddy field may be permanent as in deep water or permanently submerged fields. As a generalization, however, the aquatic phase is temporary and seasonal. The duration of this phase varies considerably from place to place and also from year to year though the general pattern remains similar at anyone site. A spatial relationship of the aquatic system always exists, however (Figure 8.3).

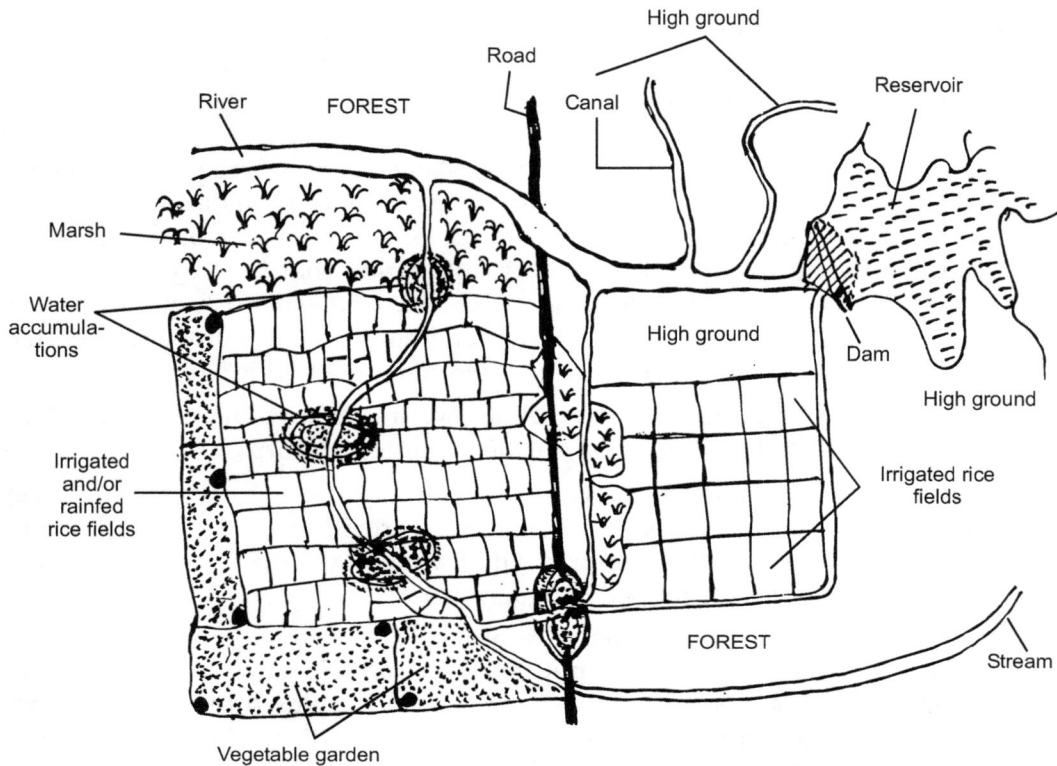

Fig. 8.3 : **Diagram showing the spatial relationship between rainfed and irrigated rice fields in a tropical region. Note that the contiguous marshes, water accumulations, sump ponds for watering vegetables and the natural streams and irrigated channels serve as refuges for organisms during the dry season.**

One of the most important characteristic features of paddy fields is the slightly inclined bottom with shallow water depth throughout the aquatic phase. Ploughing the paddy field may be done when the surface is already covered with a thin flim of water or when the soil is in the moist condition. After ploughing and inundating, the fields are then crushed by using a wooden-ribbed roller pulled by a pair of buffalo or cattle. Compaction of soft mud is done on the bunds that prevents water from seepage. This soil-water mixture is left to settle and the supernatant water is drained partially before the transplantation of paddy seedlings. Transplantation is carried out with groups of plants spaced properly at regular intervals. This provides spaces between plants where fish can move without any difficulty.

The shallow water depth (5-30 cm) brings rapid changes in water temperatures. The water becomes saturated with oxygen during the day when photosynthesis occurs at a rapid rate. The organic matter which arises from the debris of plant and animal residues are deposited on the mud water and are attacked by micro-organisms. The microbes decompose them in presence of oxygen to derive energy for their growth. This process is termed as *decomposition*. During the process of decomposition, essential elements present as compounds in the organic matter are converted into simple inorganic forms. This process of conversion is termed as *mineralization*. Mineralization process in mud water brings enormous changes in the paddy field ecosystem by releasing nutrients to overlying water that are essential for ecosystem productivity.

The physico-chemical conditions in paddy fields are detrimental to many organisms. The organisms that thrive well are eurytopic forms and also those which can aestivate in the mud or colonize the aquatic phase.

The basic requirements for maximum year-round production of paddy and fish appear to be higher in the tropics than anywhere else. The total annual solar radiation and warm to hot climates provide unmatched production potential of the aquatic habitat in paddy fields. Remarkable increases in the use of conmercial fertilizers have taken place in the past several decades, especially in Asia. The aquatic phase coupled with high yielding varieties, fertilizers, and manures have stimulated unparallel increases in crop production in Asia and will likely to continue to do so. Use of commercial fertilizers is very important for paddy production and they are beneficial to particular organisms but concern over the possible build-up of toxic metabolites from large applications of fertilizers could be harmful to fish species in paddy fields. Harmful effects of fertilizers on fish in paddy fields have prevented their more extensive use with respect to fish production.

Herbicides and pesticides are widely used in most paddy–growing areas of the world for control of weeds and insect pests and consequently, paddy–fish intergration systems are contaminated by a number of inorganic compounds that, to a greater or less degree, are toxic to fish; of couresa, their degree of toxicity depends on the type of compounds. However, disease and insect problems must be encountered with the use of resistant varieties.

The paddy fields can be considered as a modified marsh ecosystem. Agriculture practices considerably damage the natural, physical, chemical, and biological conditions of paddy fields making them less favorable for some organisms but temporarily more suitable for others.

8.8 Aquatic and Semi–Aquatic Organisms in Paddy Fields

Since agriculture practices simlify the biological community of paddy fields, the paddy plant that dominates the aquatic phase provide protection to aquatic/semi-aquatic organisms and create an ideal habitat for many organisms. If the conterminous marshes and streams are rich in organisms, paddy fields have a rich and diverse spectrum of organisms. Since paddy fields are well fertilized with different nutrient carries, many species of animals are adapted to these conditions.

There is a diverse complement of aquatic organisms in paddy fields ranging from bacteria to higher plants and all groups of fauna from protozoa to mammals. All these organisms exist temporarily. A variety of indigenous fishes harbor in paddy fields. Tropical Asia has the largest number of air–breathing fishes living in marshes and they are well–adapted for living in the shallow water of rice fields. Insects populations in paddy fields serve as an important source of food for fishes. Since aquatic organisms in paddy fields vary enormously in different regions of the world, it is very difficult to make a general statement in this regard. The following examples will illustrate this point. No fish species have been reported in many paddy fields of France but tadpole, shrimps, and dipteran larvae were abundant. This condition has not been noted in tropical paddy fields. Though mosquito larvae are present in relatively small numbers, they are controlled either by chemicals or by using mosquito larvae-eating fish especially *Gambusia affinis*. The abundance of small fish in paddy fields of Thailand keeps down the invertebrate populations and also serves as food for piscivorous fishes.

As the paddy plants grow, they shade the water and the process of photosynthesis greatly reduced. There is also an abundance of decaying organic matter and the accompanying periphyton, epizoans, and bacteria which serves as food for the fauna. The population of zooplankton in paddy fields is composed largely of benthic species or those associated with the macrophytes. Benthic species mainly dominated by oligochaetes and chironomids.

When the paddy is ready for harvest, the water level in paddy fields decreases gradually and becomes dry at the time of harvesting. Along with the reduction in the volume of water, conditions for aquatic organisms also deteriorate. In tropical paddy fields, competition for food and space and heavy predation drastically reduces the density of organisms. Fishes which are confined to small shallow waters, can be readily harvested. Fishes are also harvested in dry season. In temperate regions, winter causes heavy mortality of many organisms including fish inhabiting in shallow waters.

8.9 Importance of Fish Culture in Paddy Fields

Apart from the increased production of paddy and fish to meet the demand of ever-growing populations, an obvious link between paddy-fish cultivation and vector-borne diseases such as malaria has been noted from many countries. This farming system has been linked to increase malaria incidence in Africa, Europe, North America, South America, and Asia. The expansion of paddy cultivation dictates that vector problems will be exacerbated since increasing paddy yields are being pushed ahead whatever the cost of production.

The high temperatures of water in paddy fields, high nutrient levels, and protection offered by paddy plants afford insect vectors a very suitable habitat for their propagation, Biological methods of controlling vectors in paddy fields have become more attractive proposition. A large number of fish species have been recorded in some tropical paddy fields and their role in reducing vector populations could be crucial to control of vector-borne diseases.

The importance of paddy-fish cultivation lies in the fact that this system of farming is traditional and extremely labor-intensive and therefore employment opportunity particularly where more than one paddy crop is grown in a year. Almost every member of the farmer's family is actively engaged particularly in every day of the year, in various processes of paddy-fish farming such as field preparation, raising of seedling for transplantation, manuring, collection of fish seed and their stocking, harvesting, and other management schedules as and when required. In upland regions, the terraces have to be efficiently maintained to make the culture system economically viable in the long run.

The culture and harvesting of fish in paddy fields are significant particularly in Asian countries where most of the people live below the poverty line. Therefore, paddy farmers in these regions use their fields as a source of additional income through fish cultivation which is generally considered as an artisanal fishery in paddy fields. The existence of such fishery over a long period of time and over a wide geographical area clearly dictates that extensive paddy fields are fantastic habitats for most of the freshwater fish species that may be cultured under intensive basis and consequently, this farming practice has greatly increased significantly despite adverse environmental situations.

When paddy fields are stocked with fish, paddy plants grow well and flourished with longer panicles, and more grains per tiller. This is due to the fact that nutrients derived from the use of nutrient carriers permit better conditions for the establishment of food chain for the fish. Nutrients are extracted by bacteria, phytoplankton, and other benthic organisms which are fully utilized by fish resulting in high fish production.

Another important aspect of fish culture in paddy fields lies in the fact that due to brushing behavior of some bottom-dwelling fish, the soil of paddy fields becomes loose that helps aerate the soil, thus enhancing the decomposition of organic matters and subsequent release of nutrients from the soil to overlying waters.

8.10 Constraints of Fish Culture in Paddy Fields

Disturbance of paddy-fish cultivation by some predatory animals (such as frogs, rats, snakes, birds, and a number of aquatic insects), ignorance and reluctance of farmers to paddy-fish farming, the use of chemicals for pest control, unavailability of fish seeds for stocking in paddy fields, moderate to low production potential of fish in many paddy fields are some of the most significant worldwide constraints. These constraints are most notable in many tropical and temperate regions. Lack of proper water management provides serious limitations in almost all areas of the world but are most severe in Southeast Asia.

While the use of chemical fertilizers is being encouraged both for paddy production and the growth of plankton as well as bottom fauna which are essential for fish food, practical considerations dictate the search for toxicity whether extreme or moderate or low, particularly of nitrogen. Plant and animal manures will help supplement the manufactured fertilizers.

Multiownership of Paddy Fields is a Critical Problem

In many countries, multiownership of paddy fields is a critical problem that might inhibit such farming practices. At present, many Asian countries after their independence, have tried to improve the condition of farmers in various ways, some of which have been more successful than others. In many countries, laws regarding ownership of land have been changed rigorously to prevent the land from passing out of the farmers' hands. This ensures that the farmer does not become a landless one but cannot prevent him from running himself into debt and still being dominated by the money–lender. Law has also been directed at the landlord–tenant relationship to overcome the problem of unfair rents but this leaves unaffected the problem of credit for the running of the farm.

Lack of Knwoledge in Cooperative Farming may Cause Failure in Paddy-Fish Farming

Unsuccessful cooperative farming has been the most important constraint. Paddy fish cooperative farming system, though it is being encouraged and adopted, is less successful in many Asian countries than in Europe because the members do not clearly understand about the structure and function of the cooperative society and they do not have educational background about paddy–fish farming methods. Nonetheless, where paddy–fish cooperative farming systems have been successfully established, they have enabled for farmers to reap greater benefits from their fields and have improved farming conditions.

Training and Growing Awareness among Farmers are Essential

Though paddy–fish farming technology has been developed in many regions of the world, but in the long run, it will never improve the farming methods and the productivity of the fields unless they are adequately handled. In order to overcome these problems, better education for all categories of farmers and greater emphasis on training in paddy–fish culture practices are badly needed. At the same time, growing awareness among farmers for paddy–fish farming practices will eventually reduce for ever the different constraints and traditional fatalistic outlook of many farmers. It will encourage more modern and efficient farming activities and improve the farmers status in society.

Water Management is also an Important Criterion

One of the major ways in which paddy fields can be improved for farming is the adoption of water management strategy. Since both paddy and fish cultivation requires good quality and quantity of freshwaters, the supply of water must be ascertained where rainfall is often seasonal and unpredictable. Since water is not always available all the year round particularly during the dry season, necessary steps should be taken to prevent the loss of water when irrigation and draining programs are contemplated.

8.11 Fish Production Potential in Paddy Fields

There is practically no paddy field in the world that does not produce fish. Since artisanal fisheries and/or intensive fish farming are undertaken mainly by small and marginal farmers of many Asian countries and production statistics are highly disturbing as well as not complete, it is by no means to assess the annual fish production potential from paddy-fish integration system across the world. Though no precise data on fish yields in paddy fields exist for any country/region for over 20 years or so, production data from artisanal fisheries have been noted to be as 2 kg and 38 kg/ha/year from paddy fields in Indonesia and Malaysia, respectively. On the other hand, however, using selected varieties of common carp with artificial feeding, a record production to the tune of 1, 500 kg/ha/ year has been reported from paddy fields in Japan. Many production figures have been reported from different paddy fields in the world. A few selected yields of fish in paddy fields are shown in Table 8.2.

Table 8.2 : Fish Production (kg/ha/year) From Rice Fields. Production Figures are Generally Very Short Term or From Experimental Culture.

Country	Production	Remark
China	160-260	Single rice cropping
	70-140	Double rice cropping: Artisanal fishery
India	115	Experimental culture
Thailand	50-1,710	Recent culture
Indonesia	250 - 275	Culture
	145	Experimental culture
	2.0	Capture
Bangladesh	140 - 450	Experimental culture
Japan	700 - 1,800	Using common carp only
Russian Federation	75-150	Culture
Italy	100	Culture
Zire	700	
Tanzania	110	
Madagascar	200-250	
USA	120	Crayfish still cultured and fish culture is sporadic
Brazil	965	Experimental culture

Source : Fernando (1993).

The success or failure of fish culture in paddy fields not only determined by yield alone but also on the balance sheet between economic input and output in the long-term. A relatively low level of yield can be maintained without financial loss if the demand for fish is harmonious, the outputs are low and the income is slightly higher than the cost of production. Most of the farmers, however, produce fish from their paddy fields chiefly for domestic consumption and this includes mainly low-priced species.

Fish production potential in paddy fields requires explicit control over the entire fish-paddy growing period and the refined technology that is being widely adopted and followed economically in countries where this farming potential is quite great. This requires farm mechanization which is greatest in some countries especially Japan, China, and Philippines. However, though a viable paddy-fish culture enterprise can be maintained all the year round, fish production potential is not only vary from low to moderate but the low standard of living of farmers prevails as well.

8.12 Fish Culture in Coastal Saline Paddy Field Ecosystem

The coastal saline paddy fields in some Asian countries are extremely productive where traditional or extensive paddy-fish/prawn culture systems are adopted. Coastal saline areas in Asian countries have been estimated to be about 40 million hectare. Out of which only one lakh hectare are being considered for paddy-fish cultivation.

Paddy-Fish Culture in Coastal Regions – A Case Study in India

In paddy-fish culture enterprises, Indian coastal ecosystems achieved phenomenal success in the seventies. At present, it is a profitable enterprise with interests ranging from field preparation to income. Due to favorable environmental conditions, this farming system is not very attractive to farmers with moderate to high production potential of paddy and fish/prawn. This case study deals, in brief, with the farming success in coastal fish culture industry particularly in West Bengal, Andhra Pradesh, and Kerala. Farmers in these regions are late entrant in this field. Despite late entry, farmers made new fortunes in this respect. How did farmers achieved this success? This case study will seek for to answer this query.

1. **Production Strategy:** While the planning strategy established the foundation for paddy-fish cultivation in coastal regions, it was the culture strategy of the enterprise that actually, translated the enterprise's vision into reality, the enterprise is carried out its production objective in the ideal manner. In the succeeding paragraphs we shall succinctly examine the production strategies adopted under saline conditions.

   (i) In India, the traditional system – spread over in the fringe of Bay of Bengal in West Bengal – involves the use of adjacent canal system as the main source of impounding brackishwater prawn and fish species with the traditional wet season (July to September) paddy farming which is continued up to December – January. The production of fish and paddy in this system varies between 150 and 200 kg/ha/ 4-months and 1 tonne/ha, respectively.

   (ii) Several species of estuarine fish such as *Mugil* spp. (mullets), pearl spot (*Etroplus suratensis*), and shrimps (*Penaeus mondon, P. indicus, Metapenaeus affinis, M. monoceros,* and *M. dobsoni)* form of group of subsidiary species in several types of fish culture in coastal impoundments and ponds in Asia, and in countries such as India, paddy fields have been used for a form of extensive culture of these species for centuries. The most common traditional system of paddy-fish farming which is widely practised in the tidal zone of Kerala, India, called as *Pokali System.*

The traditional shrimp/fish-paddy culture in coastal regions of India was necessarily a polyculture system because of the inability to control the composition of the seed stock. In coastal impoundments or ponds, shrimps form only a small percentage of the harvest. Consequently, the species combinations were not always compatible, and the culture methods were unfavorable for high survival rates of shrimps. At present, however, polyculture of pearl spot/mullets and shrimps have shown the interference of requirements between the species. For example, the shallow depth of mullet ponds are unfavorable for shrimps and do not allow high stocking rates. There is considerable disproportion in the time necessary to attain the species to table size. For these reasons, it is often preferable to adopt monoculture methods in paddy fields. Although polyculture is not common, the traditional extensive culture of estuarine species such as mullets, tilapia, pearl spot, and milk fish in paddy fields still continues to be an important culture system, accounting for a good proportion of present-day production. In these systems, where natural stocking was achieved through the intake of tidal water, carrying shrimp and finfish larvae in large numbers, field designs were simple and were meant to serve largely as trap ponds or impoundments. The paddy fields are fully equipped with one or two sluice gates in the embankment. The gates are so designed that only sea water can pass but not fishes and prawns/shrimps. Sluice gates have regulating valves that control the entry and exit of sea water. Each valve measures 1.5 X 6 X 9 feet and the top end of the gate lies over the embankment. The trapped species are allowed to grow under natural conditions and are periodically trapped during every high tide.

The extensive system of paddy-fish/shrimp culture has recently undergone notable changes with a view to increase in production such as the introduction of controlled stocking of fry and enhanced production of natural food organisms through better water management and manuring by paddy stubble. Total harvest is generally carried out at the end of each crop by draining and placing a conical net having a diameter of 6 feet X 6 feet at the mouth with only 30 cm diameter at the tip gate to catch the fish/shrimp. Partial harvesting can be done after partial draining from peripheral canals or from harvest basins. While the production of fish and shrimp varies between 0.5 and 2 tonne/ha under favorable conditions, paddy production has been reported to be as about 1 tonne/ha. Traditional culture systems have, however, spread out their product strategies but strategies are not based on the technical superiority.

(ii) Entrepreneurs' paddy-fish culture strategy in coastal saline areas did not end with the establishment of adequate network on exclusively farmers lands of different sizes. Entrepreneurs are aware of the fact that proper planning holds the key to successful production of paddy and fish under similar ecological conditions. Careful site selection, field preparation, farm management, provision of right planning and continuous development are the major elements of enterprise's planning management. Farmers developed an improved culture system involves the followings.

Field areas having more than 1 m tidal amplitude with elevation between high and low tide levels are suitable for fish-prawn/shrimp culture. This system facilitates water exchange during brackishwater fish farming which is followed by frequent flushing with rain water during desalination process before plantation of paddy in *Kharif* season. As a generalization, however, silty-clay or silty-clay-loam soils of neutral pH are ideal in terms of better water retention, desalination, and low reclamation cost. Adequate annual rainfall (more than 1,000 mm) is also an important aspect to be considered for freshwater paddy-fish farming during wet season.

Field preparation involves proper construction of dykes, sluice gates, and waterways. The dykes must be wide and as strong as possible and height should be 30 to 50 cm higher than the maximum tidal amplitude/flooding level. For large paddy fields, perimeter canal type design with connecting cross trenches is desirable. The beds of all trenches are constructed in such a way that would facilitate easy and rapid drainage. Flow of tidal water in the field via adjacent canal is controlled by sluice gates of around 20 cm diameter. Wooden sluice gates containing wooden box with shutter can be kept along with the width of the dyke.

Paddy-fish-prawn farming technology in coastal saline ecosystem involves mixed or synchronous farming of paddy with freshwater prawns and fish during wet season followed by brackishwater fish culture in sequence during summer.

* *Mixed Farming*: In this type of farming, desalination process of paddy fields is adopted to reduce the level of soil salinity to a desired level. Salt accumulates in coastal paddy fields because more salts move into the ecosystem than to move out. This may be due to addition of salt-laden waters or evapotranspiration. Evapotranspiration from paddy fields leads to an increased concentration of salts and hence increases its salinity. In some areas, evaporation is so intense that the salinity of soil can reach toxic levels. However, salinization of paddy fields is more intense in summer season. In order to counteract the high soil salinity (EC varies between 15 and 35 dcm^{-1}), the coastal paddy fields have to be flushed by increasing the application rate of irrigation water or accumulated rain water at every low tide during the initial phase of monsoon. Desalination process makes the paddy fields less saline condition to a favorable level (EC values range from 3.5 to 7.0 dcm^{-1}) which is safe for *Kharif* paddy. In areas of soil salinity ranging from 5 to 10 dcm^{-1}, salt tolerant paddy cultivars of 110 – 150 days duration such as SR26-B, CSR-I, Pankaj, Lunishree etc. are suitable.

After desalination of soil at acceptable level in July/August, paddy seedlings of around 30-days old are transplanted at a space of 15-20 cm between rows and 15 cm between hills. Three split applications of nitrogen at the rate of 60 kg/ha at planting, and tillering are followed. Indian major carps, silver carp, and giant freshwater prawn (*Macrobrachium rosenbergii*) at the stocking density of about 25,000/hectare with the ratio of prawn and fish as 2 : 1 are suitable for culture with paddy.

It should be kept in mind that the paddy fields should be filled with rain water but not the tidal water and the water must be kept at the desired level. Fish and prawn are harvested immediately after the paddy is harvested in December.

* *Sequential Farming :* At the end of the mixed farming, the field is left for sun drying and then prepared for brackishwater fish farming by encouraging the development of bottom organisms generally termed as *lab lab*. The use of nitrogen-containing fertilizers such as urea at the rate of 60 kg N/ha and rice bran at the 1 tonne/ha rate over the soil permit the development of *lab lab*. During the formation of *lab lab*, the trenches are filled with tidal water to a depth of 60 cm. Thereafter, water is exchanged at regular intervals during full and new moon periods through sluice gates. In the month of March, the field is stocked with juveniles/post larvae of euryhaline prawn (*Penaeus monodon*) and fish (*Liza parsia* – a mullet) fingerlings at the rate varying between 50,000 and 75,000 numbers/ hectare. The entire stock is fattened by regular feeding and are harvested in July.

8.13 Worldwide Survey of Fish Culture in Paddy Fields

Malaya

In Alluvial and swampy regions, fish are raised in paddy fields. Paddy fields are flooded with the irrigated water. Fish are aggregated, breed, and grow in paddy fields. When fields are drained, they survive in ditches and sumps (40- 50 square metre in area and 2 m deep) which are constructed in the lowest regions of a group of fields. The sumps are shaded by banana plants and coconut palms which help concentrate the fish when it is decided to harvest them. The fish reared in paddy fields include *Trichogaster pectoralis*, *Anabas testudinus*, *Clarias batrachus*, *Channa striatus,* and *Tilapia mossambica*. The average yield of fish has been estimated to be about 140 kg/ha/6-10 months.

Indo-China

Most of the paddy fields are located in mountainous regions. The water is obtained from springs or streams. *Cyprinus carpio* is the only species cultured in paddy fields. Fry are collected from inland waters or from breeding paddy fields in March and are stocked in April at the 20,000 rate/ha. Fields are drained in July for harvesting. Simultaneous farming of paddy and fish are practised during rainy season and continued upto the end of October. The tropical climate in these regions makes the farming system more profitable to farmers. The most common fish species reared in this region includes *Channa striatus,* *Trichogaster pectoralis*, and *Clarias batrachus.*

Taiwan

Tilapia mossambica has been the most successfully propagated for culture in paddy fields compared to that of the carp. In the Northern part of the country, winter kill may happen if deep wintering ponds are not used.

The paddy field is well prepared and dressed with fertilizers. Two tonnes of compost per hectare are applied as top dressing. The field is flooded with water. After the paddy seedlings have been transplanted into the field, fry and fingerlings are stocked at the rates of 7,500 numbers and 150 kg/ha, respectively. After harvesting the paddy, fish stocks are transferred to refuge in the trenches. Though many fry are present, some of them weigh about 50 g. The field is prepared for a second paddy crop, fertilized again and flooded at a maximum level until the paddy seedlings are ready for transplantation. When the field is ready, the water is drained, paddy seedlings are transplanted and a temporary bund is constructed with mud to protect the young plants from the fish. The fish are admitted in the paddy and supplementary feeding starts that consists of rice bran and mustard oil cake (1 : 1 ratio). In Autumn, fish are harvested before or after the harvest of paddy. While the larger fish are harvested, the smaller ones may be stocked in wintering ponds until the next spring.

Although this practice leads to a greater loss of paddy due to over-population of fish (two months after stocking), it greatly contributes to the fish yield by providing a safe refuge for the fish during draining periods.

Japan

Since there are vast areas of paddy fields in Japan, it is extremely beneficial to utilize these waters for fish culture. The common carp *Cyprinus carpio* is the basic fish species cultured in paddy fields. Spawning carps (several kilograns in weight) are placed in small ponds that are filled with aquatic plants. After the eggs have been laid on these plants, the vegetation is transferred into a small paddy field (less than 0.3 hectare) where hatching takes place.

The paddy fields should be so located as to allow the supplying of fresh warm water, there should be no risk of any flooding the fields, the ditch for the incoming water supply and the drainage from the field must be covered by a bamboo mesh so that fish stocks cannot be escaped from the field.

Field preparation generally involves the construction of bunds (40-45 cm high) around each field, a ditch (30 cm wide and 40 - 60 cm deep) which is connected with an inlet and an outlet pipes, a sump (50 cm deep) near the inlet, and manuring of field water with compost and lime at the rates of 1,200 and 120 kh/ha, respectively. Manuring and liming helps promote to increase zooplankton populations and other natural foods for the fish. Paddy seedlings are transplanted in June and about 10-days later fish are stocked at the rate of 4,000-15,000/ha depending on the amount of food to be given and the size of the stocking material. Fish are harvested before the paddy harvest in Septernber-October.

When the rearing is accomplished only by natural food organisms available in the fields, the number of fish (especially carp) released in the fields should be lower. If artificial feeds are regularly supplied from outside, about 4,000 carp/ha can be stocked. If yearlings are cropped, they are stocked in wintering ponds until next spring when they are restocked for second year rearing in paddy fields to yield fish suitable for consumption.

Europe

Though the raising of both common carp and trout in ponds are common in most of the European countries and the fish rearing establishments reach a size that varies between 5 and several hundred hectares depending on the agricultural structure of the region, only common carp, *Tinca tinca,* and *Carassius auratus* are the important species cultured in paddy fields of some countries especially Italy, Hungary, and Bulgaria. In autumn, however, about two-weeks before the paddy harvest, the field water is drained and fingerlings are cropped. Sometimes they are stocked in wintering ponds until the next season to be reared in the fields for a second growing season.

In some regions, irrigated paddy fields are used for carp culture by cooperative basis to reduce the cost of production. Carps are grown together with paddy from May until the end of September. Fish are drained from the fields and are transferred to nearby stocking ponds.

USA

Paddy-fish culture is concentrated in the South Central region of the United States, Mississippi Valley, and the East of the Rio Grande. At least 1.5 million hectares of irrigated paddy fields exist and the potential of fish culture is great.

After the paddy has been harvested, the field is flooded and the water level is being raised as high as possible (60-120 cm). Fish are stocked as fry for 1-3 years. During harvesting, water is pumped out until all the fish are concentrated in a ditch situated at one corner of the field where they are cropped. Fish are raised not only for human consumption but also for the preparation of fish meal, pet food, and fish oil.

Though there is vast scope for paddy-fish cultivation, it is little doubt that a new area of intensive development can be forecast for fish culture in the vast complex of United States paddy fields. Two species such as *Ictalurus punctatus* (Channel catfish) and *Ictiobus cyprinellus* (Bigmouth buffalo) are well adapted to field conditions.

Indonesia

As a consequence of intense propaganda in the years 1930-1935, the paddy-fish cultivation has spread out the whole of Indonesia. Bulk quantities of fish are produced from vast areas of paddy fields. This is due to the fact that very favorable conditions for fish cultivation in paddy fields prevail in this country. Though common carp and tilapia are the principal fish reared in paddy fields, other species of minor importance such as *Chanos chanos, Helostoma* sp., *Osteocheilus hasselti,* and *Puntius javanicus* are also reared. These fishes are cultured in paddy fields for the yield of fingerlings which are sold to pond owners as stocking materials. The methods of fish cultivation in paddy fields are more or less similar to that of other countries.

8.14 Conclusion

Paddy cultivation which is widespread particularly in the tropics and sub-tropics provides an opportunity to produce fish either concurrently or alternately with paddy. However, it is interesting to compare temperate paddy-fish culture with that of tropical one. While in Western Europe and North America, social status, rising living standards, and emergence of leisured classes have all stimulated to practically abolish paddy-fish cultivation, this farming system in the tropics has been driven principally by the need to yield more fish and to generate employment by increasing the number of labor-hours required for farming practices. On the other hand, tropical waters are abound with fishes which feed low in the food web – phytoplankton, zooplankton, omnivores – such as carps, catfishes and tilapias, which can be readily integrated with paddy.

Cultivation of paddy as a part of the green revolution has adversely disturbed the ecology of paddy fields. The increase in disease vectors in paddy fields has stimulated the use of pesticides more and more widely. Serious problems relate to the intensive use of insecticides and pesticides which create lethal conditions for fish. At the same time, the accumulation of pesticide residues in fish flesh significantly increase the human health hazards. However, an understanding of the ecology of paddy fields could definitely help fish culture enterprises immensely.

In view of ,the above limitations of fish culture in paddy fields and further in view of the fact that the paddy-fish culture strategy has opened up new dimensions of growth in tropical and sub-tropical regions, it is imperative that on the one hand steps should be taken to secure high gains in productivity and on the other steps should be taken to ensure that these gains reach all the pockets of rural economy. Hence, the overall strategy for increasing production of both paddy and fish should consist of several elements such as (i) use of better paddy field policy, (ii) continuous improvement in yields, (iii) an adequate support by education, extension, and intense propaganda for (ii) and (iv) simultaneous attention to the needs and potential of growth in the areas concerned with different levels of development. All these elements are, however, interrelated.

It should be kept in mind that the principal crop in paddy-fish farming system is paddy and hence fish culture techniques have to be altered to make them consistent with paddy cultivation. The soil fertility in paddy fields is important to fish and paddy farming systems and at the same time, water quality variables in the fields have also to be maintained at the level which is convenient not only for the fish in particular but also for paddy field ecosystems in general.

Last but not the least, the success of fish culture in paddy fields will definitely depend on the combination of local enterprise and state-of-the-art technology based on the understanding of biological cycles in paddy fields and favorable social, economic, ecological, as well as political attitudes in general.

References

Ardiwinata, R. O. 1957. Fish culture in paddy fields in Indonesia. *Proceeding of the Indo–pacfic Fisheries Council.* 7 : 119-154.

Baharin, B. K., K. G. Aug and C. E. Tan. 1979. A review of the status of research and development activities in rice-cum-fish culture in Asia. University of Pertanian, Malayasia, 49 pp.

Coche, A. G. 1967. Fish culture in rice fields; A World-wide synthesis.*Hydrobiologia.* 30 : 1-44.

Fernando, C. H. 1993. Rice field ecology and fish culture – an overview. *Hydrobiologia.* 259 : 91-113.

Halwari, M. 1998. Trends in rice-fish farming. *FAO Aquaculture News Letter,* 18 : 3-11.

Kurihara, Y. 1989. Ecology of some rice fields in Japan as exemplified by some benthic fauna, with notes on management *Int. Revue. Ges. Hydrobiol.,* 74 : 507-548.

Li, K. 1988. Rice-fish culture in China: A review. *Aquaculture* 71 : 173-186.

Mathias, J.A., A.T. Charles and P. Hu, 1998. Integrated fish farming, *BOCC Raton, F.L* : CRS Press, 420 pp.

Ruddlek, K and G. Zhong. 1998. Integrated agriculture-aquaculture in South China. Cambridge University Press, 173 pp.

Questions

1. What are the different types of paddy-fish culture generally undertaken?

2. What are the infrastructures that are necessary for paddy-fish cultivation?

3. Describe fish production techniques adopted in paddy fields.

4. Discuss fish culture methods in high latitude terrace and valley paddy-fish culture ecosystems.

5. Discuss the habitat condition in a paddy-field ecosystem.

6. What type of habitat should be considered for high production of fish in paddy fields?

7. State the importance of fish culture in paddy fields.

8. What are the constraints that govern the success of paddy-fish farming systems?

9. Discuss the fish culture strategy adopted in coastal saline paddy-field ecosystems.

10. Discuss whether there is any prospect of paddy-fish culture systems in India.

11. Discuss the conditions of paddy cultivation in India.

12. What do you mean by wet paddy cultivation? How fish are integrated with wet paddy cultivation?

13. Give a brief account for widespread cultivation of fish in paddy fields in the tropics and discuss their importance in relation to employment opportunities.

9

Fish Culture in Cages

The practice of fish cage culture was born in the context of fish production, and it remained conterminous with fish culture development in many parts of the world particularly in southeast Asian countries. The strategy that should be adopted for successful development of cage culture must, in a very great degree, depend upon the easy availability of cheaper ingradients and skilled labors. Though cage culture has first developed in the latter part of the 19th century in Kampuchea, the cage culture strategy and development have undergone a notable change in recent years and have become more comprehensive and the strategy has spread out in a number of Southeast Asian countries. At present, the contribution of cage culture to world farmed fish production is only around 5 per cent.

While discussing the fish culture in cages in a few pages of Volume 1 (See Chapter 11), the subject of fish production from cages has been kept apart for a separate discussion. Keeping with the importance and development of cage culture in mind, the present chapter has discussed this issue at length. A number of aspects relating to fish yield from cages such as types of cage culture, advantages and disadvantages, cage construction, cage designs, site selection, management protocols, problems related to cage culture and the like have been evaluated that follows.

9.1 The Origin of Cage Culture

Cages were possibly first used by fishermen as a convenient holding facility for fish, until sufficient quantities were caught to make a journey to market worthwhile. The traditional types of holding abilities such as fish traps have been in use in many parts of the world for generations. Bamboo or wooden boxes used by fish farmers for holding fish could be considered as cages.

True cage culture is of recent origin and seems to have developed independently in Southeast Asian countries. In Kampuchea, floating cages have been in use since the end of the 19th century. Snakeheads, catfishes (*Clarias* and *Pangasius*) were held in wooden or bamboo cages, fed by trash fish and transported to the market. During the 20th century, this type of cage culture has spread to many regions of the lower Mekong delta and Vietnam.

In Indonesia, floating bamboo cages have also been in use since the early 1920s. A different form of cage culture also appeared in Indonesia around 1940 where small bamboo and wooden cages were anchored to the bottom of organically-polluted rivers and canals and stocked with common carp which fed on wastes and natural food organisms carried in the current.

Traditional cage culture is still practised in many parts of Indonesia and Indo-China. Although moderately successful, these methods of rearing fish have a localized influence and did not directly give rise to the current cage fish farming industry. Modern cages utilize synthetic mesh or netting materials and have collars fabricated from synthetic polymers and metals, although wood is widely used in many forms. In Norway, cages were being used to culture Atlantic salmon, *Salmo salar* in the early 1960s. 'The culture of tilapia, sea-bass Common carp, Indian major carp, and catfishes in cages are of more recent origin, and the work is being carried out in most tropical countries in the late 1980s.

9.2 Diversity of Cage Types

At present, there is an enormous diversity of cage types and designs. Generally, four basic types of cages are found such as (i) fixed, (ii) floating, (iii) submersible, and (iv) submerged.

Fixed Cages

Fixed cages consist of a net bag supported by posts driven into the bottom of a lake or river. They are used in some tropical countries (such as Philippines) where fixed cages have been found to be inexpensive and simple to fabricate, although they are more limited in size and shape and restricted to sheltered shallow areas with suitable substrates.

Floating Cages

The floating cage is supported by a buoyant collar or frame and can be designed in an enormous variety of shapes and sizes to suit the purposes of the farmer. Some floating types rotate about a central axis mounted on a coller, other designs are rotated by means of moving the supplementary flotation materials or by adjusting the buoyancy of the frame membranes. The more widely used non-rotating floating types can be constructed with wide or narrow collars. The wide collars are common on larger cages and serve as work platforms, thus facilitating many of the routine farm works. Simple and inexpensive flexible collar narrow cages can be fabricated using rope and buoys, but in practice they have been found to be difficult to manage. Rigid narrow collars, made of glass fibre or steel section and buoys are very popular in Western Europe. A very simple floating cage design for short-term culture is shown in Figure 9.1.

Submersible Cages

Net and rigid mesh bag submersible cages have no collar, but rely on the frame to maintain shape. The position of this type of cage in the water column can be adjusted to take advantage of prevailing environmental conditions. Some designs rely on the bag being suspended from buoy on a floating frame on the water surface. In some designs, however, it has proved difficult to maintain the shape of bag when the cage is submerged. All species of fish are not adapted to culture in submerged cages.

Submersible Cages

Whilst a number of submerged cage designs have been proposed, few have actually been built or tested. Simple submerged cages are, however, widely used in running waters in some regions of Indonesia and the Russian Federation. Most of the designs are wooden boxes with gaps between the slats to facilitate water flow, and are anchored to the substrate by stones or posts.

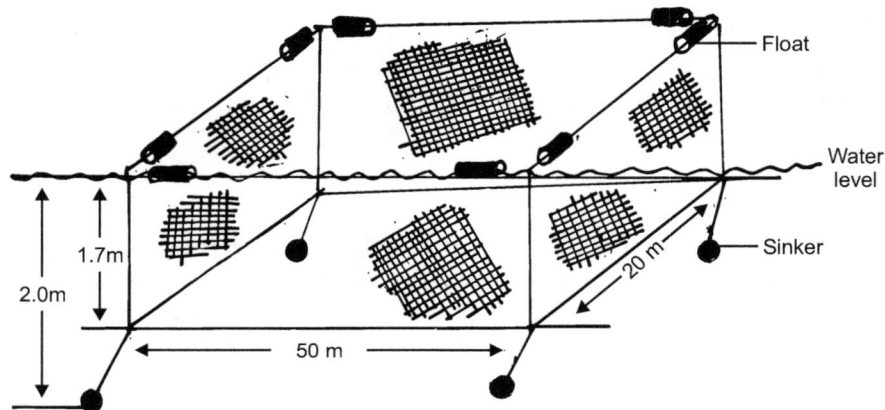

Fig. 9.1 : A floating cage

9.3 Cage and Cage Culture

Because most of the cage designs are not expensive and convenient to hold aquatic organisms in captivity, cages have been used for a variety of purposes such as to hold and transport bait fishes for tuna pole-and-line fishing. More recently, cages have been developed to hold fish for water quality monitoring of effluents. Cages have also proved invaluable in experimental work, where it is important that there are no differences in environmental conditions between groups of organisms being studied.

On the basis of food inputs, cage culture can be classified as extensive, semi-intensive, and intensive. In extensive culture, fishes must rely on available natural foods carried in the drift. Semi-intensive culture involves the use of low protein (less than 10 per cent) feedstuffs whilst in intensive culture operations, fish rely exclusively on an external supply of high protein food (more than 20 per cent).

Extensive Cage Culture

This type of culture system is restricted to freshwaters and may be practised in two types of environments such as highly productive lakes and reservoirs and water bodies which receive considerable quantities of domestic wastes or sewage. The productivity of water is dependent upon the availability of essential nutrients such as nitrogen and phosphorus, light and temperature. High nutrient loadings are likely to be highly productive. However, productivity is also correlated with latitude and between temperate and tropical zones. Hence, tropical water bodies with high nutrient loadings permit the best opportunities for extensive fish culture in cages.

Commercial extensive cage culture is at present widely practised in the Philippines although cages of bighead carp are used at several reservoirs in Singapore to control excessive algal blooms. The culture of bighead carp without supplementary feed in the United States has, however, been disastrous.

Both sewage-fed ponds, streams and rivers subject to high loading domestic wastes have been found in extensive cage culture, although there is concern about the public acceptability of fishes grown in such systems. Carps and tilapias may also be suitable candidates for extensive culture. The choice of fish can be critical. For example, in flowing water ecosystems, some planktonic species are able to survive and grow well. Therefore extensive cage culture of plankton-feeding species is likely to be impractical. The following example will illustrate this point. In Malaysia, cages were stocked with 25 g of fish at a stocking density of 15 fish/m^3. During two-month trial, 95 per cent of the carp died and the average weight of the survivors was 19.5 g.

In the Philippines, yield of tilapia by extensive cage culture is claimed to be high — up to 2.0 kg/m^3/month, although it is unlikely that yield at this level would be substantiate in the long run. Nevertheless such culture may be of value in several areas of the world. It may prove to be viable alternative to the management of a resource as a fishery and it may be judicious to carry out some extensive culture of fish in cages in conjunction with intensive lake- based fish rearing, thus reducing the environmental impact of wastes whilst increasing the profitability of the venture. It may also be advisable to switch between semi-intensive and extensive cage culture at different times of the year so as to take the advantage of periods of high plankton densities. Research and development efforts are aimed not only at determining which species perform best under which conditions, but also at the endeavors of increasing predation pressure on a particular community or size-group of aquatic organisms in order to better evaluate the potential and economic viability of this method of fish culture.

Semi-Intensive Cage Culture

Semi-intensive rearing of fishes in tropical freshwaters is the most common method of cage culture of species which feed low in the food chain such as *Oreochromis mossambicus*, *O. niloticus*, *O. aureus*, bighead carp, silver carp, and common carp are well fed with boundless variety of materials such as paddy bran and domestic wastes. Semi-intensive culture is also practised to a limited extent in Eastern Europe.

The supplementary feeds are determined by their availability and tend to be delivered on an *ad hoc* basis rather than according to any predetermined set of rules. There is little known about the importance of supplementary feeding and consequently, it is impossible to provide any stringent guidelines. Many freshwater sites have proved to be suitable for semi-intensive fish culture, although cautions may have to be taken to avoid excessive feed losses.

Intensive Cage Culture

In freshwater ecosystems, salmonoids, channel catfish (*Ictalurus punctatus*) and common carps are reared intensively. Intensive rearing of caged tilapia is practised in parts of the United States, Mexico, and South-east Asia where they fetch a high market price. Intensive fish culture is not recommended in lotic sites where feed losses can be excessive.

It should be kept in mind that whether extensive or semi-intensive or intensive, monoculture is the rule and polyeulture is the only exception. There are several reasons for this. The most important is that there are few feeding niches to exploit in cages than in ponds, even when the cage bottom is burried in the substrate. Thus natural foods are unavailable to caged benthic species. Whilst it is possible to establish a polyculture of several plankton-feeding species in extensively or semi-intensively managed cages, this type of system has not at all been evaluated in detail. Polyculture of carps in cages are practised in Hungary and China, with yields of up to 7.5 kg/m^3/year from extensive systems and up to 13.5 kg/m^3/year from semi-intensively managed cages.

9.4 Production Potential of Finfish and Shellfish in Cages

More that 130 fish species and nearly twelve species of lobster, prawn and crabs have been grown in cages on an exterimental basis. However, cages have proved suitable for both small-scale artisanal and large-scale commercial production, the materials used to fabricate the cages and the size for the rearing units being chosen accordingly. Some species such as tunas (*Thunnus thynnus, Euthynnus pelcunis*) have been found to adapt in floating cages. Bottom-dwelling flat fishes (*Scophthalmus maximus*) too, have been cultured in cages fitted with a solid bottom. Prawns are difficult to culture at high densities since they exhibit cannibalistic behavior. If bottom of the cage is, however, burried in mud, prawns can be efficiently reared in small units.

Production in freshwater ecosystems accounts for 80-90 per cent of the world finfish culture and around 95 per cent of this comes from ponds. Thus cage culture will account for about 4 per cent, or 100-140,000 tonnes of freshwater finfish/year. In cretain sectors, cage culture assumes to be an extremely important strategy. In Scotland, for example, more than 40 per cent of the rainbow trout yield is produced from freshwater cages. In China, more than 50 per cent of the common carp production is from freshwater cages.

The culture of prawns and other crustaceans in cages is still in its infancy and commercial yield is being limited to a few tonnes of penaeid prawns in several Southeast Asian countries.

9.5 Advantages and Disadvantages of Cage Culture

Advantages

1. Though some designs are fairly sophisticated, the majority of cages are very simple to construct and can be put together in a day or so, using local and unskilled labor and are also easily managed. Also, fish can be harvested easily using little more than a scoop net.

2. Cage farms can be expanded simply by adding a few more cages as experience grows and circumstances allow to do so and cages can be moved around with little restriction.

3. Cage-reared fish are often superior to fish reared in other systems in terms of condition factor, appearance, and taste.

4. Cage culture is highly important in areas where fisheries and aquaculture are on the decline, thus greatly increasing the income of fishermen families.

5. Fishermen possess most of the skills necessary to construct and operate fish cages.

6. Fish cages not only permit for a free exchange of water to maintain high levels of its dissolved oxygen but help to remove metabolic waste products of fish from cages.

Disadvantages

1. Cage fish farms have an impact on the aquatic environment due both to the presence of the cages and to the method of culture. Cages take up space which can disrupt access and make navigation difficult, reduce the landscape value of a site and alter current flows and increase local sedimentation rates.

2. Irrespective of the method of rearing, cage culture can introduce or disrupt disease and parasite cycles, change the aquatic flora and fauna and alter the behavior and disruption of local fish stocks.

3. Associated with the intensive cage culture is the release of unconsumed food and faeces into the environment, stimulating primary productivity and adversely affecting water quality. At the same time where extensive cage rearing of fishes is practised, algal populations can be over-grazed, effectively reducing primary productivity.

4. Cages are often sited in public or multi-user water bodies and farmers may be powerless to control any pollution that occurs. Pollution problem is very important for cage fish farmers in many parts of the world causing thousands of dollars of damage each year.

5. Cages are more susceptible to storm damage than ponds, tanks or raceways.

6. Cages are also highly vulnerable to poaching and vandalism which may be such a major problem as to prohibit their extensive use in certain parts of the world.

9.6 Economics of Cage Culture

Several economic evaluations have been carried out by cage culturists. Capital costs, however, vary with cage size and materials used. About 3- fold differences in costs per unit volume between two cages constructed from identical materials but differs with size. At the same time, in cage constructed from rigid copper-nickel mesh costs around three times more than a cage of similar size which utilizes a synthetic fibre net bag. Those cages designed to withstand extreme weather conditions can cost twice as much per unit volume as more conventinal designs.

Operational costs are also highly variable and are determined by species, size, method of culture, management and scale of operations. Costs vary between 10 and 40 per cent of operating costs depending very much on species. In intensive farming, feed costs can account for more than 50 per cent of operating costs compared with only about 4 per cent at semi-intensive cage tilapia operations (Table 9.1).

Table 9.1 : Operating Costs (per cent) of Small ($\bar{X}$ = 420 m²), Medium ($\bar{X}$ = 848 m²) and Large ($\bar{X}$ = 2,499 m²). Cage Fish Farms Located in Philippines (Around San Pablo City, Laguna)

Item	Small	Medium	Large
Seed	34	42	38
Feed	3	4	3
Labor*	19	11	8
Depreciation	30	31	45
Interest	10	9	5
Miscellaneous	5	3	1

* Labor costs include estimates for unpaid labor, which can account for 50 per cent of total labor costs.

Source: Aragon *et al* (1985).

Depreciation costs can be as high as 45 per cent of the operating cost, indicating a short service life for most cages, although it is argued that copper-nickel mesh cages have longer life span. There are also differences in operating costs between small and large operations for semi-intensive tilapia farms in the Philippines.

9.7 Construction of Fish Cage

The cage design must take recognition of the type of fish culture such as whether it is large commercial farm or artisanal operation. In practice, ignorance of the effects of cage size or shape on fish yield and the lack of knowledge on the interaction between environmental factors and cage structure has resulted in cage designs having largely evolved tentatively. Therefore, all cage designs deliberate more the limitations of the materials used, local skills and the prevailing socio-economic situation rather than the environmental conditions in which the cages are installed. The forces acting on the cage structure and mooring system are very complex and hard to quantify and analysis of the responses of a cage structure to these forces requires the development and use of computer models. Since the cage farming industry in Europe, Japan, and North America has expanded very rapidly and has attractecd the interests of a growing number of large multi-nationals and with facilities to carry out research and development, it is expected that in future, cages will have to be more rigorously designed, tested, and cost-effectively manufactured. This will definitely satisfy not only the demand of fish farmers but also the insurance companies and government safety standards.

Size, Shape and Materials

All cage designs should start with the bag which should be designed to meet the requirements of the fish. However, octagonal or square designs are suitable for salmon. Other less active fish such as tilapia and common carp are thought to be less particular with regard to the shape of the holding facility. Although circular cage bags are stronger and made of available netting materials resulting in the lowest costs per unit volume, several disadvantages exist. Because the surface area: volume ratio is small, water exchange is beleived to be relatively poor and the technology required to construct a supporting collar for a circular enclosure is more sophisticated than for other shapes.

The influence of surface area and depth on the yield of fish has been little studied to date. Cages of less than 1.5 m depth have been shown to retard growth of tilapia and common carp, whilst depths of greater than 10-12 m are poorly utilized by most of the fish species unless surface temperature or water quality conditions deteriorate.

Although larger cages have been shown to be better suited to the larger and fast swimming fish, they require more sophisticated materials and refined technology. At the same time, cages are also difficult to manage and require at least four farmers to perform routine operations. From fish farmer's point of view, however, the cost of bag-sized cage per unit volume is cheap and hence it is widely used. The use of large-sized cage has several disadvantages. Water exchange is likely to be poor, since surface area: volume ratio decreases along with the increase in size of the cage. Moreover, grading is also a problem for the giant cages. Therefore, it is suggested that these cages may be best suited to the culture of fish which do not require periodic sorting and harvesting.

The materials used to form the cage bag should be (i) strong, (ii) corrosion and weather resistant, (iii) fouling resistant, (iv) inexpensive, (v) smooth textured and thus non-abrasive to the fish, (vi) easily worked and repairable, and (vii) drag free. As a generalization, however, rigid and flexible material is better since it facilitates easy harvesting. It should be kept in mind that no single material possesses all the qualities noted above. Some materials are better suited to certain species, sites, and purposes.

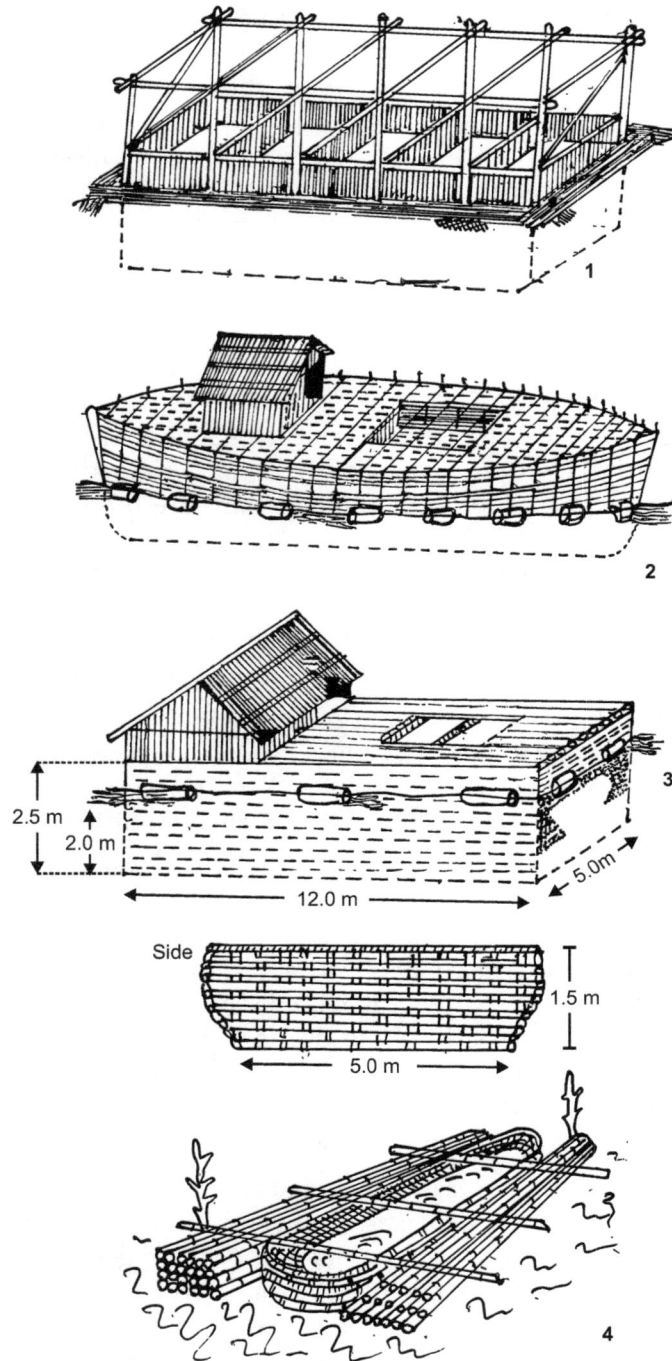

Fig. 9.2 : Traditional fish cage designs in some Asian countries.
(1) Battery of small cages in Kampuchea; (2) Boat-shaped cage in Kampuchea;
(3) Box-shaped cage in Vietman; (4) Basket-shaped bamboo cage in Thailand.

Traditional Designs

In traditional designs, the cages are fabricated using existing skills from whatever materials are in hands. Such designs are very familiar to the builders. Several natural materials such as grasses or wood, reeds, jungle vines and creepers, bamboos, hardwoods and softwoods are widely used for this purpose. Of course, some materials are rapidly decomposed in a matter of month in tropical waters, though life expectancy may be increased a little by application of tar-based substances. They are also damaged by the caged fish and predators.

Bamboo is a more suitable material being strong, cheap, widely available and easily worked with simple tools. Slats are cut and polished using a knife and either woven into matting or fixed directly to the cage frame, sometimes in combination with natural or synthetic fibres and then fixed to the cage frame. In many parts of Thailand, however, traditional cages are virtually constructed with bamboo slats that forms a basket-shaped structure (Figure 9.2). In tropical countries, bamboo-slatted cages are more superior to several species of fish such as *Clarias batrachus, Heteropneustes fossilis,* carps and prawns, and in some cases highly superior to those fitted with nylon mesh and at the same time the cost is considerably less. These cages are also labor-intensive to construct, heavy and difficult to manage but are highly suitable for use in freshwater environments.

Hardwoods are denser than bamboo and are thus stronger and highly resistant to fungal attack and to infestation by boring organisms which is very common in many tropical waters. Planks are fastened to form box-shaped constructions which are either anchored to the bottom with stones and posts driven into the subtrate or floated using additional floatation materials. Though hardwood cages cost around 15 times more than those fabricated with bamboo slits; they have a working life of about 12 years in freshwater environments.

Traditional floating cages utilize a variety of materials such as bundles of bamboos or hardwood logs and oil drums which are lashed to the sides of the structure for supplementary flotation. Simple anchoring systems such as ropes and stone weights or posts are widely used.

Modern Designs

For the construction of cage bag, flexible netting and rigid mesh materials are widely used. These materials can be fabricated from a range of man-made and natural materials and are available in a variety of forms.

1. **Flexible Mesh Materials (Netting):** Application of natural fibres for cage netting has been replaced by synthetic fibres because the natural fibre is subject to rotting and consequent loss of strength. The most common synthetic fibres used at present in the manufacture of fishing nets are polyamide, polyester, polyethylene, and polypropylene.

Though the details of preparation of yarn and the knitting is beyond the purview of the present treaties, fish farmers should understand the properties of different netting materials, so that the appropriate choice can be made. Generally, cage nets should be slightly denser than water, thus facilitates easy hanging, but not so dense as to make handling arduous or labor-intensive or to affect the flotation systems. The netting should be capable of supporting a portion of the biomass of enclosed fish when the nets are lifted for harvesting. Resistance of the netting to lateral and bending deformation, abration, and high extensibility are also desirable. Moreover, the netting must be resistant to fauling. Lastly, durability, availability, cost and maintenance requirements are extremely important considerations.

The characteristic features of netting are determined by the properties of the yarn, the type of fibre, diameter of the yarn, degree of twisting, method of fabrication (such as knotted or knotless), and mesh size. The addition of chemical stabilizers, dyestuffs, antifoulants, and stiffening compounds can also modify these properties.

2. **Rigid Mesh Materials :** Rigid mesh materials consist of plastic and metals. The mesh panels are attached to a frame and supported by a floating collar.

Rigid plastic mesh material (such as nylon) though does not corrode and is highly rot-resistant, it is prone to weathering and is more resistant to fauling. Metal meshes, on the other hand, have been tested for use in the construction of rigid cages. Since metal meshes are subject to chemical attack or corrotion when immersed in water, they are well protected using protective coatings. Metallic coating may be applied by dipping, electroplating, spraying, cementation, whereas inorganic oxide coating are formed by chemical treatment. A variety of paints, varnishes, plastics and rubber compounds also permit to isolate the metal from the corrosive environment.

In general, only three types of metal mesh are widely used in the fabrication of cages. *Firstly*, 90 : 10 Copper - Nickel wire and expanded metal mesh is most appropriate for use in fish culture. *Secondly*, zinc galvanized steel mesh is also highly effective but the longevity in a particular environment is dependent upon the type of mesh and the quality of galvanising. In general, the service life varies from 12 to 16 months, and *thirdly*, galvanized steel mesh is coated with polyvinyl chloride that is considerably less resistant to fouling.

(i) *Construction*: Rigid mesh cages usually require some sort of supportive structures, otherwise they will crack. Small designs can be assembled by binding sheets cut to the appropriate size to a previously fabricated frame. A top hinged panel is often included in rigid mesh designs. Small and simple panel frames are constructed from woods, metals, synthetic materials or fibre glass. Panels of 30 cm wide are joined to form a rectangular or circular enclosure of variable sizes. In some cases, galvanized steel reinforcements are used at corners and joints.

Synthetic fibre net bags are designed with an area of freeboard to prevent fish from jumping out. The fabrication of simple square or rectangular cage bags from fine mesh (2- 5 mm) mosquito netting, suitable for tilapia hatchery is very simple and most economical.

The nature and extent of rigging depends upon the size of the cage bag. For small bags (less than 50 m³ volume), for example nylon ropes are attached to the outside of all edges, whilst for larger designs, rigging may also have to be attached at intervals along the panels to stiff the structure and to facilitate lifting at the time of harvesting.

On the cage bag, ropes are stiched to the outside of the walls and floor of the bag every few metres. For floating cages, loops should be included not only at the corners, but also at meter interval on all ropes to facilitate lifting. The loops should either be knotted and whisped or spliced. Weights can be hung from the rigging which runs along the floor seam in order to help the bag hang properly in the water, or alternatively a metal frame can be fabricated and lashed to the exterior of the floor seam.

Cage Collar and Support System

1. **Fixed Cages:** The mesh bag is supported by bamboo posts driven into the. substrate. Bamboos are widely distributed throughout the tropics, sub-tropics and mild temperate areas of the world. The common bamboo species that are most suitable for cage construction purposes include *Dendorcalamus spinosa, D. hamiltonia, D. merillianus, Bambusa blumeana, B. vulgaris,* and *Gignathocloa gigal.* Bamboo is resistant to bending and tension forces, but due to its unique structure, it is much weaker. Although bamboos are easily split and readily fabricated into a cage mesh they suffer from one principal disadvantage such as it has a very short useful working life (18-24 months in freshwaters). It is susceptible to rotting and to destruction of boring insects and also becomes water-logged, thus losing its flotation properties.

Softwoods and hardwoods are more resistant to weathering and rotting but are much more costly. Although this material costs at least 5 times as much as bamboo, it has a life span of at least 4 years.

The posts in a fixed cage must be able to resist the static vertical forces imposed by the weight of the net and the farmers and the horizontal dynamic forces exerted by wind and currents. The posts that support the net bag must be buried securely at sufficient depth (1.5 metre) to resist tbe movement of the bag as a result of the wind, wave, and current forces acting on the structure.

If a buried depth of each post is 2 metre, a water depth of 6 metre and that the bamboos extend for 3 metre or more above the water surface, hollow stems (culms) of 11 m plus are required. Before installation, the culms should be well-cleaned to prevent the net from damage. Bamboos are floated out to the site on a raft which also serve as a construction platform. The corner posts of one side are fixed first, and a line is stretched between the two to ensure that the posts lie along a straight line. After one side has been finished, the others are completed in a similar manner. Horizontal supports are lashed to the vertical posts about 1.5 - 2.0 metre above the water level and serve not only to strengthen the structure but also as a narrow walkway. The net is then tied to the posts and the bottom rigging being secured by divers.

In the case of a shallow surface mud stratum, anchorage of the posts can be upgraded by driving bamboo pegs (25 cm diametre x 450 mm long) through internodes at the base of the culm to form two cross pieces, about 30 cm apart. Bracings may also be used at corners to strengthen the structure.

2. **Floating Cages :** A great diversity of floating fish cage designs always exist. The collar not only helps support the cage bag in the water column but also maintain the shape of the structure. Many cage collars also serve as work platforms. Both static and dynamic forces act ceaselessly on the collar. Static loads act on vertically and comprise the weight of the bag, super-structure and additional loads which may be expected during routine operation (such as weight of farm staff, feed bags, pumps, aerators etc.) and these must be kept in mind to design the flotation system.

In some cage designs, the simplest flotation system is constructed from bamboo to serve not only as a rigid frame work for maintaining the shape of the bag, but also to support the weight of the structure. In case of sophiticated commercially-produced cages, polyurethane foam-filled plastic pipes form the basic framework of the collar. Steel and aluminium alloys and steel drums are also used in some situations for strength and supplementary flotation, respectively.

If bamboo or wood is used as floats, the buoyancy is lost as a result of rotting. Rotting can, in part, be protected by impregnation with preservatives such as Coaltar Oil. Plastic or steel drums rapidly accumulate fouling organisms under certain conditions, which reduces their flotation capabilities. Consequently, high maintenance costs may negate the benefit derived in terms of time spent on cleaning. Antifouling compounds drastically reduce the rate of fouling or these floats are sometimes covered with polythene sheets which are changed as necessary. The overall robustness of a cage is, however, governed by several criteria that follows.

(i) Properties of the construction materials and size and profile of the structural members.

(ii) Strength of joints between structural members.

(iii) Degree and distribution of flexibility in joints.

(iv) Design and location of mooring points.

All structures of different types of cages are highly flexible and such flexibility particularly in the collar region is desirable that reduces both the incident forces acting on the collar and the motion of the bag. Flexible joints in the floating cages though costly, are generally recommended to keep the cage as rigid as possible;

of course, some home-built designs have utilized simple flexible joints very successfully.

In inland waters where current, wind and wave exist, unsophiticated and simple designed cages have proven absolutely sufficient. In some Asian countries (such as Philippines) a very simple design is constructed entirely from bamboo. For a 3 X 3 X 3 metre cage, suitable for rearing tilapias, 20 bamboo culms and 100 m of 4 mm polyethylene rope will suffice. The bamboos are prepared in the same way as described for fixed cages and cut into lengths. Using tyres to support each corners, 5 lengths cut from the culm bases are laid out to form each side of the collar. The pegs (25 cm long X 12 cm diameter and tapered to a point) should be cut and driven through the bamboos. 'The rigidity of each side is greatly improved by lashing a section of split bamboo across the walkway. Supports (1 m long) are lashed to each corner and handrails run between them.

* *Methods of Construction*: The buoyancy of the collar is around 112 kg which is enough to carry the weight of one man and the cage bag when it is in the water. For larger cages, bamboos are replaced by plastic or steel drums for flotation purposes and where the cages used in more exposed locations, lengths of wood and steel bolts are substituted for bamboo. Hardwoods which are used for cage construction in many regions of the southeast Asia, are denser and hence more resistant to bending. A hardwood collar may last up to 4 years if regularly treated with preservative chemicals.

The plan diagram of a typical wood and oil drum collar is shown in Figure 9.3. 1t is, however, constructed from six (7 metre each) and three (8 metre each) hardwood beams. The beams are joined by 12.5 mm diametre and 20 cm long steel bolts. Fifteen 200 litre steel drums are secured to the frame by rope that permits adequate flotation, and each compartment holds a 3 X 3 X 3 metre net. There are many variations of this design, details of which have been reported in a number of FAO publications.

In western Europe, a simple design has been developed which is extensively used in both inland and coastal areas. Although there are numerous variations of this design, it essentially consists of walkways which runs along two sides of the cage and joined together at either end by timber cross-beams. The walkways are fabricated from larch which is treated with wood-preservative compounds to improve rot-resistance. Galvanized brackets secure the corners and galvanized steel stanchions are used to the walkways and crossbeams are used to support the wooden hand and net rails.

Since larger cages are widely used to rear fish in less sheltered locations, galvanized steel or aluminium has been used to replace much of the timber and larger as well as heavier brackets and bolts used at corners. These materials, though more expensive, have a much longer life span of about 10-12 years for steel- frame design. Moreover, several designs are built into the walkways and incorporate handrails fabricated from galvanized steel. The cage super-structure thus performs a greater structural role, that makes disagreement to bending and curling.

Most of the fish cages have collars which not only buoyant the cages sufficiently but also provide a work platform. Most of the working collars have walkways that are at least 1 m wide. Others have narrow walkways on several sides and these walkways are sometimes linked to a central pontoon which facilitates the transport of items such as feed and fertilizers. In recent designs, the pontoons are well-structured and buoyant to support small vehicles.

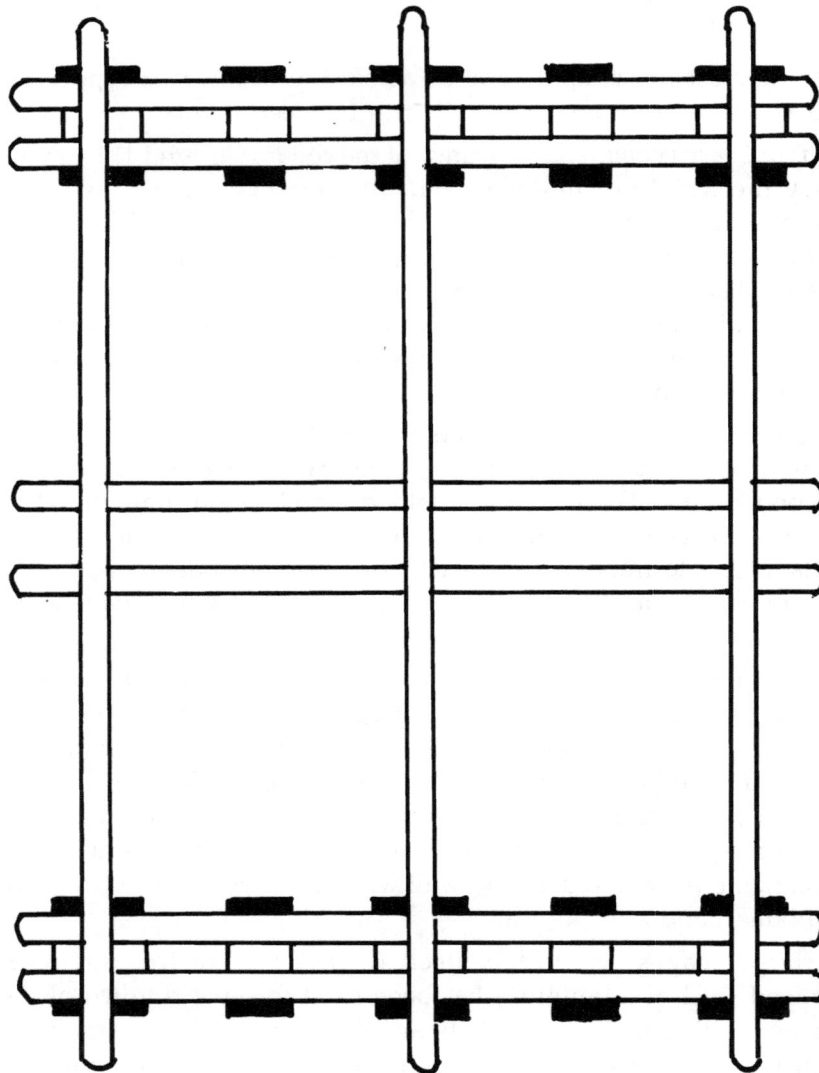

Fig. 9.3 : Typical 4-cage raft constructed from oil drums and hardwood beams (Redrawn from Mok, 1982)

*   *Cage Linkages and Groupings*: Although flexible collar and rigid collar cages are moored individually, cages are generally grouped together both for minimizing the costs of mooring arrangements and management reasons. The number and arrangement of cages grouped together thus depends on various factors such as the size of the farm, the size and nature of the site, the shape and design of the cage, and linking systems, mooring constraints and environmental considerations.

    For a large production unit, cages are grouped conveniently. On the other hand, it is not desirable to group cages for a small knit. This is due to the fact that the size, shape, depth, and physical features of the site may circumscribe the positioning of the cages and command the number of cages that can be grouped together and rectangular cages can be mustered in a variety of shapes such as hexagonal, octahedral, and circular. In certain commercial designs, the cages can only be grouped together in a vary limited number of ways.

    Though linkages and groupings of most modem cages facilitate the rapid removal of a cage from the group for repair or rapid harvesting or for re-setting, several environmental factors deserve serious considerations. This is because there are marked differences in water quality between cages and consequantly deteriorate fish production. In some Southeast Asian countries such as Philippines and Malayasia, reduced yield of fish in the central cage of a 3 X 3 groups has been noted compared to those in the outside cages. This is due to the fact that the transmission rate of water reduces from 0.8, 0.6 and 0.4 cm/second in the first, second, and third cage respectively. This situation generally hampers the production potential of fish in such systems.

    Linkages between cages should be designed so that the pitching motion is only damped whilst rolling and yawing are kept to a minimum. Not only it is difficult and expensive to construct a joint which permits universal movement, it is also desirable from the farmer or from the fishes' point of view, to facilitate free movement of the cages. On the other hand, the more rigid the linkages, the greater the force that are concentrated at these points.

    Ropes or chains are most commonly used for linkages and groupings that reduces the motion to a minimum. Rope tends to scrape fairly rapidly and so chain is preferable. Rubber tyres are also bolted between cages and act as buffers to prevent a cage from striking against another cage.

    Several more sophisticated designs of linkages which are well-suited to larger and heavier cages are constructed from stronger materials and are designed to distribute loads over a greater surface area. And due care should, however, be given to the choice of materials not only to satisfy loading criteria, but also to avoid the creation of galvanic cells.

*   *Anchor*: The simplest and cheapest type of anchor is the dead- weight or block anchor which consists of a bag of sand or stones or a block of concrete metal. An anchor is connected to a fish cage by a length of mooring line. The block will begin

to move when the horizontal component of the force exerted by the cage and mooring line equals to the frictional force between the block and substrate.

Block anchors are not efficient because they have low holding power per unit installed weight. For example, a sand bag anchor weighing 100 kilogram in water and installed on a sandy substrate, would have a holding power of between 19 and 27 kilogram, depending upon the angle between the mooring line and the sand bag. In practice, however, a concrete block anchor is more efficient since they tend to bed down into the substrate. A concrete block with a mass in water of 970 kilogram, is sitting on a sandy substrate and is buried to a depth of 10 cm.

A variety of embedding anchors are used, the holding power of which is related to its frictional resistance in soil, and so is dependent upon the mechanical properties of the soil. Anchor penetration is, however, a function of fluke shape and the angle between the flukes and the shank.

9.8 Site Selection

The correct choice of site in any fish farming operation is definitely important, since it can greatly influence the economic viability by determining capital outlay and by affecting running costs, rate of yield, and mortality factors. Water-based fish culture systems suffer in comparison to land-based operations in that there is less room to make mistakes in choice of site. Poor land-based sites can, for example, sometimes be improved by drilling boreholes to increase water supply, or by introducing filters and sediment traps to remove suspended materials. However, there is little that can be done at the cage, pen or enclosure farm if the site is too exposed and water exchange is poor, or if water quality deteriorates. Hence, the site selection for cage farming is an extremely important consideration.

Criteria for Site Selection

The criteria for selection of suitable sites for cage fish culture may be arranged into three categories (Table 9.2). The first is primarily concerned with the physico-chemical conditions which commends whether a species of fish can thrive well in an environment.

Table 9.2 : Criteria for Site Selection of Cage.
Note That Several Factors are common to More Than One Category

Category I	Category II	Category III
Temperature	Depth	Legal aspects
Salinity	Shelter	Access
Pollution	Substrate	Security
Suspended soilid		Proximity to markets
Al gal blooms		
Disease organisms		
Water exchange		

The second dictates the conditions that should be considered in order to install a cage structure successfully, whilst the third category involves those factors which determines the possibilities of constructing a farm and the profitability of the venture.

Informations regarding site selection can also be gained by discussing with local people about the occurrence of toxic blooms or pollution. Consultations with them on the establishment of a cage fish farm may also help to avoid the problem of poaching and vandalism.

9.9 Environmental Criteria for the Fish.

Water Quality*

An ideal cage farm site must have good water quality which means that it should not only be uncontaminated by toxic industrial pollutants such as ammonia, nitrite, metals, and phenolic compounds, but also that the pH, temperature, oxygen, and salinity requirements of fish species should be carefully considered. Water quality assessment and methods for specific tests have.been discussed in a number of text books.

1. **Temperature and Salinity:** When selecting a site for cage culture, the species' optimum temperature and salinity conditions should be met, since even immediately outside these optima, behavior, feeding, food conversion and growth can be adversely affected. Sub-optimal conditions can also contribute towards stress, leading to increased susceptibility to parasitic infections and reduced resistance to diseases. Rapidly fluctuating temperatures and salinities are more tolerant than gradual and seasonal changes, although some species are more tolerant than others.

    A close relationship exists between temperature regime and the amount of solar radiation received by an aquatic ecosystem. In large water bodies, due to its high latent heat[1], it does not become so cool in night hours and water currents mix up and abridge the effects of local heating and cooling. The longer wave lengths are completely absorbed in the top metre or two of the water column. If there is no mixing of surface warm and deeper cold waters, there would be an exponential fall in temperature with depth. The mixing requires energy, and the degree of mixing depends both on the energy inputs to the water body and on the density difference between surface and underlying water. In small freshwater ponds, however, diurnal variations in temperature may be pronounced. In deep freshwater lakes and reservoirs, a pattern of thermal stratification may often be observed where a layer of warm (22-23°C) and less dense surface water (epilimnion) is situated on the top of a colder (as low as only 5°C) and denser

---

* For further detail, see Alabaster and Lloyd (1980).

[1] Latent heat is the heat required to convert a solid into a liquid or vapor, or liquid into a vapor without change of temperature.

deep layer (hypolimnion). The two layers are separated by a short zone of rapdily falling temperature termed as *thermocline.*

In temperature regions, surface waters begin to heat up in the spring causing the onset of stratification which may last until the end of the summer. As the autumn advances, the surface waters cool down and the prevailing winds increase in strength and frequency until there is sufficient energy to turn the whole water body over, bringing up cooler water from below. During the winter, some lakes may freeze on the surface resulting in a second period of stratification which lasts until spring when the surface waters warm to reduce density difference and facilitate wind-induced mixing.

Deep stratified lakes and reservoirs occur throughout temperate regions and in some parts of the tropics. Water bodies which exhibits two periods of circulation per year (spring and autumn) are called *dimictic.* On the contrary, lakes which turn over once in september or october are termed as *monomictic* and are common in the UK and Western Europe. Some extermely deep lakes (such as Lake Tanganyika in Africa) require huge amounts of energy to turn over completerly, and have a restricted or partial turn over and are termed as *meromictic.*

Shallow lakes may not stratify at all, or stratify for short periods of time (day), and then turn over. Polymictic lakes as they are called, occur throughout tropical and temperate regions. Thermal stratification in lotic (flowing water) systems is less pronounced than in lentic (standing water) systems and is restricted to the slow flowing lower reaches.

Cage structures may be installed in cooling ponds or canals or in the receiving water body. The temperature of the water available to caged fish depends upon the design of the system, site (pond/canal/receiving water body) and the proximity of the cages to the power station.

2. **Oxygen :** Several environmental conditions encourage the devlopment of large populations of algae (blooms) during the warmer months of the year in areas subject to high nutrient influxes. Sewage discharges and agricultural washes may be important contributors. A sudden upwelling of nutrient-loaded water from the deeper parts of the water body during the breakdown of thermal stratification may also stimulate blooms. Enclosed water bodies with flushing rates tend to be more subject to blooms. Serious problems can occur when algal blooms (See panel 9.1) die due to sudden changes in climatic conditions affecting light and temperature, and during the subsequent decomposition, microbial respiration may remove much or even all of the dissolved oxygen resulting in fish kills.

Benthic communities may require dissolved oxygen for metabolism and the higher particulate waste loadings which are associated with the intensive cage fish farm areas may greatly increase the elimination of dissolved oxygen by the benthic microbial and invertebrate communities from the overlying water, which in turn may reduce the content of its oxygen to a large extent in and around the cages.

PANEL - 9.1

PHYTOPLANKTON BLOOMS

Blooms refers to the period occurrence of large populations of planktonic algae in fresh and marine waters when the appropriate conditions prevail such as high light and nutrient levels, warm wate and a combination of favorable hydrobiographic conditions. Blooms can drastically affect fish stocks by damaging and clogging their gills as well as their influence on the pool of dissolved oxygen. Though several species of phytoplankton are beneficial, other species are known to produce toxins which can kill a wide range of aquatic organisms. In this panel, however, emphasis will be given on the factors involved in controlling the development of toxic blooms.

During bloom conditions, the algal community becomes extremely large. Blooms of certain species seem to occur more frequently than others. In freshwaters, the most important groups are the diatoms and cyanobacteria (blue-green algae).

1. Diatoms: Diatoms occur in temperate regions most frequently during the spring and the autumn months. Blooms are present in large numbers and they cause gill damage.

2. Cyanobacteria: Cyanobacteria are very common than the former type in inland waters of both temperate and tropical areas. Since nutrients and enough light stimulate the growth of the algal population, such blooms occur in productive lakes and ponds.

Many species of cyanobacteria accumulate gas vacuoles in their cells during photosynthesis, causing the cells/colonies to float towards the surface. Common genera of bloom-forming cyanobacteria in inland waters are noted below.

Order	Family	Genus
Chlorococcalis	Chroococcacae	*Coelospharium, Gomphosphaeria, Microcystis*
Nostocales	Oscillatoriaceae	*Oscillatoria, Spirulina, Trichodesmium*
	Nostocaleae	*Anabaenopsis, Aphanizomenon*
	Rivulariaceae	*Gleotrichia*

Source : Reynolds and Walsby (1975)

Of greater concern to fish farms is the appearance of toxin forming bloom species which belongs to the genera *Mycrocystis, Anabaena, Oscillatoria* and *Aphanizomenon*. These species have been identified throughout tropical and temperate inland waters. Algal toxins are of two basic types: (i) alkaloid neurotoxins and (ii) protein or peptide hepatotoxins. The toxin produced by a particular species may vary form strain to strain thus making it practically impossible to identify a toxic bloom or the specific toxin involved unless laboratory tests are carried out following sophisticated methods.

Cyanobacterial toxins*		
Species	Name of Toxin	Structure
Lymgba majuscula	lynghatoxin A	Alkaliod
	Debromophysiatoxin	Phenolic
Schizothrix calciola	Debromophysiatoxin	Phenolic
Oscillatoria migroviridis	Oscillatoxin A	Phenolic
Nodularis spumigera	Nodularia toxin	Unkown
N. aeruginosa	Microcystin	Peptide
Anabaena flos-aquate	Anatoxin - a	
	Anatoxin - b	
	Anatoxin - c	Peptide
Oscillatoria agardhii	Oscillatoria toxin	

* Source : Skulberg et. al. (1984)

From a cage fish farmer's point of view, it would seem best to avoid site where such algal species exist, although fish mortalities as a result of these algae seem to be rare at cage farms. Extensive and semi-intensive cage operations may seem to be at risk, since the fish rely entirely or partly on the available phytoplankton. Intensive methods of cage culture are severely restricted to the tropics and sub-tropics where very little is known about the distribution and abundance of toxin-generating algae or their effect on economically important phyto-feeders such as tilapia and silver carp. Toxic species tend to occur in highly productive ecosystems and such ecosystems though ideal for extensive cage culture, are less suitable for intensive operations.

Supersaturation of dissolved gases (oxygen and nitrogen) in thermal power station effluents can cause extensive mortalities of a number of species held in cages through gas bubble disease.

To sum up, water bodies which are strongly stratified for much of the year, and/ or where algal blooms tend to develop are likely to have periodic poor oxygen conditions and such areas, if possible, should be avoided. Water areas which have good bottom currents and disseminate sedimenting wastes are highly preferable.The availability of dissolved oxygen to fish is dependent not only upon the dissolved oxygen concentration in the water, but also upon water exchange through the net cage.

3. **Hydrogen Ion Concentration (pH):** Freshwater, because of the great variability in their ionic compositio, may have pH values that greatly varies between 3 and 10. Acid water are often associated with highland lakes,or with lakes that are subject to acid rain precipitation such as those that occurs in many areas of Northern Europe or North America. Alkaline waters, on the other hand, occur where the

geology of the watershed is dominated by alkaline sedimentary rocks such as limestone or chalk.

In contrast to sea water, care must be taken with freshwater sites, since there are maked seasonal and diurnal changes, particularly in low-buffered waters. Intensive cage farming will stimulate phytoplankton production which can, due to increased photosynthetic oxygen generation, lead to elevated pH values particularly in summer. Ammonia toxicity may become a problem at such times. Problems of low pH water have been discussed in Volume 1 (See Section 12.29 of Chapter 12).

4. **Turbidity :** Although suspended solids can cause a number of problems in fish culture, their direct effects on fish reared in cages are of great concern. High level of suspended solids result in gill damage. If damage is sufficiently severe, the fish will die. Mortaility rate varies with species and with the nature of suspended materials.

At turbidity levels below about 100 mg/l there is little effect on cage fish species. Above this value, the situation becomes complicated. Although cage fish farmers are always advised to avoid areas where such turbidity levels occur, this may not be possible in riverine areas where several thousand milligram per litre suspended solids can occur during flood periods. It should be added as a note that cage fish farms are themselves a source of suspended solids.

5. **Pollution :** Pollutants severely damage the cage structure, adversely affect the cultured fish or its food, or accumulated in fish body to such an extent that it causes human health hazards. Enormous number of pollutants are entered the aquatic ecosystems and at least 1,500 pollutants in different freshwater have been estimated and it is beyond the scope of this chapter/book to discuss or list at length. The principal categories of aquatic pollutants are, however, given in Table 9.3. Therefore, extensive sampling and sophisticated methods of analysis must require to detect many of these compounds and risks may be reduced by installing cages as far away as possible from industrial contributors. But avoidance of pollution is a very difficult task, since competition for resources has been forced to develop fish farming alongside the industry or in heavily polluted areas.

Table 9.3 : Categories of Aquatic Pollutants

Acids and Alkalies

Anions (Sulfide, Sulfite. Cyanide)

Detergents

Domestic sewage and farm manures

Gases (Chlorine, ammonia)

Food processing wastes

Heat

Metals (Mercury, Cadmium, Zinc, Lead)

Nutrients (Phosphate, Nitrates)

Oil and oil disperants

Organic wastes (Phenols)

Pathogens

Pesticides

Polychlorinated biphenyl

Radionuclides

Source: Mason (1981)

Moreover, sampling of considerable quantities of toxic wastes and domestic sewage in and around the pond/lake/river are definitely responsible for a number of fish kills.

Thermal effluents may contain biocides (such as chlorine) which are used to control fouling, corrosion inhibitors, and heavy metals and therefore, care must be taken to ascertain the levels of biocides during the site selection process.

6. **Disease :** Organically-polluted water bodies are the source of varieties of disease agents. Pathoganic bacteria-induced mortalities (such as *E. coli* and *Vibrio* spp.) up to 80 per cent of cage stocks have been noted from regions where intensive cage culture is undertaken. Also, caged fish have become severely infested with the ecto-parasites (cestodes and copepod, *Argulus* and *Diphilobothrium* spp.) resulting in heavy mortalities and the eventual shut-down the farm activities. Disease risks, therefore, can be minimized or even eliminated by avoiding sites where there are parasites or other disease agents present in the wild fish, intermediary hosts, or environment which could be transmitted to the caged fish.

   Disease outbreaks still occur through introduction of diseased stocks to the farms or attraction of birds and other predators to the area concerned.

7. **Water Exchange :** Good water exchange at the cage site is essential for intensive cage culture operations in reducing the build up of waste substances and all the concomitant problems. It is dependent upon currents, temperature and topography of the area involved. Although currents at the site may appear to be ideal (at the rate of 10-60cm/ second) for cage culture, water exchange can still be low, leading to poorer than expected water quality in surface waters, and a build up of waste substances on the bottom. However, intensive cage fish farms must be established in areas with good water exchange.

8. **Fouling :** The fouling of net cages decreases the specific mesh size of the net bag, whilst increasing its surface area. The reduction in mesh size severely restricts the flow of water through the cages and thus reduces the rate of dissolved oxygen supply and waste metabolite removal which can adversely affect the fish. The additional weight of fouling on cage bags can lead to net failure, and makes net changing extremely difficult and time consuming.

   A large number of fouling plants and animals of the groups Cyanophytes, Chlorophytes, Coelenterates, Annelids, Arthropods, and Molluscs are found to be clinged to net cages after immersion for only 3-5 months. As a generalization, galvanized panels develop much less fouling than the synthetic fibre netting panels.

   Fouling is a problem at sites with low current velocities, and is modified by temperature and nutrient status. However, either such sites should be avoided, or management practices should be followed to solve the fouling problems.

9.10 Environmental Criteria For Cages

Currents

As noted earlier, better water exchange through cages is inevitable both for enrichment of oxygen consumed by fish and removal of waste metabolites. In intensive systems, currents are extremely important for the supply of natural food organisms. But excessive currents will inflict additional dynamic loadings to the cage supporting structure and moorings that may severaly affect fish behavior and contribute towards food losses from semi-intensive and intensive systems. High flow rates have also been criminated in skeletal deformities of cage reared carps.

Stream and river waters are subjected to two forces such as gravity and friction. Cage structure in lotic ecosystems is widely practised in many Southeast Asian countries where fish are reared in traditional/extensive/semi-intensive situations. In Kampuchea, for example, slow flowing reaches of the Mekong and Tonle Sap rivers have cages, while in Indonesia, rivers and irrigation canals are widely used. Though current velocities greatly exceed the normal velocity (at the rate of 10cm/second) during mansoon floods, these cages can able to withstand such harsh conditions because of their robust construction and method of anchorage.

Current velocities in lentic water bodies, being fairly uniform, rarely influence site selection.

Depth

Some cages are designed to install in shallow waters in Indonesia and Philippnes with their bottoms in contact with the substrate. The depth of water is not critical if they are covered for most of the culture period. Drastic reduction in water level could seriously reduce cage volume, thus increases in stocking density and adversely affect the water quality variables.

Fixed cages are used in shallow areas of lakes, reservoirs and rivers where the depth does not exceed 8 metre or so. Floating cages may be used virtually in any depth of water, although the cost and other problems concerned with mooring increase with depth.

As a generalization, the cages should be installed in adequate depth to maximize the exchange of water, and it is wise to keep the bottom of the cages well clear. Fishes often generate water currents which flow through cages by drawing in water during swimming and feeding activities. During these activities, water is drawn into the cage structure through the bottom panel and as the cage bottom approaches to the substruate, this flow is obstructed. In intensive cage structure, waste substances are developed resulting in localized oxygen depletion and a build up of potentially toxic compounds. Although the extent and degree of deoxygenation depends very much on how the operation of cages is managed and the nature of the site, it is better to place the cage structure at the bottom. Under some circumstances, however, noxious gases (such as hydrogen sulfide) are generated and these have been attributed to a possible causative agent of gill damage of fish. Also there can be a build-up of potentially disease-causing microorganisms in the sediments where cage structures are established.

To sum up, itis wise to avoid unnecessary risks such as dissolved oxygen depletion in and around the cage, waste sedimentation, water quality deterioration, high turbidity, occurrence of disease-causing micro-organisms, etc. by holding the fish at least 5 metre above the sediments. This is practically not possible particularly in ponds where cages are being used. In runnimg water ecosystems, suitable depth can be found by taking fluctuations in water level into consideration. In some areas of the world where water is abstracted from lakes and reservoirs for irrigation purposes, water level fluctuations are very large and in this case, floating cages are generally recommended.

Substrate

The type of substrate may vary from soft mud to rocky and may have a considerable influence on the choice of cage design. In freshwater areas with rocky substrate where it is arduous to drive stakes into the hard and rocky ground, use of floating cages are advisable.

9.11 Site Facilities and Management

For cage fish culture developments, the availability of sufficient land to construct a cage farm with almost all facilities may be an important consideration in site selection.

The safety of a cage fish farm is a buring problem in many parts of the world, since cages are often cited in water bodies which are publicly owned or have no restriction to access and are thus vulnerable to poachers and vandals. Cages located near centres of population may be more at risk. Although a number of security measures are taken to protect cages, farmers may prefer to site their cages where they can keep an attentive watch on them.

9.12 Carrying Capacity of a Cage Farm*

The carrying capacity of the site refers to the maximum level of production of fish which can be expected to suatain on an area of water body, usually calculated from food requirements. Intensive cage fish culture results in the production of wastes which can stimulate the productivity and alter the abiotic and biotic characteristics of the water body. On the other hand, less intensive methods of cage farming can result in over-cropping of algae and a fall in productivity. Both these impacts can radically alter the value of the resource. A serious deterioration in water quality will stress or even cause mortalities in fish stocks and can encourage disease organisms to thrive, whilst over-grazing of algae can result in poor growth. Consequently, the profitability or even viability of cage farming industries are seriously affected. It is, therefore, extremely important for all concerned

* For further detail, see Beveridge (1984).

with cage fish farming developments such as farmers, planners, and financial institutions that an accurate evaluation of the sustainable levels of production at a farm can be made.

A few models have been proposed for predicting the carrying capacity of different types of cage fish culture (particularly intensive cage culture of rainbow trout and extensive as well as semi-intensive cage culture of tilapia) and may be considered as guides for other species of commercial importance. But nonetheless, there are certain problems with each model that must be kept in memory.

For intensive culture, the establishment of acceptable water quality criteria presents a major difficulty. Although various agencies particularly. United States Environmental Protection Agency have presented a tentative water quality criteria these are primarily concerned with reducing nuisance algal blooms in multi-use water bodies. The establishment of management objectives for different freshwater resources make extremely difficult by the poor discernment of the relationship between algae and water quality, water quality and stress/growth/disease/ and mortality and water quality as well as the species being cultured, quality of stocks, time of stocking, and the composition of algal community. Hence, the establishment of acceptable water quality objectives is a highly quarrelsome and in view of this, the values presented in Table 9.4 must be used in the context of information collected from environmental monitoring programs.

Table 9.4 : Tentative Value for Maximum Acceptable $\overline{[P]}$ in Lentic Inland Water Bodies Used for Cage Culture of Fish

Water body	Species cultured	$\overline{[P]}$
Temperate	Salmonoid	60
	Carp	150
Tropical	Carp and Tilapia	250

Source : Beveridge (1984)

The well-established chain between water quality variables and algal densities is definitely a main foundation stone of modern water management wisdom and the relationship between water quality and fish health has been frequently drawn to the readers attention in various parts of this book.

The idea of a limiting nutrient is also a major concern with regard to the carrying capacity of a cage fish farm. While a range of nutrients are essential for algae, and if the supply of anyone nutrient is less than demand, this nutrient will definitely limits growth. In fish culture ecosystems, it is phosphorus that is limiting, since it is the rarest element with respect to algal and higher plant demand.

9.13 Management

An economic benefit can be achieved by the cage culturist who produces robust and market-sized fishes in a short time and the capability to maximize growth depends on suitable management. Fish culturists must provide (i) adequate space and food, (ii) water of high quality (particularly temperature and dissolved oxygen) for the fish, and (iii) diseases as well as parasites.

For sustained yield, the caged stock must be kept in good conditon as to minimize losses and encourage better growth. The management protocols generally involve (i) stocking the fish at densities suitable for the site, species and methods of rearing, (ii) feeding the fish in a cost-effective manner, (iii) ensuring the best quality within the cages, (iv) maintenance of cages, anchors and other apparatus, and (v) periodical checking of stocks for signs of diseases and parasites and treatment of diseased fish.

The water quality conditions and fish growth should be recorded so that management procedures can be properly evaluated, modified, and shape into scenarios if required. The choice of management is crucial to any successful cage fish farming venture. However, the general routine of cage fish farming has been discussed briefly in five sub-sections that follows.

Seed Supply and Stocking

Nutrient requirements for normal growth of the major cultured fish species are essentially known and the quantities required under standard conditions have been fairly identified. However, methods of formulating, processing, and feeding diets for a variety of commercial cage fish culture species in a variety of culture environments present a countless challenges to the progressive fish culturists.

Feed costs account for over 50 per cent of the variable costs in applying the best feeding strategy can have a significant impact on optimizing profit which is the principal objective of commercial fish culture.

1. **Seed Supply :** Juveniles of cultured fish must either be produced in hatcheries or collected from wild stocks and then stocked in cages for culture. For long distance transport to cage sites, fry or fingerlings require 1-2h starvation prior to packing. Because of the stupendous journey, fish should be carefully inspected and injured fish must be removed. Quarantine measures should also be taken prior to despatch.

    The process of capture, handling, loading, and transport are highly stressful to fish that result not only in physical damage but also increases in oxygen consumption, osmoregulatory problems and susceptibility to diseases. Some species of fish such as *Catla catla* and silver carp are difficult to transport.

    Hence during transport, handling should be kept to a minimum. In temperate regions, a few thousand of fish may be transported over short distances with few problems. Plastic bags are used which should be one-third filled with water and the remaining space is filled with oxygen prior to sealing. For larger consignments,tankers with a capacity of 10,000 litres or more are available and these are equipped with refrigeration and aeration/oxygenation devices.

    In many parts of the world particularly in Southeast Asia, fry are transported in water-filled chambers by boats. This type of transportation is, however, very popular where road access to cage fish farms is arduous.

High temperatures may aggravate transport problems. In tropical countries, transportation at night of containers with saw dust and ice (in 1 : 1 ratio) is recommended. Moreover, transportation of fish over considerable distances results in build-up of toxic metabolic wastes such as ammonia and carbon di-oxide and increased bacterial numbers which can be removed by the addition of neutral zeolite, a buffer, and antibiotics to the transport media, respectively.

2. **Stocking :** At the time of stocking, fish bags are placed in the cages and the temperatures are allowed to equilibrate prior to release. If fish are transported following adequate steps, fish can be released without any problem. However, in warm climate, transfer to cages should be carried out in the early morning or late evening when temperatures are low.

Immediately after stocking to cages, the feeding of fish is generally not recommended. Some fishes such as tilapia, catfish, and snakeheads recover quickly from handling and regular feeding can start 4-6h after stocking.

Types of Food

In fish cultures four types of food are used that follows.

1. **Semi-intensive Food:** These types of food are made from local ingredients available at low cost and contain small amounts of protein. Therefore, fish grown in semi-intensive culture conditions rely entirely on natural food which is high in protein, carbohydrate, and fat. This type of fish culture is ideal for some species who are planktivorus, herbivorus, omnivorus, and detrivorus in nature and grown in environments where the supply of natural food is enormous. Semi-intensive cage culture is restricted to freshwater fish and is widely practised in tropical and sub-tropical countries.

2. **Intensive Food :** These foods are used for all cultured species' nutritional requirements including right quantities and qualities of carbohydrates, fats, proteins, vitamins, and minerals. With the improvement of fish culture technology, however, intensive foods have been developed and improved beyond measure and a number of intensive food-stuffs are used in fish culture.

3. **Moist Food :** Many farmers use moist foods prepared from trash fish or silage mixed with a binder meal which contains protein (fish meal), carbohydrate (wheat meal), vitamins, and a bindder (cooked plantain, potato, etc.). The paste is then either fed as a moist ball or pelleted to the required size using a size die on the grinder. Though nutritional qualities of moist foods are higher than trash fish and possess high water stability, they suffer from many of the disadvantages of trash fish diets such as seasonal fluctuations in quality, high pollution, transport and storage problems. Consequently, moist diets are no longer used in many regions of the United States and Europe.

4. **Dry Diets:** Dry diets have less pollution problems, high water stability, readily eaten by most species of cultured fishes and more digestible and therefore, these

diets are widely used. The simplest method of preparation of dry diets involves the use of food mixtures which produces a wet-extruded diets and the commercial-scale operations also involve more sophisticated processing technology which results in the preparation of quality foods. In cage fish farming, however, sinking or floating feeds are used. Though moist and dry diets may be suitable for cage fish culture, there are relative advantages and disadvantages (Table 9.5) that has great practical significance in relation to cage fish culture industries around the world.

It should be added as a note that due to seasonal variations in the availability of foodstuffs and costs, their use is infrequent.

Table 9.5 : A Summary of The Relative Advantages and Disadvantages of Different Wet And Dry Diets

Type of food	Composition	Advantages	Disadvantages
Trash fish	Scraps from fish processing plants: Fresh/frozen and minced/chopped/ whole.	Cheap to purchase and palatable to fish	High moisture content, expensive to transport and store, high pollution loadings, variable quality; Coloration of farmed fish
Wet pellet	Trash fish and binder meal which contains vitamins, minerals and protein	Improved nutritional and water stabilities over trash fish	Requires regular fresh/frozen fish supply; used immediately when pelleted; limits choice of feeding system
Dry diets		Water-stable, convenient, long shelf life, low storage and transport costs	Expensive, not palatable to a few species of fish

Storage of Foods

Since foods are supplied in bulk quantities, any fish farming operations require storage facilities. Different foodstuffs must be stored in such a way that the quality of foods may be maintained because humidity, heat, insects, fungi, rodents, dirt and other contaminants can destroy or greatly damage foods, making them unpalatable, less nutritious, and even toxic to farmed fish.

Fresh and good appearance of foods are stored in cold storage for several months. Since cold storage facilities are highly sumptuous to run, an alternative is to store the food as silage which can prove cheaper than freezing and produces a highly palatable moist food. The most common method of producing silage involves the use of acid. The fish or fish offal is first chopped and a mixture of 1.5 per cent sulfuric acid and 1.5 per cent formic/ propionic acid is added to reduce the pH to less than 4. At this stage, an anti-oxidant may be added at the rate of 250 mg/l. The prepared silage is immediately used or stored in plastic bins or storage tanks.

Dry foodstuffs are packed in plastic-lined sacs or stored in hoppers. However, fish foods should be conveniently kept in clean, dry, and cool conditions in order to avoid contamination with other materials. Both humidity and temperature not only helps stimulate the growth of insects and fungi but also increases peroxidation with the consequent destruction of fat-soluble vitamins. High humidity is also conducive to the destruction of vitamin C. Insects and rodents severely damage food in large amounts than they consume through spoilage.

Faecal contamination of foods are primarily responsible for *Salmonella* infection in farmed fish. Several species of fungi generate mycotoxins and some foods contain toxic compounds which produce mycotoxicosis in fish. However, deterioration of stored dried foods and foodstuffs can be reduced if the following criteria are followed.

1. Temperatures should be kept as low as possible.
2. Maximum ventilation in the storage facility.
3. Foods should be checked before storage.
4. Galvanized iron sheeting is not recommended as the construction materials become very hot in summer season.
5. Insects and rodents should be kept under control.
6. The moisture content of dry foods and presence of moulds, insects and rodents in stored houses should be frequently checked. The shelf lives of fish foods are summarized in Table 9.6.

Table 9.6 : Summary of Storage and Shelf-life of Different Types of Foods.

Type of food	Storage and duration
Dry supplementary foods (such as rice bran, mustard/ groundnut cake, and wheat middlings)	Moisture content is more than 10 per cent and minerals are stored in cool, dry and pest-free environment. Food can be stored for several (tropical) and many (temperate) months.
Trash fish (Frozen)	Low fat content – for 1 year at - 20ºC. High fat content – for 3 months at - 20ºC
Silage	Can be stored for 6-8 months if the fat content is very low and contains adequate antioxidants.
Pelleted, commercial and intensive foods	In tropical countries, foods tend not to have high levels of antioxidants or vitamins present, hence they can be stored for 2-3 months.
	In temperate countries, food generally contains extremely high levels of antioxidants and vitamins and hence they can be stored in clean and dry conditions for 9 months or more.

To sum up, fish stocks that are fed contaminated foods may exhibit abnormal behavior, loss of appetite, poor growth, and disease infections. Therefore it is necessary to store foods and foodstuffs to retain their quality. Several simple systems of retaining the quality of foods have proved to be cost-effective and beneficial and are considered as wise practices for food conservations.

Feeding

Fish are definitely fed throughout the culture period. Extremes of heat and cold indicate that fish will not feed or feeding should be discontinued. In the Russian Federation, for example, cages of common carp are over-wintered below the ice surface and are not fed until the spring when the ice melts and the cages are lifted.

Most of the cage farms fed their fish by hands since it is cheaper and easier to do so. Trash fish is simply broadcast over the surface using a scoop and shovel. Rainbow trout farmers fed their stock on frozen blocks of minced trash fish which are floated on the water surface. Moist pelleted foods and dry supplementary foods are fed to the fish by keeping them on the top net of the cage and gently lowering the net into the water.

Hand feeding has several advantages: *firstly*, farmer can access how hungry their fish are and consequently, can adjust the amount of food they use. *Secondly*, farmers can monitor the health condition of their stock, because stressed or sick fish will stop feeding. *Thirdly*, in semi-intensive cage fish rearing, hand feeding is very effective in small farms since operations are small.

1. **Use of Feeders in Feeding:** Most of the intensive cage farmers generally use mechanical feeders for reasons of economy of labor. Mechanical feeders are of two types such as (i) demand feeders and (ii) automatic feeders.

(i) *Demand Feeder* : Demand feeders consist of a feed hopper fitted with a plate on the bottom which is connected to a pendulum rod that protrodes down into the water. The feeders are mounted on a gantry over the middle of the cage, and when the pendulum comes in contact with the fish, the plate moves, releasing small quantities of food. Water currents are also sufficient to trigger the pendulum frequently, releasing more food than the fish can encounter with. Demand feeders are, however, cheap and make food available to the fish round the clock. Demand feeders work fairly well with channel catfish, trout, and tilapia. The relative merits of using demand feeders include better food conversion ratios, high production and good fish health, less disease problems, and improved water quality.

(ii) *Automatic Feeder*: Automatic feeders are used by farmers where a measured ration is supplied at fixed intervals. As a generalization, dry pelleted automatic feeder is perhaps the most common type installed at cage fish farms. It essentially consists of a plastic feed hopper climbing a triggering device which releases a quantity of pellets each time it is activated, and are operated either by compressed air or by rechargeable batteries. Different types of feeders have, however, relative advantages and disadvantages and have been summarized in Table 9.7.

System Management

1. **Monitoring Water Quality:** The most important data that should be collected for monitoring water quality variables relevant to caged fish include dissolved oxygen, temperature, nitrogen (ammonia, nitrite and nitrate), dissolved phosphorus levels,

pH, secchi disc visibility, and chlorophyll levels. These data adequately provide the fish farmer a vivid picture of what is happening in the cage culture environment that helps farmer to understand about the dangerous levels of toxins, and the effects of the farming operation on algal populations. Water quality parameters can be measured using simple equipments or by chemical methods. Several portable test kits suitable for use by farmers, are now available.

2. **Fish Husbandry and Management:** Fish husbandry and management of resources involves sampling, grading, monitoring, disease outbreaks, harvesting, and marketing. However, fish sampling should be taken at regular intervals and weighed for monitoring the growth of stocks. This information helps make a number of decisions such as feeding strategies and harvesting time.

Table 9.7: A Summary of Feeding Methods at Cage Fish Farms

Type of feeding	Type of food	Usage	Advantages	Disadvantages
Demand feeder	Pelleted foods	Used for intensive and semi-intensive system of farming of a range of species	Cheaper than automatic feeders and less labor intensive than hand feeding; feeding rates can be adjused to suit conditions	Over-feeding must be avoided; waves can trigger some feeders at exposed sites; may not be suitable for some species; poor feed distribution pattern
Hand feeding	All types of intensive and supplementary foods	Used in semi-intensive and intensive farms	No capital expenditure on feeders; fish health can be monitored; feed can be distributed widely and fish can be fed according to appetite	Can be expensive at large farms; farmers are actively engaged in farms; sometimes may lead to over-feeding: poor feed conversion ratio values and high waste loadings
Automatic feeders				
(a) Air-operated	Pelleted dry foods	Not commonly used on cages and used for intensive culture of salmonoids	Good food distribution and less labor-intensive than hand feeding	Not suited to all cage sites; not feedback from fish to farmers and hence there is a risk of over-feeding
(b) Electricaly operatred	Pelleted dry foods	Widely used for intensive fish culture	Good food distribution and less labor-intensive than hand feeding	Not feedback from fish to farmers and hence there is a risk of over-feeding
(c) Water powered	Pelleted dry or wet foods	Generally used for salmon culture	Improved food conversion ratio values and reduced solid loadings	Very expensive to install, vitamin losses may be problems

(i) *Sampling* : Sampling involves lifting the cage structure so that the fish concentrate in a small volume of water. The fish are captured using a dip net, counted and weighed.

(ii) *Grading:* Along with the growth of fish, the stocking density in the cage increases and hence it is necessary to thin fish stocks from time to time which helps maintain optimum growth conditions and reduce disease risks. Fish are generally graded in intensive operations to yield fish of a standard size.

There are several methods for grading fish and the grading depends on the type of the farming systems. Grading of fish by eye is widely practised in less intensive culture systems and at farms where large fish are grown. In case of intensive cage fish farms and larger ones, grading of fish are routinely carried out by machine. In intensive cage fish farming systems, automatic graders are used and the grading operation is carried out either on the cage raft itself or on the board of the boat. The machines are designed to handle fish that varies between 50 g and 500 g, and can be adjusted to classify fish into 4 or 5 different sizes. Classified fish are then caught in a series of tanks placed below the machine and transferred to the appropriate cages.

(iii) *Monitoring*: Regular monitoring of fish stocks from a disease point of view is also inevitable. If something erroneous is suspected at the time of feeding, then some fish should be removed from the cage and examined whether there is any change in general appearance (deformed spine), skin (color, presence of lesions, rashes, spots, and excessive mucus), eyes (bulging and cloudy), and fin as well as tail (erosion).

(iv) *Disease Outbreaks*: In any fish culture systems, disease outbreaks always accur at a farm. Dead fish should be eliminated from cages as they may be a source of further infection and attract predators. This involves periodic (once per week) lifting of nets very carefully and removal of dead fish or scooping out any fish floating on the surface.

(v) *Harvesting*: Cage fish harvesting is either performed in batches or continually depending on how the production cycle is manipulated. Fish are harvested *in situ* or the cages are dragged to a landing stage where netting operation is carried out more conveniently. For larger cage fish farms, mechanical gears are used to lift the cages. In case of small farms, the cage structure is pulled up until the fish are assembled in a small volume of water and netted out using dip nets. If the cages do not have walkways, harvesting may become a little painstaking.

(vi) *Marketing*: After harvesting, fish are despatched as readily as possible to ensure the freshness of the product to consumers. In a number of southeast Asian countries, several species of freshwater fish such as common carp, *Labeo* spp. *Mystus* snakeheads, tilapia etc. are cultured only for live fish market. In many cases, however, fish are killed prior to despatch. Sometimes fish are gutted, descaled, cleaned, smoked, dried and frozen before despatch. Harvesting, processing, and marketing of farmed fish have been discussed at length in Chapters 22, 23, and 24.

3. **Maintenance of Cages:** Since cage structures are vulnerable to damage that can be caused by predators, high currents, drifting objects, poachers, and vandals, all materials used in the construction of cages have a limited life span. Cages, nets and moorings, therefore, must be checked with degree of regularity for signs of damage and repaired if necessary.

    Dirty nets are cleaned by soaking them for 2-3 days in a mixture of 9 per cent copper sulfate solution and 3 per cent formic acid. Chemical method of cleaning is, however, not practised now-a-days largely for economic reasons and because farmers are too much concern that the chemicals used may have dangerous effects on fish stocks. The use of antifouling compounds at the time of manufacture of cage nets have been found to be more effective.

    Mechanical method of cleaning of cage nets involves the use of hard bristle brushes or sticks. Brushes are used to remove weed and accumulated debris.

9.14 Problems in Cage Fish Culture

A number of criteria such as suitable site selection, cage construction, environmental factors, management, and other production-related criteria are the correct strategies to culture fish in cages. These criteria permit high survival, growth and production of fish. But unfortunately, it is not often possible to follow the above-mentioned criteria. It is also beyond possible ,to predict all of the problems which might appear and to examine all risks associated with the cage culture systems. In the following sub-sections, some important common problems associated with the growth of fish in cages are briefly discussed and suggestions have been noted to alleviate adverse effects.

Acidification of Freshwater Bodies

Acidification of freshwater bodies occurs naturally in many areas of the world. However, since several species of fish are not able to tolerate acid conditions, fish diversity in acidic waters is low especially in temperate regions. Therefore, it is essential that cage fish culturists must consider these implications for fish yield when selecting sites and culture methods.

If acidification of water occurs, farmers either control the process of acidification or attempt to culture acid-tolerant species. Several chemical substances have been recommended to neutralize acidic conditions of water. These substances include lime [CaO and $Ca(OH)_2$], limestone ($CaCO_3$), soda-ash (Na_2CO_3), olivine [$(MgFe)_2 SiO_4$] fly-ash, and sodium hydroxide ($NaOH$). Of them, limestone is by far the best because it is cheap, safe and effective on a wide range of lake type. Direct application of lime to the acidified water body is likely to be most cost-effective. Finely crushed limestone is sprayed with the help of a special device. Liming materials help to elevate the pH of water from around 4 to 7.7

that reduces the direct effects of low pH on fish and also to reduce the toxic effects of aluminium. Aluminium is leached in high amounts from acidified conditions of freshwater ecosystems.

Currents

For cage fish farming, suitable current conditions are essential. High currents (more than 50 cm/second) result in distortion of cage structures, excessive stress on mooring, and losses of food. High currents also affect the metabolic rate of fish, since fishes are forced to expend energy in swimming against the current. As a generalization, the flow rate of water inside the cages should not exceed 2, 35, and 40 cm/second, respectively for trout, carp, and tilapia. Fast currents are encountered in running water habitats but such condition is not desirable since high flow of water through cages makes feeding very onerous.

To maintain the cage structure in better condition under high currents, the use of floating copper-nickel expanded mesh rigid cages to culture trout, cyprinids and tilapia is recommended and has been very successful.

Disease

In fishes, two main groups of diseases are noted such as pathogenic and non-pathogenic. Pathogenic diseases are those caused by invading organisms (pathogens) which may be regarded as *parasites*. On the contrary, non-pathogenic diseases are very common and include deficiency diseases, genetic diseases, environmental diseases etc. However, the type of disease generally encountered in fish may be divided into six categories: (1) Genetic, (2) Nutritional, (3) Environmental, (4) Physical injury, (5) Parasites and micro-organisms, and (6) A combination of the above. Most of these diseases are non-communicable and if properly diagnosed, management practices are enough to eradicate the problem.

Nutritional disorders may be treated with the application of balanced and nutrient-enriched diets. Environmental hazards such as adverse light, temperature, pH, and dissolved oxygen of water can be reduced by putting the fish cages in suitable sites and good husbandry. Physical injuries can be reduced by handling fish with care and the choice of netting materials and mesh sizes.

Communicable diseases are caused by parasites, bacteria, fungi, and viruses. Fish parasites include protozoa, nematodes, acanthocephalans, cestodes, trematodes, leeches, and crustaceans. At least 20 species of bacteria have, however, been isolated from freshwater fish and are regarded as *potential pathogens*. Fungi can also invade living fishes often through lesions. At least 15 commercially important fish viruses have been identified including infectious pancreatic necrosis (IPN), viral haemorrhagic septicemia (VHS), spring virema of carp (SVC), and infectious haematopoietic necrosis (IHN).

Fouled caged structure can harbor a variety of parasites. Environmental stresses (such as low dissolved oxygen, high temperature etc.), handling and physical damage to the caged fish definitely lower the resistance to diseases. In the case of intensive cage farming where fish are normally stocked at high densities, there is every possibility to spread diseases rapidly.

1. **Prevention Treatment:** Prevention of fish diseases should always be kept in mind and some important steps should also be taken by all farmers to reduce risks that follows.

    (i) Choice of disease-resistant species.

    (ii) Proper site selection.

    (iii) Selection of robust fish for stocking,

    (iv) Daily observation of fish. Early diagnosis of fish diseases is inevitable if mortalities are to be kept to a minimum,

    (v) Avoidance of over-crowding and over-feeding.

    (vi) Monitoring and maintaining the water quality,

    (vii) Use of fresh and uncontaminated food,

    (viii) Removal of dead fish from cages and handling of fish with care.

Inspite of precautions, disease outbreaks do occur and before any action can be taken, proper diagnosis must be carried out. In many cases skilled farmers can make preliminary diagnosis and consequently medicaments for the fish are possible. Though external examination of infected fish is enough to make a diagnosis; in most cases, a post-mortem examination becomes necessary so that proper treatment can be made.

Treatment of diseased fish have to be made as early as possible and the three methods are generally followed such as (a) direct use of chemicals to the fish, (b) use of chemicals in the food, and (c) bath treatment in which chemicals are added to the water. Direct use of chemicals involves, for example, the use of drugs as a paste to cure fungal skin infections of brood fish and intra-muscular injections of individual egg-laden fish.

Some diseases such as bacterial septicemia (caused by *Aeromonas* and Vibriosis infections), internal protozoan infections (such as *Octomitus*), and some cestodes can be treated by using drugs which are incorporated in the food. When wet pellet diet or trash fish are used, drugs are generally mixed with the food prior to feeding; but when dry diets are used, drugs are coated with a gelatin mixture which are then allowed to dry.

It should be kept in mind that the use of drugs in fish food requires a correct prescription and this is important in reducing the risk to human health of developing resistant strains of pathogenic microorganisms. It is also recommended that fish are allowed to sale for human consumption for at least 20 days following drug treatment. This time lapse will permit to eliminate all drugs from their tissues.

Extremely soluble chemicals are used in bath form to treat fish with gill and skin diseases, but chemicals used in cages would be diffused by currents. To solve the problem, the net bag is lifted to concentrate the diseased stock in a smaller volume of water and enclosed in a plastic bag (in case of small farms) or diseased stocks are netted out and transferred to the treatment cage (in case of large farms) containing suitable chemicals and allowed for the recommended time before being transferred back to the production cage. During treatment, however, farmers should follow the following precautions.

    (i)   Since oxygen consumption by the diseased fish is likely to increase during treatment, an oxygen diffuser may be installed.

   (ii)   The use of an aerator helps disperse the chemical following treatment.

  (iii)   Feeding should be stopped for at least 24h prior to treatment.

  (iv)   Plastic packets should be used for treatment procedures.

   (v)   Treatment at high temperatures should be avoided. In tropical fish farming, treatment should be made in the morning provided that dissolved oxygen levels in water are satisfactory.

  (vi)   Courses of treatment should be completed, otherwise the disease may revert with greater severity.

It should be remembered that most of the diseases stem from inadequate management and that medicament should be seen as a last resort. Many fish diseases cannot be tackled by therapeutic means. Most of the bacterial and viral diseases can, however, be controlled by removal of stressing factors such as low oxygen, crowding etc. The economics of treatment should also be considered. While systemic treatment via the diet is a very costly affair in terms of drugs, bath treatments are less costly in terms of chemical but are very labor-intensive.

Drifting Objects

Cages are more vulnerable to damage by drifting objects such as different species of floating aquatic plants. These objects make holes in the cage nets and restrict water flow through the cage structure. Drifting masses of floating plants such as water hyacinth, *Pistia* etc. in inland tropical water bodies severely damage the cage structures.

These plants cover substantial portions of the water surface and when strong winds blow, floating plants can overthrow on a site and consequently the cages are damaged. To overcome this problem, a barricade is constructed which distract the weed masses and keep the cage regions very clear.

Fouling

Fouling generally occurs at all cage sites. In contrast to marine sites, freshwater ones are less worse. However, organically-rich waters exhibit high degree of fouling and the degree of fouling depends upon the species being cultured, materials used in cages, and the mesh size (knotted or knotless). Fouling reduces the actual mesh size thereby reducing the flow of water through the cage.

Several compounds such as organotin (tributyl tin oxide), copper-based substances etc. are known to contain a biocide which surrounds the net and form a thin layer of toxic material which prevents the fouling organisms from adhering to the cage nets. Application of anti-foulants to the nets involve immersion of the nets in the compounds for at least 30 minutes and subsequent drying for several hours in such a position that none of the meshes become clogged. It is recommended that for long-lasting, the cage nets must be treated with antifouling agents each year.

Planktonic organisms which foul net cages generally form the diet of cultured herbivores such as carps and tilapias and consequently these organisms are removed from cage structures. Stocking of Indian major carps, for example, at low densities (5 fish/m^3) have been found to be effective in maintaining the cleanliness of cages.

The use of fouling-resistant materials such as PVC-coated wire, copper-nickel materials etc. in the construction of the cages have proved effective to a limited extent in reducing the fouling of cages. Several companies manufacture polyethylene netting inlaid with copper wire. The use of these materials is, however, much expensive than conventional nylon bag systems. Since the copper provides limited support, the nets should be replaced every year or treated with anti-fouling substances after the copper wire has crumbled.

Oxygen

In cage fish farming, the main source of dissolved oxygen is water flowing through the cage due to external currents and the movement of the caged fish. If the cages are kept relatively free from pollutants or if there is no fouling problem, the dissolved oxygen content of water remains at saturation levels. Tropical cage fish farming frequently create difficulties by decreasing the absolute amount of dissolved oxygen and by increasing the oxygen demand of the fish.

1. **Reasons of Dissolved Oxygen Depletion:** Dissolved oxygen levels drastically fall below saturation in natural water bodies under the three conditions. First, where algal blooms are large enough, respiration of fish at night results in significant decline in dissolved oxygen level, reaching a minimum at dawn. Second, at the end of the algal bloom, bacterial decomposition of algal mass can lead to anoxia and third, upwelling of anoxic waters can occur at certain cage sites. These problems are common at both freshwater and marine sites in tropical and temperate regions and the problem can, however, be solved if cages are not installed in eutrophic or strongly stratified water bodies.

2. **Control of Dissolved Oxygen Depletion :** The problem of dissolved oxygen levels in cage fish farmers can be avoided either by increasing (i) the content of its oxygen or (ii) the rate of water flow through the cage. In practice, oxygenation is accomplished by maximizing the area of contact between water and oxygen or air either by introducing fine bubbles of gas into the water or by breaking the water into fine droplets in air. Air is delivered to the water from an air blower via the bubble diffusers while droplets of water can be generated by different mechanical devices such as submerged pumps, aerators, and surface aerators.

Submerged pumps provide excellent results with a very high oxygen transfer efficiency (2.3-3.2 kg O_2 KW/hour). This device also helps reduce water temperatures and consequently improves the content of its oxygen. Surface agitators and aerators are also effective since they aerate only the top few metres.

Oxygen may be supplied from a pressurized oxygen source. Bulk liquid oxygen is the most common source of oxygen at fish farms which is about 95 per cent pure and is stored under pressure in cylinders or takns.

(i) *System of Aeration*: An aeration system typically consists of a diesel or electrically-operated blower or compressor connected to a flexible PVC hose which distributes air to an aerator or system of aerators immersed about 3 m below the cage surface. However, there are two alternatives for aeration. First, increase in the rate of water exchange through the cage using a low speed propeller and second, the use of oxygen instead of air. Propellers generate water flow and create horizontal as well as vertical mixing and have been successful in several temperate countries. Compressors, on the other hand, are cheap and easily available to the farmers in tropical countries, though they are less efficient than those of low speed propellers.

Predators

Any animal that exists by hunting and feeding on other animals is known as a *predator*. However, many cage fish farms have a lot of problems with predators. The range of predators noted from freshwater cage fish farms is enormous and include fish turtles, lizards, snakes birds, mammals, and insects. Many of these are seasonal visitors. Cormorants (*Phalacracorax carbo*) in Scottland begin to arrive at freshwater cage fish farms in August, increase in numbers during the autumn and winter months, and disappears in April when the breeding season starts. A marked seasonal pattern in the number of herons arriving at fish farms has also been noted.

In farms where fish are reared in cages, grebes, shore birds, ducks, gulls and crows feed on scattered fish food and from beneath hoppers. Droppings of many predatory birds in surrounding waters and on floating structures of cage farms where they settle for rest or sleep, can pose human health hazards when droppings are loaded with pathogenic organisms.

Some predatory birds act a carriers of viruses. Viruses cause viral Haemorrhagic Septicemia (VHS) and Infectious Pancreatic Necrosis (IPN). These viruses have been detected in regurgitated food and in the faeces of heron, pelical, cormorant, crane and spoonbill 4-6 hours after feeding on contaminated fish.

Many species of birds and mammals exhibit diurnal patterns of activity and attack the caged fish at dusk and dawn when farmers are absent from the farm. Most of the predators try to enter the cages either from below the waterline or above. Freshwater farms are however less attracted by the wide range of predators than marine ones.

Predators create a lot of problems to fish farms such as (1) spreading diseases and damage the cage nets, (2) loss of small-sized fish, (3) Consume fish as much as they can,

(4) loss of scales and horrific wounds on the sides of the fish, and (5) severely damage the large-sized fish. Once enter into the cages, predators are not able to get out again. However, minks, otters, cormorants, gulls, herons, snakeheads, catfishes, water snakes, etc. are most common in many freshwater cage fish farms. This damage creates rupture of nets through which the total stock can vanish.

Several measures are commonly employed against predators. These include the use of (i) top nets to restrain from gulls and herons and anti-predator nets to foil underwater attacks, (ii) dogs in tropical countries to restrain from otters and rats. A number of deterrents such as flashing lights, scare crows and loud noise-emitting devices are used in many cage farms but these methods have had limited success since most of the predatory species have become habituated to these deterrents. Shooting is also adopted by many farmers against birds. Live traps are sometimes used against mink.

To sum up, it should be kept in mind that the use of flexible and rigid mesh cages are much superior in terms of resistance to damage and infiltration by predators. This is only possible if cages are fitted with adequate protection above the waterline. Last, but not the least, the poor management performance of cage fish farms in general creates frequent problems which have long-term adverse effects on fish production potential in particular.

Wastes

It is pertinent to mention that the magnitude of the sediment of waste materials (particularly uneaten food) under fish cages at intensively managed farms is higher than that of control site. The rate of waste generation is enormous due to local site conditions, food type, management, and fish species. The closer the cage bottom to the pond/lake/ river bed the greater the proportion of wastes that is likely to sediment under the cage and in areas where water currents are most predominant, the waste food particles will be removed.

The accumulation of waste materials can adversely affect fish production in a number of ways. A small increase in the rate of supply of organic material to the sediments will encourage bacterial, fungal, and micro-invertebrate activities causing an increase in the oxygen demand of sediment. In intensively managed freshwater farms where the rate of waste sedimentation is high enough, the rate of oxygen supply may be not adequate to fulfil the respiratory requirements of the macro-benthic and microbial communities. Consequently, the sediments become anoxic, the benthic community alters to low oxygen tolerant and the biological as well as chemical activity of inorganic and organic substances drastically reduced. During anoxic condition, however, several gases such as ammonia, methane, and hydrogen sulfide are generated in the sediments. These gases are known to be highly toxic to fish.

1. Removal of Waste Materials:

(i) *By Use of Sub-mersible Pumps*: Since accumulation of wastes under and around fish cages is a major threat to fish yield, the wastes must either be removed from the vicinity of the caged fish or the farm is transferred to a new site. Periodic removal of sedimented wastes using sub-mersible pumps has been successful particularly in large commercial fams in Norway. Although the method is highly effective, it is very difficult to operate and with risk. During operation, the stocked fish must be removed from the area concerned prior to pumping.

(ii) *By Use of Filter Collector*: This method involves the use of filter funnel-shaped collectors. These collectors are enclosed the cage bottom and suspend under the cages and the sedimented wastes are periodically removed by pump. However, about 50 per cent of solid wastes are removed in this method.

The removal of fish cages from waste-laden areas to new sites where dangerous problems do not arise, is very simple; of course, if the farmer has new areas suitable for the purpose.

To sum up, it is very difficult to advise farmers which method is best and cost-effective since economic benefits of waste entrapment have not been satisfactorily proven and deserve further evaluation. Prevention is, however, preferable to cure and efforts must be concentrated towards the cage fish industry to reduce waste generation rather than simply avoiding the consequences. Intensively-managed farm areas with poor water circulation may, however, develop a lot of problems. Moreover, leaving farm areas fallow for several months or years may sometimes alleviate the effect of accumulated waste materials in the sediment.

Poaching and Vandalism

Since fish are quickly removed from cages by hand nets, caged fish are more vulnerable to poaching and vandalism. Farms located some distances from farmer's residence can be easily monitored. Vandals can cause severe damage the cages and loss of fish stocks.

Prior to establishment of cage fish farms, due consideration must be given to the site selection and security must be ascertained in the area concerned. However, small cage fish farms may be established,if possible close to the farmers residence. But in case of large farms, watchman houses are constructed on the cage rafts. A number of devices such as lockable lids, alarm systems, warning lights, etc. have been successfully employed at a number of large farms especially in the UK, USA, and Southeast Asia.

9.15 Cage Culture of some Species of Fish

Previous discussions remind us that the cage culture has contributed only around 4 per cent in overall farms to world farmed fish, shellfish, and crustacean production. In some sectors, cage culture industry is extremely important and is the most economically feasible method of producing several species of fish in certain regions of the world such as China where the area of cage culture has expanded at an average annual rate of 71 per cent from 1978 to 1996 and fish production from cage culture increased by an average of 10 per cent per year. Commercially viable cage farming has also been developed in other areas of the world, but several informations or ideas on the methods of cage culture have spread gradually through a group of farmers.

Rural tanks, ponds, seasonal and perennial water bodies, running water environments, reservoirs, and minor irrigation tanks provide sizeable areas for consideration of cage culture of a variety of fish species especially catfishes (*Clarias, Heteropneustes, Pangasius Mystus, Ictalurus etc*), snakeheads *(Channa spp.),* carps *(Cyprinus carpio, Labeo rohita, Cirrhinus mrigala,),* Cichlids *(Oreochromis spp.),* white fish *(Coregonus spp.),* eel *(Anguilla* spp.), etc. There are, however, two ways by which farmers can use fish cages: (1) for growing fishes to table size (6 months culture period) and (2) for raising fingerlings (2-3 months culture period). In this section, some economically important cage farming systems have been briefly noted.

Cage Culture of Tilapia

The tilapias are members of the large tropical Cichlidae family of inland water fishes and about seventy five or so species have been reported. The group may, however, be divided into three genera *Tilapia, Sarotherodon*, and *Oreochromis*. The fish have a fine-textured, white flesh and have an excellent taste which has proved universally popular.

Around half-a-dozen species and various hybrids are currently important in commercial cage culture not only in the tropics but also in sub-tropical parts of the world. Two types of cage are in common use such as floating and fixed. Knotless nylon of variable mesh size are employed (Table 9.8).

Table 9.8 : Recommended Mesh Sizes For Cage Culture of Tilapia

Fish size	Purpose	Mesh size (mm)
Fry (less than 12g)	Nursery	1 - 3
Fingerling (12-30g)	Grow-out	4 - 8
30-200g in size	Grow-out	10 - 20
200g in size	Grow-out	20 - 25
Brood fish (150g +)	Breeding	1 - 3

Cages vary enormously in size and in general, the larger the farm, the cages that are used.

1. **Stocking:** The fingerlings required by cage farmers are between 0.5 and 20 g in size. Most of the farmers prefer small-sized fish for stocking. Stocking of tilapia in cages is carried out in the early morning or late afternoon when temperatures are slightly lower. Stocking densities range from 5 to 50 fingerlings/m^3 depending on site, cage size, feeding practices, and desired size at harvest.

2. **Feeding:** Supplementary feeding is carried out by the majority of farmers at different sites, although productive water bodies produce fast-growing fish without recourse to the use of foods. Rice bran and mustard oil cake are the most

widely used food. Other materials include chicken and pig manure, dried and shrimp meal, ipil-ipil leaf meal, and wastes from various food processing industries. Recommended feeding rates with these foods are 3-5 per cent body weight per day, depending on the water surface or moistened into balls and fed to the fish several times per day.

3. **Production:** Since there are tremendous differences in production between sites (production varies between 0.5 and 18 kg/m3/ month with and without supplementary feeding) and also year to year variation, profitability has also been found to vary enormously.

Cage Culture of Common Carp

The fish common carp enjoys global distribution occurring in the tropical as well as temperate regions of the world where it has well acclimatized itself to a variety of habitats and extremes of environment. However, common carp has been one of the most versatile fishes cultured globally and its seed production and culture methods are easier than those of either Chinese or Indian carps.

Cage culture of common carp dates back to the mid 20 th century in Europe and some Asian countries particularly in China where production has significantly increased. In China, for example, the yield of common carp reached 157 tonnes in 2009 from cage culture. At present, the principal exponents are several European and Asian countries who raise substantial quantities of common carp for increasing inland fish production.

As a generalization, a cage size measuring 2 X 1.3 X 1.9 m^2 provides an effective volume of 1.68 m^2. Cages are set in ponds, lakes, and other freshwater ecosystems. Bamboo rafts of the cages form the floats and branches of *Ipil ipil* are used to make the cage upright for suspending the knotless polythylene nets. Concrete anchors may be used to keep the cages stationary. Fry are stocked in cages at the rate of 25, 50, or 100 numbers per cubic metre. Rice bran, mustard oil cake, and commercial pellets are used as artificial food items. Food is supplied in earthen feeding trays which are suspended in each cage. With a few modifications in feeding strategies, it has been noted that fish production is highest in cages where commercial pelleted diets are used.

Cage Culture of Trout

Trout culture is not popular in the tropics but is practised more widely on commercial scale in several sub-tropical and temperate regions of the world. Cage farming of trout is, however, a rapidly expanding industry and is extensively cultured both in freshwater and salt-water environments. Though there is restriction in the use of cage structures in inland waters, considerable development is taking place in the design and construction of cage farm units. These units can be operated in more exposed areas and can withstand rough weather conditions.

Rainbow trout is the most common species widely used in cage culture. The fingerlings required by cage farmers are between 5 and 30 g in size. The smaller fish are, however, preferred by most farmers, although commercial farms operating their own hatcheries often use larger fish. Fingerlings are stocked in cages in spring and are harvested in autumn after a culture of 1.5 years or stocked in the autumn for harvest after one year.

Floating cage culture is widely practised that has several advantages over fixed cage counterparts, among which the most important is (1) the relatively high level of control the fish farmer can have on the stock and (2) the possibility of operations on the small scale with a small investment. Cages vary enormously in size and in general, the larger the farm, the larger the cages that are used.

Stocking density of fish in the cage varies from 8 to 20 kg/m^3 at harvest, depending on water circulation, cage size, feeding practices, and desired size at harvest. In general, large quantities of fingerlings that are required to obtain the above density are stocked in each cage and as they grow larger they are graded in other cages.

Quality of fry and fingerlings is a matter of concern not only to cage growers but also to all trout farmers. The economics of cage farming in different trout-producing regiqns of the world has shown that the profitability has been found to vary enormously. However, according to one estimate, a six unit cage operations could realize adequate profit to enhance rural economics more than several folds.

Cage Culture of *Pangasias pangasias*

The fish *Pangasias pangasias* is a warm water-loving genera of the family Pangasidae which is widely distributed in Thailand, Java, Malaya, Archipelago, Burma, and India. It is now accepted that there are three species within the family such as *P. pangasias*, *P. larnaveli*, and *P. sutchi* which are very suitable for culture in cages.

Since the early 1960s, cage culture of this fish has been carried out experimentally in India and adjacent countries. But inspite of high market price, tasty, and consumer demands, its culture in cages has not yet been popularized. However, fry are produced by capturing wild fish, injecting and fertilizing eggs and incubating them in hatcheries.

Generally, three types of cages are used for the culture of this species: (1) large-sized cage (2.5 - 4.5 m breath X 8 - 10 m length X 2 m deep), (2) medium-sized cage (2 m X 5 m X 1.5 m), and (3) small-sized cage (1 m X 2 m X 1 m). Small-sized cages are used for rearing of fry. The rate of stocking of fry in each small cage is about 2,000. In medium- and large-sized cages, fingerlings at the rates of 1,000 and 2,000 numbers per cage are stocked, respectively. Cages containing fish are floated at the surface of the river or in ponds. All cages are covered with straw mats or aquatic plants such as *Eichornia* sp. to protect the caged fish from the sun. Since the fish is a predatory in nature, all the fish stocked in cages must be about the same size to avoid cannibalism. The cages are made of wood or split bamboos each bamboo is about 2-3 cm width, with interstices between the bamboo of 0.5 to 1.5 cm.

Though the diet of fry consists of different species of zooplankton such as *Cyclops,Daphnia Bosmina*, and rotifers, a well-grown Pangasias may consume about 150 g of chopped trash fish per day. Alternative food for fingerlings includes the mixture of banana (two parts) and rice flour (one part). Water lettuce (*Ipomea*) and other chopped aquatic plants, soyabean waste and cooked maize are also used as food for the fish. The food is given twice daily – early in the morning and in evening.

The rate of production may be as high as 50-70 kg of fish/cage in 2-3 months. If cages are kept in running water habitats enriched with bottom fauna, fish production has been found to be highly significant. The possibility of intensive stocking, feeding, and culture under close observation makes the culture strategy more effective, attractive, and economic.

Cage Culture of White Fish

The corigonidae or white fishes are a coldwater inhabitant family which are widely distributed throughout freshwater lakes of Asia, North America, and Northern Europe. Three genera of this family have been noted:*Coregonus, Stenodus*, and *Prosopium*. However, possibly there are 20 to 30 species belonging to the genus *Coregonus*. The most three commercially important cultured species are *Coregonus lavaretus C. albula*, and *C. peled*.

1. **Biology:** Most of the species are pelagic at least for part of their life. They feed largely on zooplankton or some species adopt a benthophagic life style. Many species possess long and fine gill rackers which are efficient for filter feeding. When surface water temperatures become too high, fish may remain below the thermocline in stratified lakes. During spawning season, the fish congregate in huge numbers in the spawning grounds. Fish matures sexually at 3-4$^+$. Spawning takes place in the colder months of the year. Eggs are laid over sand gravel substrates and hatch after 100-200 days. The number of eggs released by females vary between 1,000 and 1,05,000 depending on species, age, and certain environmental factors. The fry remain pelagic for at least first year.

2. **Cage Culture:** Since the early 1970s cage culture of coregonid species has been carried out with considerable success in the Federal Republic of Germany, the German Democratic Republic, the Russian Federation, Finland, Switzerland and Poland. However, fry are produced by capturing wild fish, stripping and fertilizing the eggs, and incubating them in hatchery jars. The larvae fed on live food such as *Daphina, Cyclops, Artemia, Bosmina*, etc. before they consume dry diets. Two to 5 day-old larvae are transferred from the hatchery to cages. Generally small and floating cages are used for the purpose. Stocking densities of fry vary between 2,000 and 10,000 numbers per cubic metre.

    The extensive cage culture of corigonid fry for restocking purpose is a viable proposition, and trials with rearing of planktivorous juveniles of other species such as *Esox lucius* (pike), *Lucioperca lucioperca* (pike perch), *Tinca tinca* (tench), and *Salmo salar* as well as *S trutta* (salmonoids) have been successful. The extensive culture of corigonids for the table is not highly encouraging since growth beyond the first few months are insignificant. Moreover, the use of light may also severely affect the fish. Since only the juvenile stages are phototactic, it is logical to conclude that the survival and growth in illuminated cage coregonids are definitely significant in contrast to unilluminated cages. But continuous illumination may contribute towards stress and disease susceptibility to fish stocks.

9.16 Probable Future Expansion of Cage Industry

Cage fish culture practices have spread rapidly in some developed nations for the last 20 years. Whereas in 1950 these practices were used on only 2-5 per cent of the cultivated water areas, by 1980, this figure has risen to about 40 per cent. Cage fish culture systems in China and Indonesia are highly encouraging.

Though the cage culture has not been given due attention in many developing countries till today, potential regions definitely exist and may have great opportunities if financial supports are available among cage-growing farmers in these regions. The expansion of cage culture in areas concerned is likely to turn out well because of its possibility of high fish production.

Up to this point, attention has been given to small farms of many cage-growing regions of tropical countries. It is appropriate to have done a lot since there is scope to produce more fish from these regions through cage culture.

9.17 Conclusion

Fish culture in cages which is practised widely particularly in some tropical and sub-tropical countries provide an opportunity to raise fish and the contribution of cage culture to world farmed fish production potential is not fit for consideration. Cage culture is virtually non-existent in regions where fish production potential is great. However, cage culture practised in these countries has a localized influence and did not directly give to the current cage fish industry. The expertise of cage fish culture has been evaluated in several Asian regions but the success/failure of fish culture in cages over extensive areas has yet to be demonstrated. Cage fish culture in several countires such as India, Bangladesh and other adjacent countires has not yet received much attention among fish farmers. When the need arise, substantial quantities of fish are traditionally raised in cages.

The success of fish culture in cages principally depends on the combination of local expertise and state-of-the-art but cost-effective technology based on the understanding of biological cycles in cage culture ecosystems and favorable ecological, cultural as well as economic attitudes. Extreme variations in production potential in cage fish culture ecosystems virtually exist which are associated with the economic and ecological factors in a particular country or region.

The activity of cage fish culture on commercial basis could not create the needed impression among fish farmers in many regions of the world. The lack of attitude towards cage cultivation, unavailability of culture technology to progressive fish farmers and short-term lease of culture sites, short life span of cage materials, high depreciation of cage materials, over-engineering of the cage structure, poaching and vandalism are some of the most important reasons that should be kept in memory.

Scientists and extension personnel, must gain sound knowledge of the characteristics and potential of freshwater ecosystems where there is possibility of establishing cage culture industry. This knowledge is necessary in determining the extent to which suitable water areas should be brought into cage fish industry, but the knowledge of cage fish culture in potential regions is critical to humanity's struggle for fish production through cages. And this farming system is equally critical if we expect to use knowledge gained at one region to produce fish at other regions where more or less comparable ecosystems prevail.

There is great potential for tropical regions to produce fish in cages for at least satisfying the demand of both rural and urban populations. Whether this potential will be reached at the optimum level will depend in part on the wisdom that is used in managing cage farming strategies.

References

Alabaster, S. and R. Lloyd. 1984. water quality criteria for freshwater fish. Butterworths, London, 361 PP.

Aragon, C. T.E. B. Torres, M. M. Lin and G. L. Tioseco. 1985. Economics of cage culture in Laguna Province.*IN* Proc. PC ADRD/ICLARM. Tilapia Economics Workshop, U.P. LOS, Bainos, Philippines, August 10-13,1983. I. R. Smith, E. B. Torres and E. O. Tan (Eds.), ICCARM, Philippines.

Beveridge, M. C. M. 1984. Cage and pen fish farming, carrying capacity models and environmental impact. *FAO Fish Tech. Pap.* 255, 131 PP.

FAO.1983. Freshwater Aquaculture Development in China.*FAO Fish Tech Pap.* 215, 125 pp.

Mason, E. E. 1981. Biology of freshwater pollution. Longman Group Ltd, Harlow, (Essex, 250 PP.

Mok, T. K. 1982. Design and construction of floating cages and rafts. *In:* Report of the training course on small-scale pen and cage culture for fish. 26-31 October, 1981, Philippines, p.103-108.

Reynolds, C. S. and D. E. Walsby. 1975. Water blooms. *Biol. Rev.* 50: 437 481.

Skulberg, O. M. , G. A. Codd and W. W. Carmichael. 1984. Toxic blue-green algal blooms: A growing problem. *Ambio* IB, 244-247.

Questions

1. Trace the origin of cage fish culture.

2. Discuss different types of cages and cage designs.

3. What do you mean by cage culture? How it differs from pen culture? Mention some common economically important species of fish which are considered best for culture in cages. Why monoculture in cages is considered than polyculture?

4. Trace the merits and demerits of extensive and intensive cage culture.

5. Discuss the advantages and disadvantages of cage culture.

6. Discuss whether the use of cages is highly economical for intensive rearing of fishes. Briefly discuss why cage industry in European countries, Japan and North America has expanded rapidly.

7. What are the factors that must be taken into consideration while cage design is constructed for culture?

8. What are the differences between traditional and modern cage designs? State which design is effective for small and marginal cage fish growers.

9. Discuss different steps involved in cage construction.

10. According to some experts cage culture is a costly and wasteful business. What is your opinion?

11. What type of cage is suitable for fish culture in tropical countries?

12. Why cage fish farming has not been developed in India ? Discuss. Is there any possibility to culture fish in cages in this country ?

13. Why site selection is essential for successful fish farming? What are the criteria for selection of suitable sites for cage fish culture?

14. What are the environmental criteria that are essential for an ideal cage farm? Discuss.

15. Discuss how water currents, depth and substrate affect cage fish farming.

16. What is management? What are the management philosophy involved in cage culture industry?

17. State how feeding of cage fish is performed.

18. Discuss the steps involved in routine management of cage fish farms.

19. Discuss some common problems associated with the culture of fish in cages.

20. Discuss how drifting objects and fouling affect fish cages.

21. Write notes on the following:

(a) Carrying capacity, (b) Cage culture of tilapia, (c) cage culture of common carp, (d) Cage culture of trout and pangus, (e) Diseases of caged fish, (f) Maintenance of cages, (g) Cage management, (h) Storage of fish food, (i) Management of cage fish culture, (j) Fixed and floating cages.

10

Fish, Fisheries and Fish Culture in Mountain Regions

Cool and clean water, stunning and exquisite scenes above the clouds and dramatic surroundings the peak of mountains is a very attractive region indeed. A mountain (See Panel 10.1) is generally considered as a part of the earth's crust that has risen noticeably above the surrounding lands, exhibiting steep slopes, a relatively confined summit area, and considerable local relief. Mountains exhibit varying conditions, not from one to another, but within each mountain from base to peak the mountaineous regions throughout the world are provided with rivers and their tributaries, lakes, streams, ponds, and reservoirs. These water bodies are predominantly coldwaters and are abound with fish fauna of varied forms.

The term coldwater fish and fish culture is used for the fish inhabiting the above mentioned water bodies having a temperature not beyond 20°C; the lower water temperature may vary between 0 and 5°C or slightly more depending upon various factors such as topography, altitude, and climate. For proper distribution and abundance of coldwater fish and their culture, other factors such as chemical, physical, biological, and geochemical are equally important. As a generalization, water bodies which are situated above an elevation of about 1,600 m may be referred to as *coldwater*.

Different coldwater resources offer significant potential and scope for development of coldwater fisheries; of course, the abundance and distribution of coldwater fish species and their culture strategy are determined by the activity of streams and rivers in mountain regions (See Panel 10.2). Keeping the vast potential of these resources in mind, coldwater fisheries are necessary not only to conserve most threatened fish fauna but to produce fish as well for mass consumption in mountain regions. Several coldwater lakes have become obsolete and consequently, they are not now in a position to increase fish production. Several others are unable to increase fish production though endowed with better production potential because of the lack of technical personnel, technology, and inaccessibility in many regions.

10.1 General Considerations

Although most of the upland water resources are teemed with a variety of fish species, some regions include zones which are inhabited by both cold and warm water fish species.

PANEL - 10.1
MOUNTAINS

A portion of land surface rising considerably above the surrounding country either as a single eminence or in a range or chain is known as *mountain*. Some authorities regard eminence above 600 metres (2,000 feet) as mountains, and those below being referred to *as hills*.

Mountains are often spectacular features which rise several hundred metres or more above the surrounding terrain. Some occur as single isolated masses; the volcanic Cone Kilimanjaro (Africa), for example, stands almost 5,895 m (19,1.60 feet) above the sea level over-looking the grassland of East Africa. Others make up a portion of an extensive mountain chain, such as the American Cordillera, which runs continuously from the tip of South America through Alaska. Some chains, like the Himalayas, are youthful, gigantic mountains that are still rising, while others are very old and nearly worn down, such as Applachian mountains (USA) and Aravallis (India). Although every mountain system is unique, the similarities of systems are greater than the difference.

A mountain may have several forms such as mountain ridge, range, chain, system, group, and cordillera. Also on the basis of their most dominant characteristic features, mountain can be classified as folded mountain, volcanic, fault-block, upwarped (dome) mountains. And all these mountains have freshwater resources and most of these water bodies have, to a greater or less degree, high/medium/low productive potential.

Flood plain lakes of Kashmir, lower altitude reservoirs and foot hills of the Himalayas, for example, receive cool water from mountain rivers. However, stocking of fish and their culture at high altitudes in different countries of the world are diverse and their management strategies are extremely variable depending on the topographical and climatological variations. In mountain rivers and streams, trophic status, water temperature, dissolved oxygen concentration, velocity of water, turbidity, availability of natural food organisms, and the nature of substratum all have significant influence on the distribution of coldwater fish and their culture technology.

Coldwater aquaculture is truly emerging as a growing industry in many upland areas of the world. It has been considered as the principal industry to produce as more coldwater fish seed and fry as possible not only for increasing the natural stocks and culture but also to provide a great opportunity for yield of valuable food, employment, best use of water and upliftment of montane people. Since yield of fish from warm water resources is declining due to severe pollution, habitat destruction and the like, many countries of the world have resorted to large- scale production of commercially important species of coldwater fish.

Asia is taking the lead in the world accounting for about more than 50 per cent of the total world production of coldwater fish which is expected to be doubled by 2025.

PANEL 10.2
ACTIVITY OF STREAMS AND RIVERS IN MOUNTAIN REGIONS

In general, streams consist of three closely-related activities such as erosion, transportation, and deposition. Stream erosion is the gradual removal of mineral materials from the floor and sides of the channel. Stream transportation consists of movement of the eroded particles dragged over the stream bed, suspended in the body of the stream. Stream deposition is the accumulation on the floor of a standing body of water into which the stream empties. Erosion cannot occur without some transportation taking place, and the transported particles must eventually come to rest. Thus erosion, transportation and deposition are three stages of a single activity.

Erosion of Streams: Streams erode in various ways depending on the nature of the channel materials. The force of the running water not only creates a dragging action on the bed and banks but also causes particles to impact the bed and banks. These two activities (collectively termed as dragging impact) can easily erode alluvial materials such as gravel, sand, silt, and clay. This form of erosion is called *hydraulic action*, can excavate enormous quantities in a short time. The undermining of the bank causes large masses of alluvium to slump into the river, where the particles are segregated and become part of the streams load. This process of bank caving is an important source of sediment.

Stream Transaportation: The solid matter carried by a stream is the *stream load*. It is carried in three forms. Dissolved matter is transported in the form of chemical ions. All streams carry some dissolved ions resulting from mineral alteration. Sand gravel and larger particles move as bed load by sliding or rolling. Silt and clay are carried in suspension. This fraction of the transported matter is the suspended load.

Large rivers such as Ganges and Mississippi, carry as much as 90 per cent of their load in suspension. The Yellow river of China heads the world list in annual suspended sediment load, and its watershed sediment yield is one of the highest known for a large river basin.

Stream sediment has different fractions. However, according to the Agriculture Department of USA, sediment fractions may be classified as follows:

Clay	more than 5 µm
Silt	5 - 50 µm
Very fine sand	50 - 100 µ m
Fine sand	100 - 250 µm
Medium sand	250 - 500 µm
Coarse sand	500 - 1,000 µm
Fine gravel	1, 000- 2,000 µm

Capacity of Stream to Transport Load : The maximum solid load of debris that can be carried by a stream at a given discharge is a measure of the stream capacity. The total solid load includes both the bed load and suspended load.

The capacity of a stream to carry suspended load increases with an increase in the velocity of a stream because the swiftness of the current, the more intense the turbulance. The capacity to move bed load also increases with velocity – the fast water motion produces a stronger dragging force against the bed.

Two Approaches: Autochthonous and Allochthonous: Due to ferocity of the current in hill-streams, rolling and breaking results in weathering of rocks that causes disintegration and decomposition. The rock constituents that are formed are dissolved in water and are transported to down-streams. This phenomenon is called *autochthonous*. During this phenomenon, enormous quantities of minerals are transported by streams. In contrast to this, the term *allochthonous* denotes a deposit of materials which are developed from land washing, erosion, landslides, and debris. These materials drastically affect the fish life by reducing the concentration of plankton and bottom organisms in stream habitats.

Some Asian countries particularly India and China have joined the coldwater fisheries and aquaculture race to increase fish in mountain regions for domestic consumption.

But it is unfortunate that it has even not been possible to solve some of the basic criteria such as water/land allotment for culture and capture fisheries. Most of the farmers consider coldwater aquaculture as subsistence level with limited budget. Development of aquaculture in mountain regions is basically a capital-intensive industry and macro-planning with massive investment.

10.2 Topographical Features of Mountain Regions

The topographical features of mountain regions dictate, to a large extent, the fish production potential in that particular area where coldwater fish culture can be undertaken. Therefore, it is necessary to have a brief idea about the topographical conditions of the region concerned before implementing coldwater fisheries and aquaculture research projects. Asia covers nearly one-third of the total land area of the world and this extensive land mass is fed by more than 100 major rivers and their tributaries with diverse ecosystems. At the same time, this continent has a great diversity of coldwater fish fauna than that of the other continents. Therefore, the discussion on this subject has been more concentrated in this region. The topographical and climatological features of Asia have been summarized in several paragraphs that follows.

The total area of Asia is about 4,44,11,000 sq.km which is nearly one-third of the total land area of the earth. The variety of landforms is infinite in the continent. This continent extends from nearly 25° East longitude in the West to nearly 170° West longitude in the East. From the north, it stretches from nearly 17° North latitude to nearly 10° south latitude.

The topography of Asia has a great variety. The extensive areas are covered with young fold mountains or plateaus (See Panel 10.3) formed by old rocks or plains formed by new alluvial. According to the variety of relief, the landform in Asia can be divided into six parts such as (1) The Central Mountain Ranges and High Plateau Region, (2) The Southern Plateau Region, (3) The Northern Plains, (4) The Southern and Eastern River Plains, (5) The Coastal Plains, and (6) The Islands. In the following paragraphs, we shall discuss about the mountanious regions of Asia.

Nearly in the centre of Asia lies a large mountanious region and in between the mountains, lies some important plateaus. Most of the mountains extend from two centres or Knots termed as (1) *Pamir Knot and* (2) *Armenian Knot.*

Mountains Arise from Pamir Knot

Mountains that extend from the pamir Knot stretch eastwards and Westwards. Southeast of the Pamir extends the Himalayan range. The Himalayas include four parallel east-west extending mountains. Further to North, lies the Tethys Himalaya, to its south Himadri, further south Himachal and finally the Shivaliks. From the eastern corners of the Himalayas, five mountain ranges have extended southwards such as the Patkai, Naga, Lusai Hills, Arkan Yoma, and Pegu Yoma. The last two mountains enter the Mayanmar. Further north to the Himalayas, lies the Karakoram mountain range.

East from the Pamir extends the Kunlun and Altun Shan mountain ranges. These ranges extend into China as Nanasan, Sinlingsan and Khingan mountains. Later the Khingan mountain merge with the Stanovoy mountain in the north-east.

Stretching north-east from the Pamir, extend the Tian Shan, Altai Sayan, Yablonovy, and Stanovoy mountain ranges. West to the Pamir extends the Hindukush mountain range and further West, this range makes a curve and merges with the Armenian Knot.

South-West of the Pamir Knot extends the Sulaiman and Kirthar mountain ranges. Further west, it forms the Zagros mountain which merges with the Armenian Knot.

Mountains Arise from Armenian Knot

This knot is located to the South-east of the Black Sea. To the East of this knot, two mountains are extended such as Zagros and Elburz and to the North-west of the knot also extends two mountains such as Tauras and Pontic. The latter extends along the north of Turkey and the former stretches around the south of Turkey, extending to the shores of the Mediterranean Sea in the west. A brunch of the Hindukush mountain extends between the Caspian and Black seas and is called as *Caucasus mountain*

Rivers Arise from Mountain Regions in Central Asia

Numerous rivers arise from the extensive mountain regions in central Asia. These rivers flow in several directions according to the slope of the land. Some rivers also arise from the plateau of the south but they are comperatively smaller.

PANEL – 10.3
PLATEAUS

It is an elevated tract of relatively flat land, usually limited on at least one side by a steep slope falling abruptly to lower land. It may also be delimited in places by abrupt slopes rising to residual mountains or mountain ranges, as in the Tibetan plateau. The surface of plateaus may, however, be plain-like in quality, very flat rocky or hilly in nature, or they may be so dissected by streams and glaciers that it is very arduous to recognize their original plateau characteristics.

The most important Plateau such as Intermontane plateau and Southern plateau are abridged below.

1. *Intermontane Plateau* : In the central part of Asia, a number of plateaus such as Tibet (between the Himalayas and Kumlum), Siadom (between Altun Shan and Tian Shan mountains), Mongolian (between Yablonory, Altai, Khingan and Altun Shan mountain ranges), Sistan (between the Hindukush mountain in the north and the Sulaiman mountain in the south), and Iran (between Pontic and Tauras mountains), Bolivia, Peru, and the Mexican plateau exist. These plateaus are surrounded by mountains but most of them are desiccated with little vegetation.

2. *Southern Plateau*: In the southern part of Asia, there are three plateaus such as (i) Arabian plateau in the South-West Asia, (ii) Deccan plateau of India: To the west and the east, the plateau lies the Western Ghats and the Eastern Ghats, respectively, which bounds the plateau on both sides, and (iii) Indo-China plateau and Yaman plateau of China. Over the Deccan plateau, however, several rivers such as Mahanadi, Godavari, Krishna, Cauvery, etc. are drained from the West to the East making the slope very distinct. On the other hand, rivers such as Mekong, Salwen, Sikiang, etc. flow over Indo-China and Yaman plateaus. Most of the rivers and their tributaries are rich in fish fauna.

1. **North-Flowing Rivers:** Three main rivers such as Ob, Yeniser, and Lena arise from the central mountain region of Asia. These rivers flow to the north and not able to reach the sea as they are dammed by solid ice. Though these rivers extend long distances due to their frozen mouths, they are not used for agriculture and fisheries. At the same time, regular flood waters collect in several areas to form marsh-lands where fisheries activities may be undertaken to some extent.

2. **East-Flowing Rivers:** The four large rivers of Asia such as Amur, Hwang-ho, Yangste-Kiang, and Si-Kiang flow to the east arise from central mountain area and plateau region. They then travel long distances to flow into the Pacific Ocean. Since these rivers flow over China, the role of these rivers from fisheries point of view is, therefore, tremendous.

3. **South-Flowing Rivers :** A number of important rivers such as Mekong, Menan, Ganges, Indus, Brahmaputra, Tigris, Salween, Irrawaddy, Euphrates, etc. flow over South-Asia and drain into the Indian ocean. Most of the rivers flow through monsoon regions. Though these rivers receive snow-melt waters from high mountains, they are mainly dependent on monsoon rains.

4. **West-Flowing Rivers:** Very few rivers of Asia flow to the west. In India, the Normada and the Tapti rivers flow over the Deccan Plateau to the West draining into the Arabian Sea.

Most of the rivers are fed either by glaciers or rain waters and are abound with many varieties of coldwater fish species, some of which are important from commercial point of view. Most of the species are considered as academic and research purposes.

Climatic Conditions of Asia

Asia has a great variety of landforms and accordingly, also has a variety of climates. Latitude, topography, ocean, wind movement, and rainfall are factors that influence the climate of Asia. During summer, winds carrying plenty of water vapor move from the Indian and Pacific oceans towards the mountain region of central Asia. These winds cause heavy rainfall in North-east Asia as they are blocked by the mountains in the north. The average rain-fall is 110 cm . Since the central Asia is surrounded by mountains, the plateaus in these regions receive the least rainfall, the average being 25 cm.

In winter season due to high pressure over central Asia, cold and dry winds move to the oceans. Most of the areas, therefore, do not receive any rain. In South-east coast of India, North-west India, and Japan receive some rains. In winter, convectional rain occurs in Southern Asia along the equatorial region in Malaysia and Indonesia. The potential of cold- water fisheries in these regions is, therefore, great.

10.3 Characteristic Features of Mountain Regions

Landslide and erosion are the two principal features of mountain regions that makes fisheries and aquaculture activities more difficult. Landslide is a general term for a wide variety of processes and land-forms involving the downslope movement, under-gravity, and masses of soil and rock materials. On the other hand, erosion is the wearing away of soil and rock by mass wasting, weathering, and the action of glaciers, streams, waves, and underground water. However, soil erosion in mountain regions depends on the length and steepness of slopes. The greater the steepness of slope, other condition being equal, the greater the erosion, partly because more water is likely to flow-off but also because of increases in the velocity of water flow.

More than 65 per cent of the world's upland region has a rich and varied water potential from snow-capped mountains down to the plains of different regions. Flowing as a sheet across the surface of high land, running water picks up particles and moves them downslope into a stream channel. When rainfall is heavy, streams and rivers swells lifting large volumes of sediment and carrying them downstreams. In this way, running water erodes mountains and hills, valleys and deposit sediments in different standing water bodies.

Running water is one of four flowing substances that erode, transport and deposit minerals, and organic matter. The other three are waves,glacial ice and wind. These agents carry out the process of denundation. Denundation is the total action of the processes by which the exposed soil of the mountains are worn away and the resulting sediments are transported to the closed inland basins.

These above-mentioned characteristic features drastically alter the habitat condition that has a crucial role in abundance and distribution of coldwater fish. In some cases, the farmed fish live a healthy life; but in most situations when habitat conditions do not favor the existance of fish life, production potential of coldwater ecosystems is hampered. High oxygen content and low carbon di-oxide round the year and high transparency of water are the most important factors to be considered. High oxygen level in water accelerates to a maximum limit the physiological activities of hill stream fishes, below which their vitality significantly reduced, sometimes causing their death. Other habitat conditions include sparse vegetation resulting in the scarcity of food, presence of inorganic constituents in high amounts in bottom soil and low fertility of water and soil.

10.4 Characteristic Features of Coldwater Streams

Streams of mountain regions may be divided into three types such as foothill, mountain and high mountain. These types of streams have considerable variations in the range of their important factors such as water temperature, dissolved oxygen, biota, velocity of water (Table 10.1).

Physico-chemical Features

In mountain regions, glacier- and spring-fed streams are very characteristic in nature. This is due to the fact that there are marked differences in the physico-chemical features between this two types of streams. However, the upper parts of glacier-fed streams are characterized by very high turbidity (10-650 mg/l), dissolved oxygen content (4-14 mg/l), low phytoplanktonic and benthic community and their diversity, blustering current (3-6 m/second), steep gradient, stony substratum, and lack of nutrients. On the other hand, spring-fed streams have comperatively low water current velocity (0.5-2.0 m/second), high temperature range (12-29°C), total hardness (50-115 mg/l), total alkalinity (40-100 mg/l), and low dissolved oxygen (4-10 mg/l).

Macrozoobenthos

The population of macrozoobenthos in spring-fed streams is relatively higher than that of glacier-fed ones. Due to steep gradient and turbulent current in glacier-fed streams, most of the macrozoobenthos are washed out from the stream bed. They could no longer persist in streams for prolonged period of time. However, the dominant forms of macrozoobenthos include caddis flies, stone flies, may flies, water bugs, beetles, dragon flies, two-winged flies, bivalves, snails, segmented worms and water mites. In general, smaller streams are used as potential breeding grounds for many coldwater fishes. These streams have diverse and rich benthic communities (3,400 ± 300 individuals/m^2).

Table 10.1: Physico-chemical and Biological Features of Three Main Types of Himalayan Streams

Feature	Stream type		
	High mountain	Mountain	Foothill
Elevation (m)	1,470 and above	875 - 1,470	875 and above
Current speed	Fast	Moderate	Slow
Substratum	Boulders and rocks either pitted or smooth	Boulders and stones covered with slimy algal matter and sand patches	Pitted rocks and stones with periphytic and filamentous algae
Water Temperature (ºC)	7-18.5	7.2-20.8	18.5-28.0
Dissolved oxygen (mg/l)	10.0-12.0	8.0-11.0	7.5-8.5
Total alkalinity (mg/l)	60-90	70-95	95-120
Dipteran larvae (per cent)	3.0	13.0	9.0
Beetle larvae (per cent)	4.0	8.0	8.0
Caddish fly larvae (per cent)	31.0	37.0	25.0
Stone fly nymph (per cent)	7.0	4.0	2.0
May fly nymph (per cent)	37.0	32.0	26.0
Dragon fly nymph (per cent)	1.0	3.0	20.0

Source: Jhingran and sehgal (1978), Sehgal (1983)

10.5 Characteristic Features of Coldwater Lakes

In contrast to coldwater streams, the lakes contain high concentration of dissolved oxygen (6-20 mg/l) and rich in nutrients. Different groups of phytoplankton are dominated by chlorophyceae, bacillariophyceae, cyanophyceae, dinophyceae, cryptophyceae, and chrysophyceae. On the basis of altitude, three types of lakes have been recognized such as mountain lake, valley lake, and forest lake. The last two types of lakes are rich in nutrients and are teemed with plankton and fish. In mountain lakes, fish populations are sparse.

Most of the high altitude lakes are ultra-oligotrophic or oligotrophic, although it has been found that some high altitude lakes such as Ooty lake in Nilgiri hills, Kodaikanal lake in Palani hills, Nainital lake, and most of the valley lakes of Kashmir are being transformed into or have become eutrophic due to human settlement around these lakes.

10.6 Characteristic Features of Coldwater Fish

1. The coldwater fish exhibit great power of locomotion while Glyptothoracine catfishes and mahseer are inhabitants of the lower reaches of the hill streams, the trouts and Schizothoracids find their way furthest up the high streams.

2. Most of the coldwater fish possess structures adapted for burrowing, clinging or otherwise to withstand fast water currents.

3. Due to ferocity of the current in hill streams, certain fishes are encountered only at their substrates and such fish have become limpet-shaped such as *Balotora, Glytosternum* etc.) or have developed modified organs (*Garra, Glyptothorax* etc.) which enable them to attach with rocks.

4. Coldwater fish have acquired some modifications of the mouth suitable for reaping encrusted organisms and removing algae from the emergent or submerged rocks and boulders. The modification of the lips is well seen in snow trout, mahseer, minor carps, and catfishes. These fishes are not well adapted for feeding in muddy and deep waters.

5. Adaptation to live fish in highly oxygenated water is/are available in torrential streams of the mountains. Such ecosystem has induced structural adaptations in the respiratory organs. The gill openings have narrowed and gills are greatly reduced so much so that fishes cannot live for prolonged period in low-oxygenated water.

6. Coldwater fishes may be distinguished as (i) eurythermal (having a broad temperature tolerance range) such as common carp, *Barilius benedelisis, Chedra,* and *Schizothorax plangiostomus* and (ii) stenothermal (having a narrow temperature tolerance range) such as *Salmo trutta ferio* (brown trout) and *Salvelinus fontinalis* (easten brook trout). Possibly all coldwater fish are endowed with a faculty to tolerate temperatures nearly upto the freezing point of water. The upper and lower temperature tolerance definitely limits growth, survival and distribution of fish in space and time.

7. The common features of the fish fauna in high mountains are their low species diversity, endemism, and a number of morphological as well as anatomical adaptations which helps fish to tolerate the harsh conditions prevailing at high altitudes.

10.7 Coldwater Fish Species

Though more than 1,000 coldwater fish species are distributed around the world, all the species do not form fishery of considerable magnitude. Based on the percentage representation of the catch, the following species of fish have been recognized in different upland rivers, streams, lakes, and reservoirs of India. Most of these water bodies are fished for subsistence and some species are considered as sport as well as food fish. Some other coldwater fish which are distributed in high altitude lakes of Asia have also been noted elsewhere in this chapter. This continent is, however, supported by about 750 fish species (both indigenous and introduced), notably the family Cyprinidae and some other families. Some common coldwater fish species are shown in Figure 10.1.

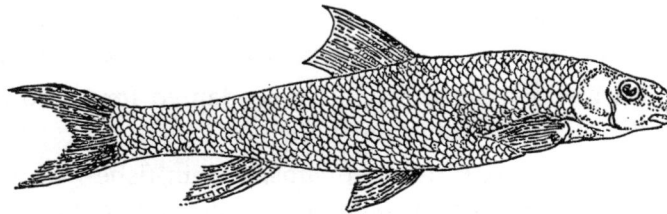

1. *Oreinus richardsoni* (Gray)

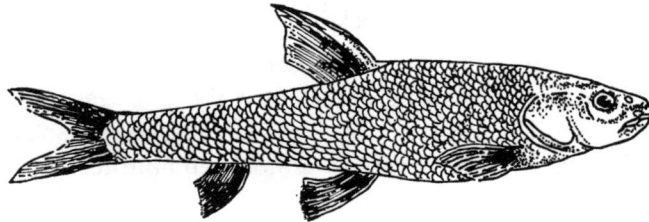

2. *Schizothorax niger* (Heckel)

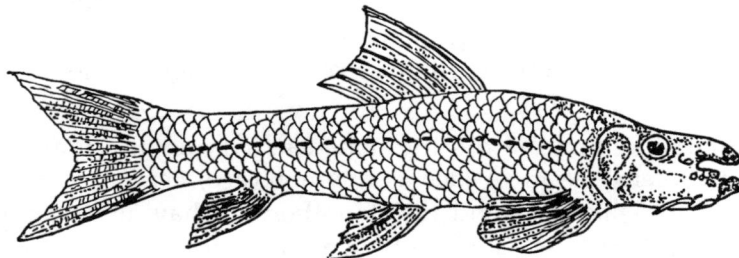

3. *Garra gotyla gotyla* (Gray)

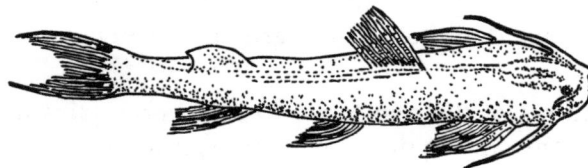

4. *Glyptothorax pectinopterus* (McClelland)

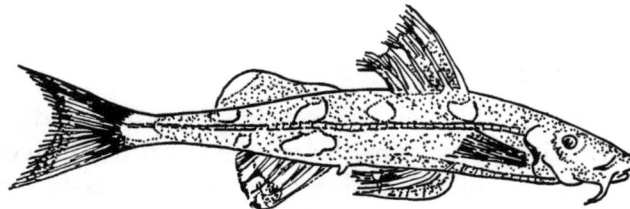

5. *Pseudecheneis suleatus* (McClelland)

Fig. 10.1: Some common coldwater fish species.

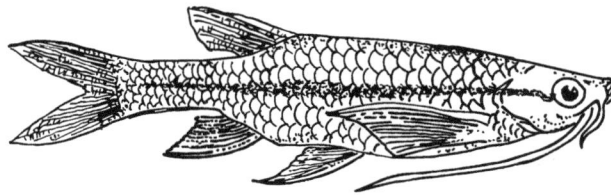

6. *Esomus dandricus* (Hamilton)

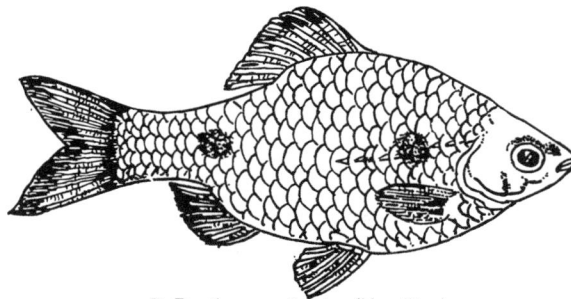

7. *Puntius conchouius* (Hamilton)

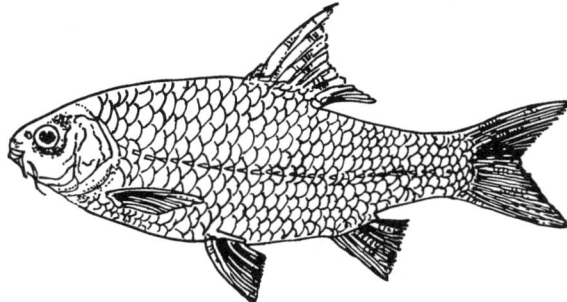

8. *Puntius sarana sarana* (Hamilton)

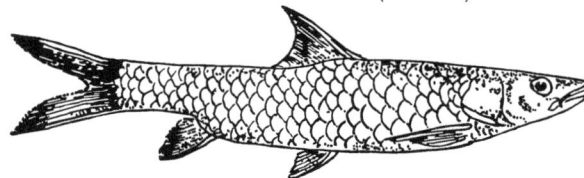

9. *Tor putitora* (Hamilton)

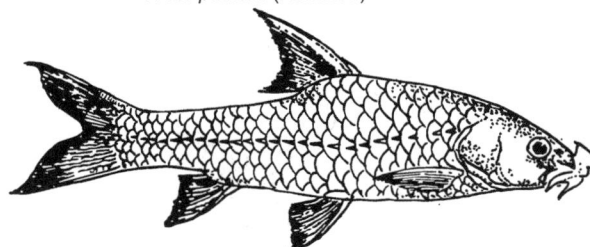

10. *Tor tor* (Hamilton)

Fig. 10.1 – *contd...*

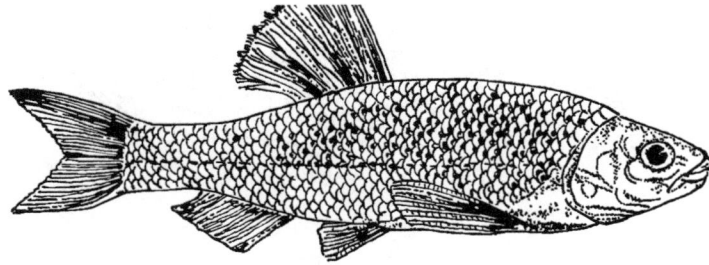

11. *Barilius bendelisis* (Hamilton)

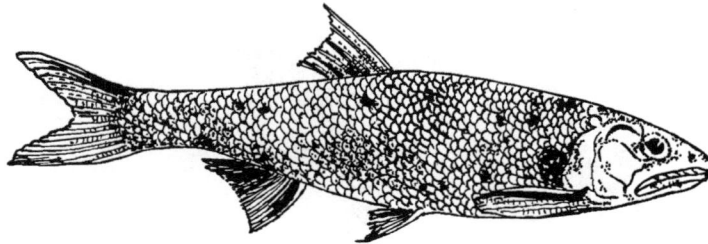

12. *Barilius bola* (Hamilton)

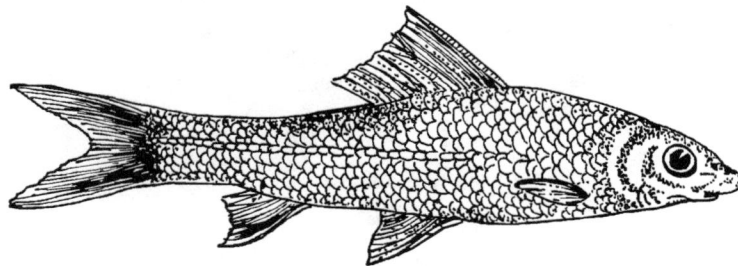

13. *Labeo dero* (Hamilton)

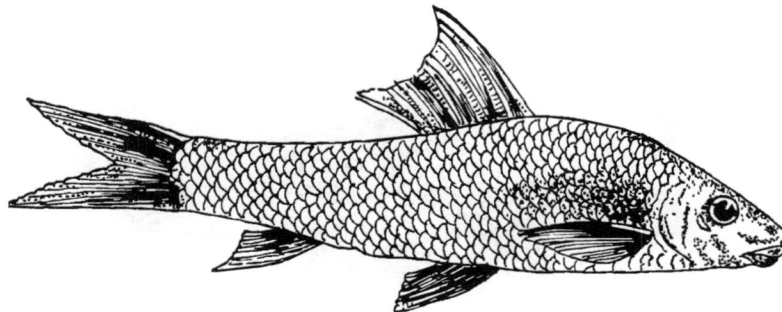

14. *Labeo pangusia* (Hamilton)

Fig. 10.1 – *contd...*

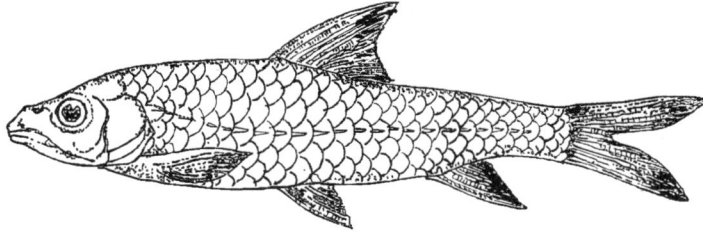

15. *Tor mosal* (Hamilton)

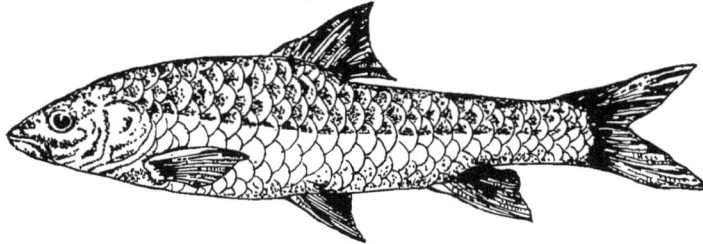

16. *Tor khudree* (Hamilton)

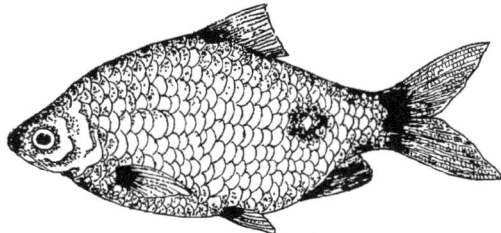

17. *Puntius javanicus* (Hamilton)

18. *Noemacheilus poonaensis* (Menon)

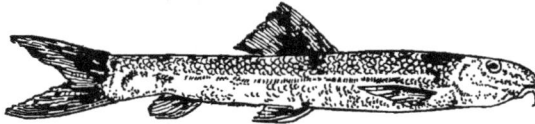

19. *Noemacheilus anguilla* (Annandale)

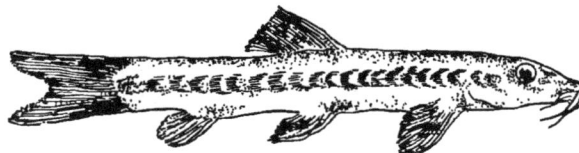

20. *Noemacheilus monilis* (Aora)

Fig. 10.1 – *contd...*

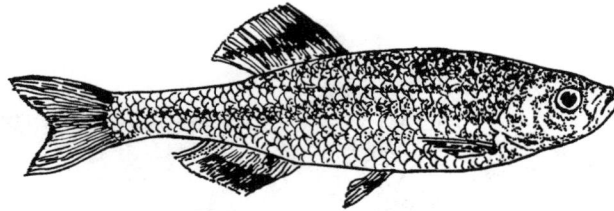

21. *Danio aequipinnatus* (McClelland)

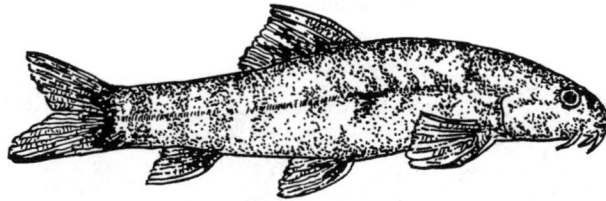

22. *Noemacheilus horai* (Menon)

23. *Noemacheilus kangrae* (Menon)

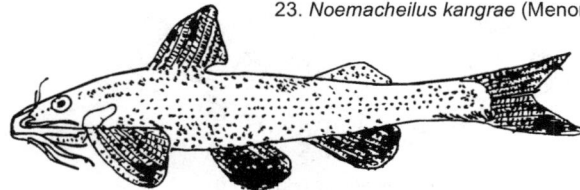

24. *Glyptothorax trilineatus* (McClelland)

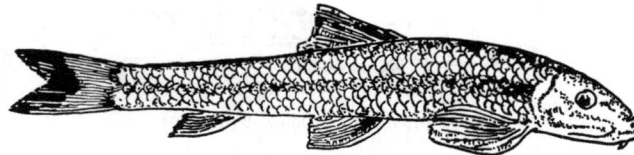

25. *Garra hughi* (Silas)

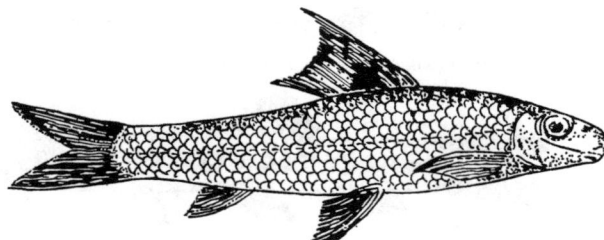

26. *Crossocheilus latius* (Hamilton)

Fig. 10.1 – *contd...*

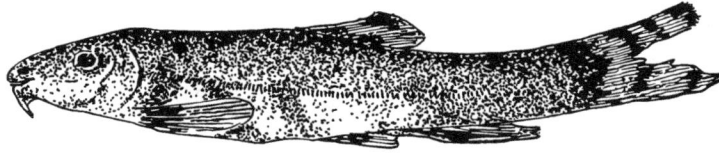

27. *Botia almorhae* (Gray-Young)

28. *Lepidocephalus irrorata* (Hora)

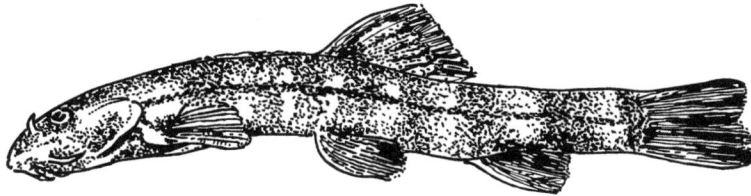

29. *Noemacheilus kangjupkhulensis* (Hora)

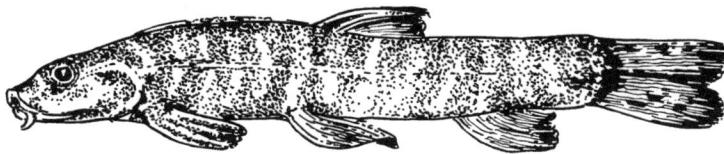

30. *Noemacheilus manipurensis*

31. *Salmo qairdnerii* (Richardson)

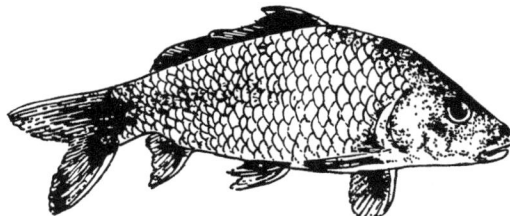

32. *Cyprinus carpio* (Linnaeus)

Fig. 10.1 – *contd...*

A. Tor and related species (Order: Cypriniformes, Family: Cyprinidae)

1. *Tor putitora* (Ham.) (Golden mahseer) 2. *T. tor* (Ham.) (Deep-bodied mahseer) 3. *T. mosal* (Ham.) (Copper mahseer) 4. *T. khudree* (Sykes) (Decan mahseer) 5. *T. progeneius* (Mc Clelland) 6. *T. mussullah* (Ham.) (High-backed mahseer) 7. *Neolissochilus hexagonolepis* (Mc Clelland) (Chocolate mahseer) 8. *N. crynaadensis* (Day)

B. Snow-trout and related species (Family: Cyprinidae)

1. *Oreinus richardsoni* (Gray) 2. *O. kumaonensis* (Menon) 3. *Schizothorax currifrons* (Heckel) 4. *S. esocinus* (Hackel) 5. *S. hugelii* (Hackel) 6. *S. longinnis* (Hackel) 7. *S. nasus* (Hackel) 8. *S. niger* (Hackel) 9. *Diptychus maculatus* (Steindacher) 10. *Lepidopygopsis typus* (Raj.) 11. *Pychobarbus conirostris* (Steindacher) 12. *Schizopygopsis stoliczkae* (Steindacher) 13. *Racoma labiatus* (Mc Clelland) 14. *R. progasta* (Mc Clelland)

C. Carp species (Family: Cyprinidae)

1. *Labeo dero* (Ham.) 2. *L. bata* (Ham.) 3. *L. calbasu* (Ham.) 4. *L. pangusia* (Ham.) 5. *L. angra* (Ham.)

D. *Barilius* and related species (Family: Cyprinidae)

1. *Barilius benelisis* (Ham.) 2. *B. bola* (Ham.) 3. *B. canarensis* (Jerdon) 4. *B. shacra* (Ham.) 5.*B. tileo* (Ham.) 6 *B. vagra* (Ham.) 7. *Amblypharyngodon melettius* (Valenciennes) 8. *Danio aequipinnatus* (Mc Clelland) 9. *D. dongila* (Ham.) 10. *D. devario* (Ham.) 11. *D. naganensis* (Chaudhuri) 12. *D. neilgherriensis* (Day) 13. *Esomus barbatus* (Jerdon) 14. *E. dandricus* (Ham.) 15. *Raiamas bola* (Ham.) (Indian trout) 16. *R. guttatus* (Day) 17.*Rasbora rasbora* (Ham.)

E. *Puntius* and related species (Family: Cyprinidae)

1. *Puntius chola* (Ham.) 2. *P. arelius* (Jerdon) 3. *P ticto* (Ham.) 4. *P. conchonius* (Ham.) 5. *P. dorsalis* (Jerdon) 6. *P. filamentosus* (Valenciennes) 7. *P maheoda* (Val.) 8. *P. sophore* (Ham.) 9.*P. wageni* (Day) 10.*P. hexastichus* (Mc Clelland)

F. *Glyptothorax* and related species (Fam : Sisoridae)

1. *Glyptothorax reticulatum* (Mc Clelland) 2. *G. pectinopterus* (Mc Cle.) 3. *G. conirostris* (steindachner) 4. *G. cavia* (Ham.) 5. *G. modraspatanam* (Day) 6. *G. telchitta* (Ham.) 7. *G.gracile* (Gunther) 8. *G. arnandalei* (Hora) 9. *G. indica* (Talwar and Jhingran) 10. *G. kashmirensis* (Hora) 11. *G. lonath* (Sykes) 12. *G. platypgonoides* (Bleeker) 13 *G. prashadi* (Mukherjee) 14. *G. stoliczkae* (Steindachner) 15. *Erethistes pussilus* (Muller and Trochel) 16. *Gagata cenia* (Ham.) 17.*G. gagata* (Ham.) 18. *G. sexualis* (Tilak) 19. *Laguvia ribeiroi* (Hora) 20. *Sisor rhabdophorus* (Ham.) 21. *Pseudecheneis sulcatus* (Mc Cle.)

G. *Garra* and related species (Family: Cyprinidae)

1. *Garra gotyla gotyla* (Gray) 2. *G. annandole* (Hora) 3.*G. bicornuta* (Rao) 4.*G. gravelyi* (Annandale) 5.*G. hughi* (Silas) 6. *G. lamta* (Ham.) 7. *G. kempi* (Ham.) 8. *G. mcclellandi* (Jerdon) 9. *G. mullya* (Sykes) 10. *G. naganensis* (Hora)

H. *Crossocheilus* and related species (Family : Cyprinidae)

1. *Crossocheilus latius latius* (Ham) 2. *C. latius diplocheilus* (Heckel) 3. *C. latius punjabensis* (Mukherjee).

I. *Schizothorax* (Snow trout) (Family: Cyprinidae)

More than 30 species of snow trout have been reported from different regions of Asia especially Indian sub-continent, China, Central Asia, Mayanmar, Kazakhastan, Pakistan, Afganistan and others. Schizothoracines are highly valued fish, most of them are important sport and food fish. Some major commercial fish species include *Schizothorax niger. S. curvifrons, S. longipinnis, S. micropogon, S. plagiostonus S. richardsoni* etc.

J. *Noemacheilus* spp. (Family: Balitoridae)

More than 60 species have been reported from different small spring-fed hill streams in many regions of Asia.

K. *Balitora, Homaloptera* and related species (Family: Balitoridae)

1. *B. brucei* (Gray) 2. *B. mysorensis* (Hora) 3. *Bhavania australis* (Jerdon) 4. *H. bilineata* (Blyth) 5. *H. rupicola* (Prasad and Mukherjee) 6. *H. montana* (Herri)

L. *Botia* spp. (Family: Cobitidae)

1. *B. almorhae* (Gray) 2. *B. berdmorei* (Blyth) 3. *B. dario* (Ham.) 4. *B.* birdi (Chaudhuri) 5. *B. histrionicus* (Blyth) 6. *B. straitus* (Rao)

M. *Amblyceps* spp. (Family: Amblycipitidae)

1. *A. mangois* (Ham) 2. *A. laticeps* (Mc Cle.)

N. *Mystus* and related species (Family: Bagaridae)

1. *M. cavasius* (Ham.) 2. *M. bleekeri* (Day) 3. *Aorichthys aor* (Ham.) 4. *A. seenghala* (Sykes) 5. *Batasio batasio* (Ham.) 6. *B. travancoria* (Hora and Law)

O. *Channa* spp (Family: Channidae)

1. *C. gachua* (Ham.) 2. *C. orientalis* (Bloch and Schnelder) 3. *C. punctatus* (Bloch)

These species have been reported in foot-hill streams.

P. *Mastacembelus* spp. (Family: Mastacembelidae)

Among different species, only two species such as *A. pancalus* (Lacepede) and *M. armatus* (Lacepede) are commercially important and have been reported both in turbid hill-streams of uplands and foot-hills.

Q. Common carp (Family: Cyprinidae)

1. *Cyprinus carpio var. communis* (Linnaeus) 2. *C. carpio* var. *nudus* (Linn.) 3. *C. carpio* var. *specularis* (Linn.)

R. *Xenontodon* and related species

1. *X. cancila* (Ham.) 2. *Esox cancila* (Ham.) 3. *Belone cancila* (Ham.)

S. *Lepidocephalus* and related species

1. *L. guntea* (Ham.) 2. *L. annandalei* (Choudhuri) 3. *L. berdmorei* (Blyth) 4. *L. irrorata* (Hora) 5. *L. menoni* (Pillai and Yazdani) 6. *L. thermalis* (Valencinnes) 7. *Pangio pangio* (Ham.)

8. *P. longipinnis* (Menon) 9. *Somileptes gongota* (Ham.) 10. *Neoencirrhichthys maydelli* (Banarescu and Nalbant).

T. Trout and Salmon (Family: Salmonidae)

Trouts are usually found in the temperate and artic regions. Trouts have drawn serious attention to anglers, ichthyologists, and fish culturists because of their commercial value, delightful taste and attractiveness. They are, represented by fourteen species distributed all over the world. The trouts comprise six main species as noted below:

1. Brown trout, *salmo trutta trutta* (Linnaeus)

2. Rainbow trout, *S. gairdnerii gairdnerii* (Richardson)

3. Cutthroad trout, *S. clarkii* (Richardson)

4. Eastern brook trout, *Salvelinus fontinalis* (Mitchell)

5. Lake trout, *Salvelinus namaycusk* (Walbaum)

6. Dolly Varden trout, *S. malma* (Walbaum)

Trout populations are inhabited both in freshwater and seas. Differences in the body form and color of the marine and freshwater habitats are very distinct. Since these fishes have been transported from their native place to many countries of the world, they have now been widely distributed and are intensively cultured on commercial basis with artificial feeding. The following species have been successfully introduced to the Indian mountain and hill streams.

1. Rainbow trout 2. Brown trout 3. Eastern brook trout 4. Golden rainbow and tiger trout (A hybrid between female brown trout and male brook trout) 5. Land-locked variety of Atlantic salmon, *Salmo salar.*

Among the salmons, two species are European in origin such as Atlantic salmon, *Salmo salar* (Linnaeus) and Danube salmon, *Hucho hucho* (Linn.). American salmons are represented by the following six species.

1. Pink salmon, *Onocorhynchus gorbuscha* (Walbaum)

2. Chum salmon, *O. Keta* (walbaum)

3. Coho salmon, *O. kisutch* (Walbaum)

4. Masu salmon, *O. masou* (Brevoorf)

5. Sock-eye salmon, *O.* nerka (Walbaum)

6. Chinook salmon, *O. tshawytschu* (Walbaum)

It is not possible in one chapter to do justice to the variety of both coldwater and warm water fish species; only a few representatives of coldwater fish have been somewhat summarily noted. For more detailed information about this aspect, the readers should consult specialized books dealing with the subject concerned. There is, unfortunately, no single book that provides a more extended survey on the entire subject.

10.7 Impact of Man's Activities on Coldwater Fish Culture

In the history of man's intervention on coldwater resources, there are now few regions in the world uninjured by man-made environmental changes. About 95 per cent of rivers across the world are considered as afflictively altered. Even the'Virgin' area of the polar regions or tropical Africa, and the United States are subject to such phenomena as acid rain, deposition of pesticides and the like-all of these are derived from man's activities.

Types of Effect

There are two basic types of impact under the general idea of water use : (1) Indirect impacts and (2) Direct impacts.

1. *Indirect Impacts :*

   (i) *Agricultural Activities :*

     * *Nutrients* : The fishery may be benefitted when increased concentrations of nutrients lead to increased levels of production, although shifts in species composition may follow sustained eutrophication. In cases where nutrient load is high enough, primary production as well as macrophytes may lead to reduced production and loss of fish throughout the periods of lethally low dissolved oxygen levels.

     * *Suspended Solids* : Turbidity due to the carriage of silt in the water mass may obstruct light from the water column. This will reduce the ability of the plant to yield food through photosynthesis and thus lower the levels of primary production. Silt may also create provision that will cause and possibly death of fish. The settlement of solids in areas where water flow decreases will affect the physical condition of the stream bed configuration. This may change the water courses and obliterate spawning and feeding areas to the detrimental of fish production.

     * *Toxic Materials* :Pesticides can be leached from the soils or washed off vegetation and thus pass into the drainage waters and kill or reduce fish life.

   (ii) *Civil Construction Work*: Construction of dams, bridges, roads, and their associated earthworks can have devastating effects on the natural aquatic ecosystems of the area concerned. The basic problem lies in the disturbance due to the movement of solid materials in large amounts and the exposure of soils to erosion and dispersal by waters. This will be not unimportant at times of intense rainfall and can lead to drastic changes in the topography of the swelling waters and the condition of the substrate. The debris (from fine silt to pieces of rock) may block waterways and drainage channels. Near the construction site, there may be devoid of aquatic life.

   (iii) *Industries*: Industries has long been considered that the water courses are the cheap and most convenient way to get rid of unwanted solid and liquid. In the past, but to a lesser extent now, this has lead to the total obliteration of fish life in many water bodies due to constant discharge of toxic chemicals from the industrial activities close to them.

It should be kept in mind that some discharges have concealed effects. It requires a small discharge of some highly toxic substances to move down a water course to have a momentous effect on fish life.

(iv) *Urbanization* : Water transportation has always been a traditional way of eliminating human sewage wastes from many urban areas where the rainfall is exuberant. These waters discharge high organic loads but usually with little toxic chemicals. Development of residential areas in thickly populated mountain regions also discharge large quantities of nutrient- rich water from washing and cooking.

In addition to the domestic waste, water discharge and urban built-up areas have a rainfall run-off loads and paved surfaces. This run-off may contain chemicals, especially heavy metals and organic micro-pollutants in suspension or solution from the wash-off of road surfaces.

(v) *Deforestation*: Unparalleled increases in human populations have placed great strain on deforestation in many upland areas of the tropics. The soil cover on steep slopes is generally fairly thin but support the growth of vegetation. If the vegetation is destroid for agriculture or other developmental activities, the soil will no longer be protected by a leafy canopy from the driving rain. As a consequence, some of the soils are washed away into the ponds, rivers, streams, and reservoirs which tends to become chocked with silt.

(vi) *Acid Rain*: Acid rain is the precipitation that is acidified by sulfuric and nitric pollutants. However, it is a serious global environmental threat produced by natural forces and, to a greater extent by humans. Rain water is slightly acidic, but acid rain is more acidic because of the presence of sulfuric and nitric acids. Winds can carry the pollutants that produce acid rain over great distances, causing widespread acidification of ponds, lakes, and lands and consequently killing many species of fish and tree.

The impact of acid rain on aquatic environment depends on the altitude, latitude, and the chemical composition of the soil as well as the direction of the prevailing storm track. At lower altitudes, the effects of acid rain on streams and lakes are less severe. The water flow-off has the capacity to buffer by soil, because there is wide variety of rivers and streams at lower altitudes than in mountainous regions.

The effects of fish population have been observed in many lakes and streams. Fish have, for example, disappeared from the upland lakes and streams of the Adirondack mountains of eastern New York State. Fish kills were observed particularly in the spring when the snow melted; the acid precipitation that collected in it over winter months was released rapidly into the streams and rivers, suddenly lowering the pH. At a pH of 4.5 fish begin to disappear.

Similar pH drops and spring fish kills have been noted in southern Scandinavian mountain regions where storms laden with pollutants move from Germany and Britain. As the winds reach the Norwegian mountains, the region is suppressed with snow and acid rain. Out of 265 lakes in Norwegian basin, only 91 contain fish.

Acid rain disrupts the ecological balance on aquatic ecosystems. High concentrations of hydorgen ion can have a direct toxic effects on aquatic life. The common problem associated with low pH is the impaired ability to balance sodium, chlorine and hydrogen. Ion regulation problems have been documented for fish and benthic invertebrates. Many species of fish feed on benthic invertebrates for at least part of their life history. Declined in species richness of benthic invertebrates with reduced pH in many upland lakes have been noted.

2. Direct Impacts :

(i) *Dams* : Because most of the large rivers of steep gradient do not have falls, dams are essential to establish the vertical drop required to spin the turbine of electric power generators. Globally, dams fragment the streams, leaving less than 20 per cent of the length of rivers in a natural condition. Although dams offer important services such as hydro-electric power, flood protection, and water storage for agriculture as well as aquaculture, a growing body of experts indicates that these structures have also lead to widespread degradation of riverine fisheries.

Dams and the resultant man-made lakes generate a complex web of impacts which affect biological and physical components of the ecosystem. The principal effects of these activities include the following:

* Obstructin to upstream migration of brood fishes

* Loss of fry during passage through or over dams

* Gas supersaturation

* Delay in migration for spawning

* Reduction in survival of egg/ fry to gravals

As a consequence of dam construction, the alteration in the ecosystem will inevitably create a change in the dominance of coldwater fish and their food organisms. The following examples will indicate how fish stocks disappear from natural resources due to dam construction: salmon population disappeared due to dam construction from the Dordogne River, France, soon after the first dam were built on the lower reaches between 1840 and 1905. Similarly, the stock of mahseer and snow trout and their fisheries in the Himalayan rivers (such as Yamuna, Ganga, Alakananda, and Bhagirathi), India, have suffered considerably from constructional activities of quite a number of projects concerned. Extinctions have often been associated with the accidental or deliberate introduction of exotic species aimed at enhancing fish populations of newly established aquatic environments.

(ii) *Road Construction*: Construction of road projects in mountain and hill regions has resulted in considerable damage to the river and stream ecosystems. Blasting of rock structures for road construction and dumping of spoils lead to the movement of hugh amounts of debris downward in the river. According to one estimate, the construction of one kilometre road requires removal of about 40,000 - 80,000 cubic metres of debris in addition to the increase of sediments in waters.

(iii) *Agricultural Activities*: Recent increases in agricultural activities in mountain regions have, to some or large extent, aggravated the freshwater ecosystem generally. Most of the agricultural lands are situated near or on the bank of rivers and streams from which agricultural washes along with sediments are drained into the river systems. The terraced nature of agricultural lands on the mountain slope encourage rapid erosion of the top soil.

(iv) *Dumping of Sewage* : Rapidly growing populations coupled with the tourism development have caused an increase in the organic load of sewage. Since there is no proper treatment facilities of raw sewage in mountain regions, sewage dumping definitely causes physical, chemical, and biological deterioration of the river water.

10.8 Integrated Watershed Management is Inevitable for Fish Culture

As a generalization, watershed regions at high altitudes cover about 55 per cent of the total surface area of the earth. In India, mountainous region accounts for about 24 million hectares of the total geographical area of the world. Integrated watershed management in most of the Asian countries is, however, an age-old practice. During 1990s, special emphasis has been given towards various aspects of watershed development such as agro-forestry, agri-horticulture, creation of operational research, and treatment of flood prone rivers and streams in hill and mountain regions. In most countries of the world, different watershed management programs have been brought under a single umbrella by constructing different rural development projects, national watershed development project and international funding agencies under the Ministry of Agriculture. Recently, some advanced technologies have been adopted for watershed management programs such as data-base creation and compilation, socio-economic evaluation, watershed hydraulics, agro-farming systems, planning of aquaculture and fisheries, remote sensing and computer modelling.

Research on Watershed Management : Indian Context

During 1980s, research thrusts on watershed development in India were based principally on bio-physical and agronomical features such as soil loss, reduced run-off and increased crop yield. Thereafter, advances in aquaculture and fisheries sectors and their importance on human nutrition across the world have changed the entire watershed management scenario. The importance of aquaculture and fisheries in watershed regions has recently been realized by establishing research organizations on planning, designing and methods of implementation of water harvesting technologies in mountain and hill regions to implement capture and culture fisheries projects. The fastidious insufficiency of research inputs for the development of different sectors concerned through watershed management was perceived by the Indian Council of Agricultural Research (ICAR) and accordingly, Central Soil and Water Conservation Research and Training Institute (CSWCRTI), Dehradun (India) has been established in 1974 to sustain integrated management along with fisheries by watersheds or catchment basins. This institute is entrusted with the objective of integrating fisheries and aquaculture particularly in mountain regions of the country. At the same time, the National Research Centre on Coldwater Fisheries (NRCCWF) has been set up to explore the potential resources of mountain rivers and streams in the country.

10.9 What is Watershed Management*

Watershed management can be defined as the judicious use of water and land resources for sustained production with minimum hazards to natural resources. It is essentially a conservation phenomenon of mountain water and soil for aquaculture and agriculture activities. Naturally created or newly constructed water bodies in watershed areas can be profitably used for fish culture.

Huge quantities of water are available on the mountain region each year for possible harvest and conservation in ponds, lakes and small reservoirs. Conservation of water for fish culture coupled with crop production, horticulture, animal husbandry , and poulty can be effectively executed in well-managed mountain ecosystems (Figure 10.2)

10.10 Water Harvesting in Mountain Regions For Fish Culture

For harvesting flow-off water in mountain regions, different land management measures such as contour bunding and trenching, half moon terraces, bench terraces, land levelling and grassed waterways are adopted. These measures will improve the soil profile and, in turn, enrich the ground water table in valley and foot-hill areas. For harvesting water, ponds are constructed in areas whecre water can be trapped on gravety flow in the valley. These ponds may be either seasonal or perennial depending on the water collection capacity of the watershed. According to an estimate, however, about 25 per cent of the rainfall can be collected into one hectare pond from every 10 hectare area of a watershed.

Fish culture ponds that are located in the foot-hill of a watershed may generally receive 75.3, 22.4, and 2.3 per cent water from sub-surface flow, direct rain, and surface flow, respectively. In highly porous soils, ponds can be lined with low density polyethylene flim, clay materials or bentonite. These methods help to reduce the loss of water through seepage. Through faming system research programs, however, more than 90 per cent rain water can be harvested for the purpose.

---

* For further study, see Ghosh (2002).

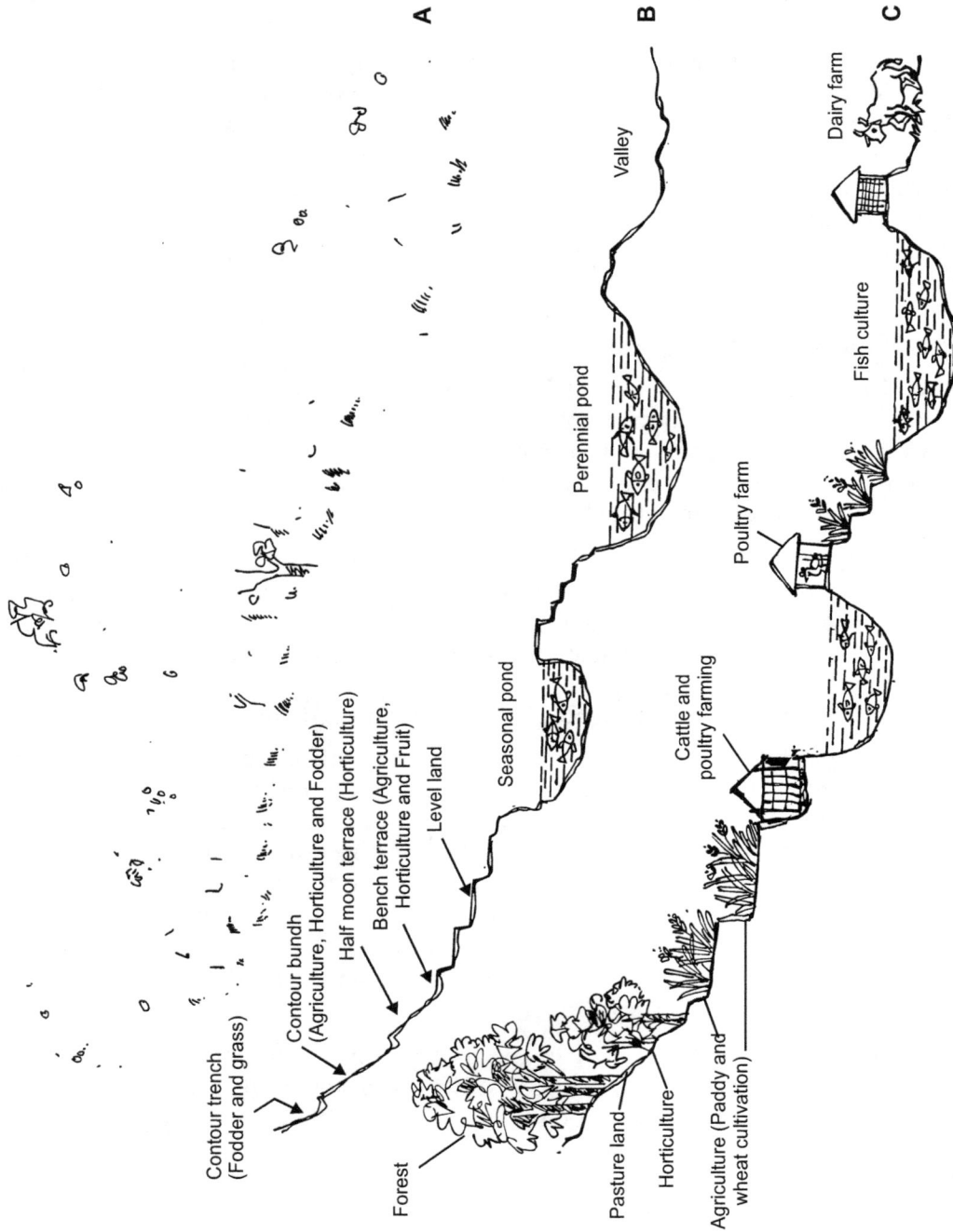

A

Contour trench
(Fodder and grass)

Contour bundh
(Agriculture, Horticulture and Fodder)

Half moon terrace (Horticulture)

Bench terrace (Agriculture,
Horticulture and Fruit)

Level land

Seasonal pond

B

Valley

Perennial pond

Forest

Pasture land

Horticulture

Agriculture (Paddy and
wheat cultivation)

Cattle and
poultry farming

Poultry farm

Fish culture

C

Dairy farm

Fig. 10.2 : (A) An unmanaged watershed; (B) After soil and water conservation measure; (C) Integrated with plants, Cattle, poultry, fruit, padi and fish. Though implementation of this strategy involves heavy investment and skilled technicians, farmers' community will definitely receive high profits through such an integrated fish farming sytem.

10.11 Integrated Faming in Watershed Areas

Integrated farming system is essentially the practice of different faming components including animals, plants, and fish by recycling waste substances from one component to other to enhance production potential with as minimum inputs as possible.

Advantages of the System

1. The waste materials from different farming components can be utilized for fish culture. While the silt and water are used for crop production, a substantial amount of crop by-products can be used to produce food for poultry, fish, and livestock. Animal manures can be used for fertilizing fish ponds and crop yield.

2. The system is likely to offer a potential source of employment and revenue to lower income populations without damaging the environment.

3. In an intergrated faming system, the upper one-third area of the hill should be utilized for forest and silviculture. The middle portion is used for horticulture on the half moon terraces and contour bunds. The lower one-third region is suited for horticulture and agriculture on the bench terraces and levelled lands. Below these areas, livestock and poultry farming systems can be adopted.

Management

Research has demonstrated that suitable management of watersheds for integrated farming is a critical factor in mountain regions not only for fish production but also for sustaining productivity. Even though water and soil of the mountains differ from those of plains considerably, some plain-zone management principles are quite pertinent for fish culture in mountain regions.

Activities of aquaculture developnent have stressed the critical need for the characterization of waters and soils. The pollution status, down-stream flooding and landslides can well determine the probable success of fisheries projects in hill and mountain regions. Lack of productivity can be precisely identified, as can be the virtual for run-off and erosion problems. The time and endeavour being dedicated to ecosystem characterization are not significant compared to the demand for this kind of knowledge.

Promising Alternatives

Soils in upland areas must have cover with plants to check erosion. For this purpose agricultural development activities must be encouraged. If erosion occurs, the potential freshwater areas are subject to siltation, making fish culture virtually impossible. This turn of result may be checked by land-use planning. Treatment of watershed development from valley to ridge with bioengineering estimates will afford water resources and harvesting for agriculture and fisheries. Harvesting of water and its recycling for better use should be comprehended as one of the effective means of encouraging sustainable integrated farming in mountain regions.

Efforts are being made to develop viable devices for integrated farming systems that would permit fish production. Some development activities have shown that modest impacts of nutrient carriers, reclamation of problematic soils and low productive waters help sustain fish yield in mountain regions. Use of organic manures along with better means of management may be effective in stabilizing production. And in some regions of upland areas, farmers near the small streams have adopted integrated farming schemes and are pleased with its greater stability.

The next alternative closely related to the development of aquaculture is the establishment of coldwater hatcheries. Except a few experimental hatcheries such as State Fisheries Departments at Chamoli, Bhawali, Bhimtal, Dehradun, Kalyani, Pithoragarh in Uttar Pradesh and Pathlikool in Himachal Pradesh, at Avalanchi (Nilgiri Hills) in Tamil Nadu and a few others in rest of the Himalayas, the existing seed availability is not adequate to meet the requirements of stocking materials in reservoirs, lakes, and rivers.

10.12 Problems of Coldwater Fisheries

Development of coldwater fisheries faces problems that are common in most of the tropical countries. These include (1) Protection of native fish species and establishment of good sport and recreational fishery, (2) Habitat protection against pollution, (3) Catch regulation to prevent over-fishing, and (4) Enforcement of anti-poaching measures.

Limited Factors

1. **Low Fertility:** A deficiency of soil nutrients is the most significant constraints. Most of the water bodies are also deficient in nitrogen, phosphorus, and calcium. This constraint is notable in most of the tropical mountains. Nutrient deficient provides serious limitations to fish production in these regions.

2. **Low Temperature :** Special problem relates to mountain regions is characterized by low temperature of water. Temperate and sub-temperate climates with a marked winter significantly affect the growth of fish. Generally, the coldwater fish species thrive well in the temperature range of 5 to 22°C. The temperature range sometimes goes down to sub-zero. During winter season, most of the areas of the mountain environment is covered with ice which becomes more difficult for aquaculture development. To overcome the problem, there is need for better fishing methods. Fishing units require better fishing technology and better mobility to improve their efficiency. This includes provision of better fishing boats, drilling machines for drilling ice for winter fishing from ice, and transport vehicles. Therefore, winter is a severe limiting factor for coldwater aquaculture because mass mortality of fish occurs during winter kill.

3. **Low Organic Materials :** The low organic materials (Biomass) in mountain ecosystems is a dominant component that drastically affect the nutrient cycle. oligotrophic status and other factors of the ambient environment such as turbidity, moderate to high current of water, nature of substratum etc. have an impact on feeding, growth, and reproduction of coldwater fish.

4. **Inaccessibility to Mountain Regions :** Mountain regions are characterized by difficult terrain and inaccessibility to areas where potential capture and culture fisheries exist. This factor is very common in mountain regions and must take this factor into consideration.

5. **Soil Erosion:** The loss of soil from upland agricultural and non-agricultural lands is a serious problem across the world. Several billion metric tons (Mg) of soil are moved annually by soil erosion from altitude regions. The load of sediment carried by some of the world's major rivers is enormous as shown in Table 10.2.

   Soil erosion has serious implications for fish production in mountain regions such as the shortening of the life span of coldwater resources by inputs of sediment from eroded lands. Soil erosion is, however, most certainly a concern of not only for fisheries and aquaculture in hill areas but for agriculture as well.

Table 10.2 : Annual Sediment Loads of Some Major Rivers

River	Country	Annual sediment load (million Mg)	Erosion (Mg/ha drained)
Ganges	India	1,450	270
Brahmaputra	India	1,590	285
Amazon	Brazil, Peru	360	15
Kosi	India, Nepal	170	550
Mekong	Thailand, Vietnam	170	40
Nile	Egypt, Sudan	110	12
Yellow	China	1,600	480
Red	China, Vietnam	130	215
Mississippi	United States	300	95

6. **Poor Capital and Economic Sustainability :** Backwardness in rural areas, of small water/land holdingls and economic instability of the rural people definitely interrupt the effective implementation of aquaculture programs in these regions. To overcome these difficulties, loans with low interest and subsidies should be granted to farmers. At the same time, capital and economic instability can be regulated through advancement of low input integrated farming systems.

7. **Distribution and Freezing of Water:** Irregular distribution of freshwater areas and river/lake freezing to the bottom in winter and harsh winters are very important constraints which do not permit fish culture.

8. **Production of Fish Feed:** The most important problem that is related to coldwater fish culture is the non-availability of balanced and quality food and effective feeding strategy. Farmers generally use locally available cheap food ingradients such as rice bran, oil cakes, trash fish, boiled rice etc. with improper food management strategy stances which makes fish culture more difficult. Development of quality food for fish using locally available food ingradients is badly needed.

9. **Breeding and Seed Production:** Non-availability of brood fishes in high altitude regions is the main impedement in augmenting fish production. Although it has been possible to produce seeds of some species of coldwater fish such as snow trout and mahseer, seed supply falls much shorter than demand. However, effective extension programs should be implemented right now for proper distribution of hatchery-produced seed to fulfil the requirement of farmers. Construction of fish seed farms will not only bring fish ponds under culture but to recover the diminishing trend of fish stocks from rivers/streams as well.

10. **Over-fishing:** The most important problem stemming from mountain aquaculture and fisheries is the over-fishing of coldwater resources. Several species of commercially important fish such as *Tor* spp., *Raiamas bola, Racoma progasta, Botia* spp., *Balitora brucei* and many other species have declined considerably as a result of over-fishing. Over-fishing has been identified as one of the most significant issues in almost all the potential resources and efforts to increase production and profit from upland coldwater resources have included the stocking of water bodies, disease control, improvement of breeding techniques, and habitat improvement. Fish stocks should be enhanced by regular stocking. Progressive fish culturists should be realized the potential danger of over-fishing and suitable guidelines must be adopted to regulate the use of gill nets of smaller mesh.

11. **Habitat Destruction, Pollution and Siltation:** The use of explosives for development activities and pesticidal plants for fishing not only kill the food fish but also pollute the water bodies to a large extent, thus adversely affect the quality of ecosystems. Most of the coldwater lakes, ponds and reservoirs are subject to severe pollution, eutrophication, and siltation. These phenomena drastically hamper the fisheries and aquaculture activities at large.

12. **Environmental Degradation :** Environmental degradation continues to play a key role in coldwater aquaculture across the mountain region. Mahseer farmers in India, for example, have suffered dramatic loss in production due to construction of hydro-electric projects and dams on the rivers Ganga, Yamuna, Alakananda, and Bhagirathi which has resulted in stress to feeding and breeding migrations and complete destruction of spawning grounds. Trends in intensification of coldwater resource use and yield are increasingly apparent in Asian hill stream aquaculture, rising concerns over the approaching release of waste substances.

Other environmental degradation phenomena include habitat destruction, pollution, siltation, human settlement, excessive run-off loaded with silt and others that has already been noted in previous paragraphs.

Different environmental issues with respect to coldwater aquaculture development are receiving considerable attention in the countries concerned, and efforts are under progress to assure more sustainable development of the hill and mountain regions.

13. **Financial Constraint :** Finantial constraint is one of the most significant problems in the development of coldwater fish culture in tropical mountain regions. Heavy

investments are obviously needed for commissioning different projects in mountain regions to meet the vast infrastructural gaps in communications, transport power, and generation for the development of stable fish culture round the year. It would take a considerable time for these areas to build up an adequate resource based on their own.

The guiding principles in different plan periods on which the mountain area development programs are based include the promotion of a source, basic life support system of farmers, and judicious utilization of hasvested water in a total perspective embracing the complement of interests in fish culture in both hills and plains. Effective programs on fish culture seem to be very much suitable for the mountain's economy especially for small and marginal farmers. A rigorous drive for several years would suffice in bringing home the returns.

Suggestions for Better Water Harvesting Management

There is urgent need to keep water to the most efficient use in order to maximize production per unit of water area. The steps required to be taken to achieve this objective are noted below.

1. Efficient management of harvesting water by encouraging farmers to layout the pond structure properly.

2. Fish culture in terraced ponds should be planned in such a manner that it would be possible to make the best use of water. By adopting such an approach, it would be possible to adopt integrated farming programs. There is considerable scope for increasing gross cultured area in the productive zone. The approach to improve the farming patterns in water harvesting area will pay high dividends in the long run. Such approach should be adopted in new aquaculture projects right from the beginning.

3. A major deterrent in the spread of fish culture practices under water harvesting management strategy is the lack of cooperative effort among farmers. A fish culture unit which is established by small farmers may be viable but in many cases, it cannot be economical unless operated on a large-scale where several farmers are actively involved.

4. An increasing awareness to obtain efficiency in the utilization of water resources involves the integrated effort of devoted workers including the specialist, the engineer, the policy maker and the farmer. Their active involvement will certainly assure the success of watershed mangaement projects. Some such projects have been implemented successfully in many regions of the world. This approach will help to conserve thousands of cubic kilometres of freshwater areas which still remain unutilized. The challenge before us is to develop a comprehensive and integrated water management progrms to reduce the possibility of droughts and floods.

5. Proper use of the available water requires attention. Water stored in tanks or reservoirs should be conveyed as and when required through certain distances before it reaches the areas where fishes are cultured.

To sum up, proper management of harvested water requires action at levels higher than individual farmer and action is also required at planning as well as policy levels. Integrated detailed micro- and macro-planning of water resources and at the same time institutional reforms have to be carried out. Detailed designs have to be carried out well ahead of time. It does not involve heavy investment but much better management of existing infrastructures and the adoption of modern techniques to improve water harvesting strategy that is required for significant yield of fish in mountain regions where potential coldwater exists.

10.13 Potential of Tropical Mountain Aquaculture and Fisheries

One cannot assist but be optimistic about the future of mountain aquaculture and fisheries in the tropics. Hilly regions of the tropics constitute about 53 per cent of the total area. Almost all the upland areas in the tropics possess abundant water resources that could be utilized for aquaculture and fisheries. The principal requirement for year round fish production seems to be higher in the tropics than anywhere else. Total annual solar radiation and cold to moderately hot climates provide good production potential.

A variety of coldwater fish species can be used for culture on commercial basis. The available aquaculture technologies for warm water species can also be improved and extended in coldwater climates. The potential for fish culture and capture from tropical mountain regions is immense.

10.14 Research on Coldwater Fish Culture in the Tropics

It is unfortunate that very little informations have been available regarding activities of coldwater aquaculture in mountain regions of many tropical countries. Research on coldwater aquaculture, their potential and prospects for fish production is not significant compared to that of other regions of the world. Knowledge on their potentialities makes it possible to identify only the topographical and climatological variability. More intensive study will indisputably exhibit different resources in regions where only a few can be explored. Adequate knowledge on the coldwater aquaculture and the adoption of the state-of-the-art management practices *vis-a-vis* training, education, and research might well have prevented much of the project's recurring and non-recurring costs.

There is good probability that makes intensive research and extension works more effective and at the same time will definitely identify complexities of ecosystems in mountain regions similar to those known for temperate zone counterparts. The role of genetic and biotechnological implications should also be apprehended. The existing technological infrastructures to support integrated coldwater farming system in some tropical regions are, however, essentially worthless from aquaculture point of view and it is very arduous to receive the profits derived from fisheries and aquaculture activities. Unsatisfactory research data on the sector concerned has embarrassed the setting up of systematized aquaculture and fisheries in mountain and hill regions.

A major breakthrough is being achieved by fishery scientists in the field of integrated aquaculture technology in many hill regions for conservation, development and management of glamarous species of coldwater fish. This will definitely open up the possibilities of several-folds increase in fish production. And the establishment of coldwater fish seed farms in suitable places will enable a round-the-year rearing of fish through continuous supply of seed.

Increase in farming activities and seed production potential of coldwater fish species through co-ordinated research projects will undoubtedly facilitate the coldwater culture revolution in mountain regions for the present and the future. With the culmination of all the efforts, coldwater aquaculture strategies may not be far off.

All coldwater fishery research and development efforts concentrate on mahseer, brown trout, rainbow trout, carp, and schizothoracine production for markets and maintaining healthy stocks in streams, rivers, and lakes for anglers to improve the economic status of the mountain regions. Since the price of some coldwater species is extremely high both from sport, nutritional, and commercial view-point, their restoration and conservation through collaborative research projects deserve serious consideration.

10.15 Prospects of Coldwater Aquaculture

Since integrated aquaculture in mountain regions is the main sector for rural economy, the coldwater resources would undoubtedly provide great opportunity for development of the sector concerned. The percent catch of different species of fish of the river Alaknanda at different catching centres in the Garwal region of the Himalayan mountain has been summarized in Table 10.3. Note the high proportion of the catch at 4 to 7 regions that are mainly dominated by *Barilius* sp., *Labeo* sp, *Tor* spp. *Schizothorax* sp., *Oreinus* sp., and *Garra* spp. On the other hand, on comparison of the percent catch of snow trout from the foot hills and the snow-fed streams, it is evident that Schizothoracins contribute substantially to 80-90 per cent of the total in the river Alaknanda while *Labeo* spp. and Mahseer in the foot hills contribute to 20-45 and 20-75 per cent of the total catch, respectively. These regions are, however, highly productive. With adequate management including periodical stocking and conservation strategies, these regions can be made quite productive. However, large farming areas of mountain regions are located far away from coldwater hatchery centres, and as a corollary, fish production significantly reduced. The important water resources in Indian upland areas are derived from various watersheds of the upper stretches of three major river systems such as Ganges, Indus and Brahmaputra which are originated from different glaciers of the Himalaya regions and the watersheds of Krishna and Cauvery rivers and Nilgiri streans in Western Ghats of Peninsular India. All these regions are abound with fish which can be readily integrated with crop and livestock production. In contrast to tropical regions of the world, tropical upland aquaculture is characteized by being an integral part of the prevailing farming systems.

Table 10.3: Average Catch Composition of Important Fishes of the Alakananda at Different Catching Regions During Winter season

Species	Catching region						
	1	2	3	4	5	6	7
Orenus sp.	94	93	88	78	74	66	57
Schizothorax sp.	5	4	4	5	5	2	0.2
Tor sp.	–	–	1	2.5	4	4	9
Labeo sp.	–	–	–	2	3	4	6
Garra sp.	–	–	1	2	1	2	5
Puntius sp.	–	–	2	2.5	3	3	3
Pseudecheneis sp.	1	3	3	4	4.5	4.5	2
Glyptothorax sp.	–	–	–	2.5	3	5.5	7
Barilius sp.	–	–	–	1	2	7	7
Crossocheilus sp.	–	–	1	2	1	2	4

1. Vishnuprayag; 2. Birahi; 3, Nandprayag; 4. Karnprayag; 5. Rudraprayag; 6. Srinagar; 7. Devprayag.

Source: Badola and Singh (1981)

Preference of Option

There are four possibilities that planners and aquaculturists may follow to increase fish production in high altitude regions.

(1) Intensify production in water areas already under culture, (2) Construction of new ponds for culture operation, (3) Increasing fishing intensity from rivers, streams, and reservoirs, and/or (4) To maintain capture-culture balance in medium-sized watershed areas. The preferences of option, however, depend upon the climatological and geographical conditions of the hill and mountain regions. In tropical countries, the physical potential for increasing water under cultivation is great although most of the mountain regions are not accessible to adopt modern farming. Moreover, the cost of managenent of watershed areas, clearance of water areas, transplantion of the needed inputs, and distribution of seeds from hatchery units are very high. While these areas may be more fully utilized for integrated aquaculture, the cost of making then suitable for capture-culture fishery and the environmental consequences of habitat destruction as well as topographical characteristics make this preference of operation extremely questionable.

Development of Artificial Propagation

The artificial propagation of coldwater fish involves collection and selection of broodfish from natural populations, stripping and artificial fertilization, water hardening process, hatching of eggs, running water and sprinkler systems for oxygenation, emanating of swim-up fry and their initial feeding, and raising of fry as well as fingerlings on formulated diets. However, to proceed from fish production common in subsistence aquaculture to those dictated by fish protein requirements, dramatic increases in supplies of fry and fingerlings of certain coldwater fish such as *Schizothorax* sp., *Tor* spp., and *Cyprinus carpio* have occurred. From the year 1970, fry production through artificial propagation have been quite successful (Table 10.4) and mahseer eggs, for example, have been supplied by Tata Electric Company, Lonavala, Maharashtra, to different regions of the country where conservation and managemet strategies of this species have been undertaken on commercial basis.

**Table 10.4: Production Potential of Mahseer (*Tor mussullah* and *T. khudree*)
eggs, fry, and Fingerlings at Tata Electric Companies Fish Farm**

Year	Number of fertilized eggs produced (X 100)	Number of fry and fingerlings produced (X 100)
1970	14	10
1975	22	131
1980	417	180
1985	651	361
1990	464	290
1994	480	288
1996	620	341
1998	436	292
2000	410	314
2005	440	325
2010	500	535

Source: Modified after Ogale (2000)

1. *Success in Artificial Propagation* : During the mid 1990s, millions of fry of *Tor putitora* have been produced at Bhimtal hatchery in Kumaun. Similarly, artificial propagation of snow-trout in Kashmir, Kumaun, and Garwal regions is highly encouraging. Breeding of common carp in Jammu, Himachal Pradesh, Punjab, North- eastern hill areas, and Nilgiris has been found to be pertinent. Trout propagation in several streams of Garwal himalayas (such as Asiganga, Mandakini, Pinder, and upper reaches of the river Bhagirathi and Alaknanda) have also given encouraging results. These achievements have resulted in a demand for information and technology on methodology to explore the possibility of producing more fish seeds and their large-scale distribution to other mountain regions. Considerable attention should be given to artificial propagation and culture of coldwater fish species for production. But unfortunately, the run-off and rainfed ponds in mountain regions are not being utilized in full potential. Research and development efforts on extensive and intensive coldwater aquaculture have demonstrated the response that can be obtained from farmers in mountain regions, but their financial and social issues have interrupted more extensive use of coldwater resources.

10.16 Fishery Potential in High Altitude Lakes of Asia

Most of the high altitude lakes (See Panel 10.4) of Asia are rich in different species of coldwater fish. Fish species are extremely suitable for coldwater farming in mountain lakes.

Many mountain areas have been noted as suitable conditions (such as lakes in Mongolia, Northern areas and North-West Province of Pakistan, several Himalayan regions of India, many coldwater lakes of Georgia, Azerbaijan and Abkhazia and biggest potential for farming of trout, Altai Osman (*Oreoleuciscus warpachowskii*), Siberian roach (*Rutilus rutilus lacustris*), pike (*Esox lucius*), Perch (*Perca fluviatilis*), burbot (*Lota lota*), and Amur catfish (*Paras-iuros asotus*). Depending on the climatic variations they are termed as *coldwater fish* for water having a temperature range of 6-20°C. Some warm-water fish species can be considered for cultivation in cooler lakes where the water temperature occasionally goes down below 20°C. In Table 10.5, some important species of fish have been noted which are, for example, suitable for culture in upland areas of India.

PANEL - 10.4
GENERAL CONSIDERATIONS OF A LAKE

A lake is generally regarded as any inland body of standing water, larger and deeper than a pond. The term lake, however, includes a wide range of water bodies, ponds (which are small and shallow), marshes, and swamps with standing water can all be embraced under the definition of a lake. Lakes receive water input from streams, overland flows and ground waters and so are included as parts drainage systems. Lakes are landscape characteristics but are not generally considered to be landforms.

Lakes are very important from food production view-points. They are commonly used as sources of freshwater for agricultural and domestic purposes as well as food particularly fish. Lakes are also important for recreation and source of natural beauty.

Where lakes are not naturally present in the valley bottoms of drainage systems, lakes are constructed as needed by establishing dams across the stream channels. Some natural lakes are small ponds and serve to fabricate farms, while others encompass hundreds of square kilometers. In some regions, the number of artificial lakes is large enough to have momentous effects on the hydrologic cycle of the region.

One of the most important aspects of lakes is that similar to other small water areas, the productivity of lakes remains for short period of time. Most of the lakes accumulate inorganic sediments carried by streams, entering the lake and organic matter derived from plants and animals in the lake. Eventually, they fill up, forming a boggy wetland with little or no free water surface (Figure 10.3).

Most of the high altitude lakes are fed by melt waters of high glaciers and snow-fields in the elevated Hindu Kush, Pamir, Tien Shan, and Himalayas. Lakes persisted through thousands of years as oases for plankton, fish, and other aquatic organisms which are directly related to the food chain or/and food web. But in the last 40 years or so, most of the lakes once teemed with fishes of commercial/recreational importance, have lost their economic values. The volume of their waters has decreased by more that 30 per cent. The catch of commercial fish has drastically reduced; of course, commercial hunting and trapping of fish have almost banned from several productive lakes.

What is the ground for this drastic reduction in fish population? The answer is not complex. The reserve of lakes' water significantly reduced, lakes' water is lost by evaporation, and withdrawal of water from lakes for irrigation and domestic consumption drastically reduced the level and volume of water.

The future of most of the lakes seems to be dreadful indeed with the present reduction of freshwater inflow into the lakes, draw-down of water level, pollution, and eutrophication of lakes due to human interference, the productivity of lakes will soon attain the level of unproductive. Without the sacrifice of agricultural production and other activities for water to satiate the lake, or the importation of additional water from vast distances through special distribution networks, there is little that deserve serious considerations to rescue the productivity of lakes.

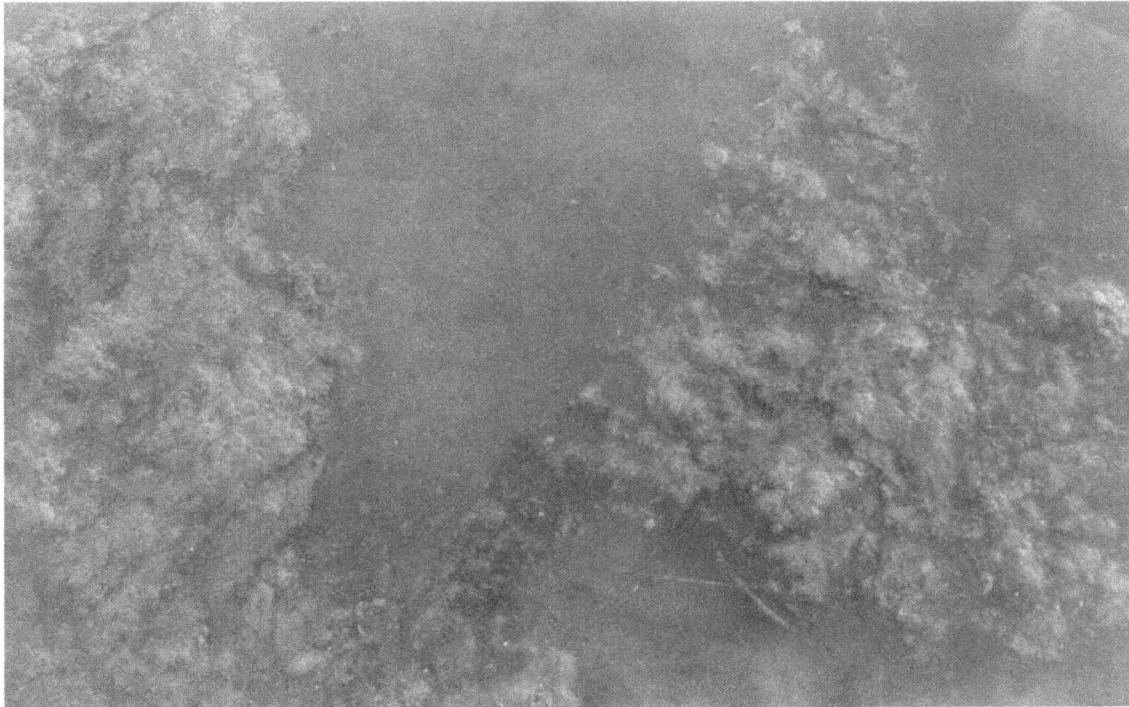

Fig. 10.3 : A high altitude freshwater pond in India. Vegetation is slowly growing inward from the edges, and eventually the pond will become a bog.

Table 10.5 : Some Important Fish Suitable for Culture in Water Harvesting Ponds Under Indian Conditions

Region	Species
I. Coldwater	*Salmo trutta fario, S. gairdneri gairdneri, Schizothorax richardsoni, S. cumaonensis, S. esocinus, Tor tor, T. putitora, T. mosal, T. khudree, Neolissochilus hexagonolepis, Berilius bendalisis, Labeo dero, L. dyocheilus, Garra gotyla, Glyptothorax pectenopterus, Puntius sarana, Nemacheilus* spp.
II. Warm water	*Catla catla, Labeo rohita, Cirrhinus mrigala, Cyprinus carpio, Ctenopharyngodon idella, Hypophthalmichthys molitrix, Labeo bata, Aristicthys nobilis, Clarias batrachus, Heteropneustes fossilis, Channa striatus, C. marulius*
III. Warm water fish elevated to coldwate	*Labeo rohita, Catla catla, Cirrhinus mrigala, Clarias batrachus, Cyprinus carpio*

Intensive culture of some species of trout,*Tor tor, T. putitora, Labeo dero, L. pungassia, Cyprinus carpio, Catla catla, Hypophthalmichthys molitrix, Acrossocheilus hexagonolepis, Barilius spp, Bagarius spp.* perch, tench, pike, *Glyptothorax* spp, and *Leuciscus schnidti* in some tropical lakes at altitudes ranged from 1,900 to 2,500 m above the mean sea level (amsl) has given encouraging results. It is significant to note that a number of water bodies in the form of lakes at still higher altitudes such as Pamir lake and Karakul lake (3,960 m), lake Zorkul (4.126 m), lake Yashilkul (3,730 m), lake Bulunkul (3,740 m), and 1ake Savezskoe (3,220 m) have also been found to be suitable for culture of some coldwater fish such as *Schizothorax intermedius* (snow trout), *Noemacheilus stoliczkai* (Tibet stone loach), *Schizopygopsis stoliczkai* (False osman), *Carassius auratus gibelio* (Giebel carp), and *Coregonus peled* (White fish). Most of the lakes are of tectonic origin and some are formed due to land-slides or moraine.

There are large number of lakes in the Himalayan regions. All mid-altitude lakes are rich in fish. Except some lakes such as Wular, Dal, Nainital which are eutrophic, most of the mid-altitude lakes exhibit either mesotrophic or oligotrophic to ultra-oligotrophic. A variety of fish can be grown successfully in these lakes for food and anglers.

Some mid-altitude lakes such as Sevan (1,950 m amsl), Karagel (3,195 m amsl), Khanchali (1,930 m amsl), Sagama (2,000 m amsl) in the Caucasus and lake Nainital (1,937 m amsl), Bhimtal (1,330 m amsl), Sattal (1,290 m amsl) and Naukuchiatal (1,220 m amsl) in the Himalayan region have been under major stress due to human intervensions. Though steps have been taken up for conservation of lake habitats, still damage is on the increase because of the continuous eutrophication of these lakes. High concentrations of phosphorus (3-25 mg/l), ammonia-nitrogen (10-325 mg/l), nitrate-nitrogen (5-20 mg/l), and silicate (80-3,400 mg/l) have been noted in these lakes which definitely indicates an increasing pollution. At the same time, the detection of bottom anoxia due to organic matter production, concomitant with the production of methane and hydrogen sulfide, most of the lakes may exhibit massive fish kills particularly in winter. However, the potentiality of these lakes can be increased by continuous monitoring of the input of toxicants and prohibition of fishing along with regular stocking with fingerlings. Though fish production potential of mid-altitude lakes in Asia is quite great, in majority of the lakes, fish production capability through management efforts is high at a much higher cost.

Fishery Potential in Lakes of Nepal

Several lakes such as Phewa, Begnas, Rupa and Reva provide good habitats for a wide variety of coldwater fish of the genus *Barilius barna, B. bendelsis, B. varga, Chagunis chagunio, Crossocheilus latius, Danio devario, D. rerio, Esomus dandricus, Neolissochelus hexagonolepis, Schizothorax macrophthalmus, S. nepalensis, S. richardsoni, Tor tor, T. putitora, Glyptothorax pectinopterus, Pseudechensis sulcatus* etc. These lakes are very productive and can support an annual fish harvest of more that 150 tonnes with adequate management.

In addition to native fish species mentioned above, several exotic fish of food and sport value have been introduced into these lakes. These include rainbow trout, brown trout, common carp, silver carp, grass carp, bighead carp, *Carassius carassius, Puntius gonionotus*, and tilapia.

Fishery Potential in Lakes of Caucaus

In the countries of the Caucasus, about 40 coldwater lakes have good fishery potential. The total potential area has been estimated to be about 12,000 ha in different altitudes ranging from 800 to 2,000 m amsl. All lakes are considerably smaller than the Lake Sevan which is most suitable for fish production. Some important coldwater lakes, their area, altitudes and production potential are shown in Table 10.6.

Table 10.6 : Fish Production Potential in Some Coldwater Lakes of Caucasus

Lake	Area	Production (tonne/ha/year)	Altitude (metre above sea level)
Sevan	1,245 Km²	10,000 - 16,000	1,898
Karagel	17 ha	5 - 10	3,195
Arpagel	2,200 ha	20 - 65	2,020
Khanchali	1,375 ha	260 - 290	1,930
Tabiskuri	1,440 ha	90 - 115	1,990
Sagamo	496 ha	40 - 50	1,995
Bazalati	125 ha	5 - 10	880
Madatapa	885 ha	5 - 10	2,108
Paravan	3,680 ha	350 - 370	2,060
Bebesini	160 ha	4 - 10	1,550
Geigel	80 ha	3 - 8	1,570

Lake Sevan has native fish fauna which are represented by three species such as trout (*Salmo ischchan*), Khramulya (*Varicorhinus capeta*) and Barbel (*Barbus lacerta goktchaicus*). In order to enhance fish catches, more fish species such as *Coregonus lavaretus ladoga* and *C. lavaretus maraenoides* (White fish), and *Carassius auratus gibelio* (Giebel carp) are stocked at regular intervals.

Some commercially important fish species stocked in different productive lakes include Sevan trout, Paravan race of common carp, Khramulya, Chub (*Leuciscus cephalus*),

Coregonids, Rudd (*Scardinius erythrophthalmus*) perch, tench, Wels *(Alburnus charusini)* and several other fish species.

1. **Morphometry and Limnological Features of the Lake Seven :** It is a large and high mountain freshwater lake with a mean width of 19 km. Though the lake is fed by 28 small rivers, there is only one outflow. Snow-melt and spring floods are the main source of water. The lake bottom is sandy or sandy-clay. The Cape Noraduz and Artanish Peninsula have sub-divided the lake into two regions such as major Sevan and minor Sevan.

   The lake Sevan is enriched with different species of phytoplankton (more than 20 species of blue-green algae, the most common being *Oocystis* sp., *Gloeococcus schroeteri,* and *Anabaena lemmermanii*), diatoms (dominated by *Asterionella formosa, Stephanodiscus hantzschii, S. astaea, Melosira italica,* Peridinium sp., *Mougeota* sp and *Closterium* sp.), zooplankton (dominated by rotifers, copepods, and cladocera), and zoobenthos (dominated by oligochaets (*Tubifex tubifex, Potamothrix hammoniensis*), leeches (*Herpobdella octoculatus, Hellobdella stagnalis, Glossiphonia complonta*) bivalves *(Pisidium casertanum and P. nitidus)* chironomid larvae (*Chironomous plumosus, C. markosjani, Procladius sp.* and *Glyptotendipes*), snails (*Valvata piscinalis, Lymnaea stagnalis, Radix avata),* and gammarids *(Gammarus lacustris).*

2. **Fluctuation of Fish Populations in Lake Seven :**

   (i) *Reasons for reduction in fish stocks:*

   * Eutrophication of the lake as a result of human activities around the lake such as draining of lake water for hydro-electricity and accumulation of pollutants from the human settlements.

   * Increase in water turbidity due to presence of dissolved and suspended organic and inorganic solids, algae and other suspended substances as a result of soil erosion, agricultural flow-off, and industrial as well as sewage discharges.

   * Increase in dissolved oxygen concentrations and occasional deoxygenation of bottom water layers.

   * Death of heavy algal blooms resulting in oxygen depletion. All these situations have dramatically altered the abundance in fish population: trout stocks declined sharply, the introduced whitefish increased from 166 tonnes in 1960 to 850 tonnes in 1975, and Khramulya retained its original stocks. Due to poor condition of spawning, Sevan barbel has become rare.

   * The impact of drawdown, water level fluctuations and water quality variables have considerably damaged most of the breeding grounds of all trout sub-species. Trout stocks have declined as a result of increased discharge of agricultural chemicals and minunicipal pollutants in rivers and streams drained from the crop fields and diversion of water by irrigation. Other important causes of extinction of the trout include: (i) intensive fishing pressure on fish stocks, (ii) loss of original spawning grounds along the lake shore, (iii) alteration in substrate quality, (iv) reduction in

submerged aquatic macrophytes, (v) changes in fish food organisms, (vi) bottom water hypoxia, and (vii) poor efficiency of artificial reproduction of fish.

3. **Hatchery Production and Stocking:** To enhance the abundance of fish species in the lake, several hatcheries have been established for producing fingerlings (3-5 g) on a large scale for release in streams and rivers for feeding the lake. However, the survival of the hatchery-incubated eggs is very successful (The survival rate varies between 70 and 80 per cent).

Though the reduction in trout stocks is severe, the management strategies still consider the lake as a source of fish for releasing in other lakes/water bodies. Trout has been introduced in the Karelia, Ukraine, and Kyrgyzstan. Millions of eggs of white-fish are collected from inshore waters and transferred to lakes of Urals, Central Asia, Georgia, Siberia, Abkhazia, Azerbaijan, and St. Petersburg and the Moscow regions.

The successful introduction of Coregonids and Gegarkuni trout in the lake Issyk-Kul, for example, has given conditions in the new environment suited for their high survival and growth. Such introduction is, however, now being considered as an important step towards preventing the trout from extinction.

Transplantation of trout to selected lakes in high mountain in Armenia such as in Kari (12.5 ha, 3,185 m), Akna (50 ha, 3,030 m) and Sev (190 ha, 2,666 m) has shown record production.

Fishery Potential in Lakes of Western China

The three provinces of Western China such as Xinjiang, Qinghai and Xizang encompass several countries such as Tajikistan, Afganistan, Kyrgyzstan, Kazakhastan, Mongolia, and the Russian Federation in the north and West and Nepal, Bhutan, India and Pakistan in the South. The entire region is mountainous, bounded by the Karakoram to the West and the Himalayas to the South and divided by the Kumlum mountains, separating Xinjiang from the Qinghai-Xizang plateau (average altitude is more than 4,000 m).

Though the climate in this region is harsh and the landscape is rugged, many large rivers such as the Brahmaputra, Mekong, Yangtse, and Yellow originate from the Himalayan mountain. Run-off amounts to about 75 per cent of China's total water resources. There are many diverse freshwater and saline water bodies and the pollution of water is practically nil due to less population density. Consequently, the region is teemed with a variety of coldwater fish species.

1. **Environment:**

(i) *Climate*: The climate in this region is harsh and is strongly influenced by altitude. On the Qinghai-Xizang plateau, the mean minimum monthly temperatures range from - 10 to - 15°C, and the mean summer maxima of 15-20° C. In Xinjiang, average temperatures range from - 18 °C in January to 28°C in July.

(ii) *Limnology*: A number of lakes in Western China have fishery potential and a comprehensive report on the limnology of some important lakes have been given in FAO Publication (No. 385, Page 241-243, 1999). Many lakes may be ephemeral or permanent depending on local precipitation and their proximity to rivers. Most of the lakes are dominated by sodium chloride. Salinities vary from crystalline brines to freshwater. In shallow lakes, the water is mixed and has no persistent thermal stratification (Polymictic), but the deeper lakes tend to be dimictic (lake with two annual overturns). Depending on the nutrient concentrations, the lakes range from ultra-oligotrophic to eutrophic.

2. Biological Features for Fish Yield :

(i) *Qinghai Lakes* : All lakes support three principal groups of natural food organisms such as phytoplankton, zooplankton and bottom fauna. About 78 species of phytoplankton (dominated by more than 20 genera of Bacillariophyceae and Chlorophyceae), 60 species of zooplanktom (dominated by Copepoda, Cladocera and Rotifera) and 32 benthic species (dominated by Chironomidae) have been noted in Qinghai lakes. Fish biomass equivalents (kg/ha) of benthos, plankton, and macrophytes in different lakes of Western China are shown in Table 10.7. In general, the potential fish biomass is measured by assuming biomass-conversion efficiencies of 1/6 for benthos, 1/2 for zooplankton, and 1/3 for phytoplankton. A phytoplankton biomass of 300 kg/ha, for example, would have an estimated equivalent fish biomass of 100 kg /ha.

Table 10.7: Equivalent Fish Productivity (Kg/hectare) of Plankton, Benthos, and Macrophytes in Four Lakes of Qinghai Province

Item	Province			
	Qinghai	Gyaring	Ngoring	Keluke
Phytoplankton	4.350	0.015	0.060	7.500
Zooplankton	22.500	0.075	0.105	8.250
Benthos	7.500	10.715	10.675	8.250
Macrophytes	0.000	134.025	20.750	182.250

Source : Walker and Yang (1999)

(ii) *Xinjiang Lakes* : Xinjiang, about 157 species of phytoplankton (including 60 Bacillariophyceae and 53 Chlorophyceae) have been reported. The total number of zooplankton species noted was 76, including 10 Copepods, 25 Rotifers, and 26 Protozoans. The benthos included 130 species and dominated by molluscs, insects and oligochaetes. The estimated potential of fish biomass is 55 kg/ha.

(iii) *Xizang Lakes* : In Xizang province, different lakes are also teemed with 460 recorded species of protozoa, 208 rotifers and 60 crustaceans and the estimated potential of fish biomass is 50 kg/ha.

The foregoing discussion dictates us that most of the lakes in these regions are enriched with natural food organisms and, therefore, are quite productive and moderate to high fish production is possible if developmental strategies are implemented and enforced; of course, sustained yields require careful management, administration, development of expertise, selective breeding and regular stocking.

3. **Fish Fauna :** In Xinjiang, there are 90 species of fish, of which 50 species are native and 20 (12 introduced and 8 native) are commercially important. In Xizang, there are 56 species, of which 10 species (all are indigenous) are commercially important. In Qinghai, 66 species (48 native and 18 introduced) of fish are inhabited in different aquatic environments. Of them, 9 species are considered as most important food fish. Some species and varieties have been stocked to other regions of China. The roach (*Rutilus rutilus*) and oriental bream (*Abramis brama orientalis*), for example, have been introduced to many provinces of China and the naked carp (*Gymnocypris przewalskii*) has been translocated to several lakes of Mongolia.

4. **Aquaculture :** Development of aquaculture has become established in Western China since 1980. In Qinghai and Xinjiang provinces, extensive aquaculture utilizes both reservoirs and lakes for stocking with a variety of fish.

   In Xinjiang, some natural lakes are used for fish culture by stocking fingerlings of several species of fish; of which, three species such as *Aspirohynchus laticeps Schizothorax biddulphi* and *Hedinichthys yarkamtensis* have become well-adapted and now dominate the catch.

Boston lake in Xinjiang has an area of one lakh hectare with a mean depth of about 3.0 m. Since the temperature of water varies between 10 and 15° C, the lake is very suitable for the culture of cyprinids. The lake has the estimated potential of fish biomass of about 100 kg/ha. The lake complex is now supported by 24 resident species.

To increase the catch potential in Boston lake, fingerlings of black carp, bighead carp, silver carp, common carp, grass carp and crusian carp are stocked with degree of regularity and the total catch is highly encouraging. Millions of fingerlings of grass carp, silver carp, *Tinca tinca, Perca fluviatilis* and *Rutilus rutilu*s are produced from artificial spawning and they are stocked at regular intervals. Perch (*Perca fluviatilis*) has dominated the catch (80 per cent of the total) and the total annual yield is about 500 ± 50 tonnes/year.

The lake Keluke in Qinghai province is mainly dominated by common carp (about 70 per cent of the total catch), grass carp and crucian carp with the hope of achieving annual total catch of at least 400 tonnes/year.

Due to economic reforms, aquaculture has expanded to a considerable extent in Xinjiang and Qinghai provinces where about 5, 000 hectare of ponds exist. Many hatcheries and fingerling farms have been established with a view to producing and supplying most of the demand for stocking to different lakes and reservoirs. Fingerlings are widely used for stocking. Cage and raceway culture techniques are extensively used for trout farming and the peak yield has been recorded as 115 and 165 kg/m^2 for rainbow trout in cages and raceways, respectively.

Fishery Potential in Mongolia

In Mongolia, more than 3,000 lakes are larger than 10 ha, 27 are larger than 5,000 ha and 4 are larger than 10,000 ha.

1. **Hydrobiological Features***: Almost all lakes are teemed with rotifers, copepods, chadocera, diatoms and blue-green algae. The rich benthos is represented by many species. The major macrophytes are represented by water milfoil, duckweed, smart weed and pond weed. Most of the lakes are important for commercial fisheries.

2. **Fish Fauna****: A large number of fish in Mongolia exhibit great adaptability to a variety of environments. The same species may thrive well both in saline and freshwaters and cold as well as warm waters. They also reproduce both in lakes and in rivers. A total number of 64 fish species and sub-species have been reported from different water bodies of Mongolia. They belong to 11 families: Petromyzonidae (1 species), Acipenseridae (2 species), Salmonidae (2 species), Coregonidae (5 species), Thymallidae (4 species), Esocidae (2 species), Cyprinidae (41 species), Siluridae (1 species), Gadidae (1 species), Percidae (1 species), and Cottidae (4 species). However, some commercially important species of fish are noted below :

   1. Taimen, *Huch taimen*

   2. Lenok, *Brachymystax lenok*

   3. Siberian whitefish, *Coregonus lavaretus pidschian*

   4. Arctic eisco, *Coregonus autumnalis migratorius*

   5. Northern whitefish, *C. peled*

   6. Mongolian Grayling, *Thymallus brevirostris*

   7. Kosogol Grayling, *T. arcticus, T. nigrescens*

   8. Pike, *Esox lucinus*

   9. Amur Pike, *E reicherti*

   10. Siberian roach, *Rutilus rutilus lacustris*

   11. Altai Osman, *Oreoleuciscus warpachowskii*

   12. Amur Wild Carp, *Cyprinus carpio haematopterus*

   13. Giebel Carp, *Carassius auratus gibelio*

   14. Amur Barbel, *Hemibarbus labeo*

   15. Siberian Dace, *Leuciscus leuciscus*

   16. Amur Catfish, *Parasilurus asotus*

   17. Burbot, *Lota lota*

   18. Perch, *Perca fluviatilis*

3. **Fishery Status in Mongolia :** Commercial fishery generally focusses on large lakes of adequate depth. However, fish yield, annual catches and the most important commercial fish species are shown in Table 10.8.

---

* For further detail, see Dulma (1979) and Williams (1991)
** For further details, see Shatunovsky (1983)

Table 10.8 : Commercial Fish Catches (in tonne) and Fish Production
(kg/ha/year) From Some Major Lakes of Mongolia

Lake	Area (1,000 ha)		Fish production	Catches (tonne)	Major commercial fish species
	Total	Used by commerical fisheries			
Ugii	2.5	2.0	35	60-90	Roach, Perch, Pike, Burbot.
Kharus	186	50.0	5-10	100-150	Altai Osman, Mongolian grayling.
Khyargas	141	85	10	200-400	Altai Osman.
Khubsugul	261	40	15	250-400	Lenok, Grayling, Crusi an carp, Burbot, Roach.
Dood-tsagaan	10.5	8.0	25	160-200	Whitefish, Lenok, Grayling.
Airag	14	7.0	10	50	Mongolian grayling, Altai Osman.
Khar	58	20	10-15	90-200	Mongolian grayling, Altai Osman.
Durgan	31	11	8-10	250	Altai Osman.
Khuisiin Naiman	0.3	0.3	30	20-50	Northern whitefish.
Khoton	6.0	3.5	10-20	50	Mongolian grayling, Altai Osman.
Khorgon	7.7	3.0	10-20	50	Mongolian grayling, Altai Osman.
Dayan	7.0	5.0	5 -10	30-50	Mongolian grayling, Altai Osman.
Achit	30	17	5-10	100-160	Mongolian grayling, Altai Osman.
Tolbo	19	7.5	7-12	80-100	Mongolian grayling, Altai Osman.
Khongor-Ul en	0.6	0.6	5-10	20-50	Mongolian grayling, Northern whitefish, Vendace.
Uureg	23.0	7.0	5-10	50-80	Altai osman.
Khunguin Khar	6.5	4.5	10-15	50-70	Altai Osman.
Ulaagchny Khar	9.0	5.0	10-30	50-100	Northern Whitefish, Arctic cisco.
Tel men	19.5	8.0	5-10	20-40	Altai Osman.
Buir	62.0	25	10-15	80-200	Amur wild carp, pike, Crucian carp, Catfish.
Khukh	10	8.0	5-10	20-50	Crucian carp.
Been-tsagaan	24	11	5-10	80-100	Altai Osman.
Sangiin-dalai	17	11	5-10	30-60	Altai Osman.

Source: FAO (1999), Page 230.

In lakes with outlets to the Arctic ocean (such as lakes Vgiy, Dood, Nuur, Buyr, Hovsgol Nuur, Terhiyn Tsagaan and a number of medium-sized lakes connected to the rivers Shargin, Hegiyn and Shishhid) the commercially important fish species include Siberian Whitefish, Arctic Grayling, burbot, pike, catfish, perch, crusian carp, Amur Wild Carp, Siberian dace, and lenok.

Darhat basin is an important area for fisheries development. This basin has extensive network of streams, rivers and lakes with considerable fisheries potential. Commercial fisheries exist on lakes mentioned above where some highly priced fish such as burbot, perch, lenok, roach, Arctic grayling, etc. are harvested at regular intervals.

In Mongolia, Central Asian Internal Basin contains 32 per cent of its water resources and includes several lakes of interest for commercial fisheries such as Heh, Har and Har, Us, Dayan, Horgon, Hoton, Tolbo, Dalay, Boon, Orog, Tsagaan, Durgan, Ayrag and Hyargas with the major fish captured being Altai Osman (80 per cent of total) and Arctic grayling (20 per cent of total). Annual fish yield from some other lakes such as Durgan, Nogoon, Har, Bayan and Telmen varies between 15 and 245 tonne/year.

The largest lake of Mongolia is the flow-through freshwater lake Buyr (615 Km^2) which has the highest fish yield than other lakes. The lake is inhabited by over 40 commercially valuable species of fish especially Amur barbel, Catfish, carp, pike, lenok, chinese carp, Crusian carp, Redfish, look-up, and Arctic grayling.

During the period 1956-2000, the lake Buyer was heavily fished and annual fish catches reached about 500 tonne. Gradual decline in species diversity of fish indicate that the water body is being overfished.

4. **Problems of Inland Fisheries in Mongolia :** Since a number of fish species such as Cristo and Northern Whitefish have well adapted to different Mongolian lakes, much attention has been given on expanding fish stocks through stocking of high value fish species in different lakes where potential fisheries exist.

It has been estimated that about 3,000 tonne of fish could be harvested per year on sustainable basis but irregular distribution of water bodies, freezing to the bottom in winter and harsh environment do not permit efficient culture. The low water temperature makes the culture of some fish species such as common carp uneconomic as it grows very slowly to achieve table size. Culture of rainbow trout requires considerable investment in pond design, construction of water supply to provide the required 3 m depth at which water current would be maintained under ice. Partial or complete drying out of lakes (particularly in the valley of the lakes in the Central Asian Basin) places further constraint on the development of fisheries in some lakes. Gradual drying up of lakes has resulted in a virtual disappearance of lakes Taitsin Tsagaan and Orog. Some lakes also periodically dry up.

Fishery Potential in Lakes of Central Asia

High altitude water bodies in Central Asia such as lakes of Pamir and Tien Shan are characterized by high transparency, high depth and low water temperatures. All these parameters indicate the lakes of low productive. However, lake Issyk-Kul in Kyrgyzstan is situated at an altitude of 1,610 m.

1. **Pamir Lakes:** Many water bodies in the Pamir mountains have been found to be suitable for fish culture. The Pamir covers an area of 63,700 Km^2, of which high mountains cover about 40, 000 Km^2. The Pamir is a high mountain region of Central Asia with an annual rainfall varying between 550 and 1,250 mm. Minimum snow falls and indistinct winter and summer are characteristic in this region. More than 840 lakes are located in the valleys of mountain rivers. The area of these lakes vary between 1 and 38,000 ha and are situated in altitude, ranging from 3,220 to 4,260 m. The water is well oxygenated, thermally stratified, and alkaline in nature.

(i) *Physico-chemical and Biological Features* : In winter, most of the lakes start freezing and ice starts melting in May. Summer temperatures vary between 12 and 25° C. The lakes are dominated by calcium carbonate.

Most of the lakes have sandy/stony bottom, with stones or sand covered by gray/black clay sediments. Zooplankton, phytoplankton, and benthos reach a biomass of 92, 200, and 16 g/m^2, respectively. In shallow regions, benthos is dominated by chironomids and oligochaetes with a standing crop of 86.5 kg/ha. Diatoms, blue-green, and green algae, cladocera, and macrophytes of the genus *Chara, Myriophyllum, Potamogeton,* etc. are the most common forms.

(ii) *Important Fish Species* : Though high altitude lakes of Pamir have a medium to low productivity, the lakes are inhabited by only a small number of fish species. The native species are dominated by snow trout (*Schizothorax intermedius*), false osman (*Schizopygopsis stoliczkai*), and Karakul stone loach (*Noemacheilus lacusnigri*), and Tibet stone loach (*N. stoliczkai*). The introduced species are Northern Whitefish (*Corgonus peled*) and giebel carp (*Carassius auratus qibelio*).

(iii) *Fish Production*: Fish production figures indicate that false osman is the most important fish species harvested (90 per cent of the total catch) from Pamir lakes followed by snow trout and giebel carp. Tibet stone loach and Karakul stone loach though common in many lakes, have no commercial importance. Absence of commerial fisheries in many lakes and irregular exploitation of fish in others are factors responsible for the variation in catches. Fish production in these lakes is on the decrease. However, some important productive lakes and their production potential are shown in Table 10.9.

The total annual commercial catch from these lakes has been estimated to be about 195 tonne. Fish production could be increased through regular stocking and introduction of fast-growing siberian whitefish (*Coregonus lavarettus*) and Northern whitefish. Some other productive lakes are also to be stocked with whitefish so that a total catch of about 400 tonne/year may be obtained. Most of the lakes

Table 10.9: Production Potential in Some Pamir Lakes

Lake	Area (hectare)	Average production	
		Total (tonne)	Kg/hectare
Yashilkul	3,000	100	46
Zorkul	2, 300	60	30
Turumtaikul	535	24	43
Bulunkul	340	10	33

Source : Savvai tova and Petr (1999)

provide good conditions for fish feeding on zooplankton and their utilization in full potential may increase the yield of fish considerably by 20 kg/ha.

2. **Issyk-Kul Lake :** The lake is fed by 102 rivers and streams. Rivers are fed by melt-water from glaciers and snow above 3,300 m. The lake is 180 km long and 60 km wide with a mean depth of 275 m. About 38 per cent of the total lake area where the depth varies between 1 and 10 m, is considered as the most important fishing zone.

In summer, the surface water temperature in the central area reaches 18°C and in winter, it is above 4°C. The temperature of water may drop by 12°C down to 50 m depth and a further 2°C to the depth of 200 m. The water is well-oxygenated because it is regularly mixed by strong winds.

The bottom configuration is highly variable and consequently this has resulted in a diversity of habitats.

(i) *Fish Fauna*: The indigenous and introduced fish fauna comprises 10 and 11 species, respectively (Table 10.10). However, the fish dace (*Leuciscus schmidti*), scaleless Osman (*Byptichus dybowskii*), stone loach (*Noemacheilus strauchi ulacholius*), gudgeon (*Gobio gobio*) and snow trout (*Schizothorax pseudoksatensis issykkuli*) are very common in the lake which forms the major component of capture fisheries.

In order to establish diverse stocks of fish, the introduction of several species such as pike perch, bream, giebel carp, and common carp have been very successful. For high density stocking, advanced fingerlings are generally recommended since the higher salinity of the lake has a negative impact on the survival of fry.

(ii) *Production of Fish* : Since catches of commercial fish such as common carp, scaleless Osman, snow trout, pike perch, sevan trout and Issyk-kul dace started declining for the last 60 years or so, pre-spawning stocks are now being considered as the main fishery. At present, however, annual yield of fish stock to the tune of 350 tonnes have been noted. But this production is not enough to meet the local demand. For this purpose, total fish catch to the tune of about 700 tonnes/year may be increased by annual stocking of 1-2 million fingerlings of different species of fish particularly sevan trout, snow trout, rainbow trout, common carp, and dace.

Table 10.10 : Different Species of Fish Cultured in lake Issyk-Kul

Fish	Common name	Native	Introduced
Salmo ischchan issyko gegarkuni (Lushin)	Sevan trout	–	+
Leuciscus bergi (Kachkarav)	Issyk-Kul dace	+	–
Phoxinus issykhulensis (Berg)	Issyk-Kul minnow	+	–
Gobio gobio latus (Anikin)	Issyk-Kul gudgeon	+	–
Dyptichus dybowskii (Kessler)	Scaleless Osman	+	–
Cyprinus carpio (Lin.)	Wild carp	–	+
Nemacheilus strauchi ulacholicus (Anikin)	Spotted stone loach	+	–
N. dorsalis (Kessler)	Grey stone loach	+	–
Aspius aspius iblioides (Kessler)	Asp	+	–
Coregonus lavaretus (Lin.)	Tench	–	+
Tinca tinca (Lin.)	Rainbow txout	–	+
Coregonus lavaretus (Lin.)	Whitefish	–	+
C. autumnalis migratarius (Pallas)	Artic cisco	–	+
Abramis brama orientalis (Berg)	Aral bream	–	+
Carassius auratus gibelio (Bloch)	Giebel carp	–	+
Alburnoides taeniatus (Kessler)	Atriped bystranka	–	+
Pseudorasbora parva (Schlegel)	Stone moroco	–	+
Stizostedion lucioperca (Lin.)	Pike perch	–	+

Souce : Savvai tova and Petr (1999)

(iii) *Environmental Considerations of the Future of The Lake* : Drastic alteration in the environment of the lake ecosystem indicates that the climatic factors are involved in the fall of water level. Gradual retreat of glaciers in the catchment, lower precipitation in the basin, and withdrawal of water for irrigation are the factors responsible for drastic change in water level.

The increase in agricultural activities, industries and human settlement around the lake has led to an increase in pollution of the lake. Though a large volume of water (about 1,740 km^2) in the lake may have considerable diluting and well-mixing capacity which is able to oxidize organic matter inputs to the lake, shallow areas that are important for feeding and spawning of several species of fish, are subject to eutrophication.

The lake is oligotrophic of low productivity and has a low carrying capacity for fish. Consequently, it is not advisable to stock fish at high densities. Since the lake is very important for recreation, stocking of sport and recreational fish is badly needed to meet the demand of the tourists. At the same time, possible care should also be taken not to damage the native fish.

Fishery Potential in Lakes of Kazakhastan

The Kazakhastan Altai, Dzhungarskiy Alatau and Northern Tien Shan regions have a number of medium/large-sized lakes as yet unaffected by human activities. There are some lakes such as lake Balkhash, Alakol lakes, and lake Zaysan which have good fishery potential. These lakes are, however, dominated by a number of commercially important fish species such as common carp (*Cyprinus carpio*), marinka (*Schizothorax argentatus*), perch, pike perch (*Silurus glanis*), gibel carp (*Carassius auratus*) silver carp, rainbow trout (*Oncorhynchus mykiss*), sevan trout (*Salmo ischchan*), roach (*Rutilus rutilus*), grass carp, and tench (*Tinca tinca*).

1. **Lake Balkhash:** From 1905 to 1955, the lake received a number of commercial fish species particularly pike perch, common carp, bream, tench, crusian carp, grass carp, and silver carp from different river basins. Such introduction of fish stocks along with other fish species (such as Chinese carp) has resulted in a substantial increase in aquaculture production.

2. **Lake Alakol:** The Alakol lakes are dominated by marinka, perch and cmmon carp. During 1930-1990, about 1,30,000 tonne of common carp were harvested from the Alakol lakes. During 1960s, more than 6,000 adult pike perch were stocked but unfortunately, this introduced species began to destroy the perch, common carp and stone loach. From 1970 to 2000 about 15,000 tonne of pike perch were harvested which is distressed atonement for the 3,800 tonne of carp harvested annually before the stocking of pike perch. To solve this problem, about 20,000 bream of different age groups were stocked during the mid 1980s. This has dramatically altered the Alakol lakes into a pike perch-bream water area. At present, the Alakol lakes have led to a rapid sequence of changes in the dominant fish species, with the native fish stocks declining considerably and with little oscillations in the total catch.

3. **Introduction of Trout and Coregonids in Tien Shan Regions :** The introduction of trout in some lakes of Tien Shan has increased fish production considerably. Trout joined with the local fish fauna such as scaleless Osman and Tibetian stone loach where they starts feeding and exhibited high growth rates. Similarly, the introduction of a number of coregonids such as *Coregonus albula* (European cisco), *C. autumnalis* (Arctic cisco), *C. lavaretus* (Lake chud white fish),*C. muksum* (muksum), *C. lavaretus sevanicus.* (Sevan white fish), and *C. peled* (Northern white fish) has been very successful. However, fish stocks are regularly increased by releasing hatchery-produced fry and fingerlings. Among all the species stocked, the Northern white fish has been found to be best because of its rapid growth, excellent taste and high market potential. The average weight of the Northern white fish in different water bodies of Kazakhastan varies from 160 to 680 g and 200 to 750 g at the age of 2+ and 3+, respectively.

Fishery Potential in Some High Altitude Lakes in India*

1. **Lake In Northern India:** A number of lakes along with reservoirs, streams,rivers, and ponds constitute the main aquatic biotype in the Himalayan region. Natural lakes in the Himalayan belt extend from Kashmir in the north-west to Assam and other hill states in the north-east. This entire region is extremely diverse and ecologically dynamic.

   Generally speaking, the aquaculture and fishery in these regions are poorly developed principally due to inaccessibility and difficult terrains. However, some important Himalayan lakes of commercial importance are given in Table 10.11.

   (i) *Fishery Potential in Kashmir Regions :* Most of the flood plains and high altitude lakes in Kashmir regions have good production potential. Fish fauna in different lakes are dominated by several major species such as *Schizothorax niger, S. micropogon, S. cuvirostris, S. planiforms, Schizothoraichthys esocinus, Labeo dero, L. dyocheilus, Puntius cocchonicus, Crossocheilus latius, Glyptothorax kashniriensis, Gambusia affinis,* and *Cyprinus carpio.* Since 1960s, these species have flourished in almost all the lakes in this region, but due to excessive fishing, most of the fish stocks have diminished considerably. Consequently, fish production in these lakes now fluctuates between 10 and 35 kg/ha/ year.

   (ii) *Fishery Potential in Lakes of Uttaranchal : In Garwal region,* lakes Nainital (Figure 10.4) Sattal, Bhimtal, Khuptal, and Naukuchiatal are situated at an altitude varying from 1,200 to 1,950 m. All lakes are small and the surface area varies between 13.5 and 72.5 ha.

   * *Water Chemistry*: The water of these lakes is alkaline (pH varies from 7.2 to 9.3). The water undergoes thermal stratification in spring and mixing occurs during winter. The lake Nainital is highly eutrophic because of the presence of calcium, silicate, ammonia and nitrate-nitrogen, and phosphate in high amounts. Naukuchiatal and Bhimtal are mesotrophic and have moderate concentrations of nutrients. Other lakes are poor in nutrients.

   Eutrophic condition of the lake Nainital clearly depicts an increasing pollution which has resulted in an increase in biological production leading to a higher organic matter production. This has resulted in anoxic conditions in the bottom water layers. The lake also accumulates detritus due to human activities around the lake. At present, however, a number of measures have undertaken to arrest further deterioration of the lake which would definitely prevent the fish stock from mass mortality.

   * *Fish Species and Production*: Both Garwal and Kumaon lakes are inhabited by several species of fish such as *Tor tor, T. putitora, Labeo rohita, Catla catla, Cirrhinus mrigala* silver carp, grass carp, common carp, and *Schizothorax richardsonii.* The dominance of fish species, however, differs considerably in different lakes of the Garwal

---

* For further detail, see FAO Technical Paper (No.385), 1999. p. 64-121. Vass (1998).

Table 10.11 : Some Important Lakes in the Indian Himalayas

State/Group/Lake	Altitude (m)	Water quality
Uttar Pradesh / Uttarakhand		
1. Kumaon lakes		
* Nainital	1938	Eutrophic
* Bhimtal	1931	Mesotrophic
* Naukuchiatal	1220	As above
* Khurpatal	1670	As above
* Sattal	1286	As above
Jammu and Kashmir		
1. Flood plain lakes		
* Wular	1537	Eutrophic
* Dal	1585	Eutrophic
* Manasbal	1587	Mesotrophic
2. Mountain (Glacial) lakes		
* Gangabal	3570	Oligotrophic to ultra-oligotrophic
* Konsernnag	3670	As above
* Vishansar	3817	As above
* Kishansar	3677	As above
* Alpather	3200	As above
* Nundkul	3550	As above
* seshnag	3570	Ultra-oligotrophic
* Tarsar	3713	As above
* Tulian	3750	As above
3. Forest (alpine) lakes		
* Nilnag	2180	Mesotrophic
4. Brackishwater lakes		
* Tso Morari	4541	Oligotrophic
* Pangong Tso	4329	Oligotrophic
5. Siwalik lakes		
* Mansar	666	Eutrophic, Mesotrophic
* Surinsar	605	As above
Himachal Pradesh		
* Chandratal	4270	Oligotrophic
* Surajtal	4800	Oligotrophic
* Khajjir	2060	Meso-oligotrophic
* Renuka	875	Meso-oligotrophic

Source: Raina and Petr (1999)

Fig. 10.4 : **Front view of the lake Nainital. It is one of the most important lakes in Urrarkhand (Garwal region) and is highly attractive to tourists. The lake is mainly dominated by several species of coldwater fish such as Common carp, Mahseer and Schizothoracines. Developmental activities especially involve pollution control and stocking of fingerlings of fish at regular intervals not only to increase their numbers but to restore the lake fishery as well.**

region. Nainital, Naukuchiatal, and Bhimtal lakes are, for example, dominated by mahseer, *Schizothorax* sp. and common carp, the latter two species being abundant in Sattal. Silver carp and grass carp are abundant in Bhimtal and substantial quantities of fish are harvested from this lake. Indian major carps are dominated the catches in Sattal. Regular stocking of these lakes with fingerlings have been proposed in order to increase fish yields which could be increased to the tune of about 50 kg/ha/year.

During 1990s, fingerlings of Common carp, grass carp, silver carp, and mahseer were introduced into Nainital, Bhimtal and Naukuchiatal where they have become well-adapted and now dominate the catch. At present, fish production potential in these lakes has expanded considerably following put and take fishery and these lakes are consistent producer of mahseer, Common carp and grass carp. But mahseer fishery has become highly endangered as a result of combined effects of winter mortality of fish and increased inputs of nutrients from human settlements.

* *Restoration of Lake Fishery in Garwal/Kumaon Regions*: Restoration of lake fishery in these regions through state-of-the-art technology is a recent issue. Though a ban on fishing for mahseer has inflicted, still the rate of deterioration of lakes

(particularly Nainital) is on the increase. After the detection of near-bottom anoxia, with the presence of hydrogen sulfide and methane the lake Nainital is now suffering from massive fish kills. Gradual increase in tourism and other developmental activities have been witnessing a dramatic change in Kumaon lakes. These lakes have been so badly damaged that it is very arduous to recover their pristine condition.

Substantial increase in lake fish stocks and control of pollution requires continuous hatchery production of stocking materials in huge quantities. At the same time, continuous monitoring and sustained commitment to management of these lakes will obviously reduce the risk of water pollution considerably.

2. **Lakes in Peninsular India:** The upland areas in peninsular India are dominated by the presence of the Eastern and Western Ghats, the Aravalli Range, the Satpura and the Vindhya mountains. Among these mountain ranges, the Western Ghats are the most important from coldwater fisheries point of view. The Ghats have a number of lakes and reservoirs. Some of the most important lakes are the Ooty (2,500 m, 34 ha) in Nilgiri Hills, the Kodai Kanal (2,285 m, 26 ha), the Berijan in the Palni Hills, the Divicolam (1,985 m, 6 ha), and the Letchmi Elephant (1,880 m, 2 ha) in the Munnar High Range, and the Yercaud (1,340 m, 8 ha) in Shevaroy Hills. All these lakes receive huge quantities of water each year during monsoon months. These waters are utilized in full potential for fish culture.

(i) *Physico-Chemical and Biological Features* : Lakes in peninsular India range from oligotrophic to eutrophic and from monomictic to polymictic types. The chemical composition of lake waters is as follows: temperature, 15-28°C; pH, 7.3-8.5; dissolved oxygen, 6-10 mg/l; total alkalinity, 80-160 mg/l; silicates, 0.45-0.65 mg/l and ammonia-nitrogen, 0.40-8.0 mg/l.

Some lakes have stony bottom with stones covered by clay sediments. Several species of macrophytes such as *Eichornia, Azolla, Typha, Scirpus, Hydrilla* and *Potamogeton* dominate the lake habitat. Protozoans, rotifers, and diatoms are the most common zooplankton. The limnetic zone is poor in zooplankton but rich in phytoplankton. The littoral zone is rich in periphyton biota which are associated with macrophytes.

(ii) *Fish Production*: Most of the lakes in the Western Ghats are very suitable for the culture of coldwater fish species. However, fish culture in these regions is confined to fish farms managed by the states Tamil Nadu, Karnataka, and Kerala. These states are involved in the production of Indian trout (*Raiamas bola*), rainbow trout,brown trout, tiger trout, Kokunce salmon (*Oncorhynchus nerka) Tor khudree* and conmon carp.

In Ooty and Kodai Kanal lakes, many discrete pollution sources have been registered and water quality does not permit the fishery standards. Though these lakes are considered as tourist spots, they have lost their importance for recreational and commercial fisheries as they have become silted and polluted by waste materials inundating from channels passing along the land bodering on these lakes.

Though these lakes nave a medium to high productivity, fish catch potential is either low or moderately high and does not satisfy the demand of markets. The introduction of rainbow trout and common carp in lakes of Nilgiris, Palni Hills, and Mannar High Range has been very successful and in several lakes they have developed self-sustaining populations. These lakes are regularly stocked with hatchery-produced fry/fingerlings for sustained production.

* *Fish Production From The Hill Zone of Karnataka*: The hill zone of Karnataka (90 Km from East to West and 400 Km from North to South) is located in Western Ghats. The elevation ranges from 700 to 1,300 m. Huge quantities of waters are available during rainy season for possible return to fish culture and the annual rainfall varies between 950 and 3,700 mm. There are more than 8,760 tanks in this zone having a total waterspread area of about 12,100 ha which is more suitable for fish farming. On the basis of altitude and elevation, the hill zone has been divided into six regions (Table 10.12). Note that medium elevation and medium to high rainfall belts have high fish production potential. Fry and fingerlings of common carp, Indian major carps, mahseer, catfishes, and murrels are extensively used for culture in these regions. A number of fish seed rearing tanks have been established. The annual total production of fish seed in these zones is about 17 lakhs, with a potential capacity to the tune of 30 lakhs and the zone is able to supply most of the regional demands for stocking. Until recently the peak production in this zone has been estimated to be 2,840 tonne/year (data for the year 2009) and it is expected that this production figure will be reached to the level of about 4,500 tonne/year by 2020. However, fish culture has potential for expansion, particularly in regions 4, 5, and 6 where production is on the increase at an annual rates of about 33, 31, and 26 per cent of the total yields respectively. Following the adoption of integrated watershed management technology, there is likely to be fundamental changes in the structure of the regional coldwater aquaculture and fisheries.

Table 10.12 : Different Regions of the Hill Zone of Karnataka, Their Agro-Ecological Features and Fish Production Potential

Zone	Physio-graphical	Altitude (m)	Rainfall (mm)	Total water area (ha)	Fish production (Average)
1.	High elevation and very high rainfall belt	300	More than 3000	950	170t/year
2.	High elevation and high rainfall belt	900	2000 - 3000	1522	120 t/year
3.	Medium elevation, very high rainfall belt	700 - 900	More than 3000	1460	46 t/year
4.	Medium elevation and medium rainfall belt	700 - 900	1000-2000	2980	830 t/year
5.	Medium elevation and high rainfall belt	700 - 900	2000 - 3000	635	710t/year
6.	Low elevation and medium rainfall belt	700	990 - 2000	1307	630t/year

10.17 Culture of Some Coldwater Fish Species

The culture of coldwater fish is considered largely to yield fingerlings for stocking lakes, reservoirs, and rivers predominantly for sport and recreational fisheries. Until recently, trout, common carp and some other carps, mahseer, and schizothoracines were produced for stocking and culture. There has been a considerable progress in hatchery breeding of some sport and food fishes such as *Neolissochilus hexagonolepis, Tor putitora, T. khudree, Schizothorax richardsonii* and *Garra gotylla, Schizothorax argentatus* (marinka)*Thymallus arcticus* (Arctic grayling), common carp, leather carp, scale carp, grass carp, whitefish, and trout in suitable areas of the world. There is a general consensus that protection of catchment areas should be considered first following the use of good, unpolluted and clean water for culture. At the same time, fishery laws and regulations must be implemented which will help to preserve the most threatened fish species. Regular stocking of hatchery-produced fry/fingerlings should be made. In the following sub-sections, the culture of some species of coldwater fish has been briefly described.

Culture of Mahseer (*Tor putitora*)*

As in the case of major carps, the culture of mahseer involves several steps that follows.

1. **Collection and Management of Broodstock:** Normally broodfish of mahseer are about 3 years old and a female can be expected to lay about 6,000 eggs/kg of body weight. Though broodfish can be obtained from sanctuaries, reservoirs, rivers, and lakes, cultured broodfish grown on a diet rich in protein in a healthy and clean environment should be prefered for best results. At the onset of natural breeding, the fish ascend shallow streams to spawn and in this moment the mature spawners are collected with care by gill nets of different dimensions that varied between 38 m X 8 m and 75 X 10 m in depth with mesh sizes of varying between 75 and 125 mm. The nets are fixed during the night and removed early in the morning.

   After collection, broodfishes are stocked in stocking ponds and in suitable stocking densities. The natural food organisms produced in a well-managed pond can sustain the stocking density of 300 kg/ha, but daily supplementary feeding at the rate rate of 5 per cent of body weight fortified with animal protein and minerals is necessary for gonadal development.

2. **Stripping Operation:** Broodfish can be held in freshwater for stripping. When eggs and milt ooze out on slight pressure on the belly, the broodfish are ready for stripping. The dry method is generally followed to ensure high rate of fertilization. Soon after the eggs are stripped in a plastic container, the gravid male is at once stripped and the milt is spread over the eggs. After a few minutes, a small quantity of clean water is added to ascertain complete fertilization of eggs.

---

* For further study, see Jhingran and Sehgal (1978).

The fertilized eggs are adhesive in nature and require water hardening. For this purpose, clean water is added to the basin so as to keep the eggs submerged for at least 40 minutes. The color of eggs ranges from pale yellow to bright orange having a diameter of about 3.0 mm.

3. **Breeding Under Controlled Condition:** The stages involved in induced breeding include: stocking and upkeep of broodfish, formation of a breeding set, collection of glands and preparation of pituitary extracts, injection of broodfishes, production of spawn and fry and their rearing which is quite similar to that adopted in breeding programs of Indian major carps.

For spawning, each healthy and mature brood fish is injected with a single dose (female at the rate of 0.6 mg/kg, male at the rate of 0.2 mg/kg) of any ovulating agent (such as ovaprim). Both male and female fish are then released into a breeding hapa and also directly into the adjoining cement cistern. The level of water is raised gradually by at least one foot after the release of the injected fish. A spray of water is provided to maintain optimum conditions for spawning. The success of induced spawning of mahseer under controlled conditions is, however, highly encouraging.

4. **Incubation of Fertilized Eggs and Rearing :** The fertilized eggs are kept in a hatchery tray (@ 5,000-6,000 eggs/tray) with running water facilities and continuous stirring of water @ 2 l per min. for oxygenation. Occasional flushing of hatching stocks with malachite green @ 1 : 2,00,000 for about 30 minutes is done for prophylaxis against fungal attacks.

The incubation period ranges from 80 to 120 hours at a temperature range of 20-25°C and takes 10-12 days for complete adsorption of yolk-sac. During hatching, the mortality may occur which is due to (i) injury to fertilized eggs during handling, (ii) sudden incursion of silt-laden water, (iii) occasional high fluctuation in water, (iV) Occurrence of White-spot disease, and (v) clumping of eggs.

The hatchlings measure about 2.5 to 8.0 mm and lie at the bottom sideways. While survival of hatchling varies between 85 and 98 per cent, the survival of swim-up fry varies between 70 and 80 per cent.

5. **Development of Hatching Systems For Mahseer Eggs :** Though several systems have been developed for hatching mahseer eggs, two systems as developed by the National Research Centre for Coldwater Fisheries (NRC-CWF), Bhimtal and Tata Electric Company (TEC), Lonavala, may be conveniently adopted. Their design of hatcheries and nursery techniques are very simple and have been the most successful. Mahseer seed production techniques as followed by the NRC-CWF will be briefly considered first.

(i) *Availability of water*: Clean and unpolluted water having pH and dissolved oxygen concentration of 7-8 and 7-9 mg/l, respectively must be considered. The temperature of water varies between 20 and 25°C. Water may be collected from either rheocrene or limnocrene type of springs or from streams with very low silt load and nutrients. Hatchery water, in particular, should be of highest quality. The quantity of water required at various stages of seed rearing should be as follows:

Water flow (l/minute)	Rearing capacity
1.0 - 1.5	Incubation and rearing of 2,000 eggs/hatchling at 20-25 °C
3.0 - 4.0	For 2,000 fry (3-month old) at 20-27°C
4.0 - 6.0	For 1,500 fingerlings (4-8 month old) at 20-30° C

(ii) *Overhead Storage Tank* : The capacity of each tank should be at least 50,000 L and placed at a height of 5 m from the ground. Before lifting with a vertical pump, sedimentation of water, if necessary, is done in separate tanks.

(iii) *Hatchery Trough* : These are made of galvanized iron sheet or fibre glass of 220 cm X 60 cm X 30 cm size with separate outflow and inflow arrangements. The outflow is so designed that only the bottom water is removed first. Each unit is provided with aeration devices in the form of showers.

(iv) *Nursery Tank Units*: These units are made up of fibre glass or Polyvinyl Chloride (PVC structure) (120 cm X 70 cm X 40 cm) having a capacity to raise 10,000 - 15,000 fry per tank with shower arrangements to ensure the supply of oxygenated water.

(v) *Rearing Tanks*: These are also made up of fibre glass or GI sheet with an area of 2 m^2 (depth 0.5 m) in which about 8,000 fry/tank can be reared.

The flow-through hatchery generally has a capacity to sustain about 0.3 million fertilized eggs and can generate about 0.25 million fry. The rearing units should be installed on the iron stands about 0.5 m above the soil. The hatchery is so designed that it can be operated with minimum manpower.

* *Role of TEC in the Production of Mahseer Seed nd Fry*: Extensive studies on the distribution, biology, and fishery of commercially important mahseers have been made by TEC. This has lead to the development of refined techniques of breeding, larval rearing, and cultural practices at TECs hatchery farm. Techniques are now capable of generating fry and fingerlings of all the desired species of mahseer.

The most simple system for hatching of mahseer eggs developed by TEC involves cement cisterns, wooden floating trays and perforated pipes. Pipes provide oxygen-rich water directly into the trays containing fertilized eggs. However, the wooden hatchery trays used are 56 cm X 56 cm X 10 cm deep with a suitable 1 mm plastic or velon mesh stretched properly and fixed to form the bottom of the tray. Eight such trays can be floated in a rectangular cement tank (2.5 m X 1.5 m X 0.75 m). About 30,000 eggs can be conveniently accomodated in each tray. Installation of four such hatchery tanks are highly economic.

About 480 litres of water is necessary per hour for each cement tank which amounts to 11,520 litres per day. Therefore, the total water requirement for the hatchery consisting of four tanks would be approximately 50,000 litres per day.

Depending on the maturity of brood fish, the rate of fertilization varies between 95 and 100 per cent. The incubation period varies from 80 to 115 hours at 19-26°C. The yolk sac is completely absorbed 10-12 days after hatching. In the running water hatchery, the average survival of swim-up fry is 95 per cent. The newly-emerged fry are released in cement tanks at the stocking density of 1,200 fry/m² for 10-15 days. They are daily fed on mixture of powdered mustard oil cake and rice bran (1 : 1 ratio).

6. **Rearing of Fry in Nursery Ponds:** After collection, fry are transported and stocked in nursery ponds at the rates varying between 8,000 and 10,000 numbers/0.2 hectare. The methods of preparation of nursery ponds are similar to that followed in major carp culture ponds.

    Feeding is done at 20 per cent of the body weight (6-8 times daily) with thick solution of emulsified eggs yolk and or goat liver. Feeding is continued for about a week which is followed by artificially formulated feed mash for 30 days. Dead fry are removed along with unutilized feed every morning to avoid putrifaction. Care should be taken to prevent any deterioration in the physico-chemical factors of water and soil. At the sane time, incidence of parasites/diseases must be controlled at once to ascertain high survival.

7. **Artificial Diet:** The main ingredients used to formulate the feed mash include soyabean meal, silk worm pupae, rice starch, yeast extract, and casein fortified with minerals and vitamins. These ingredients in requisite amounts are crushed, powdered and sieved before mixing. The mixture is soaked in water to make small balls which are kept in clean dishes at 3-4 different places in rearing tanks.

8. **Transplantation of Advanced Fry to Grow-out ponds :** It is desirable to stock the grow-out pond with equal-sized advanced fry (fingerlings) for commercial purpose. Fingerlings are also released into the natural waters to increase their populations. For culture purposes, however, stocking densities that vary from 3,000 to 8,000 fingerlings/ha have been recommended but densities higher than 5,000 /ha result in several management problems such as oxygen and pH imbalance and occurrence of diseases. A stocking density of 5,000 fingerlings/ha has been found to be suitable for moderately productive ponds. Though nutrient carriers are added to accelerate the growth of natural food organisms, use of artificial food is equally important.

    During the entire rearing period, it is recommended to maintain some flow of clean freshwater in nursery and rearing ponds. All the ponds should have proper draining facilities. Thirty per cent of the total water volume should be changed at least twice a week.

Culture of Schizothoracines

The culture of schizothoracines has not been widely practised. Though this group of cyprinids exhibits a gradual decline in catches along the entire length of the Himalayas due to environmental degradation, over-fishing, and interaction with the common carp.

Induced breeding and artificial fertilization of different species of schizothoracines such as *Schizothorax plainfrons,S. niger, S. plagiostonus Schizothorachthys esocinus* esocinus, and some other forms have been found to be successful and the hatching rate varies between 20 and 60 per cent. Also, further rearing of their fertilized eggs to produce fry and fingerlings using artificial diet in terraced ponds need further elaborate and systematic investigations. Recent trials with refined hatchery technologies may provide a basis for expansion of culture of these species in near future.

Culture of Common Carp

In the Himalayan regions, two German phenotypes of common carp (such as, *Cyprinus carpio communis* and *C. carpio specularis*) are considered for extensive culture in ponds and for stocking in lakes. Mirror carp is an ideal fish for culture in hilly regions where it thrives very well and breeds. The seed of mirror carp is produced in fish farms located in many parts of the world where hatchlings are raised to fry and fingerling stages.

The fish successfully mate and produce offspring in cement tanks, earthen ponds and in rectangular cloth containers (called as *hapas*) fixed in ponds. Breeding and spawning techniques of common carp have been described in Chapter 7.

The culture of common carp in terraced paddy fields in some upland regions has focussed on increasing fish stocks with fingerlings at the rate of 6,000 number/ha. Using cattle manure @ 2 tonne/ha and supplementary feeding with rice bran and mustard oil cake (1 : 1 ratio) at 4 per cent body weight, fish production to the tune of about 400 kg/ha is possible during the paddy growing season.

Culture of Trout*

The trouts were introduced in different countries of the world from their native place principally to encourage sport fisheries. However, trout introduction has helped the species to become well-adapted in most of the coldwater enclosures of mountain regions in Asia. Apart from sport fisheries, trout culture is, at present, being considered as a high-value commercial species for the table.

The culture of trout (and salmon) (Salmonidae family), though has a long history in North America and Europe is a very recent issue in fresh coldwater regions in many Asian countries. The main aim of early trout and salmon culturists was to produce early fry through hatchery for stocking in new regions or to increase fish population in the native habitats or to maintain sport fisheries. Recently, it has been pointed out that the possibility of using the current farming technologies for high production of these species has been fully realized. Production potential of trout and salmon in farmed ponds is very high and considered as a luxury food which is beyond the reach of common people. But nonetheless, trout and salmon farming has expanded in many parts of Europe and North America.

---

* For further detail, see Gall (1992), Laird and Needham (1988).

1. Types of Trout

(i) *Rainbow Trout* : Among different types of trouts, the rainbow trout (*salmo gairdnerii*) has the greatest importance in pond culture. This trout was introduced since 1870 from the Pacific Coast drainages to all continents except Antarctica. Its distribution extends from low altitude to higher elevations. Trout culture is undertaken in upland areas of many tropical and sub-tropical countries in Asia. Several local varieties have been developed, some of them considered as species or sub-species. Several strains have also been produced through mass selection and cross-breeding to upgrade culture qualities.

(ii) *Brown Trout* : The brown trout is the native species of Central and Western Europe. Though it has been introduced into many countries of the world for stocking natural waters to develop sport fisheries, its commercial cultivation has not been developed to any appreciable extent due to their slow growth rate, poor food conversion efficiency and stringent water quality parameters.

(iii) *Brook Trout* : The brook trout, *Salvelinus fontinalis*, native to North America, has also received considerable attention to culture in different countries where the temperature of water varies between 10 and 14°C. Inspite of its rapid growth rate and good market value, it has not been possible to culture them in ponds due to several reasons such as susceptibility to water pollution and infectious diseases.

At present, large-scale cultivation of rainbow trout in ponds has become an important issue across the world since both small- and large-sized fish and their processed itens have enormous market potential. Two important varieties, however, such as steelhead (sea-moving variety) and a land-locked freshwater form. The former grows rapidly in saltwater, and attains to an average weight of 4.5 kg/year. The freshwater form which attains to an average weight of 3.5 kg under suitable conditions, is widely considered for commercial culture.

2. Major Structures for Land-based Trout Farms

As a generalization, four major structures for land-based trout farms are ponds, raceways, tanks, and cages. For brief description on these structures, their design and construction noted below.

(i) *Water For Trout Culture* : For successful rearing and restocking of trout, the water should be as clean as possible throughout the year. Any fluctuations of environmental factors must be precisely evaluated before any investment is undertaken. If streams and river waters are used, physical and chemical inspections are inevitable. Though pH should vary between 7.0 and 8.3 that does not preclude the possibility of any dangerous effects on culture strategy, care must be taken to avoid ammonia toxicity. In addition to pH, dissolved oxygen and temperature control are the most important critical factors especially during the early rearing stages.

(ii) *Pond Culture* : Commercial trout farming in ponds originated in Denmark and the pond designs are based on the 'Danish earth pond' system. This system allows for fish to be cultured under less than optimum conditions as the flow of water currents tends to be not regular. However, the ponds are rectangular in shape (30 m X 10 m) with the bottom slopping towards the outlet. A depth of about 1.8 m and 1.0 m is maintained at the lower and upper ends of the pond, respectively. About 1,200 kg of trout fry are stocked in such a pond.

(iii) *Raceway Culture* : In North America, trout is cultured in raceway where fry are reared for re-stocking in sport waters. The structure of raceways consists of long continuous channels or series of channels divided by cross walls. They are constructed of cement concrete or earth, situated either above the ground or sunk in the ground. The channels are narrow, with a depth and width of about 1 m and 2.4 m, respectively. A plentiful supply of clean, unpolluted and oxygenated spring water through the channels is necessary for raceway culture. Constant flow and temperature of water must be maintained. Stocking density varies between 4 and 6 kg/m^3 with a water exchange of 2.5 litre/min./$m^{3.}$

Though the raceway designs have the advantage of holding the capacity of fish in large numbers in comparison to earthen ponds, the most important disadvantages that appears in raceway culture include the following:

* Raceways are made up of concrete and when the stocking density attains to the maximum, the bellies of the fish can be damaged by scrapping. This causes access for bacteria which results in outbreak of diseases.

* Since the volume of water is not fully utilized by the stock, it results in growth differentials.

(iv) *Tank Culture* : Tank culture of trout is common in Europe and United Kingdom. Tanks are 1.5 m in deep and 5-10 m in diameter. Tanks are constructed on the ground. The outlet pipes are placed in trenches and connected to outlet sumps. Many tanks are provided with a central fish grading arrangement. Each tank has a separate outlet pipe which is connected to a separate main pipe, leading to a sump where the fish is graded. A tank has, however, to satisfy some requirements that follows:

* The tank should be uniform in shape to allow a smooth flow of water from inlet to outlet.

* The movement of water helps spread fish throughout the entire available volume.

* It must be placed easily on any suitable place.

* It must be durable and weather-proof.

Other management schedules are given in section 10.18.

(v) *Cage Culture* : The culture of trout in cages has received considerable attention in recent years. There is little difference between the designs of cage used in protected areas of the sea and in freshwater. Floating cage culture has the advantage of high level of control on the stock of fish. Also, there is possibility of a small-scale operation with a small investment. Fingerlings of rainbow trout are stocked in cages in springs and are harvested in the autumn after a culture of one year. A fingerling of about 60 g weight can attain to a size of about 2.5 kg within one year.

(vi) *Mixed Culture* : Though monoculture of trout is most common, intensive farming systems are generally adopted in most situations to make the cultutre system economically viable. Under suiatable climatic conditions, two fish species (such as channel catfish, *Ictaluris punctatus*, and the trout) are cultured in raceways or ponds. The trout is stocked for 5 months from November to March and the channel catfish for 7 months from April to October. Thus, fish production continues round the year. In some European countries, trout is stocked at the rate of 15 per cent of the carp stock in carp ponds to feed on the carp hatchlings. The carp fry are produced by wild spawning. In this system of farming, per hectare yield of marketable trout has been reported to be about 45 kg per year.

Culture of Rainbow Trout

The culture of rainbow trout involves the development of brood stock management, artificial propagation, incubation and rearing of hatchlings, rearing of fry, and culture in grow-out ponds, tanks or raceways.

1. **Development of Brood Stocks and Their Management :** Brood stocks of trout are collected either from natural waters or from farms. Farmers have produced strains of trout in their own ponds which are characterized by rapid growth, better food conversion and early or late maturity. Selective breeding of trout has been executed in many areas. Consequently, strains have been developed with high percentage of spawning at an age of two years and increased production of fingerlings. Fast-growing strains are now available for culture. The progeny of cross-breed trouts are often sterile, but in some cases the females or males are infertile. Fully fertile cross-breeds are obtained when crossed with closely-related species. Use of sex-reversal all-female offspring for culture has been found to grow at rapid rate. Functional males are produced when male hormone, 17-methyl testosterone, is used with feeds at the rate of 3 mg/kg. A suitable stock of brood fish is reared in brood ponds. A density of about 8,000/ha is recommended and fishes are fed on both natural food organisms and artificial feeds. But artificial feeding is reduced before spawning. Both males and females of 3-4 years are considered as suitable for breeding operations. Substantial quantities of milt and eggs can be produced from large-sized brood fish. Large-sized females have the capacity to generate in large numbers eggs and hatch into sac-fry and hatchlings (alevins).

2. **Artificial Propagation :**

(i) *Stripping and Fertilization*: Several methods are employed for trout egg collection such as one-man method, two-men method, incision method, Australian method, and Swedish methods. Under Indian condition, however, two-men method is generally employed for stripping rainbow and brown trouts. In this method, one man holds the fish and other performs egg collection operation.

Egg collection method involves gentle squeezing the abdomen of the gravid female. Gentle squeeze allows the ova to liberate without any difficulty. The spent females are immediately transferred to the recovery ponds/tanks for further rearing for the following season. To ensure maximum fertility, dry spawning techniques are carried out where no water is allowed to drip into the bowel.

After collecting the eggs and milt in the spawning tray, the products are well mixed with the help of a quill. In most part of the world, dry method is commonly employed.

For mono-sex culture of females, all female eggs are obtained by fertilizing normal female eggs with milt from sex- reversal masculanized females. The testes are removed by cutting the abdomen, kept in a dry container cooled on ice, and divided criss-cross for the milt to drain out into containers. An equal volume of standard extender fluid is then added and mixed into the milt at a ratio of about 1 : 5 (milt: extender). The testes are rinsed in this solution and then discarded. Before mixing the milt with the ova, the milt must be left at 0°C for one hour to allow the spermatozoa to equilibrate with the extender fluid. One tea-spoonful of milt is enough to fertilize eggs obtained from two females (each of about 450 g).

Fertilization takes place immediately after the addition of equal volume of water. But it should be kept in mind that if water is allowed to contaminate the bowel before the milt is added, the water will be absorbed by the eggs and expand. This effectively seals the micropyle (a small hole of the egg through which the sperm enters to fertilize the egg nucleus) resulting in poor or no fertility.

If milt is to be preserved for prolonged period, a specialized antibiotic extender (particularly benzyl penicillin potassium salt) is generally used.

3. **Egggs of Trout:** Rainbow trout eggs have about 300 degree-day hatch duration. This denotes that from fertilization to hatching, the hatch will begin on the 30th day at a constant temperature of 10°C. Any fluctuation in temperature during this period will affect this timing. After spawning, the eggs should be measured in trays or incubators with a minimum flow of water (10 litre/min./10, 000 eggs at 10°C). The eyes will eye-up at day 18. On days 20-25, they are shocked and picked. Shocking makes the eyed-ova easier to detect for picking.

Eggs which are collected from a dependable supplier do not require much picking because dead or low-quality eggs are removed by the supplier. Upon arrival, however, eggs should be disinfected by gently agitating the eggs in the disinfectant bath containing iodophor for 5-10 minutes only.

After disinfection, the eggs are kept in hatching trays. For gentle movement of eggs, the flow of water is allowed which ensures an adequate supply of oxygen to the embryo and prevents the eggs from sticking to each other on to the tray.

4. **Incubation and Rearing of Hatchlings:** Though various types of incubators are used for hatching trout eggs, troughs and California baskets are the best system used by many trout farmers. In this system, the eggs and alevins can be easily observed and the entire process of incubation can be monitored. The troughs are 40- 50 cm wide, 20 cm deep, and 4 m length. Rectangular baskets are kept in these troughs, with their top edges placing on the sides of the trough. The basket rests about 3 cm above the bottom of the trough. The bottom of the basket is perforated through which only hatchlings or alevins are passed through water below, but the spherical eggs are retained. The baskets are arranged in such a way that water will be forced trough the mesh to aerate the eggs. Water flows from the top end of the trough under the basket and passes up through the perforated base and passes to the next basket until it reaches the end of the trough. A trough would require a continuous water flow of 5,000 litre/day for every 10,000 eggs.

If large number of fertilized eggs are used for incubation within a limited space in the farm, battery incubators are used. These consist of vertically-arranged series of trays held in guides. The inner part of the tray has a perforated base where the eggs are spread. Water is allowed to flow into the outer part of the top tray. Water is passed through the edge of the inner tray and into the side of the next tray.

The hatching time of trout eggs varies with the temperature of the water that ranged from 21 days at 15°C to 100 days at 4°C. The eyed eggs (the eyes can be seen through the egg shell) are tough and can able to withstand handling and transport. The hatchlings remain in the hatchery baskets until they reach the swim-up stage and all the yolk materials have been absorbed. The egg shells are removed from the basket with a suction device.

5. **Rearing of Fry:** The swim-up fry are transferred to concrete or fibre-glass rearing tanks. A regular current of water is maintained in these tanks which helps maintain a uniform distribution of fry. Circular tanks can sustain fry in large numbers. Both square and circular tanks with rounded corners are commonly used for rearing of trout and salmon fry. Each tank has a diameter of 2 m or 2 square metre in size, with a depth of 50 cm. An elbow pipe is delivered so as to create a water circulation. A drain pipe is fitted in the centre of the tank with a flat screen leading into a sump below the tank or a vertical cylindrical screen round a central drain pipe. The drain pipe is fitted to an elbow pipe on the side under the tank which is used to control the level of water in the tank.

(i) *Feeding*: Soon after the absorption of the yolk-sac, they are ready to start feeding. Food particles are very fine and should be made available at frequent intervals under a low light intensity, preferably over a long day feeding regime. Food particles can, however, lead to problems with gill diseases or similar ailments unless the farm is maintained at a very high level.

During fry rearing, specially-prepared starter feeds are given using automatic feed dispensers by which optimum feeding conditions can only be achieved when the fingerling stage is attained. At the initial stage, feeding is done five or six times a day and as they grow older, feeding is reduced twice a day. In cases where natural food organisms are nor available for fry, conventional feeds consisting of liver, meat and trash fish are mixed together and finely ground to a size that can be consumed by the fry. One to 2 per cent salt is used to improve the binding quality of the mixture. Feed ingredients are mixed in a mechanical mixture for right consistency.

Several types of feeders are available in the market ranging from the simple demand feeder to highly sophisticated and computerized feeding devices. Simple and very easily operating demand feeders tend to aggregate the fish in one region. While compressed air feeders are designed to spray the food across the water surface to provide an even distribution, other electronic designs can spray the food in a pre-allocated trajectory. Computerized devices vary from simple variable designs to highly sophisticated ecological analysis units.

6. **Grading of Fish:** If fish stocks are allowed to grow at a uniform rate, grading of fish is inevitable which is done with degree of regularity. As a generalization, stocking of more or less same-sized fish should be the target where the stock is allowed to feed at the optimum rate. Along with the increase in growth of fish, it is further inevitable to sort out the stock but pre-marketing grading can effectively spread the stock for the entire culture period. In a commercial trout farming system where stock is fully utilized for continuous production throughout the year, an efficient level of control is required.

About one-month prior to the start of the spawning season, fully-mature male and female brood fishes should be hand-sorted with the least possible stress and should be kept into separate ponds or tanks.

7. **Culture in Tanks and Raceway :** Tanks that are used for fry rearing, can also be used for the culture of growing yearling or two-year-old trout. Tanks having good circulation of water may contain about 30 kg fish/m^2 water. Circular tanks (5-12 m in diameter and 1.2 m deep) and long raceway type tanks (30 m long, 3 m wide and 1.2 m deep) are widely used. In circular tanks with a flow of 10 litre of water/ minute, 1,000 fry are stocked per square metre. For high density culture of rainbow trout, silo type of tanks may also be used. A unit of 2.3 m diameter and 5 m in high with a flow rate of about 28.5 litre/second has been found to carry about 2,820 kg trout (27.5 kg/m^3 or 136 kg/m^2 per second of water flow) without any difficulties.

Raceways are extensively used for fry and fingerling production in Central and North America and Europe. Each concrete raceway system may be up to 500 m long, 10 m wide and 2 m deep and is divided into several segments with arrangements for aeration. Fish are cultured in raceway systems for 4-5 months and fish are graded and sorted by automatic equipment and shifted to different raceway segments.

8. **Specialized Recirculation:** In modern intensive trout culture, it is economically advantageous to control the ecosystem very accurately. The initial operation of the trout farm may be negated by a cold-spell unless the water can be controlled and the temperature is maintained precisely.

Reciculation technique may be adopted in the hatchery which involves the total control of an egg incubation system (Figure 10.5). In this case, no filtration is required since the metabolic load on this system is very less.

An ultra-violet sterilizing unit is assigned to the system efficiently since 90 per cent of the total water is passed through more than once. At the same time, sterilizing unit needs a fast-flow heated water to control a precise temperature and preclude the possibility of building up of micro-ecosystems (Figure 10.6).

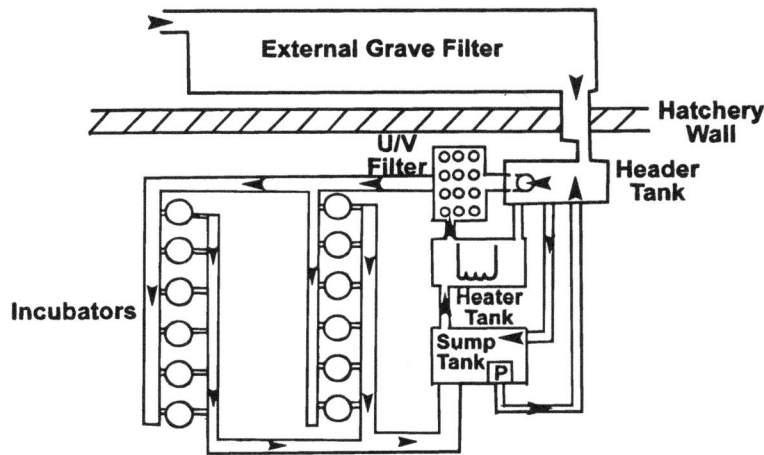

Fig. 10.5 : Recirculatory system in a clean freshwater trout hatchery. Such hatchery has a high production capacity (up to 4 million eyed ova per 21 days at 10ºC)

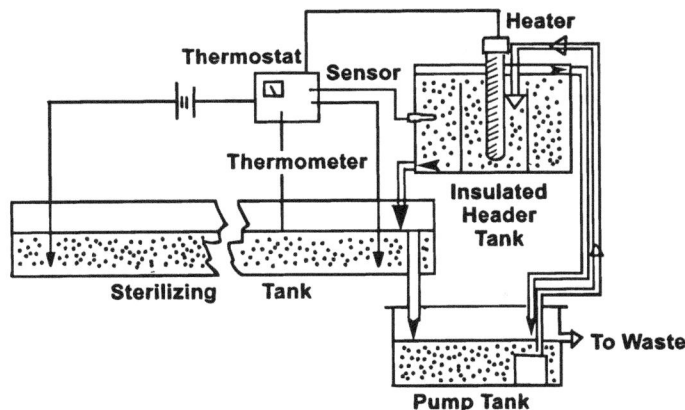

Fig. 10.6 : Diagram showing a system for the sterilization of salmon and trout ova.

10.18 Carp and Trout Culture in Europe

In Europe, the rainbow trout is the principal species used for culture in ponds. Besides trout, the European countries also offer an opportunity for developing intensive culture of common carp. Corresponding to its native habit in rapidly-flowing mountain brooks, it requires deeper ponds (1.5 - 2.0 meters) with unpolluted coldwater (temperature of water not exceeding 20°C) having sufficient oxygen. The trout pond should be small in shape and in construction, it resembles to an artificially sectionalized brook. In trout culture, the final return depends on the selection of the spawners. At the time of sexual maturity (January-April), the eggs are obtained by stripping young females (2,000 eggs/kg female) and are mixed with milt collected from young males. After fertilization, the eggs are hatched in troughs through which cool and oxygen-rich water is circulated. The hatching time depends largely upon the water temperature. As soon as the hatched fry can swim, they are immediately placed in rearing ponds where feeding is executed.

Pond Structure

Pond structure is generally determined by the agricultural structure of the region. In the Eastern parts of Europe such as Poland, Hungary, and Yugoslavia, carp and trout are cultured in large-sized water areas (several hundred hectares). In the Southern parts of West Germany, small-sized ponds (4-6 hectares) are considered, while in the Northern parts 15-20 hectares are very common.

While the minimum size of a trout culture pond varies between 1 and 3 hectares, those for carp are larger, usually 20-25 hectares. Since trout production depends very much on the marketing possibilities, some culture units are utilized for breeding of stockfish, particularly brook trout and rainbow trout.

Feeding of Carp and Trout

On the basis of artificial food, the culture of carps and trout is undertaken for high production. In some cases, farmers depend upon the natural food production of the pond. Under natural conditions, carps feed on bottom organisms and also consume phyto- and zooplankton in the water. Inspite of their plundering nature, trout feed on minor animals during their first year. In intensive fish culture ponds, trouts exhibit rapacious habit in their early stage, but as they grow, fish consume nothing else than fish. Artificial feeding of carp is supplementary as a substitute food, in contrast to the trout which entirely depends on artificial food.

The organic nutrients such as protein, fat, and nitrogenous substances are the principal constituents of artificial food for carp and trout. In addition to these ingredients, flavor substances and vitamins are also taken up by the fish. A great many high class protein-rich meal, bone meal and blood meal are also consumed by the fish. Animal and vegetable waste substances are sometimes used as food. Trout food, in addition to principal nutrients, should be supplemented by mineral substances to stimulate the intestinal activities.

Pond Management

It is the most essential means of increasing production which involves fertilization, manuring, and liming. Nitrogen and phosphorus fertilizers are extensively used for carp production. A quantity of 200-400 kg lime/ha, 30-60 kg N/ha and 30-50 kg of phosphoric acid/ha leads to a yield increase of about 100 kg of fish flesh/ha.

Mechanical preparation of pond bottom soil involves the use of organic fertilizers in the form of green manure (such as oats, barly, rye etc.) and cattle urine. This method is widely used in Eastern Europe for rearing fry and adult fish. In many cases, plants are allowed to grow at margins of the cultivated pond and are utilized as manure. During fish culture operations, cattle and ducks are allowed to graze there which provides additional manure.

Carp and Trout Production in Some European Countries

1. **Germany:** Production statistics of carp and trout in Germany is impressive. During 1990s, carp production varied between 12,800 and 15,900 metric tonnes from a total area of about 58,850 ha. Similarly, the trout production also varied between 5,785 and 14,000 tonnes.

2. **Yugoslovia :** The carp and trout production in this country amounts to about 15,000 metric tonnes and out of this production, about 5,550 metric tonnes are produced from farmed ponds. Substantial amounts of frozen trout are exported to Italy, France, Switzerland and the United States.

3. **Poland:** In Poland, about 78,000 hectares of fish culture ponds exist. These ponds account for 30 per cent of the total yield of freshwater fish. If these ponds are well-managed, annual yield to the tune of about 23,000 metric tonnes is possible.

4. **Bulgaria and Rumania :** In these countries, trout and carp are cultured in ponds to a small degree. The total yield reported from different establishments varies between 1,500 - 1,900 and 3,500 - 4,800 kg/ha, respectively.

5. **Belgium :** In this country, there are small pond areas (8,000 ha), of which a large part (about 6,300 ha) is used for carp culture. The remaining small water areas are utilized for trout. A large part of the yield (about 9,540 metric tonnes in 1990) is exported to Holland, France, and Geramany.

6. **Denmark, Sweden and Norway:** In these countries, trout culture is more important than carp. In 1956, the yield of trout in Denmark was about 4,000 metric tonnes which increased in the level of 19,800 metric tonnes in 2008. In recent years, there has been rapid expansion in freshwater-based farms for cultivating rainbow trout in Norway and Denmark, although sea trout and brook trout are also cultured. Substantial quantities of trout are exported to Germany, United States, United Kingdom, and other European countries.

7. **France :** Though carp culture in France dates back to the Middle Ages, trout culture is more important than carp and about 2,200 metric tonnes of trout are produced annually.

8. **Russian Federation:** The culture of carp is highly important in the Russian Federation. Maximum yields of 4-5 metric tonnes/ha are reported from various fish culture establishments. Though trout yield in the Moscow region is very low (450 metric tonnes/ year), the yield of carp is highly encouraging (it is more than 18,000 metric tonnes, data for the year 2007).

10.19 Coldwater Fish Culture in the United states*

Several species of coldwater fish (See Panel 10.5) especially Walleye and Northern Pike are considered as a major part of the fishery resource in the United States, and many progressive fish culturists receive prime consideration for their management and development.

Propagation for expanding the ranges of coldwater fish and their culture had started in the late 1870s. Since 1940, the construction of lakes and reservoirs throughout the United States has resulted in intensive propagation and management of coldwater species outside their natural ranges. Many states have established supplemental stocking programs because the demand has been high. Most of the species are top predators and are useful in managing over-abundant populations of forage fish. At present, however, many states are actively involved in coldwater fish production programs, their processing and marketing to a large extent. Utilization of coldwater fishes in full potential seems to be hovering.

Farmed coldwater fish are cultured intensively followed by their over-wintering in ponds, and returned to raceway culture systems in the spring. Stocks are readapted to intensive culture conditions and to dry diets when returned to the raceway.

Practical culture requires facilities for year-round growth. Maintenance of water temperature between 18 and 22°C through the development of recirculatory system is the most immediate requirement for post-fingerling culture. Quality diets and effective disease control, at the same time, would invariably improve the rate of coldwater fish production considerably.

10.20 Pen and Cage Culture of Coldwater Fish

Pens can be regarded as transitional structures between cages and ponds so far environmental and stock control are considered. However, fish pens are more or less similar to duck fences and their size varies from 50 to 1,000 square metre. The pen may be a single or double-layered enclosure and take different shapes such as circular, rectangular, square, or hexagonal. For the construction of pens, split bamboo, poles of casuaria, nylon ropes, nylon nets of smaller mesh size (0.5-1.0 cm) etc. are generally employed.

Cage culture is a method of farming in a particular type of rearing facility. Rearing facilities for fish can either be land-based or water-based, the former type includes ponds, raceways, silos and tanks, and the latter comprises enclosures, cages and pens. The terms

---

* For further detail, see Kendall (1978).

PANEL - 10.5
SOME COLDWATER FISH SPECIES IN THE UNITED STATES

Common Name	Scientific Name
Arctic grayling	*Thymallus arcticus*
Black bullhead	*Ictalurus melas*
Bluegill	*Lepomis macrochirus*
Brook trout	*Salvelinus fontinalis*
Brown trout	*Salmo trutta*
Burbot	*Lota lota*
Common carp	*Cyprinus carpio*
Cisco	*Coregonus artidii*
Flathead minnow	*Pimephales promelas*
Flathead catfish	*Pylodictis olivaris*
Freshwater drum	*Aplodinotus grunniens*
Golden shiner	*Notenigonus crysoleucas*
Gold fish	*Carrassius auratus*
Lake trout	*Salvelinus nameycush*
Lake whitefish	*Coregonus clupeaformis*
Large mouth bass	*Micropterus salmoides*
Muskellunge	*Esox masquinongy*
Rainbow trout	*Salmo gairdnerii*
Sockeye salmon	*Ictiobus bubalus*
Stripped bass	*Morine saxatilis*
Walleye	*Stizostedion vitreum*
White bass	*Morone chrysops*
Yellow bass	*Morone mississippiens*
Yellow perch	*Perca fluvescem*

---

enclosure, pen and cage seems to be virtually synonymous and can often be used interchangeably. Four basic types of cages are in vogue such as fixed, floating, sub-mersible, and sub-merged. But in mountain regions, fixed and sub-merged cages are widely used in running water environments.

Though enclosure culture of some coldwater fish such as *Salmo salar* and *S. gairdnrii* have been found to be suitable in many European countries, a few experiments have been conducted on pens and cages in some coldwater lakes and reservoirs under tropical conditions with limited success. However, enclosure culture can be undertaken in valley lakes of Kashmir, Kumaon and in other suitable coldwater lakes of Asia.

Fishes of different feeding habits (such as herbivores, detrivores and carnivores), fast-growing and resistant varieties (resistant to fluctuating environmental conditions) are selected for culture. They are stocked in separate enclosures to avoid competitions among them. In mountain regions, however, common carp, trout, mahseer, and schizothoracines are cultured in enclosures over extensive areas. Suitable design and construction of cages/pens and species selection for stocking should be evolved to standardize this culture technology in mountain waters.

10.21 Conclusion

Most of the water bodies in mountain regions represent a good potential for coldwater fisheries and aquaculture. While high-priced fish species can be stocked in some streams for anglers, commercially important species of fish stocks can either be increased or new species should be introduced into the cold-water lakes and reservoirs. Selective fishing from introduced fish should be done with the protection of native fish stocks. Pond / enclosure culture is the best method of protecting and producing native fish stocks and may provide a basis for expansion in the future. Development of induced breeding techniques for coldwater fish should also be executed to save them from extinction and their stocking with degree of regularity will definitely increase their population in water bodies in which they are threatened either by competition bteween native and introduced species or by overfishing. Minimizing human impacts resulting from the regulation of streams and rivers, pollution, habitat destruction, prohibition of over-exploitation, and control of introduction of foreign species are some of the important strategies for maintaining and conserving healthy indigenous fish stocks and developing sport and commercial fisheries.

The remoteness of the water bodies in mountain regions limit to implement management practices and development of sustained fish culture and is more expensive too and, at the same time, complex than regular calendar management of fish farms situated in plains and therefore, requires considerably greater expertise for successful operation and high production. The scarcity of native fish species, withdrawal of water for irrigation and difficulty in winter fishing from ice further limits the development of coldwater aquaculture and fishery. These requirements will need to be integrated with those of other consumers to reduce clashes over water use. The most significant constraints inflicted on high altitude climate and coldwater bodies are characterized by low productivity than warm waters and reflection of the short growing season. The fate of future aquaculture depends critically upon resource management since natural fish stocks are vulnerable to over-fishing and restoration is a long process.

For developing a succesful coldwater aquaculture program, economic threshold must be determined, environmental factors in the ecosystem must be elucidated, and high production technologies must be devised. This requires trained personnel who can obtain the data inevitable for analysing the problems. Once the strategy has been developed and tested, operation on a commercial scale requires competent field personnel who can translate the research findings into use for farmers.

Many mountain water bodies have potential for aquaculture expansion although this will require consistent financial and technical supports. Solid knowledge on culture and management technologies imparted by national and international experts would definitely provide shape and momentum to fishermen communities. If these constraints and problems noted earlier are overcome, there is likely to be essential changes in the structure of the mountain fisheries and aquaculture. Moreover, establishment of basic regional centres, selective breeding and stocking are necessary to guide the development activities which would obviously encourage a higher level of sophistication in coldwater aquaculture management.

While several warm water fish species are widely used for farming in low and/or medium altitudes, market-sized cold water fish in higher altitudes have been largely limited to trout and common carp.

References

Badola, S. P. and H. R. Singh. 1981. Fish and fisheries of the river Alaknanda. *Proc. Nat. Acad. Sci. India*. B 51: 133-142.

Desai, H. R. 2000. Reproduction biology and spawning ecology of *Tor tor* (Ham.) from the river Narmada. In: Coldwater Aquaculture and Fisheries (Eds. H. R. Singh and W. S. Lakra), pp. 235-252.

Dulma, A. 1979. Hydrobiological outline of the Mongolian lakes. *Int Revueges. Hydrobiol*. 64(6): 709-736.

Food and Agricultural Organization (FAO). 1987. Coldwater fishery: fact-finding and project idea formulating mission to mountaneous regions of Bhutan, India and Nepal. Report based on the work of *T. Petr* X. Lu and K. G. Rajbanshi, FAO, Rome, 63 PP.

Food and Agricultural Organization (FAO). 1990. Rainbow trout culture in Kabul, Afganistan. Project findings and recommendations. AFG/86/0l3, FAO, Rome, 8 P.

Food and Agricultural Organization (FAO). 1995. Coldwater fish culture in Azad Kashmir, Pakistan. Project findings and recommendations. Terminal Report for Project FI : DP/PAK/88/048, FAO, Rome 15 P.

Food and Agricultural Organization (FAO). 1999. Fish and Fisheries at Higher Altitudes. FAO Fisheries Technical Paper (No. 385).

Gall, G. A. 1992. The Rainbow Trout. Elsevier Scientific Publishers, New York.

Ghosh, G. K. 2002. Water of India: Quality and Quantity. APH Publishing Corporation, New Delhi, India, 500 PP.

Greenberg, D. B. 1960. Trout Farming. Chilton and Company, New York.

Jhingran, V. G. and K. L. Sehgal. 1978. Coldwater Fisheries of India. *J. Inland Fish Soc. India* 239 pp.

Joshi, P. C. 1991. The degrading fish habitats of the Kumaun Himalaya. Pp. 313-320. *In*: Ecology of Mountain Waters (Eds. S. D. Bhatt and R. K. Pande). Ashish Publishing House, New Delhi.

Kendali, R.L 1978. Selected coldwater Fishes of North America Proc. of the Symposium held in St. Paul, Minnesota, March 7-9 1978. Trans. Amer. Fish, Soc, Washington, DC.

Kulkami, C. V. and -5. N. Ogale. 1992. Conservation of the mighty mahseer of India. Tata Electric Company, Mumbai, p. 1-35.

Kulkarni, C. V. 2000. Artificial propagation of *Tor Khudree* (sykes) and *Tor tor* (Hamilton). *In*: Coldwater Aquaculture and Fisheries (Eds. H. R. Singh and W. S. Lakra), Pp. 203-218. Narendra Publishing House, Delhi.

Laird L. and T. Needham. 1988 . Salmon and trout farming. Ellis Horwood Ltd. Ltd. England, 271 PP.

Ogale, S. N. 2000. Mahseer hatchery: Planning and management. *Fishing Chimes*. 19 (10 & 11) : 69-73.

Raina, H. S. and T. Petre 1999. Coldwater Fish and Fisheries in the Indian Himalayas: Lakes and Reservoirs. FAO Fish. Tech. Rep. No. 385, p.64-88.

Savvaitova, K. A. and T. Petre 1999. Fish and Fisheries in Lake Sevan, Armenia, and in some other high altitude lakes of Caucasus. FAO Fish Tech Rep P. No. 385, P. 279 -304.

Sehgal, K. L. 1989. Conservation and management of aquatic ecosystem of Indian uplands. pp. 331-362 *In.* Management of aquatic Ecosystems. Society of Bioscience, Muzaffarnagar, India.

Sehgal, K. L. 1990. Ecology, abundance and distribution of fish in the North Western Himalayan streams. pp. 287-300. *In*: Management of aquatic Ecosystems. Society of Bioscience, Muzaffarnagar, India.

Sehgal, K. L. 1983. Fishery resources and their management. pp. 225- 273. *In*: Studies in Eco-Development, Himalayan Mountains and Men (Eds : T. V. Singh and J. Kaur), Print House, Lucknow, India.

Sharma, A. P. 1991. Evaluation of ecosystem characteristics and trophic status with reference to fishery potential of Kumaun lakes. pp. 149-173 *In*: Ecology of the Mountain Waters. (Eds. S. D. Bhatt and R. K. Pande). Ashish Publishing House, New Delhi.

Shatunovsky, M. I. 1983. The Fishes of the Mongolian People's Republic, Nauka, Moscow, 278 PP.

Singh, H. R. et. al 1991. Hill stream fishery potential and development issues. *J. Inland Fish Soc. India*. 23 (2) : 60-68.

Singh, H. R. and N. Kumar. 2000. Some aspects of ecology of hill-streams: stream morphology, zonation, characteristics, and adaptive features of ichthyofauna in Garhwal Himalaya. pp. 1-18. *In*: Modern Trends in Fish Biology Research, Narendra Publishing House, Delhi.

Vass, K. K. 1998. Himalayan lakes: present status and strategies for management. 1B: Advances in Fisheries and Fish Production (Vol.2). (Eds: S. H Ahmad). Hindustan Publishing Corporation, New Delhi, India, P. 113-126.

Vass, K. K. 2000. Breeding biology of trout. pp. 155-168. *In*: Coldwater Aquaculture and Fisheries (Eds. H. R. Singh and W. S. Lakra), Narendra Publishing House, Delhi.

Walker, K. F. and H. Z. Yang. 1999. Fish and fisheries in Western China. *FAO Fish Tech. Paper* No. 385. pp. 237-278.

Williams, W. D. 1991 Chinese and Mongolian saline lakes: A limnological review. *Hydrobiologia*, 210 : 39 -66.

Questions

1. What do you mean by coldwater fish culture ? How this culture differs from that of warmwater culture strategy ?

2. Discuss why coldwater aquaculture has become a growing industry in many upland areas of the world. State why coldwater fish culture strategy differs in different countries.

3. Mention the topographical features of mountain regions. Why topography is so important in fish culture in upland areas?

4. Discuss the characteristic features of mountain regions?

5. What are the features of coldwater streams?

6. What are the characteristic features of coldwater fish?

7. Trace the impact of human intervention on coldwater fish culture.

8. What is watershed management? Discuss the role of integrated watershed management in fish culture of upland areas.

9. What are the advantages of integrated farming system in watershed areas? What steps should be taken to adopt such farming in the area concerned?

10. What are the problems associated with the development of coldwater fisheries in tropical countries?

11. What are the steps required for better water harvesting management?

12. Is there any scope to develop coldwater fish culture in tropical mountain regions? If so, suggest some important steps that should be taken into consideration.

13. What are the developmental steps that have taken place in the field of artificial propagation of coldwater fish species?

14. What are the possibilities that should be followed to increase fish production in high altitude regions?

15. Discuss the prospects of integrated aquaculture in mountain regions.

16. Suppose you will have to construct an one hectare integrated fish farm in upland areas. What strategies should have to be taken into consideration to make the farm viable and profitable?

17. Most of the coldwater lakes are suffering from human activities. What steps should be taken to conserve different lake habitats and to restore their pristine condition?

18. Write notes on the following:

(a) Fishery potential in lakes of (i) Caucasus, (ii) Western China, (iii) Mongolia, (iv) Central Asia, (v) India.

(b) Pen and cage culture of coldwater fish.

(c) Carp and trout production potential in Europe.

(d) Raceway culture.

(e) Artificial propagation of coldwater fish.

(f) Problems and prospects of coldwater fish culture in mountain regions.

(g) Usefulness of water harvesting for fish culture in mountain regions.

(h)   Management of coldwater lakes for fisheries and aquaculture.

(i)   Direct effects of man's activity on coldwater fisheries.

(j)   Indirect effects of man's activity on coldwater fisheries.

(k)   Distinguish between coldwater and cool water.

19.   Among different important lakes discussed in section 10.16, which lakes are very important from fishery point of view? Suggest your comments in this connection.

20.   What are the steps that should be taken for the culture of mahseer ? Briefly discuss the culture of mahseer.

21.   Why trout farming is so important in different regions of the world? Briefly discuss the different methods of trout culture.

22.   Discuss how trout and carp are cultured in Europe.

11

Fish Production from Reservoirs

Although fish culture in different inland waters especially ponds and lakes was very popular as far back as agricultural records can be traced, reservoirs have now been extensively exploited for little more than fifty years. Reservoirs are regarded as an economic necessity to many fish farmers. Since reservoirs are the source of a variety of food fish for millions of people, their judicious management practice is inevitable for sustainable capture and culture fisheries. Fish production from reservoirs should be increased in more rapid rate at which world population is increasing. Fish production from most of the reservoirs is merely holding their own. In selected reservoirs, fish production is declining.

It should be remembered that the ancient civilizations of some countries grew vigorously because reservoirs used for various activities. And in course of time, the importance of reservoirs has been realized by millions of people. (See Box 11.A)

11.1 Fish Production From Reservoirs Involves a Multitude of Responsibilities

The progressive fishermen in a modern fish culture farm hold a multitude of responsibilities. They have to plan, direct and control the production effort of the reservoir. The task of fisherman does not stop with the achievement of fish yield. They are also responsible for bringing in the required profits. In addition, they are also responsible for producing robust fish not only for the market but also for the table. In fact, a modern fisherman has to produce healthy fish rather mere profit. Reservoirs should be guided and managed in such a way that its production becomes market-related and profitable.

Fish farmers must bear the brunt of the responsibility of reservoir fisheries development with constant assiduity for achieving the goals. They assist the farm by and large in managing production and market potential of reservoirs and in developing yield forecasts and sales budget. In addition, they have to develop the marketing program and achieve the forecasted yield by implementing the management strategy.

Building the Yield Potential

It is the responsibility of cooperative societies to build the yield potential. The formation of cooperative societies is required to ensure that the yield potential is maintained in compact condition, capable of effectively implementing the marketing program of the fish harvested from reservoirs and marketing policies and strategies of

---

BOX - 11.A
DAMS AND RESERVOIRS IN HISTORY

Dam is a barrier constructed to hold back water and raise its level, forming a reservoir or preventing flooding. However, humans have used dams to trap and store freshwater in reservoirs for more than 5,000 years. The earliest recorded dam is believed to be a masonry structure and 49 feet (15 metres) high built across the Nile River in Egypt (2900 BC). Modern dams are generally built of earth fill, rock fill, masonry, or monolithic concrete. The ancient civilizations of Egypt, Assyria, Mesopotamia, and China grew and flourished in past because construction of dams and reservoirs allowed for irrigation of agricultural lands, control of seasonal floods, and water storage during dry weather.

In the past, no special attention was given towards the development of fisheries in reservoirs. At the end of the 20th century, when humans have realised the importance of both natural and artificial reservoirs on food production, experts and scientists with the introduction of modern technology, have begun to seek ways to produce and harvest fish stocks from reservoirs for millions of people.

Humans today depend on dams to store water for electric power generation, irrigation, drinking water, and flood control in addition to fishery activities. In the middle of the 20th century, engineers constructed many dams and reservoirs for various activities. Although reservoirs produce aquatic animals of commercial importance, it does not mean that all reservoirs have production potential. At present, many reservoirs are well-known to produce fish to a greater or less degree.

Worldwide, there are over 45,000 dams and 50,000 reservoirs of different categories in 150 countries of the world at the end of the twentieth century. Dams and reservoirs built before the 1030s were constructued with little knowledge of how structures can be desgined to resist failure. The new technology has led to an era of construction of dams and reservoirs that was lasted until the present.

---

the reservoir fisheries. In addition, cooperative societies are also required to provide assistance in managing the other aspects of the yield program.

Ensuring Yield and Managing Yield Fluctuation

Cooperative societies or groups of fishermen also play a role in ensuring the yield of the reservoir, they foster an atmosphere of yield for the reservoir. In a reservoir ecosystem, there exists yield fluctuation; so are technology and management methods. In managing yield fluctuations, the society depends largely on the refined technology. In fact, the reservoir ecosystem needs extensive management to cope with the hassles associated with fluctuations of yield potential.

Human Resource Development

The human resource development is also a major part of the reservoir management. Cooperative societies are definitely responsible for providing opportunities to fishermen, their training, development, and motivation. However, the cooperative societies should maintain the production force to serve the purpose.

The various issues involved in fish production from reservoirs and management strategies are presented below in a nutshell.

1. Definition of reservoirs
2. Significance of reservoir fisheries development
3. Classification of reservoirs
4. Fish production strategy in Indian reservoirs
5. Productivity of reservoirs
6. Importance of reservoir ecosystem for fish yield
7. Reservoir fisheries

In the succeeding pages we shall discuss elaborately each of these issues one by one.

11.2 Definition of Reservoirs

Water bodies that are created by dams built of earth work with engineering precision across the perennial or large seasonal rivers or streams, ,using concrete masonry or stone for power supply and large-scale irrigation or flood control purpose are termed as *reservoirs.* Therefore, it can be said that man-made lakes that are created by artificial impoundments are regarded as reservoirs. Man-made lakes are sometimes designated as tanks and hence precluding them from the estimates of reservoirs. Though there is no vivid definition for a tank, ponds and tanks are interchangeable expressions. This expression is prevalent in West Bengal and Orissa. In some states such as Andhra Pradesh, Tamil Nadu and Karnataka, tanks are regarded as irrigation reservoirs including small- and medium-sized water areas. From fish production point of view, however, the distinction between reservoirs and tanks seems to be impertinent.

As a generalization, water bodies of more than 200 hectares in area are termed as *reservoirs.* The water bodies that are considered as reservoirs depend on various factors such as water storage capacity, irrigation etc. In China, for example, those holding more than 100 million cubic metre of water are termed as *large reservoirs,* 10 to 100 million cubic metre as *medium,* and 0.1 to 10 million cubic metre as *small reservoirs.* In Russia, water bodies up to 10,000 hectare area are considered as *small reservoirs* whereas in USA, water areas ranging from 1 to several hectares are termed as *small reservoirs.*

11.3 Significance of Reservoir Fisheries Development

Reservoir fisheries development is one of the most important components of fish production policy and water management. It is not enough if the existing production strategy is appraised of reservoirs adequately, effective measures must be seriously considered and brand discussions are taken wisely. For sustained yield, an organization or a cooperative society has to look beyond its existing production levels. A progressive society has to consider new developmental strategy as a cardinal element for the production policy of reservoirs.

Fish Yield Becomes Necessary for Meeting the Protein Requirements

Fish yield is the essence of protein requirement in adequate amounts. This is practically feasible if fish production is undertaken in reservoirs. In an age of technological and scientific advancements, changes are a natural outcome: change in yield potential and change in expectations as well as requirements. Any production has to be vigilant to these changes. Consumers always seek quality and healthy fish and more value for money. A cooperative society or any other organization who possess interest to reservoir management has to respond well to these requirements and these responses take the shape of new dimensions. Through such a response, the cooperative society reaps a good deal of benefits.

Sustained Yields Become Necessary for Making More Profits to Cooperative Societies

Sustained yields become necessary from the profit angle too. Varieties of fishes that are already established in any reservoirs, often have their limitations in enhancing the profit level of the society. Fish production potential of many reservoirs is declining gradually simply due to lack of adequate management planning.* It is, therefore, essential for cooperative societies to yield more healthy fish by stocking adequate seeds following the adoption of continuous stocking and harvesting strategy. Continuous production from reservoirs becomes part and parcel of the growth requirements of the cooperative society in general and farmer in particular and more profits come to the society only through state-of-the-art technology.

11.4 Classification of Reservoirs

Before we proceed with the chapter on fish production potential from reservoirs, it is essential to understand properly the terminology — Classification of Reservoirs. As noted in Volume 1 (See Section 20.12), reservoirs are generally classified as large (more than 5,000 hectare), medium (1,000 to 5,000 hectare), and small (less than 1,000 hectare). While the first two groups involve capture fishery strategies leading to stock the fish that are able to breed if suitable conditions prevail and get acclimatized through auto-stocking,

---

* Production history has an abundance of instances of production failures on account of the lack of proper development planning. It abounds in success stories emanating from good planning for development. It is quite natural since it is planning which provides the framework for all development strategies – strategies on production, strategies on investments etc.

the third group involves production-related innovations; it gives rise to yield robust fish for mass consumption and for marketing.

A growing body of research on the subject concerned has shown that India has 19,134, 180, and 56 small, medium, and large reservoirs respectively and with a total water surface area of 1,485,557, 5,27,541, and 1,140,268 hectare respectively; of course, with great differences in fish production potential.

11.5 Fish Production Strategy in Indian Reservoirs

Fish production from Indian reservoirs achieved a phenomenal success when the project on reservior fisheries development was launched in the 8th five-year Plan (From 1992-1993 to 1996-1997). At present, most of the Indian reservoirs afford appreciable quantities of fish with a turnover of about Rs. 46 crores per year (data for the year 2008). Reservoirs produce fish for regular supply in markets.

Fish production from Asian reservoirs is generally undertaken through a well-knit strategy and cooperative societies. The formation of cooperative societies is definitely the key to fish production strength and farmers' livelihood at large and forms an important part of the country's fish production strategy. The cooperative societies are, however, managed by expert personnel and fish culturists. They should come across several latest technologies relating to production strategy stances. Different strategy stances denote the particular approaches taken by a given cooperative society to suit particular production requirements. Whatever the strategy stances the society adopt, the core steps in the strategy will always consist of fish production from *reservoirs* through system management and marketing.

Production of huge quantities of fish from reservoirs by exploiting the innovations brought about by technology is not only a utopian view but also is a job normally undertaken by well-equipped societies who are actively engaged in reservoir fisheries management in particular and mega development in general. No wonder, sustained yields are common mostly in advanced countries. In contrast to these countries, developing nations adopting the techniques established in reservoirs as successful strategies is a difficult task as it involves adequate financial support, precise control over the entire capture/culture period, monitoring, and survellance on all categories of reservoirs for successful development. Reservoir fisheries development is, however, especially striking because it has been achieved in a production and marketing industry. Serious attention towards the development of fisheries is a recent issue. Despite the late activity, the national government has adopted state-of-the-art technology leaving traditional norms.

Expansion and Growth of Indian Reservoirs

Truely speaking, the 1990s were a decade of expansion and growth for Indian reservoir fishery projects. It sets in motion expansion plan in moderate succession for achieving a capacity of fish production to the tune of at least 150 kg/hectare. By the beginning of the new millennium, several reservoirs were producing and marketing several thousand kilograms of edible fish and by-catch, thus becoming the important national player in fish production strategy.

Some authorities claim that on the basis of rating, Indian reservoirs, on an average, are virtually at the bottom in competitiveness among some other tropical reservoirs. Most of the reservoirs are not able to successfully produce fish. The country has only a few reservoirs which can claim production standards. And the possibility of the potential of Indian reservoirs to become a national production strategy for successful value-added aquatic species has not been exploited at all in full potential.

Reservoir fisheries have come up with a plan to invest crores of rupees for wide expansion and technological upgradation of their managing, stocking, and harvesting strategies. This will possibly expect to boost the production capacity from the existing average 5 kg/ga to about 150 kg/ha. Most of the productive reservoirs are supported by a wide-service network. Government intervention following the adoption of state-of-the-art technology and management protocols have set a new incline for fish production and marketing within the country.

Our discussion on fish production strategy from reservoirs are amplified through a study under Indian conditions that follows in this chapter.

11.6 Productivity of Reservoirs

The water and soil quality variables are, in part, responsible for a marked increase in production capacity of a reservoir. Therefore, the geoclimatic features definitely play a crucial role in imparting the production potential of any reservoir. It seems quite plausible that Indian reservoirs are distributed over various types of terrains and they receive drainage from various catchment areas.

Geoclimatic Features

About 50 per cent of the total land mass of India lies in the tropics, while the rest is situated above the tropic of cancer (parallel of latitude 23°26' North). The climate varies from the warm tropical in the south to the temperate in the north. The main soil types are alluvial, deep and medium black, red and yellow, lateritic, desert, saline, forest, and hill. Almost all forms of vegetation such as tropical evergreen, tropical moist deciduous, littoral and swamp, Montane sub-tropical, Himalayan and alpine are common in various parts of the country. The landscapes include mountains, alluvial plains, riverine wetlands, plateau, deserts, coastal plains and deltas.

The physiographic division of the country are the Himalayas, the Indo-Gangetic plains, the Vindhyas, the Satpura, the Western Ghats, the Eastern Ghats, coastal plains, the deltas, and the riverine wetlands. The country receives, on an average, 105 cm of rainfall every year. The temporal and spatial distribution of rainfall exhibits wide variations within the country.

Of the total area of the country (3,287, 728 km^2), more than one million km^2 area receives inadequate rainfall. Large rivers such as Krishna, Godavari, Pennar and Cauvery pass through extensive tracts of low rainfall areas and therefore carry less quantity of water than the rivers passing through high rainfall areas such as Northeast and the West coasts. Substantial amounts of rainfall (about 1,000 cm) are received by the Khasi and Jaintia hills in the Northeast and the rate of precipitation in the Brahamaputra valley has been estimated as 200 cm. In the West coast, heavy rainfall occurs along the slopes of the Western Ghats. Similarly, the Indo-Gangetic plains and the Himalayas also receive rainfall in substantial amounts. People living in high rainfall areas have a tendency to store water by creating barricades across the streams and tributaries for agricultural activities since time immemorial. But at present, the construction of dams for agricultural production and hydro-electric power generations have drastically changed the past scenario. Such activities have become a common feature not only in India but also around the world.

11.7 Reservoir Ecosystem

Reservoir ecosystem is essentially a combination of fluviatile and lacustrine systems and therefore, it has certain characteristic features of its own. During the months of inflow and outflow, the reservoir exhibits a lotic environment whereas in summer when the inflow into and outflow from the reservoir shrinks, a more or less lentic condition prevails in most parts of the reservoir. Water renewal pattern is one of the most unique features of the reservoir.

During the monsoon months, huge quantities of water are drained into the reservoirs and all outlets are opened, resulting in total flushing. This process ejects the standing crop of biotic communities. Sudden fluctuations also affect the benthos by exposing the substrata. The nutrient inputs from the allochthonous source determine the water quality, nutrient regime and the basic production potential. Deep draw-down, wind-mediated turbulence are some of the features that makes the reservoir ecosystems more attractive. Some important salient factors which are related to the increasing /decreasing the productivity of reservoir ecosystems should be kept in mind (See Panel 11.1).

Several factors are primarily responsible for the formation of reservoir ecosystems and have been noted in several sub-sections.

Climatic Factors

Reservoirs are exposed to a wide range of climates from the temperate Himalayas in the north to extreme tropical in the southern peninsula that has great practical significance in determining the productivity of large bodies of water. Also, the latitudinal location is equally important that determines the quantum of solar radiation available for primary productivity. About 0.2 to 0.7 per cent of the incident solar energy is fixed as chemical energy by the producers in different reservoirs which are located between 11° 25′ N and 31°25′ N latitude.

PANEL 11.1
FACTORS INVOLVED IN THE PRODUCTIVITY OF RESERVOIR ECOSYSTEMS

A. Factors augmenting the productivity of reservoirs :

1. Development of shoreline.
2. Low mean depth (less than 18 m).
3. Existence and extent of marginal vegetation.
4. Nutrient enrichment during floods.
5. Moderately developed macrophyte community.
6. Abundance of periphyton.
7. Well-developed plankton and benthos.
8. Introduction of fishery which are adapted to lentic conditions.
9. Implementation of modern fishing gear.
10. Enforcement of fishery regulations.
11. Conservation of fish biodiversity.

B. Factors decreasing the productivity of resevoirs :

1. Low transparency due to high turbidity.
2. High men depth.
3. Erosion of watershed areas in reservoirs.
4. Reduction in quantity of water flowing into the reservoir.
5. Wide fluctuation of water levels.
6. Pollution in the watershed.
7. Presence of unbalanced fish population.
8. Exposure of fish breeding grounds during draw-down.

It is important to note that air temperature and wind velocity play a key role in the productivity of a water body. The wide seasonal variations in air temperature is the predisposing factor in the thermal features of the north Indian and peninsular reservoirs. On the other hand, the southern reservoirs are characterized by the narrow range of fluctuations in air and water temperatures. This phenomenon prevents the formation of thermal stratification. As a generalization, the thermal gradient occurs when high air temperature warms up the upper layer of water. In peninsular India, the air temperature remains high round the year. During summer, high temperature at the bottom does not offer any scope for thermal resistance by the warmer upper layers. Thermal stratification is important because in this situation, the water above and below thermocline does not mix up thereby nutrients at the bottom layer are locked up. A warm bottom layer facilitate rapid decomposition of organic matter, thereby increasing the process of nutrient release.

The deep basin of some reservoirs (such as Nagarjunasagar) does not favor the formation of thermocline. The continuous draw-down from deeper layers, the wind and the wave facilitate the mixing up of water column in most of the Indian reservoirs. However, some upper peninsular reservoirs undergo transient phases of thermal stratification during the summer stagnation depending on the depth, water abstraction, and the wind. Wind helps distribute the heat and equalize the temperature in the water. Wind-induced turbulance is definitely important in churning of the reservoirs and hence facilitates the nutrient availability at the trophogenic zone.

Morphometric Factors

The term morphometry refers to the process of measuring the external shape and dimensions of landforms, living organisms, or other objects. In case of a reservoir ecosystem, however, morphometry is the function of the height of the dam and the topography of the impounded areas. The mean depth is the most important morphometric factor that determines the productivity of reservoirs. The shallow regions of the reservoir waters are highly eutrophic, facilitating greater mixing and circulation of heat and nutrients and hence higher productivity. Deep regions serve as a nutrient sink at the bottom, where organic matter accumulates in large amounts and thus nutrients become unavailable at the photosynthetic zone. Medium-sized reservoirs have mean depth and that varies between 2.3 m (Poondi reservoir) and 24 m (Bhatghar reservoir) . Small reservoirs have a mean depth that also varies between 2 m (Vidur) and 15 m (Badua). Mean depth of the hydel reservoirs of the mountain slopes is high compared to that of irrigation reservoirs of the plains and plateaus. The two largest impoundments in the country such as Gandhisagar and Hirakud have low mean depth of 12 m and 11.5 m, respectively. On the other band, Idukki — the hydel reservoir of the Western Ghats which is one-tenth of Gandhisagar, has mean depth of 30 m.

The mean depth does not always exhibit any direct correlation with productivity. Shallow reservoirs such as Vidur do not support rich plankton community. Some other shallow reservoirs such as Govindgarh and a Kulgarhi in Madhya Pradesh also exhibit oligotrophic nature. On the contrary, deepest reservoirs such as Govindsagar have high production potential (about 100 kg fish/hectare).

Hydro-Edaphic Factors

Deficiency of nutrients and other minerals in many reservoirs of Kerala indicates the oligotrophic nature. This is due to the fact that the rivers of Kerala drain humus and lateritic soils from the hills of Western Ghats which are deficient in calcium, nitrogen, and phosphorus. Those reservoirs which receive drainage passing through agricultural area have higher levels of hardness and alkalinity. Similarly, reservoir waters are soft with less mineral salts due to geo-chemical reasons. Soils of Madhya Pradesh are, in general, deep black, medium black, mixed red, and shallow black in colour. Such soils are low in nitrogen and phosphorus.

Many reservoirs in Tamil Nadu and Andhra Pradesh are characterized by hard water due to presence of calcareous rocks and limestone underlying the water course. The acidic condition of the water of the North-eastern reservoirs such as Barapani, Nongmahir etc. is attributed to the acidic soil of the reservoir bed.

1. **Nutrient Budget :** Though most of the tropical reservoirs are low in phosphate and nitrate, fish production potential does not seem to be an indicative of low productivity. Reservoirs such as Gandhisagar, Bhabani sagar, and Amaravathy are apt examples where phosphate and nitrate levels are sufficiently low and moderate to high primary productivity has been noted. However, the reservoirs of Rajasthan and Madhya Pradesh have high levels of phosphate that varies from traces to about 15 mg/l.

In addition to the nutrient budget, specific conductivity (35-65 μ mhos), total hardness (15-50 mg/l), and total alkalinity (20-100 mg/l) strongly supports very rich in plankton community and and good stock of fish. It must be kept in mind that voluminous data on physico-chemical factors of water and soil partaining to different reservoirs lead to the conclusion that none of the edaphic, morphometric and water quality parameters can be used as a dependable yard-stick to signify the productivity with degree of accuracy. Physico-chemical features of Indian reservoirs are, however, shown in Table 11.1

Table 11.1 : Physico-Chemical Features of Indian Reservoirs

Feature	Range	Productivity		
		Low	Medium	High
A. Water	6.5-9.2	6.0	6.0-8.5	8.5
pH	6.5-9.2	< 6.0	6.0-8.5	> 8.5
Alkalinity (mg/l)	40-240	40	40-90	90
Nitrate (mg/l)	Trace-0.90	Trace	Upto 0.2	0.2-0.5
Phosphate (mg/l)	Trace-0.40	Trace	Upto 0.1	0.1-0.2
Temperature (ºC)	12-31	< 18	18-22	> 22
B. Soil				
pH	6.0-8.8	< 6.5	6.5-7.5	> 7.5
Available phosphorus (mg/100g)	0.47-6.1	< 3	3-6	> 6
Available nitrogen (mg/100g)	13-65	< 25	25-60	> 60
Organic Carbon (per cent)	0.6-3.2	0.5	0.5-1.5	1.5-2.5

Biotic Communities

A community is a group of interacting species populations living together in any particular aquatic ecosystem. The living part of the ecosystem could be observed as a dynamic unit with a web of energy flow and an exchange of matter. Consequently, reservoir ecosystems supply many different food items. Productive reservoirs are, however, generally teamed with different groups/species of plankton, macrophytes, bottom organisms, and fish.

1. **Plankton:** The group Cyanophyceae or blue-green algae forms an important plankton community. The presence of different species of Cyanophyceae especially *Microcystis* in productive reservoirs of Gangetic plains, Deccan plateau, Orissa and Tamil Nadu is worth mentioning. High intensity of sunlight, extensive catchment area, and agricultural land favor the development of algal blooms. In high altitude reservoirs such as Gobindsagar, the community of macrophytes is replaced by Dianophyceae. Oligotrophic reservoirs of the Northeast have Chlorophyceae-dominated plankton community.

(i) *Plankton Pulses* : Plankton pulse is the rhythmical throbbing of the population of both phyto- and zooplankton in water in different times of the year. Most of the reservoirs have three plankton pulses that harmonize with the post-monsoon (september-November), winter (December-February), and summer (March-May) seasons. The monsoon season (June-August) highly disturbs the standing crop of plankton. So long as the dam outlets are closed, the allochthonous nutrients favor the growth of plankton communities. In winter season when the water becomes clean, the plankton community progresses through successions to reach a point of highest development. In contrast to deep reservoirs, the shallow and nutrient-rich ones sustain a permanent plankton bloom.

In many situations, plankton pulses may exhibit 'antibiosis' which refers to complete or partial inhibition or death of one form of life by another through generation of some substances or ecological conditions as a result of metabolic pathways. In this case, none of them derives any benefit. Antagonistic substances are reported in some plankton pulses. As for example, in culture populations of *Chlorella vulgaris*, some substances accumulate which inhibits the growth of diatoms, *Nitzschia frustrulum*. Blooms of blue-green algae especially *Microcystis* are widely known to generate toxins which causes death of fish. Many examples exist in this regard and there has been accumulating day by day much informations on this aspect, detailed account of which is beyond the scope of the present chapter. Interested readers are referred to consult books on microbiology.

2. **Macrophytes** : As a generalization, Indian reservoirs do not exhibit the population of macrophytes except in some cases where there are severe pollution, low water renewal, and aging of reservoirs. Microphytes are, however, restricted to isolated patches of *Vallisneria*, *Hydrilla* and mats of *Spirogyra*. Some irrigation reservoirs in Uttar Pradesh and Andhra Pradesh are known for luxuriant growth of

macrophytes. Old reservoirs such as Vanivilas and Markonchalli in Karnataka have a luxuriant growth of macrophytes.

The qualitative and quantitative distribution of macrophytes depends, to a large extent, on the physiographic conditions and soil types. Thick macrophytes comprising the littoral, submerged and emergent types are very important characteristic features of most irrigation tanks of coastal plains. In the region between plateau and hills which are characterized by the presence of red and lateritic soils, the tanks are fertile and aquatic weeds are sparse. Those tanks which are located in black soil regions, submerged and emergent weeds grow profusely.

Although macrophytes provide suitable habitats for insects and molluscs which substantially contribute to the species diversity of the reservoir, the presence of macrophytes is not desirable from fish production point of view for the following reasons. *Firstly*, floating weeds utilize the solar radiation for photosynthesis and make it available to the phytoplankton populations, *Secondly*, submerged weeds offer shelter for weed fishes and minnows. These fishes emulate with carp as food, *Thirdly*, huge quantities of nutrients are robbed by different types of macrophytes, and *Fourthly*, luxuriant growth of macrophytes causes high rate of decomposition of dead plant materials which helps deplete the content of its oxygen, thus creating an anaerobic condition which is further aggravated if floating vegetation is covered over extensive unmanageable water areas. And as a corollary, such covering helps prevent light from penetration into the water body. Under these conditions, however, fish mortality in summer months is very common in several reservoirs. Thus macrophytes really create a lot of hassles associated with the reservoir fisheries management. Some common and important macrophytes which are grown in different reservoirs are listed below.

TYPE	SPECIES
A. Submerged	*Ceratophyllum sp., Potamogeton crispus, P. natans, P. pectinatus, P. nodosus, Najas minor, N. graminaea, Hydrilla verticallata, Vallisnaria spiratis, Chara intermedia, Nitella* spp.
B. Floating	*Eichornia sp. Lemna sp.* and *Azolla* spp.
C. Emergent	Ipomea reptans, I. aquatica, Polygonum glabrum,
	Typha elaphanta, T. angustata

3. **Benthos:** In the benthic habitat, fishes are found in the bottom. Plants are found only in shallow areas of the reservoirs though they are also abundant in different depths where light penetration permits their growth. The bottom itself may be rocky, high fertility of soil, variable slopes, different types of substrates, mud, silt, and debris. This condition provides suitable habitats for many species of bottom animals and plants. Several aquatic macrophytes also act as a favorable criteria for the development of benthos.

Several reservoirs in Karnataka such as Tungabhadra, Vanivilas Sagar, Markonahalli, Krishnaraj Sagar, some small as well as shallow reservoirs of the gangetic basin have dense populations of benthic organisms and their growth and developnent in relation to fish production are very striking. Deep reservoirs such as Rihand have,however, barren benthic community. The groups of benthic organisms are represented by Chironomids, Molluska, and Annelids.

* *Productivity of Benthic Community*: The Productivity of benthic communities depends on the food available and is related to local primary productivity. Shallow water areas of reservoirs usually have the highest productivity and as a generalization, the biomass of fish declines considerably along with the increase in depth and distance from shore area. Thus very deep reservoirs are populated by small number of fish. And consequently, the major fisheries developed by bottom-dwelling fish are highly restricted to shallow-water areas of the reservoirs.

Benthic animals especially fish, crustacea, and molluska have a variety of food items available to them. The organic matter generally termed as *detritus* is found in the sediments, and much of it is derived from the plankton. This material is made more worth-while by the presence of bacteria on its surface. Large living plants of the bottom of the reservoirs are not used for food, but provide a hard surface on which small algae and bacteria develop. These organisms are consmed by harbivorous fish and mollusks. Large-sized aquatic plants, however, only become food for them when they are dead and undergo decomposition.

4. **Fish Species :** In general, Indian reservoirs are teemed with a variety of fish species. Large reservoirs, however, harbor more than 70 species of fish, out of which about 45 species significantly contribute to the commercial fisheries. Although Indian major carps occupy a dominant role among different commercially important species of fishes, a number of exotic species, catfishes, featherbacks, air-breathing fishes, murrels, and weed fishes have contributed substantially to the commercial fisheries. Various species of fish that follows are dominated in Indian reservoirs.

GROUP	SPECIES
1. Indian major carp :	*Labeo rohita, L. calbasu, L. fimbriatus, Crihhinus mrigala, Catla catla*
2. Minor carp:	*Cirrhinus cirrhosa, C. reba, L. kontius, L. bata, Puntius sarana, P. dubius, P. cannacticus, P. dobsoni, P. chagunio, Schizothorax plagiostomus, Osteobrama vigorsii*
3. Catfish:	*Arichthys aor, A. seenghala, Wallago attu, Pangasius pangasius, Silonia silondia*
4. Featherbacks:	*Notopterous notopterous, N. chitala*
5. Air-breathing Fish:	*Clarias batrachus, Heteropneustes* fossilis
6. Murrels:	*Channa marulius, C. striatus, C. gachua, C. punctatus*
7. Exotic Fish:	*Oreochromis mossambicus, Silver carp, Cyprinus carpio communis, C. carpio specularis, Gambusia affinis,*

8. Weed Fish: *Ambassis nama, Esomus dandricus, Aspidoperia morar, Amblypharyngodon mola, Puntius sophore, P. ticto, Oxygaster bacaila, Laubuca laubuca, Barilius barila, B. bola, Gadusia chapra, Osteobrama cotio*

(i) *Fluctuations of Fish Species* : Drastic fluctuations of fish species in many reservoirs are the result of a series of human intervensions and natural changes of the ecosystem such as water currents, turbidity and the loss of breeding grounds. A state of chronic overfishing now exists in the world's reservoirs. The increased landings have reflected the development of fishing fleets and the targetting of one fish species after another.

Fishermen maintain income by catching more rather than fewer fish. If this reaches the point where fish are caught before spawning, a catastrophic decline in numbers will occur and local extinction may result.

Accidental or deliberate introduction of exotic species have led to drastic change in the species set-up in several reservoirs. For examples, introduction of silver carp, common carp, and tilapia respectively in Gobindsagar, Krishna Sagar, and Amaravathy reservoirs have drastically altered their community structure.

11.8 Impact of Reservoir Formation on Indigenous Fish

NEGATIVE IMPACT

A decline in pristine condition of many Indian rivers have already been noted due to changed ecological conditions especially silt deposition and stratification of water body. At present, many native species (see Panel 11.2) which inhabits in river ecosystems are on the decline. High reproductive potential of introduced fish such as tilapia, common carp, and silver carp have further posed serious threat to the indigenous species.

Prior to construction of Hirakud reservoir, the river Mahanadi had about 105 species of fish which at present, has declined to 40. Similarly, several native fish species such as *Labeo pangusia, Puntius sarana P. dubius, P. borcallus* borcallus etc. have either partially disappeared or declined drastically from Tungabhadra and Nagarjuna Sagar reservoirs in the Krishna river system due to the absence of their fluviatil environment and the changed trophic structure. At one time, these reservoirs flourished a variety of fish species such as *Labeo fimbriatus, L. calbasu* and *Tor khudree*. But due to habitat destruction, these species have declined considerably over the years and the vacant niche has been filled up by minnow-predator combination.

POSITIVE IMPACT

A number of fish species have the ability to adapt to the reservoir ecosystems where they find congenial habitat and flourish there. This favorable condition is established soon after the holding of water in reservoirs. However, all species are not considered as commercial rather, a very limited number of species such as *Pangasius, Osteobrama vigorsii,*

PANEL 11.2

The following indigenous species of fish are on the decline :

1. The mahseer, snow trout, and *Labeo dero* of the Himalayan streams.
2. The catadromous eel of all major river systems.
3. *Punctius sarana, Tor tor, T. mosal, Labeo fimbriatus L. calbasu,* of the Mahanadi river.
4. *Cirrhinus cirrhosa* and *Labeo konitus* of the Kauvery river basin.
5. *Puntius dubius, Puntius sarana, P. porellus, Labeo calbasu, L. fimbriatus,* and *Tor kudree* of the Krishna river system.
6. The mahseers, eels and *Osteobrama belangiri* of the Northeast rivers.

and *Salmostoma phulo* support a flourishing dry fish trade in Tungabhadra and Nagarjunasagar reservoirs. This indicates that many reservoirs act as sanctuaries for several species of fish such as *Bariliu bola* in Tilaiya (Damodar), *Mystus krishnensis, Osteobrama vigorsii* in Nagarjunasagar (Krishna),*Tor kudhrii* and *T. mussuliae* in Shivajisagar (Krishna), *T. putitora* in Pong (Beas) and Vallabhsagar (Tapti).

11.9 Strategies Involved in Fish Production From Reservoirs

An explicit assessment of fish production from Indian reservoirs is fallacious. Statistical evaluation of the voluminous data received so far on fish production has shown that production potential per hectare basis is not attractive and most of the data generated is so far are of historical importance. At the same time, the production figures available on most of the reservoirs are not reliable and accurate due to several factors as noted below:

1. Unorganized market channel since markets are controlled by illegal money lenders.
2. Ineffective cooperative set-up.
3. Diverse royalty/crop-sharing systems practised by different state governments.
4. Inadequate and poorly trained manpower at the disposal of cooperatives/state governments to collect catch data which is beyond possible to cover the entire reservoir.

Due to several reasons such as streamlined machineries to record catch stastistics, size of reservoirs, and the large number of remote landing centres, an effective monitoring of catch is practically impossible. However, on the basis of the recorded data, fish production potential of 422 reservoirs has been presented in Table 11.2 Fish production figures of small reservoirs of Andhra Pradesh are very impressive (about 188 kg/ha), followed by those of Kerala, Madhya Pradesh, Tamil Nadu, Rajasthan that varied from 46.43 to 53.5 kg/ha. and medium-sized reservoirs of Rajasthan, on an average, produce fish to the tune of 24.5

Table 11.2 : Fish Yield (Kg/ha) in Different Categories of Reservoirs in India

State	Small Reservoirs			Medium Reservoirs			Large Reservoirs			Pooled Reservoirs		
	Number	Production	Yield	Number	Production	Yield	Number	Production	Yield	Number	Production	Yield
Tamil Nadu	52	760	48.50	8	269.0	13.74	2	294.0	12.66	62	1,323.0	22.63
Uttar Pradesh	31	168	14.60	13	156.0	7.17	1	50.0	1.07	45	374.0	4.68
Andhra Pradesh	37	2,224	188.00	29	306.0	22.00	3	800.0	16.80	69	4,330.0	36.48
Maharashtra	6	72	21.1	12	313.5	11.83	4	794.0	9.28	22	1,179.6	10.21
Rajasthan	78	970	46.43	17	600.0	24.47	2	120.0	5.30	97	1,690.0	24.89
Kerala	7	118	53.50	2	17.3	4.80	-	-	-	9	135.0	23.37
Bihar	25	22	3.91	3	7.2	1.90	1	1.0	0.11	28	30	0.054
Madhya Pradesh	2	24	47.26	20	624.9	12.02	3	1,184.0	14.53	25	1,833.1	13.68
Himachal Pradesh	-	-	-	-	-	-	2	1,453.0	35.55	2	1,453.0	35.55
Orissa	53	349	25.84	6	163.0	12.76	3	925.0	7.62	62	1,437.0	9.72
Total	291			110			21			422		
Average			49.90			12.30			11.43			20.13

Production in tonne
Source : Sugunan (1997)

kg/ha,while Tamil Nadu, Maharashtra, Madhya Pradesh and Orissa, about 50 per cent of this yield has been recorded.

On the basis of catch statistics of 422 reservoirs, the national fish production rate has been estimated as 20 kg/ha. Applying this production data into 1,485,557 ha small, 527,541 ha medium and 1,140,268 ha of large reservoirs, their current production rate can be estimated as 74,129, 6,488, and 13,033 tonnes, respectively. An average increase in yield rate up to 100, 75, and 50 kg/ha in respect of small, medium and large reservoirs (See Table 20.15 of Volume 1) would definitely assure fish production at the rates of 148,556, 39,565, and 57,013 tonnes. This will increase the production rate by at least three times.

11.10 Management of Reservoir Fisheries

In contrast to some other Asian countries such as Indonesia, Thailand, Sri Lanka, and China where fish yield is high (yield varies between 50 and 285 kg/ha in different reservoirs), fish production in Indian reservoirs in very low (about 20 kg/ha) that can be attributed to inadequate management inspite of the availability of scientific manpower in fisheries sector of the country. Fish production in reservoirs is essentially deducible in nature and hence the quintessence of management strategy lies on full utilization of natural stocks. Nonetheless, the ecosystem management provides different degrees of freedom for stock manipulation. The extent of human intervention in ecosystem management can be used to differentiate between capture and culture fisheries.

While fish culture systems provide avenues for the nation as a whole to monitor and alter the habitat variables and the biotic communities at will, this freedom will attenuate as we proceed from aquaculture to the culture and capture-based fisheries. In contrast to small water bodies, the large ones have little room for altering the habitat variables. And the scope for effective changes in fish populations is limited to stocking which has no practical significance in relation to fish production.

Management norms in capture and culture-based fisheries vary greatly depending on the type of reservoirs. In general, large and medium reservoirs are used for capture fisheries and the management protocols are based on the principle of stock manipulation, effective conservation measures, gear selectivity, and adjustment in fishing efforts. On the contrary, small reservoirs are effectively utilized as culture-based capture fisheries and the main emphasis is given on stocking at regular intervals, fattening and harvesting the stock. For small reservoirs, however, effective management must be carried out to achieve satisfactory results.

Stocking

Selective stocking must be ascertained not only for setting up species diversity but also to fill the vacant ecological niches and to ascertain the utilization of food reserves. Stocking is, however, carried out at the time of trophic disruption. Failure in this management protocol causes the proliferation of trash fishes by utilizing fish food organisms. This, in turn, may create a good habitat for catfishes, minnows, and murrels particularly in large reservoirs. As a consequence, a long food chain is established. Though these reservoirs harbour good standing crops of plankton and benthos, these food chains are not utilized by fish biomass.

Since large and medium reservoirs are developed on the principles of capture fisheries, it is desirable to stock the species that are capable of breeding and ultimately acclimatized in reservoir ecosystems through auto-stocking. This is inevitable to meet the long-term objectives of obtaining a sustained yield. However, continuous stocking the reservoirs with fish involves heavy costs and also create many practical difficulties in producing the stocking material in adequate quantities.

Strong currents completely expunge the carp eggs from shallow regions to deep zones of the reservoir leading to their destruction and consequently recruitment failure is not uncommon. Also the annual fluctuations in the magnitude of monsoon floods significantly affect the breeding and stocking of carps.

The relative abundance of natural food organisms generally determine the stocking rate. Utilization of natural food organisms by the fish stock in full potential increases the carrying capacity of reservoirs and hence summons for a higher population density which is achieved by adding species to the original fish stocks. Also, information on differences in the growth rate of native and introduced species and factors involved in attaining the introduced species for the table will provide an insight into the production dynamics. However, a number of principles should be strictly followed while selecting the fish for stocking.

1. The cost of stocking and managing the species should be such that the operation is economically viable.

2. Availability of stocking materials with minimal transportation cost.

3. The stocked fish should be fast-growing.

4. The fish should find the ecosystem ideal for growth and reproduction.

5. Suitable size should be maintained for fingerlings to achieve better yield.

6. Stocking should be made in such away that natural food organisms of the ecosystem are fully utilized.

1. **Stocking Measure:** Generally fingerling stages of a species or combination of species are considered for stocking without any definite density levels or ratios. The rate of stocking and the species combination are, however, determined by their availability.

Most of the Asian reservoirs are represented by a wide variety of biotic communities that essentially consist of phytoplankton, zooplankton, and benthos. These communities are primarily shared by the carps and trash fishes. This focuses on the need for controlling carp minnows and weed fishes. For the ecosystem-related management strategies stances, due attention should be given to trophic strata in terms of alloted, unalloted, and vacant niches. The detritus and the grazing food chains are the two princinal pathways through which energy is locked in the fish flesh.

Before the development of induced breeding technology, reservoirs were stocked with fry/fingerlings collected from rivers. At present, however, most of the gangetic and peninsular reservoirs are largely dominated by the Indian major carps and to a less extent, the minor carps. But since Indian major carps are less suited to utilize phytoplankton in full potential, stocking of reservoirs with silver carp has been found to be suitable for efficient utilization of phytoplankton. In reservoirs where huge quantities of macrophytes are grown, some water-loving fish such as *Ctenopharyngodon idella*, *Puntius pulchellus* and *P. Javanicus* should be stocked to check the growth of macrophytes. The common carp though stocked in reservoirs, is not considered as suitable species because of its browshing behavior in the sediment that makes the water body more turbid. Mahseer, silver carp, grass carp, *Tinca tinca*, and *Carassius carassius* are recommended for stocking in high altitude reservoirs. It should be noted that for establishment of a variety of fish species and to utilize all natural food resources, stocking materials should be diversified. The deleterious effects of reservoirs following the stocking of introduced species need to be assessed before their extensive stocking is permitted.

2. *Impact of Stocking*: Suitable stocking is generally performed for sustained production. This production is possible if the stocked fishes survive, grow and trapped in the fishing gear. However the spawning program in large and small reservoirs can be successful if the stocked fish are able to breed successfully in reservoir ecosystems. This breeding will definitely permit auto-stocking. But in many situations, this strategy does not exhibit any significant increase in production potential of reservoirs inspite of the persistent stocking, thereby furnishing the expenditure tending to destruction of stocked-fish. Recapture of fish from large and small reservoirs are not certain in many cases on account of several reasons such as fish escape through outlets of the damand frequent failure of fish breeding programs due to erratic monsoon. Unless the stocked fish propagate themselves, the capture fishery potential in these types of reservoirs will never be sustained.

In contrast to medium and large reservoirs, sustained stocking of small reservoirs has been more effective in augmenting the yield. This is due to the fact that small reservoirs have the advantage to monitor and manipulate the stocked fish comfortably. In general, periodical stocking and harvesting schedules are the most fundamental cues for fish production in small and shallow reservoirs such as Aliyer (Tamil Nadu), Meenkara and Chulltar (Kerala), Markonahalli (Karnataka), Gulariya, Bachhra and Baghla (Uttar Pradesh) and Bundh Beratha (Rajasthan) where the yield of major carps has increased by more than 60 per cent.

Removal of Predators and Weed Fishes

It is very difficult to control/manage the predatory and weed fishes from medium and large reservoirs since these fishes cause extensive damage to commercial species. To solve the problem, however, repeated use of gill nets of suitable mesh size, use of long lines and traps are recommended for their control. In some countries, poisoning in selected areas is executed that has limited use in India due to multiple use of reservoir waters. Bottom trawling is, however, more effective to catch about 80 per cent of these unwanted fishes from many reservoirs.

Introduced Fish and Their Role in Indian Reservoirs

During 1970s, the Indian major carps and exotic carps have firmly established themselves in reservoir ecosystems and in open water systems as a result of their introduction either accidentally or deliberately. Having acclimatized themselves to reservoir ecosystems, they have stabilized as a natural fishery to a considerable extent. Consequently, their contribution to reservoir fish production has substantially increased. This strategy of fish production has created severe competition among different fish species for food, space, and shelter, forcing many indigenous fish species to live in an inimical situation. Obviously, competition plays a crucial role in the survival and growth of many indigenous fish species (Chart - 11.1).

CHART - 11.1
COMPETITION

1. Intra-specific and Inter-specific Competition:

Different species of fish live together influencing each other's life directly or indirectly. Such processes as growth, nutrition, and reproduction depend very much upon the interactions between the individuals of same species called *intra-specific competition*, or between those of different species called *inter-specific competition*. The relationship between species may be beneficial to both, harmful to both, or harmful to one and neutral for the other.

2. Negative Interaction:

Although there are various types of interactions among different organisms in different ecosystems, negative interactions exist in reservoir ecosystems. These interactions, as a generalization, include the relations in which one or both the species are harmed in any way during their life period. Such types of associations may be referred to as *antagonism*. Negative interactions are, however, generally classified into exploitation and competition.

(i) *Exploitation* : In case of exploitation, one species of fish harms the other by making its direct or indirect use for food or shelter. Predatory fish is free-living which catches and kills another species of fish for food. They feed upon the adults or larvae or even eggs of their prey. Different species of herbivorous and bottom-dwelling carps receive varying degrees of harmness as the result of browsing and grazing on benthic community by several species of catfish. Predacious fish consume much quantities of

---

other seeds of commercial fish species as food. Moreover, the grazing and browsing may bring about marked changes in the composition of other fish that are not so highly competitive.

(ii) *Competition* : Competition takes place when individual species attempt to obtain a resource that is not adequate to support all the individual fish species seeking it, or even if the resource is sufficient enough, fish species harm one another in trying to obtain it. The resource may be food, space, habitat and hide from predators. It is an important area of population ecology in both situations and involving populations of same species as well as interacting populations of different species.

A number of carnivorous fish that harbours in most reservoirs generally consume insects and other small animals including small fish for their food. They are well adapted in remarkable ways to attract, catch and digest their rations.

Intra-specific competition is often called *scramble competition* and is an important density-dependent factor controlling populations. At high density, native fish populations drastically reduced because over-crowding leads to fewer successful matings due to interference between fish species. These encounters are the principal factors resulting in decrease in native fish population density. Subsidiary effects involve increased egg mortality and increased competition of young fish for food at high density. The mortality of eggs is due to continuous knocking the eggs by adult fish so that fertilized eggs are not transformed into hatchlings or fry.

In contrast, inter-specific competition sometimes called as *contest competition* or *interference competition*. According to Lotka- Volterra mathematical model, three types of outcome that follows may be predicted.

* Only one species survives, and growth of the surviving species to its carrying capacity level is slower than if the second species had been absent.

* When inter-specific competition is less intense than intra- specific one, both species coexist for long or short period of time. The population will reach the carrying capacity level.

* If one species persists at higher density, the other is eliminated. This is the case particularly when the populations have negative effetcs on the growth of each other, but inter-specific competition is stronger than intra-specific one.

While a few of the introduced fish have been found to be highly profitable and regarded as a strategy to boost fish production from reservoirs, some other exotic species in open waters have evolved a lot of debate. In the yesteryear between 2004 and 2008, there is a growing concern over the possible dangerous impacts of some of the introduced species that follows on the indigenous species diversity.

1. **Tilapia:** Tropical reservoirs have afforded an ideal habitat for tilapia and consequently, the fish has established a secured position in these waters. The

presence of large-sized tilapia (1.0-2.5 kg) in many south Indian reservoirs has virtually eliminated many indigenous species including carps. Sizeable populations of tilapia have flourished in some reservoirs of Tamil Nadu since the last forty years (from 1965 to 2005) and still contributing substantially to commercial catches. The catch potential, however, fluctuates between 20 and 70 per cent of the total fish biomass.

Though tilapia is a prolific breeder, the fluctuation of water levels has drastically affected the breeding pits and at the same time the weed and predatory fishes severely destroyed the population of young tilapia. These two factors act in concert and undoubtedly helps check the excessive growth of tilapia population in reservoirs.

2. **Silver Carp:** Although silver carp did not get established as a breeding population in any reservoir, its performance in some reservoirs such as Gobindasagar is striking. In this reservoir, the fish has established himself as a breeding population and consequently reached a plateau of production (production increased from 160 tonnes in 1970-1971 to 1,520 tonnes in 2005-2006). The silver carp is a cold-water fish and when introduced into a warm-water environment, consumes natural food organisms in high quantities and grow at a rapid rate. However, the high performance of this fish under Indian conditions is highly significant than in other Asian countries. In China, for example, the fish mature at 5^+ whereas in India, it breeds just at the age of one year under optimum environmental conditions.

It is significant to note that despite its entry into a number of reservoirs, silver carp has not been able to acclimatize anywhere in India except Gobindasagar reservoir. Since the reservoir has temperate climate which is close to the original habitat of the fish, the fish seems to have found a congenial habitat for growth and reproduction.

(i) *Interaction Between Silver Carp and Catla* : There is a strong interaction between silver carp and the native fish *Catla catla*. Since the two species share a common niche and compete with each other for food in the reservoir ecosystem, the silver carp is extremely deleterious to the population of catla (and also other carps) in some reservoirs of Himachal pradesh where the population of major carps has declined considerably. And since the silver carp hampers the growth of catla, caution must be taken before extensive stocking of silver carp in Indian reservoirs is permitted.

3. **Common Carp:** Similar to tilapia, common carp has also acclimatized in Indian reservoirs. But it seems reasonable that the common carp is not a suitable fish for stocking in Indian reservoirs for various reasons. Since the fish is indolent in habit, the possibility of its survival in a predator-infested reservoir is very unsatisfactory. Also, the sluggishness and bottom-dwelling habit of the fish makes the fishing less effective as they are not able to catch with gill nets. Like silver carp, this fish also competes with *Cirrhinus mrigala, C. cirrhosa,* and *C. reba.*

The minor carp, *Cyprinus carpio specularis* , has a great significance in relation to reduction in the survival rate of several species of indigenous fish after its introduction in some upland reservoirs. Minor carp, for example, has caused serious damage to the indigenous snow trouts such as *Schizothoraichthy niger, S. esocinus*, and *C. curvifrons* in Gobindasagar reservoir and *Osteobrama belangeri* in Loktak lake of the Northeast (State Manipur). This situation has also occurred in warm-water reservoirs of the plains where other commercially important species of fish abound.

Other exotic species such as *Hyphophthalmicthys mobilis* (Bighead carp), *Cirrhinus molitorella* (Mud carp), *Mylopharyngodon piceus* (Snail carp), and many other catfishes have been considered for stocking in Indian reservoirs. Though these species possess propensity to cause severe damage to native fish, steps must be taken before their stocking is permitted; otherwise, all native forms will likely to be damaged and the time will come when many indigenous species will be listed in the Red Data Book of India.

Fertilization of Reservoirs for Sustained Yield

In India, the use of organic manures and chemical fertilizers in reservoirs to augment their productivity has not received much attention. Multiple use of water and the resultant conflict of vested interests among the various water users are the prime factors that has prevented the adoption of management strategies. The last three decades (1970-2000) were the dawn of fertilization programs in some reservoirs such as Vidur in Maharastra, Naktra in Madhya Pradesh, Krydemkulai and Nongmahir in the Northeast. The fertilization programs are still continuing with degree of regularity and the results have been successful. A 50 per cent increase in fish yield along with 3-fold increase in the average weight of Indian major carps were achieved.

In other Asian countries, management strategies of reservoirs involving fertilization/ manuring and feeding exhibited significant production that varied between 1,500 and 6,500 kg/hectare. It should be kept in mind that high production of fish is possible particularly in small reservoirs where fertilization and feeding strategies can be undertaken on economic basis. On the other hand, the cost of fertilization, manuring, and feeding schedules involved in medium- and large-sized reservoirs will definitely negate the benefit achieved.

Pollution

Similar to other freshwater resources, reservoir ecosystems are sensitive to changes induced by a variety of factors such as pollutants, domestic Wastes, industrial effluents, pesticides and heavy metals, and siltation. Consequently, reservoir ecosystems have become stressful and the pristine condition has been damaged to a large extent. Also, industrial society not only makes radical changes in water flow by construction of engineering works but also pollutes and contaminates most of the reservoirs with a large variety of wastes.

The sources of reservoir pollutants are many and varied. Industrial, chemical, and thermal power plants dispose of toxic compounds which are deposited in reservoirs through surface flow-off. Also, toxic substances are carried by upstream rivers and their tributaries loaded with organic and inorganic wastes. These consequences drastically affect the biological productivity of reservoirs.

1. **Thermal Pollution :** This is the most important form of pollution of many reservoirs.Thermal pollution refers to the discharge of heat into the reservoir ecosystems from combustion of fuel and from conversion of nuclear energy into electric power. Thermal pollution of reservoirs takes the form of discharges of heated water/effluent into reservoirs via streams which can have drastic effects on fish life. The impact may be quite large in a small area.

   The principal ecological consequences of a heated water discharged into the aquatic resources are (i) increase in water temperature and (ii) change in chemical composition and subsequent alteration in life history of aquatic communities. Discharge of heated water may elevate the temperature of water by 6 to 8 °C which may cause mortallty of fish and fish food organisms. Temperature of water also exerts direct influence on chemical toxicity. The formation of a thick covering layer with fly ash on the reservoir bed is, however, one of the most deleterious impacts of thermal pollution. Fly ash covers over extensive areas of the bottom, covering the layer of the reservoir bed, resulting in partial or complete extermination of benthic communities. Deposition of a thick layer of fly ash at the reservoir bottom over the years may restrain the release of nutrients from soil to water phase, thereby affecting the productivity.

2. **Industrial Effluents:** Chemical industries generate a variety of wastes, both organic and inorganic, which occurs as a mixture with a range of component concentrations. However, industrial effluents cause pollution hazards in many reservoirs. Effluents are drained into rivers and then reservoirs where they accumulate in quantities that are locally damaging or lethal to fish communities in particular and to reservoir ecosystems in general.

   Several instances of ecosystem-damage, fish kill and replacement of uneconomic fish by economic ones following discharge of effluents have been noted. Discharge of wastes into reservoirs such as Rihand, Hirakud, Sandynulla and Bhawanisagar) poses a threat to fish health.

3. **Pesticides and Heavy Metals :** Toxic and hazardous substances such as pesticides and heavy metals are drained into the reservoirs through effluents and surface flow-off. Since these substances are highly persistent in soils, reservoir ecosystems are easily contaminated and the entire bio-geochernical cycle of reservoirs is damaged. The hassles are aggravated when hazardous substances get biomagnified in fish body which otherwise persists in water and soil in low or moderate concentrations.

Though hazardous substances have been detected in significant amounts in almost all the riverine ecosystems of the world, such reports from reservoir ecosystems are very rare. It is significant to note that most of the reservoirs are situated in relatively remote areas which are far away from industrial and agricultural activities and therefore, the opportunity of severe contamination of reservoirs is less. However, a relatively few reservoirs have been noted to contaminate with hazardous wastes to a greater or less degree.

4. **Siltation:** Excessive siltation is a common problem in reservoirs. This leads to drastic decrease in the water-holding capacity. Siltation also hampers the reservoir productivity by affecting biotic communities. Erosion of top soil in the catchment area is the factor that substantially contribute to increase in the sediment load in rivers. Generally dragging and impact can easily erode alluvial materials, such as gravel, sand, silt, and clay. This form of erosion can excavate enormous quantities in a short time. The undermining of the banks causes large masses of alluvium to slump into the river, where the particles become part of the stream's load. Heavy erosion and high sediment load are characteristic features of rivers.

Suspended particles tend to settle down in the reservoir causing many problems. A growing body of research has shown that Indian rivers, for example, sustain about 2,050 million tonnes of silt, of which nearly 490 million tonnes are deposited in the reservoirs. As a consequence, the storage capacity of reservoirs drastically reduced. Destruction of breeding grounds and retardation of the overall productivity of reservoir ecosystems are the most serious consequences of soil erosion.

To sum up, although changes in the reservoir ecosystem due to excessive siltation and pollution have received much publicity, hassles of capture-culture fishery balance in most of the reservoirs are likely to be more serious and perhaps more expensive to solve. If this situation continues, the potentiality and importance of fisheries in many reservoirs will be lost in future. And in some other reservoirs, fish production will be reduced to a minimum.

There is uncertainity regarding forecasting about the future of siltation and extent of pollution in reservoirs as these processes are on the increase and make it difficult to predict. And consequently, effects of pollutants and siltation on the reservoir ecosystems by and large are fairly unpredictable.

Pen Culture

So far environmental and stock manipulations are concerned, pens are considered as transitional structures between cages and ponds. While pen culture practice of certain species of finfish and shellfish such as Yellow tail, *Chanos chanos* (Milk fish), *Mugil* sp., *Anadara granosa*, *Meretrix* sp., *Penaeus Japonicus* sp. etc. in some sountries especially Japan, Philippines, Norway, and India have got great extent of ability to carry into effect, attempts to bring forward these systems have not encountered with much success in many other countries of the world. This can possibly be attributed to the difficulties in the use of state-of-the-art techniques and the involvement of high cost of earthwork constructed in confined waters and water management. Decades of experience coupled with scientific study clearly indicates that the success of fish culture in pens, to a large extent, depends very much on the hydrological conditions of the region concerned. The design of the structure and working procedures should be based on the satisfactory knowledge of floods, waves and currents, water and soil qualities, and occurrence of predatory animals.

Fish culture in pens (See Panel 11.3) of the yesteryears between 1999 and 2005 over unmanageable extensive areas is highly remarkable. Since it is one of the intensive culture systems fostering fish production to a large extent, the pen culture has become very popular in several tropical countries and this system of farming gives cue to produce more fish if fish culturists work in concert. And it is hoped that such culture system in shallow areas of reservoirs and lakes will definitely explore the possibility of producing robust and disease free fish for the current and future generations.

1. *Relevance of Pen Culture in Reservoir Management and Constraints*

Pen culture has a special relevance of reservoir management since it has been recognized as a means to rear the fingerlings for stocking. As a matter of fact, enormous quantities of fingerlings are required for stocking reservoirs and it is practically beyond possible to raise the stocking materials from land-based nursery farms. Hence pen culture is inevitable for high production. Nonetheless, pen culture technology has not been developed anywhere in most of the Asian countries. The factors that hamper the successful adoption of pen culture technique include (1) weed infestation and harvesting problems, (2) unavailability of suitable pen materials, (3) wave and wind actions, and (4) water retention time. The water retention time is one of the most important factors for pen nurseries in reservoirs since the rearing has to be completed before the water level in the pen drops below the critical limit. The water retention time in reservoirs with high drawdown is very limited. This condition makes the pen nurseries more eutrophic.

Crafts and Gears

1. **Crafts:** Craft refers to a boat or ship used in fishing or other purposes. A saucer-shaped craft, called *Coracleas* is widely used in harvesting the fish stock from reservoirs of peninsular India. It is made of a split bamboo frame which is covered with the buffalo hide. The craft is very simple, inexpensive, durable, and is able to operate skilfully and carefully in waters having many small waves. The craft is able to adapt many different functions or activities such as laying and lifting of nets, transport of fish and other farming materials.

   Wooden boats are also widely used for fishing in a number of reservoirs, particularly in northern parts of the country. Flat-bottomed and locally-fabricated boats are used in some reservoirs such as Rihand, Gandhi sagar, Gobindasagar, Hirakud etc. Bengal type boat, popularly known as *Dinghy*, with additional facility of setting sails for wind propulsion is also used in many Indian reservoirs.

PANEL-11.3
FISH CULTURE IN PENS

Fish culture in pens involves the following points.

1. **Site Selection :** Careful site selection and proper pen design are the two essential prerequisites to make the pen culture practice viable. In general, well-protected shallow water areas are more suitable for the installation of pens. Areas used for inshore fishing and boating should be avoided. The following criteria may be considered while selecting a site for pen culture.

   * Areas where culturable seeds for stocking in pens are abundant.

   * Depth of water column where pen is installed should vary between 1.0 and 2.0 m.

   * Stagnant or slow-flowing unpolluted waters should be used. Turbid water should be avoided.

   * Bottom soil of the site should be muddy, clay, clay-loam or sandy-mud with gentle slope and detritus.

   * Places exposed to high winds should be avoided. Gentle wind helps water circulation through pens.

   * Places near the disposal points of sewage or waste must be avoided.

   * Areas where excessive siltation and decompostion of organic matter occurs, should be avoided.

   * Speed of water flow should be 0.2-0.3 m/ second in order to bring water heavily loaded with natural food organisms and oxygen into the pens and remove fish wastes out of pens.

   * Dissolved oxygen, temperature and pH of water should be 6.0 mg/l, 24-35°C and 7.5-8.5, respectively.

2. **Pen Materials and Construction :** Fish pens resemble to duck fences and their size veries from 30 to 1,000 m². Pens installed in freshwater areas may be single or double-layerd enclosure and take different shapes such as rectangular, circular hexagonal or square. However for easy sampling and harvest, rectangular pens are generally recommended.

   For the construction of pens, poles of split bamboo, wood, iron polyethylene (nylon) nets of smaller mesh size (0.5-1.0 cm) nylon rope etc. are commonly used depending on the environmental conditions, size of pens and species of fish cultured. Coir ropes and synthetic twines are used as fastening materials. Iron mesh (1 cm x 1 cm) is used for covering the pen.

   The poles prior to their use are coated with coal tar or antifouling materials. This prevents the poles from attacking with biofoulers.

Split bamboo poles having 2.5 m high and 5-15 cm wide with pointed tips at the bottom are arranged horizontally. Poles are placed at an interval of 1 meter if bamboo screens are used. In cases where synthetic nets are used, poles are fitted at 2 meter intervals. An inter-space of about 1 cm between the bamboo slits of the mats is cardinal for free exchange of water through pens.

Poles are driven into the mud at a depth of 50 cm. A 2 mm polythelene rope serves as a head rope and a foot rope. The head rope is attached to a nail driven at the top of the poles. The laterite stones are attached to the foot rope at an interval of 1.0 m. The stones along with the foot rope and net are anchored about 30-40 cm in the mud. A slackness of 1 m for every 10 m of net is permitted so that the net is not held in tension. A scare-line is fixed to the polythelene twine at an interval of 1m inside the ecclosure about 50 cm above the bed of the pen so that the fingerlings would not dash against the net. The pen walls have sufficiently high free board over the normal water level.

3. **Species Suitable For Culture in Pens :** Fishes which are detrivores or herbivores, fast-growing and tolerant to fluctuate environmental factors should be selected for culture. Though catfishes, murrels, prawns, crabs etc. are cultured, Indian major carps and some species of exotic carps are highly suitable under Indian conditions and can be cultivated in pens either as individual species or in combination with other species.Polyculture of carp species along with prawn especially *Macrobrachium rosenbergii* and *M. malcolmsonii* can also be undertaken in pens.

4. **Stocking Density :** Continuous exchange of waters through pens definitely satisfy the basic requirements of growing fish. The possibility of using supplementary feeds should not be ruled out. The stocking densities followed in pens vary between 10 and 100 number m^2. For growing Indian major carp in pens, 250 fry are generally stocked in each 100 m^2 area and uniform size (100-150 mm in length) should be stocked. The rate of stocking of carp in a 100m^2 area is generally recommended as follows.

Species	Number
Catla catla	100
Labeo rohita	75
Cirrhinus mrigala	50
Ctenopharyngodon idella	25
Total	250

Fingerlings grown in pens can be used for stocking in different culture areas such as large-sized ponds, lakes and reservoirs.

5. **Management :** Management protocols in pens are similar to that of pond fish culture and involves liming, fertilization and manuring, feeding, disease control,

harvesting and adoption of precautionary measures. However lime and cattle manure are used in a 100 m² area at the rate of 3 and 30 kg/month, respectively. Cheap and easily available common food such as mustard oil cake, rice bran, snail meat, kitchen waste, poultry offals etc. are widely used. Highly nutritious food items such as fish meal, de-oiled rice bran, soya meal supplemented with vitamin-mineral mixture if available, may also be used for sustained production.

Harvesting methods are very much dependent on the rearing techniques. When harvesting is done at low water level or when a portion of the stock is harvested it is necessary to use dip cast nets. Some fish culturists use seine or drage net (mesh size of 1 cm or less) for total harvesting.

6. *Advantage of Pen Culture :*

   * Higher yield in limited spaces is possible in pens owing to continuous movement of water with rich supply of natural food organisms and oxygen.

   * Since fish are able to save energy without spending much energy during movement, better growth rate and higher yield could be expected in pens.

   * Pen culture in running water communities has special advantage since dissolved oxygen is not a limiting factor and accumulation of faecal matter of fish in intensive stocking causes no harm. Since toxic metabolites especially ammonia are flushed regularly, there is no possibility of fish mortality.

   * Since there is plenty of water, pen culture can be undertaken throughout the year.

   * In contrast to freshwater ponds, the occurrence of fish disease and parasites in pen culture systems are very much limited.

   * Pen culture does not involve any complicated design. As soon as pens are installed it is ready for full-scale operations.

   * Pens may serve as fish seed stocking centres.

   * Fish harvesting is not at all a problem in pens.

7. **Constraints :** Inspite of several advantages, a number of constraints exist that should be kept in mind when fish culture in pens has to be planned.

   * Biofouling of bamboo mats and polythylene nets in pens with algae, molluscs etc. is a common feature. These organisms drastically reduce not only free exchange of water but also affect the longevity of pen structures. Several species of aquatic insects harbour in pens and cause damage.

   * Algae and macrophytes in the vicinity of pens may decay and pollute water over extensive areas.

   * Periodical wear and tear of pen materials and theft of materials make the pen culture costly.

   * Heavy rains in momsoon season and high wind action may damage the pen and at the same the entire culture area is innundated.

These constraints could, however, be overcome by adopting several judicious management practices. The most important management practices include : (i) periodical cleaning of net enclosures or bamboo mats with a wire brush, (ii) rearing the fish in pens only during non-flood season, (iii) regular checking of pen structure (iv) temporary increase in height with nets during rainy season, and (v) to avoid fish mortality in pens at high temperature (above 35°C). Sites should be selected for pen culture where at least 1m depth of water is available. Moreover, a small patch of floating aquatic weeds may be kept inside the pen structure that makes the water body extremely suitable for fish survival.

Besides using indigenous and simple crafts, mechanized/motorized crafts are also used in some reservoirs for fishing at high rates. Use of mechanized crafts is a key factor for effective fishing. But their extensive use in reservoir waters is not widely accepted because they are very expensive for fishermen, though mechanised crafts are inevitable for fishing in large and medium reservoirs.

2. **Gears:** In most reservoirs, entangling-type of gill-net without a foot rope is commonly used. However, the underwater obstacles severely restrict the use of active gear and the choice is limited to simple gill-net. Shore seines of various dimensions and mesh sizes are also used in many reservoirs. Though a variety of other fishing gears such as long lines, hand lines, pole and line, cast-nets, dip-nets etc. are in active operation in Indian reservoirs, their contribution to the total catch is not significant nonetheless.

Other Management Schedules

1. **Timber in Reservoir Bed:** Most of the reservoir fishery experts are aware of the fact that the presence of timber in any reservoir bed severely restricts the use of active fishing gears to exploit fish stocks in full potential and for many other management schedules. Another body of experts perceive that it is necessary to retain at least non-commercial timber for a variety of purposes such as providing habitat for fishes and peryphyton development.

2. **Exploitation System :** Both government and cooperative societies are actively involved in harvesting, processing and marketing operations. The type of operations, however, vary from one reservoir to another. Therefore, some sort of uniformity in fishery regulations among different types of reservoirs must be maintained.

    Exploitation systems on commercial basis followed in different regions can be broadly divided under four types: (i) departmental fishing, (ii) lease by auctioning, (iii) issue of licences to cooperative societies or individuals, and (iv) fishing on a royalty basis.

11.11 Evaluation of the Economics of Reservoir Fisheries

Statistical evaluation of the economics of fish production from reservoirs is a very difficult task due to the multiplicity of agencies involved in reservoir management. In general, development of reservoir fisheries is primarily based on capture fisheries following the common property norm. Capture fisheries strategy of the resevoir is congenial to the industries such as coal, oil and ore where the production potential depends on the state-of-the-art technology adopted and the quantum of labor and capital deployed. But the renewable nature of the resource and the complicated biological principles embraced in the ecosystem management bestow a great challenge to the reservoir fisheries management personnel. Though human interference is less intense in reservoir ecosystems, drastic fluctuations in fish production still exists even if the adoption of state-of-the art technology and other management protocols are kept constant. Hence, yield-function relationship is fastened to be very complicated. A firmness must be ascertained in production potential of reservoir ecosystems by adopting improved technology along with good returns to fish farmers.

Owing to over-concentration of fishermen in reservoir fisheries managanent, too much anxiety in their profit level is the most important scenario. At the same time, reservoirs that are low in productivity, do not encourage the farmers to obtain sufficient revenue. For sustained economic development, however, four objectives such as conservation, increase in production potential, employment opportunities, and the best use of reservoir resources through adequate management would have to be kept in mind. Whilst serious fishermen work hard for producing more fish and encounter actively with fish merchants for economic deliberations, the legal obligation of both government and non-government organizations should never be ignored to ascertain that the economic advantage can be achieved by farmers' community.

Reservoir fisheries economics are primarily concerned with the impact of the economy on reservoir ecosystems, significance of the ecosystem to the economy, and the appropriate way of controlling economic activity so that a balance is achieved among production, consumption and market. For most aqua-products in a modern reservoir fisheries economy, however, we must rely on markets to match producer costs with consumer demands.

11.12 Strategy Needed for the Sustainability of Reservoirs in Fish Production

In view of the deficiencies experienced in the developmental system in the past, it is necessary to give a fresh thought to the strategy that may be developed in future. Both national and sub-national governments have, in collaboration with various research organizations, cooperative societies, and local fishermen group evolved a definite strategy in different plan periods as to how reservoir development strategy could be formulated and implemented. In this effort, the systematic survey of reservoirs for development through application of the state-of-the-art technology, their infrastructural gap in the process of development, need for human resource development — all have helped both government and non-government agencies to become fully aware of the constraints of fish production from reservoirs. The collaboration has revolutionized the process of development in respect of increasing productivity of several Indian reservoirs, setting up of infrastructure for marketing the aqua-products etc. and affording minimum needs to the fisherman families. However, for improving the reservoir ecosystems and for increasing fish yield, a number of strategies that follows may be suggested.

1. Breeding grounds and sheltering areas of fish have to be identified for providing protection to fish stocks.

2. Cooperative societies and private enterprises could perform an effective development by advising the fisherman communities on the use of technology they adopt.

3. For proper assessment of reservoirs, the availability of commercially important species and their abundance in different seasons of the year should be made by applying geographic information system and remote sensing tools. These techniques have, however, been described in Chapter 2.

4. Adoption of breeding and larval rearing techniques for propagation of high-value fish species under controlled conditions is a valuable task for conservation of fish stocks and for increasing fish production from reservoirs.

5. Regulation of fish catch and mesh size of gears would facilitate the survival and growth of young fish.

6. A fishing holiday of at least 3 years following a short holiday during the peak spawning season is inevitable. This strategy has given encouraging results.

7. Active involvement of cooperatives towards fish production have given reassuring results. Reservoirs situated in Rajasthan and Madhya Pradesh have, for example, been managed well by the Tribal Area Development Federation and Matsya Vikas Nigam, respectively.

8. Introduction of compatible species, legislation to prohibit the use of non-selective gears, enforcement of closed seasons, manipulation of fish population, and water regulation are some of the cues for reservoir fisheries development.

9. Adoption of suitable models for determination of the key management factors of culture fisheries such as size of fish at capture, species selection, stocking size, and stocking density.

10. Raising fish seed for regular stocking.

11. Screening of exotic species, minnows, and large-sized catfishes at regular intervals.

12. Development of cage and pen culture in medium- and large-sized reservoirs.

13. Modernization of crafts and gears for effective fishing.

14. Loss of fish through sluice gates should be prevented.

15. Fertilization programs in littoral zones of small- and medium-sized reservoirs may be carried out to increase fish yield.

Major Constraints to Development

1. **Lack of Technology Clout:** The state-of-the-art technology has been a major plus point of reservoir fisheries development and is the strongest attribute to the profile of reservoir productivity. While technology is not trying to rectify the situation at present, the fact remains that acquiring technology clout takes time. And the consequence of staying with obsolete technology or no technology at all has been severe on the development front.

2. **Capital Inadequacy:** In most of the cases, it has been the second major weakness of reservoir fisheries development. And this came to the force in the 1990s. The weakness is related to the need for upgrading the scale of operations and the establishment of research agencies and farmers' co-operative societies. The capacity of reservoirs to produce more fish has to be expanded to remain the productivity round the year. This calls for hefty and new investments, which most reservoirs do not command. Majority of reservoirs are too neglect to create adequate reservoirs over time. They did not have the clout to raise the required capital from the sale of fish because of low productive potential of most reservoirs. And cooperative societies in such a financial plight obviously face a threat in this context.

3. **Vastness of Reservoir Area :** Vastness of reservoirs is one of the most important constraints to reservoir fisheries development. Small-sized reservoirs can be managed to some extent but large-sized ones are by no means easy for most development authorities. Pollution of water over extensive areas and difficulties in management strategies pose problems.

4. **Low Productivity:** Another major factor bring forth by the developnent constraint relates to low productivity. Many Indian reservoirs have been reported as low productive and this is coming to the fore.

Low productivity is essentially a failure to yield fish of commercial importance in huge quantities on account of some edaphic factors. It can, however, be manipulated to some extent if reservoir fisheries management staffs work harmoniously. But unfortunately, most of the reservoir fisheries developnent authorities do not follow any major development tasks in concert with degree of regularity that can increase the productivity of reservoir ecosystems. Catch statistics have shown the average yield of about 50.0, 12.3, and 11.4 kg/hectare respectively for small, medium, and large reservoirs that are not significant to meet the demand of markets.

(i) *Analysis of Productivity Development* : Productivity development analysis is the process of gathering and analyzing informations relating to the productivity. Involved in the processes are the tasks of monitoring the changes in the productivity taking place and forecasting the future position in respect of each of the factors.

Productivity development analysis involves a diagnosis of the mega development as well as the development that is specific to the given productivity. Under mega development, it studies the production-related ecological factors and government policies. It also studies factors relating to the productivity and factors relating to production of fish biomass.

11.13 Potential of Tropical Reservoirs

Unparallel increase in human populations has placed great strain on the traditional fish culture systems in many areas of the tropics. Millions of people who depend upon culture and capture fisheries for most of their food represent a dramatic increase in numbers during the past three decades. To provide fish protein for populations, fishery scientists have been forced to culture fish in small reservoirs and capture from medium- and large-sized reservoirs. Fish cultivation in several small-sized tropical reservoirs is more successful than the seemingly more efficient temperate zone counterparts. Lack of adequate management, financial assistance and at the same time, pollution have resulted significant decrease in fish production in most Asian reservoirs and will likely to continue to do so if these factors persist.

One cannot help but be optimistic about the future of reservoir fisheries in the tropics. The basic requirements for maximum year-round production appears to be higher in several tropical regions than anywhere else. Warm to hot climate and total annual solar radiation provide unmatched photosynthetic potential. Substantial quantities of fish can be grown in some particular reservoirs of the tropics. Already fishery scientists have noted some small- and medium-sized reservoirs which are rich in fish. Means of controlling pollution and the use of artificial food as well as fertilizers/manures in some reservoirs are becoming more common in tropical areas. The potential for fish production there is enormous.

11.14 Research on Tropical Reservoir Fisheries Development

Unfortunately, little is known about tropical reservoir fisheries and their management. Research on this aspect, their properties, nature, and their capability for production or resources for use, development is not singnificant compared to that of temperate reservoirs. Adequate knowledge on their characteristic feature makes it possible to identify only the potential categories of reservoirs.

The little we have learned about some small-sized tropical reservoirs is highly encouraging. While some reservoirs are excellent for fish yield and these reservoirs respond well to modern management protocols and mechanization (Table 11.3), other medium- and/or large sized reservoirs have resulted catastrophic decline in fish production. Knowledge on the nature of the reservoirs and the adoption of management practices to keep fish biomass might well have prevented much of the projects Rs. 10 billion loss.

There is a good likelihood that more intensive study will identify the complexities among tropical reservoirs similar to those known for temperate ones. Some of the tropical reservoirs are easily managed and fish yields are highly encouraging. At the other extreme are reservoirs that are not highly productivity and they produce very less quantities of fish (Table 11.4) and hence such reservoirs are essentially worthless from fisheries point of view. Unfortunately, these water areas are extensive.

Table 11.3 : Fish Production Potential in Some Small Reservoirs of India with Modern Management Protocols

Reservoir	State	Production (Kg/ha)
Yerrakalva	Andhra Pradesh	152
Markonahalli	Karnataka	63
Aliyar	Tamil Nadu	194
Tirurnoorthly	Tamil Nadu	182
Meenkara	Kerala	108
Chulliar	Kerala	316
Gulariya	Uttar Pradesh	150
Bachhra	Uttar Pradesh	140
Baghla	Uttar Pradesh	102
Bundh Beratha	Rajasthan	94

Source : Nath and Das (2004)

Table 11.4 : Some Unproductive Reservoirs in Some States of India

State	Size	Name of Reservoir	Production (Kg/ha)
Orissa	Medium	Mandira	53.0
		Gohira	0.5
		Satandi	8.7
		Upper Kolab	11.0
		Kalahandi	2.7
		Bhatrajore	0.97
	Large	Hirakund	6.6
		Rangali	0.50
		Balimela	15.70
Madhya Pradesh	Medium	Kharkora	1.5
		Monohar Sagar	2.0
		Darri	2.5
		Tighra	3.0
		Kodar	8.0
Kerala	Medium	Peachi	4.5
		Malampuzha	5.0
Rajasthan	Large	Mahibajaj	3.0
		Kadana	8.0
	Medium	Meja	1.0
		Mangalsar	5.0
		Urvania	6.0
Himachal Pradesh	Large	Pong	20.0
		Govindsagar	36.0

Contd.....

Freshwater Fish Culture Vol. 2

State	Size	Name of Reservoir	Production (Kg/ha)
Bihar	Large	Tilaiya	10.0
		Maithon	5.0
		Panchet	8.0
		Kansjor	1.0
	Medium	Nandini	1.0
		Konar	20.0
	Small	Batane	5.0
		Malayjalashaya	3.0

Source : Compiled by the author from different published literatures

11.15 Conclusion

Reservoirs are most important source of kaledoscope variety of fish that provides nutrition to the human society and employment opportunity to fishermen. In India, several small- and medium-sized reservoirs have high fish production potential and substantial quantities of fish are harvested every 2-3 weeks interval. Judicious use of reservoirs following adequate planning and programming for sustained yield are, however, sound practices that have characterized by successful production systems through decades. Although reservoir fisheries development is cost-effective and requires refined technology, it is not less important today in view of the high protein demands to feed an ever-increasing population. Robust and disease-free fish should be produced from productive reservoirs to the possible mangnitude where sound management will permit.

Most of the reservoirs are less important in producing robust fish due to contamination of water with a large variety of wastes. Reservoir ecosystems contain both inorganic and organic chemicals at levels that are baleful to fish and man.

High producivity of many Asian reservoirs vividly indicates that there is potential for the tropics to produce adequate fish for current and future generations. Whether this potential will be reached will depend in part on the wisdom that is used in managing reservoirs. And this wisdom will be based largely on the service of devoted reservoir fishery scientists in the region involved.

References

Jhingran, A.G. 1990. Recent advances in the reservoir fisheries management in India. *In.* Reservoirs Fisheries of Asia. Sena, De-silva (Ed). Proceeding of the second Asian Reservoir Fisheries Workshop, Hangzhou people Republic of China, 15-19 October, 1990.

Nath, D. and A.K. Das 2004. Present status and opportunities for the development of Indian riverine, reservoirs and beel fisheries. *Fishing chimes.* 24 (1) : 61-67

Sugunan, V.V. 1997. Reservoir Fisheries of India. FAO, Rome, Italy, 423 p.

Questions

1. Sum up the impact of the exotic fishes on the existance of native species in Indain reservoirs.

2. Evaluate critically the nature of reservoir ecosystem and its impact on fish production.

3. What strategies would you suggest to make reservoirs effective in production?

4. Fish production from reservoirs embraces a number of responsiblities. Discuss.

5. How will you classify the reservoir ecosystems ? Discuss the significance of reservoir fisheries development.

6. What are the problems of reservoir fisheries development ?

7. Discuss how reservoir ecosystems are polluted.

8. Discuss how pen culture in reservoir ecosystems is undertaken.

9. Discuss different factors involved in increasing and decreasing the productivity of reservoir ecosystems.

10. State the importance of stocking in reservoir fisheries management.

11. Write a short note on the following :

    (a) Plankton, (b) Benthos, (c) Macrophytes, (d) Hydro-edaphic factors, (e) Siltation, (f) Crafts and gears, (g) Pollution of reservoir waters, (h) Nutrient budget, (i) Importance of pen culture in reservoir ecosystems.

12. What are the strategies that are necessary for the sustainability of reservoirs in fish production?

12

Role of Periphyton and Microalgae in Fish Culture

Judicious use of chemical fertilizers and organic manures are the two important management schedules for sustained yield of farmed fish in ponds. These nutrient carriers help to develop a variety of periphytic and microalgal species either in water or over the substrata that are present in pond bottom. For production of herbivorous fish in ponds through different farming systems (such as extensive, intensive, semi-intensive, and super-intensive), primary producers must be developed since producers are directly consumed by herbivorous fish. These fishes normally consume benthic/periphytic/epiphytic algae rather than phytoplankton populations. Development of sessile algae over substrata per unit of water area is several times higher than that of phytoplankton and fish are more competent to skim periphyton as food. Moreover, association of sessile algae is more perpetual than that of phytoplankton populations. Therefore, the potential of periphyton-based fish culture systems is enormous in many regions of the world.

12.1 What is Periphyton?

The term periphyton originated in 1960s from Greek (*peri*='around' + *phyton*='plant') which refers to freshwater organisms attached to or clinging to plants and other objects projecting above the bottom sediments.

Simply defined, periphyton is the total assemblage of sessile or attached organisms on the most diverse aquatic supports necessary for increasing fish production. Periphyton contains algae, protozoans, bacteria, fungi, rotifers, annelids, insect larvae, and crustaceans. Traditional fish culture in some Asian countries such as Philippines and Indonesia principally depends upon the exploitation of periphyton in full potential as food for fish which is termed as *lab-lab*. Other countries such as India, Bangladesh, Cambodia, Sri Lanka, and West Africa adopt periphyton-based fish culture to a large extent.

12.2 Species Composition in Periphyton

The abundance of various forms of organisms depends very much on the habitat they have assemblaged. Almost all well-managed ponds harbor a large number of periphytic organisms. For successful develpment of periphytic organisms, several factors appear to be of major importance. Different factors include the type of substrate, nutrients, light, and regular monitoring of culture ecosystems.

Growth of a periphyton layer over substrates generally begins with the gradual deposition of organic substances especially mucopolysaccharides on which bacteria are grown. In course of time, algae are developed and later on different species of green algae with long strands are developed. Periphytic organisms are attached to the substrate in various ways such as cushions of filaments (aquatic mosses, algae and sea weeds), muscular suction pads (snails), adhering to the substrate (insect larvae), stalks with sticky ends (ciliates), and sticky capsules (bacteria and blue-green algae).

Species composition within the periphytic congregation is extremely high. In some fish ponds, only a few species of algae dominate the periphyton gathering while in others more than 75 species of algae are developed.

12.3 Biological Cycle of Periphytic Biomass

When a fish production system is based on the growth and development of periphytic biomass, understanding of its biological cycle is inevitable. The origin of this cycle is in the mineral nutrients dissolved in pond water. Nutrients are derived from fertilizers and manures which are applied for sustained production or come from substances carried to the water by exogenous detritus. With the help of sunlight, aquatic plants transform these nutrients and carbonic acid in water into organic matter which forms macrophytes, planktonic algae, and periphyton. Detritus are consumed by small organisms which ultimately serves as food for larger aquatic animals which, in turn, are consumed by fish.

The cycle of fish production only through fertilization and manuring is governed by the following links: nutrients, plant production, intermediate animal consumption, and production which leads to the final production – the fish biomass. In the last stage, mineralization mechanism is accompanied by bacteria which permits the return in which all dead components of organic matter are re-ingested into the biological cycle.

12.4 Principal Groups of Nutritive Fauna in Periphyton

The principal constituents of nutritive fauna found in well-fertilized ponds belong to several animal groups such as crustaceans, rotifers, worms, molluscs, and insects.

Crustaceans

Crustaceans are the most important group which forms the food for fish. The small-sized animals are the main zooplankton for fry and young fish. The large-sized forms are consumed by adult fish. The most common and important forms include the cladocera (*Daphnia magna, D. carinata* etc.) and copepoda (*Cyclops* spp. *Mesocyclops* etc.) which are found among the vegetation or they remain as floating/drifting condition in a body of water.

Rotifers

These are small planktonic animals and are found in prodigious quantities in ponds. They are found in great numbers on the bottom of ponds where the aquatic macrophytes are grown profusely. Some forms of rotifers are dominant periodically (such as *Brachinous, Anuraca, Conochilus*), some other forms are not permanent (such as *Asplanchna, Synchaeta*) while others such as *Polyarthra, Rattulus,* etc. are irregular but permanent. Rotifers are consumed by aquatic animals which are used as food by fishes.

Worms

A few species of worms play an important item for fish as food. Worms vary in size. They run the gamut of different sizes such as from almost microscopic to 30 cm in length. Development of worms takes place in the mud especially when it is rich in organic matter. Some forms are associated with aquatic plants.

Insects

Insects form one of the most important natural food organisms for farmed fish. Insects are consumed by fish as larvae or nymphs. The most common groups of insects include Plecoptera, Odonata,Megaloptera , Trichoptera, Ephimeroptera, Diptera, Coleoptera, and Hemiptera and all are found in natural ecosystems both as larval and adult stages. They either burrow in the pond mud or swim in water or live on the submerged plants or on the bottom.

Molluscs

Though a variety of species live in ponds, a few of them form important food items for fish. They live along with submerged plants. The most important species which are used as for fish include *Lymnaea stangnalis, L. ovata, L. auricularia, Planorbis* spp., *Vivipera bengalensis, Physa fontinalis,* and *Valvata piscinalis.*

12.5 Principal Groups of Flora in Periphyton

In addition to faunal composition, a number of algae and macrophytes also constitute the base of periphyton. However, the algae are also able to synthesize protein and oils from carbohydrates which helps to manufacture soluble forms of nitrogen and other minerals in water in which they grow and reproduce.

The nutrient substances which are stored as food reserves in the form of polysaccharides, vary from one group to another and therefore, provide important data for basic grouping of algae. Polysaccharides are generally termed as *starch*. Various kinds of starches are found in different algal groups such as cyanophycean starch (found in Cyanophyceae), floridean starch (found in Rhodophyceae), laminarin (brown algae), paramylon (Euglenoids), leucosin (Xanthophyceae), and Cyanophycin – a protenaceous substances (found in blue-green algae). Fats also occur as reserved food in appreciable amounts in the cells of Bacillariophyceae, Xanthophyceae, and Chrysophyceae. The detrimental factors which favor the growth of algae are temperature, light, proper supply of oxygen and carbon di-oxide and essential elements. Of twenty five thousand species of algae that are known to exist, about 50 per cent of them consist of only single-cell species.

Large-sized, multicellular algae tend to be thread-like or flattened in shape. A well-managed fertilized freshwater fish culture ecosystem is teemed with innumerable species of algae which form the food for herbivorous fishes.

12.6 Role of Environmental Factors for Periphyton Development

When the fish attains the juvenile stage, the principal work involves the supply of natural food organisms for larvae and fry of fish. The provision of natural food organisms is always a condition for successful fish culture in ponds. Periphytic biomass, however, require minerals, vitamins, nitrogen, and phosphorus for their development which, in turn, establish the productivity of fish culture ecosystems.

Nutrients

Inorganic nutrients such as calcium, nitrogen, phosphorus, and potash encourage the growth and development of periphytic biomass. Research data has shown that maximum periphytic biomass to the tune of 3.5 mg/cm^2 can be achieved through pond fertilization by urea, triple superphosphate and cattle manure at the rates of 150, 150, and 4,500 kg/ha, respectively; of course, the degree of productivity of periphyton depends upon the type of substrates used during pond preparation.

Minerals

In fish culture ponds, minerals especially calcium, phosphorus, potassium, cobalt, and zinc are required by fish for various life processes. In contrast to marine fishes, freshwater forms live in a dilute medium and must receive minerals from the diet. If ponds are periodically fertilized with nutrient carriers along with the mineral premix containing a variety of other minerals, periphytic organisms will utilize these essential minerals for their growth and development.

Light

The quality and quantity of light falling on the ecosystem and photoperiod are equally important to periphyton growth and may exert considerable influence on fish growth. The onset of development of periphytic organisms is often related to temperature, but may also be influenced greatly by photoperiod, particularly in climates that are characterized by little or no seasonal fluctuations in water temperatures. The photoperiod requirements for periphyton should be determined so that adjustments can be made at appropriate time.

Development of periphytic biomass is controlled by the intensity of light that penetrates into the water column. The standing crop of periphyton greatly reduced at greater depths of the ecosystem. Light is less important for organisms other than algal communities although most of the algal communities extract nutrients from the habitat.

Suspended Solids

Suspended solids are made up of sediment particles, organic materials, phytoplankton and other living micro-organisms. The higher the concentration of suspended solids in the pond water, the more is the turbid condition and consequently, the penetration of light decreased. In a highly turbid water, the productivity of periphytic biomass greatly reduced as a result of shading.

Particulate matters that are suspended in waters provide a vast amount of surface area for the gorwth of bacteria and fungi, and could increase a variety of diseases in aquaculture systems. Suspended particles can also absorb and adsorb different chemicals such as phosphates. Therefore, fertilization may be less effective in turbid water, not only because of shading, but also because nutrients may not be available to incorporate into the periphytic biomass.

12.7 Type of Substrates Used in the Culture of Periphyton

Huge quantities of periphytic biomass are developed in fish culture ecosystems where different types of materials are used as substrates such as polyvinyl chloride pipes, plastic sheets, and custom-designed materials (also termed as *quamats*). For pond culture, bamboo poles, tree branches, paddy straw or sugar cane bagasse bundles, jute sticks, ceramic tiles and plastic pipes are also widely used. Hanging of woven bamboo mats in the water column have given encouraging results. Bamboo mats provide sufficient space for development of periphytic biomass. substrate type has a distinct effect on the composition and density of periphyton. As a generalization, substrates made of bamboo have resulted in the development of better quality periphyton in high densities than on other materials. The reason for these differences in the densities of periphyton may be attributed to leaching of nutrients, emission of any bad odor or relative roughness of the surface. However, the energy values of periphyton assemblaged in bamboo mats, sugarcane bagasse, and polyvinyl chloride pipes have been estimated as 24.2, 20.7, and 18.3 KJ/g, respectively.

12.8 Nutrient Quality of Periphyton

Physical and chemical aspects of soil and water other than temperature may have direct impact on the nutrient quality of periphytic biomass in ponds. The nutrient components of periphyton are protein, fat, and carbohydrate. The protein, fat and carbohydrates contents have been estimated as 511, 27, and 273 g/kg of periphyton, respectively and the calorific value in these nutrient groups varies between 6.7 and 25.5 KJ/g. In general, the proximate composition of periphyton is quite similar to that of other natural food organisms but the presence of organic matter in high amounts due to accumulation of mucopolysaccharide matrix (80 per cent of total organic matter) makes the periphytic assemblage more palatable to fish.

Practical fish diets are composed of vitamins, minerals, and fibre in addition to the energy-bearing materials. Minerals and vitamins are essentially important for fish nutrition, but requirements for many of them are poorly known for several aquacultural species. More informations on the nutritional aspects are known about the salmonoids than any other groups of fish, though considerable informations have been obtained on channel catfish and a few other varieties.

In addition to nitrogen, phosphorus, calcium, magnesium, minerals, and vitamins, periphytic biomass also contain several amino acid residues. A protein molecule may contain thousands of amino acid residues, but in most of the cases only about 20 different amino acids are actively involved. However, floating microbial mats grown on grass silage contain amino acid profiles of proteins that promote excellent growth of fish (especially tilapia) in grow-out ponds with adequate management.

The amount of periphyton which are grown on different substrates and required for fish is extremely variable but the most important aspect is to provide nutrition to farmed fish. The optimum quality of periphyton for a fish species alters due to age, condition of fish and variation in the environment. In general, rapidly-growing fish needs considerable quantities of periphytic biomass than do older ones. Moreover, bottom-dwelling fish requires more periphytic biomass than surface-feeders.

12.9 Fish Response to Periphyton

Development of nutrient-rich aquatic ecosystems through fertilization and manuring is fundamental to high survival, growth, and yield of farmed fish and as a principle that drives ecological engineering design, it creates greater efficiency and productivity of culture ecosystems. The application of ecological engineering principles to enhance ecosystem quality criteria can not only reduce fish production costs but increase fish yield as well. The use of various types of substrates suitable for development of periphyton assemblage is most important aspect for fish growth. Substrates help generate natural food organisms in large volumes. Since periphytic loading on substrates is the critical factor contributing to ecosystem productivity, periphyton production is inevitable.

Periphyton play many essential roles in fish production. It is a producer of dozens of nutrients responsible for fish growth. Periphyton has more than one role in fish culture. It increases water quality, survival, growth, yield and health status of fish. Consequently, the efficiency of fish culture systems is enhanced. The use of substrates is in part responsible to reduce the need for artificial food and therefore periphyton can be regarded as an alternative to supplementary feeding. A number of biochemical processes such as photosynthesis, protein and lipid synthesis take place in periphytic biomass. Many species of fish respond well to periphyton. As with other natural food organisms, periphyton may be present in large quantities in the substrate and exert no harmful effect on fish.

A number of trials with periphyton were conducted to evaluate the suitability of different substrates and fish species. A comparative evaluation of the efficiency of polyvinyl chloride pipes, sugarcane bagasse and bamboo poles as substrates for increasing the yield of Mahseer (*Tor khudree*) in concrete tanks has shown that bamboo poles are very important for high production. Use of sugarcane bagasse at the rate of 2,800 kg/ha as a substrate exhibited higher production in ponds that are stocked with *Labeo rohita*, *Catla catla* and *Cyprinus carpio communis* provided with or without artificial food. Production of mahseer and *Labeo fimbriatus* (fringe-lipped carp) fed on only periphyton was found to be significantly higher (yield ranged from 30 to 75 per cent) than the yields achieved without food or periphyton.

Experimental Trial

Experiments conducted in ponds stocked with common carp, tilapia (*Oreochromis mossambicus* and, *O. nilotica*), and *Labeo rohita* using sugarcane bagasee, paddy straw or water hyacinth have shown that higher yield (40 to 50 per cent) was obtained when sugarcane bagasse were used as substrate. Moreover, production of bluegill (*Lepomis macrochirus*) has been found to be linearly increased with increasing in surface area of the substrate. Similarly, production of nile tilapia in ponds with microbial mat innoculating with blue-green algae was higher than that of commercial food items.

Significant responses of different parameters of farmed fish to periphytic assemblage using specific substances are becoming more wide-spread as periphytic-based fish culture has expanded over extensive areas. Responses of fish survival, growth and reproduction of Indian major carps to sugarcane bagasse and bamboo slits, for example, have been noted in regions with neutral to alkaline pond soil. Biodegradable substrates such as farm wastes which have been used in several tropical fish ponds for significant production has received considerable attention in recent years specially in areas where quality supplementary food items are not available to farmers. These types of substrates have produced good response to fish growth. This is due to the fact that biodegradable substrates harbor periphytic species in large numbers because these substrates provide a better surface structure for them and the periphytic species extract nutrients for their growth. These examples clearly illustrate the need for these substrates in fish culture if optimum yields from ponds are to be maintained.

12.10 Nutrient Availability to Periphyton

Major sources of nutrients and the general reactions that makes them available to periphyton and microorganisms are summarized in Figure 12 .1. Both organic and inorganic forms are the primary source of nutrients. The nutrients are used by periphyton in life-supporting processes. Nutrients and minerals are removed by fish harvest and as a corollary, the soluble nutrient pool will be reduced if it is not replenished with nutrient carriers nutrient deficiencies will result. These nutrient carriers, though contemplated primarily for supplying nutrients and minerals, are important source of nutrients and minerals.

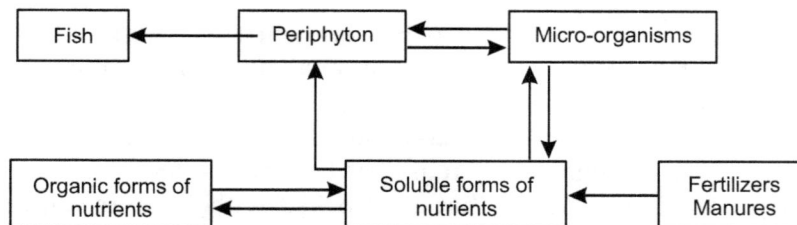

Fig. 12.1 : Nutrient sources of periphyton and micro-organisms and their utilization by fish

12.11 Benefit of Microbial Bioflim

In order to produce robust fish, higher immune responses to fish and their protection against diseases, development of the pond microbial bioflim is indispensable. Of several beneficial effects reported on fish, the most important effect can be mentioned here.

Development of a Stable Oral Vaccine

The most significant contribution of the microbial bioflim to fish health is their ability to develop stable oral vaccine. Microbial bioflim principally consists of bacteria. The bioflim cells are provided with glycocalyx matrix that has been well demonstrated to be resistant to the action of antibiotics and chemicals. This property has been fully utilized for the development of stable oral vaccine which is supposed to be very important for fish culture. By consumption, these periphytic components are broken down, exerting higher immune response to the posterior part of the alimentary canal (hind gut)and kidney and provide protection against diseases. Development of bioflim in ponds can boost immune response to fish against pathogenic micro-organisms. Stability of bioflim vaccine in fish ponds has been well demonstrated by antigen localization through specific antibody (monoclonal antibody) produced by hybridoma cells which are derived from the fusion of a myoloma cell with an antibody-producing B-lymphocyte.

12.12 Problems in Periphyton-Based Fish Culture

Deficiencies of natural food organisms/periphytic species have been noted in most of the fish ponds of the tropics and sub-tropics. Of course, in some areas, the extent of these deficiencies is not severe. Some experiments have suggested that there is scope for increasing pond productivity by using different substrates, but their availability determines whether such fish farming strategies could be adopted.

Periphyton-based fish culture in many countries of the world has not been utilized the desired technical supports, which are necessary for sustained growth and yield of fish biomass. This type of culture comprises a package of emperically-developed practices. Fish culturists must have to receive the benefit derived from system-related research on this aspect. Through proper pond management, it is desirable to produce periphyton suitable for propagation and those suitable for the fish. Under pond management, farmers have by which methods by which water and soil quality characteristics optimum for the culture of periphytic organisms can be evaluated.

The quality of pond water is of utmost importance in periphyton- based fish farming. Water exchange during fish culture operations, if possible, provide a good environment from that found in static ecosystems. While constructing a periphyton-based commercial fish farm, choice must be made on which sufficient clean water is available.

Mineralization of organic detritus at the pond bottom is comparatively a slow process due to low oxygen content. The process of decaying excessive organic matter at the pond bottom leads to organic pollution. This is more so in case of intensive and super-intensive culture systems when fish stocks are regularly fed with periphyton in addition to artificial feeds. This can be rectified by reducing accumulated metabolites in pond bottom, and by supplying oxygenated water or by aerating the pond bottom with the help of aerators. However technological inputs in periphytic-based fish culture can bring about essential changes in increasing the population of periphytic organisms and improving the growth and health of fish.

Periphyton-based fish farming has not received much attention in several Asian countries. In some developed countries, however, experiments on this type of culture system had been conducted with varying production results. In India, major carps, common carp and tilapia are some of the most common species and their culture have been tried. The relatively long time period (4-10 months) required to attain fish for the table, pollution of water and soil and reluctance to adopt such culture system by farmers have hampered the development of this type of fish farming.

In areas where there is scope for adopting periphyton-based fish culture, frequent harvesting of fish from ponds may create severe problems. During netting, substrates obstruct the net which makes the removal of fish from the pond troublesome and the cost of netting may negate the benefit achieved.

12.13 General Considerations of Microalgae

Algae of microscopic dimensions that ranges from 1-2 microns are termed as *microalgae*. These include dinoflagellates, diatoms, coccolithophorids, silicoflagellates, and blue-green algae (Chart 12.1). These are widely distributed in oceans, lagoons, freshwater ponds, lakes, and rivers. Several species of microalgae are known to grow well in mass culture and are extensively used as live food for various species of fin-fish, shell-fish and zooplankton (rotifers, copepods and cladocerans). Similar to higher plants, these algae bear a light-trapping green pigment termed as chlorophyll, which are essential for synthesizing organic compounds of high potential energy. Microalgae grow both in laboratory and in outdoor hatchery systems under controlled conditions. Also the growth, development and quality of microalgae entirely depend upon several factors such as light, temperature, nutrients, carbon di-oxide, and salinity. Microalgae have different biological compositions which are generally considered as indices of nutritional value and consequently, the type of microalgae to be used should be recommended for the purpose according to the specific nutritional requirements of fish. Thus a specific microalgae may be suitable for one kind of fish but not for another. Consequently, the development and quality of a particular microalga can affect the success or failure of aquaculture.

The growth and development of microalgae and periphyton are no doubt vital because they largely control the productivity of fish culture ecosystems by supplying nutrients, minerals, vitamins, and amino acids. These ingredients take active part in the nutrition and health status of fish as well as soil micro-organisms. This nutrition and health status involves breakdown of organic substances in the digestive system of fish.

CHART - 12.1
SOME IMPORTANT GROUPS OF MICROALGAE

Microalgae are microscopic photosynthetic organisms. These are primitive with a simple cellular structure and a large surface to volume body ratio, which gives them the capacity to take the nutrients in large amounts. In contract to higher plants, microalgae are more efficient converters of solar energy because of their simple cellular structure. In addition to this, because they grow in aqueous suspension, they have more efficient access to water, nutrients, and carbon di-oxide. And for this reason, microalgae are capable of supplying adequate amounts of nutrients to fish.

Biologists have cataegorized microalgae in a variety of classes, mainly distinguished by their life cycle, pigmentation, and basic cellular structure. In terms of abundance, however, four important groups of microalgae have been noted below.

1. **The Diatoms (Family : Bacillariophyceae) :** The diatoms constitute a big isolated group of mostly one-celled algae which are infinite varieties of forms and often of exquisite beauty. The single cells may form filaments and colonies. These algae dominate the phytoplankton of the oceans, but are also found in fresh and brackish waters. Diatoms are mostly free-floating, while some remain attached by a gelatinous stalk. About 1,00,000 species are known to exist around the world. All cells store carbon in a variety of forms. The main storage substance for diatoms is silica.

2. **The Blue-Green Algae (Family : Cyanophyceae) :** The blue-green alage are a small group of primitive algae characterized by the presence of a blue pigment called *phycocyanin* in addition to chlorophyll (together making a blue-green color). Some species are truly unicellular, while in others the daughter cells form a chain of cells called *filament* or a flat or spherical colony. Much similar to bacteria in structure and organization, these algae play an important role in fixing nitrogen from the atmosphere. About 2,000 known species are found in a variety of habitats; of course, a great majority of them are freshwater dwellers. The cell-wall is made of cellulose and carbohydrate occurs in the form of glycogen.

3. **The Green Algae (Family : Chlorophyceae) :** The green algae are characterized by the presence of green pigment called chlorophyll. They are mostly freshwater algae and starch is the principal storage compound in these algae. About 6,600 species are found in different freshwater habitats.

4. **The Golden Algae (Family : Chrysophyceae) :** This group of algae is similar to diatoms. They have more complex pigment systems and can appear brown, yellow, or orange in color. About 1,000 species are known to exist, primarily in freshwater ecosystems. These algae produce carbohyrates as storage compounds.

Improvement of pond ecosystems is a critical component of fish production systems. For growth and development of natural biotic communities in the ecosystem, the supply of nutrient carriers must be ascertained. In contrast to unmanaged pond ecosystems, a well-managed one provides suitable condition for the development of essential natural biota. For most aquatic microalgae and periphyton, the supply of nutrients in water and soil must be kept above the desired level. In turn, the concentration of waste products and that of other potentially toxic metabolites must not be allowed to build up in ponds.

12.14 Factors Necessary for the Growth of Microalgae

Many factors of the environment are responsible for the growth of microalgae. Among different factors, the most important ones are the availability of light, temperature, nutrients, carbon di-oxide, and salinity. These effects are briefly outlined that follows.

Light

It is already known that light is an absolute requirement for plants and animals. Most forms of aquatic life depend directly or indirectly upon the metabolic products of photosynthetic organisms. In fish culture ecosystems, these primary producers are algae and their growth as well as development is restricted to the upper layer of waters through which light can penetrate. Solar radiation is the primary source of energy to heat waters. But oxygen, carbon di-oxide, clouds and dust particles intercept the sun's rays and absorb, scatter and reflect most of their energy. Only about 40 and 75 per cent of the solar radiation actually reaches the earth respectively in cloudy humid and in cloud-free dry regions.

Little amount of the solar energy is reached in the earth's surface that actually results in increase in the temperature of water. The energy is used to evaporate water. At the same time, very limited amount of solar energy also becomes the main source of radiation available for photosynthesis. The photosynthetic activity is extremely limited to the region where enough solar energy can penetrate. The depth of the photic zone greatly varies depending on local conditions such as latitude, clouds, season, and the turbidity of water – all of which are directly related to the microbial growth. In general, the photosynthetic activity is limited to the upper region of the water body (upto 50 m). The process of photosynthesis involves use of the underwater light energy by microalgae and their chlorophyll (Chlorophyll a, b, and c) and production of oxygen as a waste product as well as and organic compounds (carbohydrates) from carbon di-oxide and water.

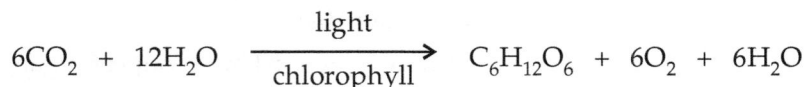

$$6CO_2 + 12H_2O \xrightarrow[\text{chlorophyll}]{\text{light}} C_6H_{12}O_6 + 6O_2 + 6H_2O$$

This oxygen is utilized for respiration and growth of microalgal cells. Through respiration, this reaction is reversed. Oxidation of carbon compounds generate carbon di-oxide and energy. While carbon di-oxide is consumed by microalgal cells during photosynthesis, energy is utilized by fishes for various metabolic functions. The general reaction is noted below.

$$C_6H_{12}O_6 + 6O_2 \xrightarrow{\text{Enzymes}} 6CO_2 + 6H_2O + \begin{array}{c} 686 \text{ K Cal} \\ \text{(Energy)} \end{array}$$

The presence of oxygen allows microalgal cells to use the more efficient aerobic respiration. In a sequence of cellular biochemical reactions, food (carbohydrates) is completely broken down into carbon di-oxide and water releasing far more energy (686 KCal) than that gained through anaerobic respiration.

Temperature

The temperature of surface waters varies from near 0°C in polar regions to 40°C in equatorial regions. In most of the well-managed tropical ponds, the range of temperature generally varies between 15 and 30°C — a condition favorable for the growth of several species of aquatic microalgae. Most types of microalgae have field temperature preferences of about 22°C circumscribed by upper and lower temperature preference of 30 and 17°C, respectively. This is due to the fact that along with the increase in temperature of water, there is a corresponding decrease in dissolved oxygen content. Consequently, deficiency of oxygen drastically affect respiration which, in turn, limits microbial growth. At low temperature, however, microalgae are not destroyed but the growth rates significantly reduced.

The optimum temperature for the growth of *Chlorella* and *Nannochlorpsis* has been found as 25 and 15°C, respectively. Mass culture of many species of microalgae at different water temperatures has a commendable growth rate at temperatures below 30°C (Table 12.1).

Table 12.1: Growth of Microalgae at Different Temperatures.
+ , Growth observed; –, No growth; S, Slow growth; x, Not reported

Species	Reported range of temperature (°C)	Growth response at different temperatures (°C)				
		25.0	30.0	30.5	33.0	34.5
Thalassiosira pseudonana	15-24	+	+	+	-	-
Chaetoceros muellari	10-35	+	+	+	+	+
Isochrysis galbana	12-28	+	+	-	-	x
Nannochloris atomus	15-20	+	S	-	x	x
Tetraselmis succica	14-34	+	+	-	x	x
T. chuii	12-32	+	S	x	x	x

Source : Nelson *et al* (1992)

Rivalry for Nutrients

The way in which aquatic organisms may be detrimental to the growth of microalgae, at least temporarily, is due to the competition for available nutrients. Aquatic organisms can rapidly absorb essential nutrients into their own bodies in order that the slow-growing microalgae can use only what is left. Nitrogen is the element for which competition is greatest, although similar competition occurs for phosphorus, sulfur, potassium, calcium, and the micro-nutrients. This aspect has been considered in Chapter 6 of Volume 1.

1. **Detrimental Effects of Temperature on Microalgal Growth :** Under extreme conditions of water temperature and poor ecosystem management, aquatic microalgae may deplete the limited oxygen supply. The deficiency of oxygen may affect the growth of microalgae in two ways. *Firstly*, the algae require a certain minimum amount of oxygen for normal growth and nutrient uptake and *Secondly*, oxidized forms of several elements such as nitrogen, sulfur, manganese, and iron will be reduced by further microbial action. In case of nitrogen and sulfur, some of the reduced forms are gaseous and may be lost through various ways. In acidic ecosystems, the reduction of iron and manganese may produce soluble forms of these elements in toxic quantities. Thus, nutrient deficiencies and toxicities can result from the same basic set of conditions.

12.15 Detection of Microalgae in Water

The size of microalgae is so small that the measurement of their concentration is a very laborious job, time consuming, and the count varies greatly. To solve these problems, the Central Institute of Fisheries Technology (CIFT) has developed a special device termed as *Microalgae Concentration Monitor* for instantaneous detection of the concentration of microalgae of various species up to 15 million/ml ± 0.1 million with digital display of data. Higher ranges can, however, be determined after dilution.

12.16 Competition Among Microalgae

In addition to competition between microalgae and higher animals, there remains in pond bottom an intense inter-microbial rivalry for food and growth. When fertilizers are added to ponds, the vigorous autotrophic organisms compete with each other for this source of food. Some algae dominate initially because they reproduce rapidly and prefer simple compounds. When these simple compounds are broken down, other types of algae become more competitive. Undoubtedly, such competition for food is the general rule and not the exception in culture ecosystems.

12.17 Food Production in Ecosystem

It has been well established that certain microalgae can produce food items of superior quality which are used as food for fishes. Because these items are beneficial to farmed animals, different nutritional products have revolutionized to combat malnutrition in humans and to produce robust fish from ponds. Several quality food products are now available in the market, which can be produced by microalgae. Thus, the ecosystem harbors not only harmful organisms but also microalgae that are the source of minerals and vitamins. This observation has been marked an epoch-making progress in fisheries science.

To meet the demand of food-fish, development of micro-algae in large amounts are necessary. And for this purpose, manuring and fertilization programms must be executed in freshwater ecosystems at regular intervals where there is scope for implementing periphyton-based fish culture. The major strategy should be to increase productivity of periphyton-based fish culture ecosystems. Productivity can be increased by three ways such as (i) appropriate monitoring and manipulation of nutritional requirements for micro-algae, (ii) bio-manipulation for superior genetic strains of micro-algae and (iii) increase in awareness among farmers about the importance of micro-algae in fish culture.

12.18 Effects of Agricultural Practice on Microalgae

Alteration in ecosystems affects both the number and kinds of aquatic microalgae. Certain management strategies at the time of fish cultivation drastically alters the aquatic environment in two ways. *Firstly*, the quality and quantity of microalgae is markedly reduced and *Secondly*, the number of species is also reduced. Aquatic ecosystems are contaminated by a number of pollutants that, to a less or greater degree, are toxic to aquatic microalgae and fish.

The overall levels of microalgae in different ecosystems are generally not seriously affected by pesticides. However, the algae responsible for reduction in pesticide levels in water and soil are sometimes adversely affected. Insecticides affect this process more than do most herbicides, although some of the latter can reduce the number of microalgae. Research reports on these aspects suggest that some microalgae are able to absorb pesticide molecules by reducing the activity of micro-organisms that are competitors of pesticide-reducing microalgae. These observations clearly explain the complexicity of the life of periphyton-based fish culture ecosystems.

The negative effects of pesticides on microalgae are ephemeral. and after a few days or a few weeks, the number of organisms generally recuperate. But the exceptions are common enough to command caution in the application of chemicals over extensive areas.

While agricultural practices have variable effects on different species of microalgae, a few generalizations can be made. Some agricultural practices generally reduce the total microalgal populations and the species diversity. Other practices such as adding fertilizers, manures, and lime will generally increase the activity of microalgae. Drastic changes in agricultural practices have ecosystem effects on aquatic microalgae by and large that can affect fish and fish food organisms. The effects of pesticides, both positive and negative, afford the perfect examples of this kind of information. Therefore, the effects of agricultural practices must be realized carefully investigated.

12.19 The Value of Microalgae

Though more than forty species of microalgae have been reported from different regions of the world, most of them grow well in sea water, a few species have been isolated from freshwater environments. However, all species of microalgae are not used successfully in fish culture. The most limiting factors for successful application of microalgae relate to their digestibility, size, nutritional value, and toxicity to fish species. Since different microalgae have different nutritional constituents and values, their suitability lies on the specific nutritional requirements of finfish and shellfish.

In terms of food value, microalgae are in no way inferior to other forms of natural food organisms. They are highly palatable, easily digestible, and highly nutritious. Different types of microalgae vary in their palatability and nutritive values. Microalgae are teemed with protein (30-70 per cent), carbohydrate (2-32 per cent), fat (1-22 per cent), vitamins (such as thiamine, riboflavin, pyridoxine, cyanocobalamin, nicotinic acid, pantothenic acid, choline, inositol, tocopherol, beta carotine, essential amino acids, and minerals (especially phosphorus, calcium, silicon, potassium, copper, cobalt, manganese, zinc, magnesium, iron, and sodium). And the presence of these ingradients in microalgae has great practical significance in relation to fish nutrition.

12.20 Response of Aquatic Animals by Microalgae

Different species of microalgae (Table 12.2) have wide applications as food for different marine and freshwater culture animals. Most of the species are developed in marine habitats. They are completely digested by aquatic animals and are consumed in dried, raw or pelleted forms. If microalgae are difficult to pellet or if dried forms are unstable in water, it is necessary to add any suitable binding agent to make the product more effective. Binders are required for most purified diets; in pelleted diets, the starch serves as the binding agent. Gelatin, carboxymethyl cellulose and a variety of other binding materials have been successfully utilized in diets used in farmed animals. The importance of microalgae used as food for freshwater fish species is, however, still unclear and further investigations are necessary to evaluate their role in fish nutrition.

Table 12.2: Major Classes and Genera of Microalgae Culture as Food for Aquaculture Animals

Class	Genera	Application
Bacillariophyceae	*Skeletonema* sp. and *Thalassiosira* sp.	PL, BP
	Phaeodactylum sp.	PL, BP, ML, BS
	Chaetoceros sp.	PL, BP, BS
	Bellerochea sp. and *Actinocyclus* sp.	BP
	Nitzschia sp. and *Cyclotella* sp.	BS
Haptophyceae	*Isochrysis* sp.	PL, BP, ML, BS
	Dicrateria sp., *Cricosphaera* sp. and *Coccolithus* sp	BP
Cryptophyceae	*Chroomonas* sp., *Cryptomonas* sp. and *Rhodomonas* sp.	BP, BL
Chlorophyceae	*Carteria* sp., *Chlorococcum* sp. *Chlamydomonas* sp., *Chlorella* sp. *Scenedesrnus* sp., *Brachiomonas* sp.	BP, BL, BS, FZ, ML
Cyanophyceae	*Spirulina* sp.	PL, BP, BS

PL, Penaeid shrimp larvae; BL, Bivalve mollusc larvae; ML, Freswater prawn larvae; BP, Bivalve mollusc post-larvae; BS, Brine shrimp (*Artemia*); FZ, Freshwater zooplankton.

Source: Selected data from Ojha and Howard (2000)

12.21 Culture Prospects of Microalgae

The exercise adopted for the culture of microalgae in future is likely to rely on the degree of economic development of the country in question. In under-developed nations, labor-intensive methods may continue to be executed not only to algae currently cultured but to culture of new species as well. The culture of microalgae for wide use in fish nutrition during the entire culture period definitely requires refined technology, intensification of culture techniques, and use of new species to culture.

Though culture of microalgae for aquaculture nutrition has greatly expanded in several developed countries, the demand for them as fish nutrition will definitely increase in Asian aquaculture sectors in the future. As noted earlier, marine microalgae are now being widely used in aquaculture of Europe, Canada, USA, and Japan.

12.22 Production of Microalgae

Large-scale production of microalgal biomass generally depends on continuous culture during daylight. Operation method involves feeding of fresh culture medium at a constant rate and the same quantity of microalgal broth is withrawn continuously. Feeding is discontinued during the night but mixing the broth must be continued to prevent the biomass from settling. About 25 per cent of the total biomass produced during daylight may be lost during the night as a result of respiration. The extent of this loss depends on the intensity of light under which the biomass is grown, temperature fluctuations, and the temperature at night.

Production Methods

Generally two methods are followed for large-scale production of micro-algae.

1. *Raceway Pond Method* : This method is a closed-loop recirculation channel which is about 0.3 m deep. A paddle-wheel aerator is used which helps mix and circulate the algal biomass. Flow is guided around bends by baffles kept in the flow channel. Raceway channels are constructed in concrete or compact earth and lined with plastic. During daylight, the culture is fed continuously in front of the paddle-wheel where the flow is starts. On completion of the circulation loop, broth is harvested behind the paddle-wheel. To prevent sedimentation, the paddle-wheel is operated all the time

2. *Photobioreactor Method* : Photobioreactors permit single-species culture of microalgae for prolonged periods. Photobioreactors have been successfully used for producing microalgal biomass in large quantities. A tubular photobioreactor essentially consists of an assemblage of straight transparent tubes which are made of glass or plastic. The solar collector tubes are 0.1 m or less in diameter. Tube diameter is limited because light does not penetrate deeply in the culture broth which is essential for high biomass productivity of the photobioreactor. Microalgal broth is circulated from a reservoir to the solar collector and vice-versa.

To sum up, raceways are less expensive than photobioreactors because the cost of construction and operation of the system is less. Although raceways are low-cost, they have a low biomass productivity compared with photobioreactors.

Since microalgae are capable of producing about 30 times the amount of oil per unit area of land compared to terrestrial oilseed crops, microalgal biomass contains significant quantities of proteins, carbohydrates and other nutrients. Hence, after the extraction of oil from microalgae, their residual biomass can be used as fish/animal food in full potential. Production of low-cost microalgal biomass, however, requires improvement to algal biology through genetic engineering.

12.23 Problems of Microalgae in Fish Culture

Increase of microbial species in large numbers in fish ponds is a problem of varying intensity in almost all systems of fish culture all over the world. But the condition is very severe enough in tropical and sub-tropical fish farms, particularly in undrained ponds. Microalgae in ponds and enclosures often lead to oxygen depletion as a result of dead and decaying algal mats. Mass mortality of fish can occur under these conditions. In cage culture, thick growth of algae on the net cages reduce water exchange capacity and thus affect the water quality within and around the cages. .

Mortality of fish stocks in farmed ponds due to the growth and development of toxic algae have not widely been noted, though certain losses of fish may be ascribed to this problem. Muddy flavour of some fish (such as catfish, common carp, *Cirrhinus mrigala*, and tilapia) can results from chemicals produced by certain species of blue-green algae. Production of fish with unpleasant smell due to liberation of toxic chemicals is very common in many ponds. At present, there are no management strategies that fish culturists may ascertain that fish crops will be free from unpleasant smell. Extensive culture systems seem to be more likely to develop the algae responsible for bad smell than do intensive systems.

12.24 Conclusion

Periphyton and microalgae are becoming increasingly important to world aquaculture as nutrition and health status of fish greatly increased. It is confirmed that various nutrient elements are essential for fish production over extensive freshwater areas under culture on commercial basis and suggest that attention towards development of these important food organisms in ponds will likely to increase in future.

Development of biological cover in ponds by using specific substrates and through fertilization is responsible for increase in immune response and protection of fish against pathogens. Several species of algae and bacteria also play a special role in providing nutrients,minerals, and vaccines for farmed fish. Competition among micro-organisms and fish and between these organisms is inevitable for some nutrient deficiencies. Adequate and regular management of farmed ponds will replenish such deficiencies. Requirements of periphyton and microalgae for fish growth and development of ecosystems are factors involved in determining the success of fish culture.

References

Ben-Amotz, A., Tornabene, T. G. and Thomas, W. H. 1985. Chemical profile of selected species of microalgae with emphasis on lipids.*J. Phycol*, 21, 72-81.

Borowitzka, A. M. 1988. Vitamins and chemicals from microalgae. *In:* Microalgal Biotechnology. (*Eds* A. M. Borowitzka and L. J. Borowitzka), Cambridge University Press, pp.153-196.

Brown, M. R. 1991. The anino acid and sugar composition of 16 species of microalgae. *J, Exp. Mar. Biol. Ecol.*, 145 : 79-99.

James, C. M., Al-Hinty, and Salman, A. E. 1989. Grcwth and omega-3 fatty acid and amino acid composition of microalgae under different temperature regimes. *Aquaculture*, 77 : 337-351.

Lin, D. S., Ilias, A. M. *et.al.* 1982. Composition and biosynthesis of sterols in selected phytoplankton.*Lipids* 17:818-824.

Nelson, J. R., Guarda, S. *et.al* 1992. Evaluation of microalgal clones for mass culture in a sub-tropical green house bivalve hatchery :growth rates and biochemical composition at 30°C *Aquaculture*, 106: 357-377.

Ojha, J. S. and Howard, G. A. 2000. Microalgae for aquaculture nutrition. *Fishing Chimes*, 20 (9) : 15-23.

Webb, K. L. and Chu, F. E. 1983. Phytoplankton as a food source for bivalve larvae. *In:Proc. Second Inter. Conf. on Aquaculture Nutri* (*Eds. Pruder G.D et.al*), *World Mariculture Soc. Spec. Publ.* No.2, Lausiana State University, Lausiana, PP. 272-292.

Whyte, J. M. C. 1987. Biochemical composition and energy content of 6 species of phytoplankton used in mariculture of bivalves. *Aquaculture*, 60: 231-242.

Questions

1. What is periphyton? Why periphyton-based fish culture is so important?

2. Which species constitute periphyton? What types of fish culture ecosystem form periphytic organisms? What are the factors responsible for the development of periphytic organisms?

3. Why the study of biological cycle of periphytic biomass is necessary? Discuss.

4. Discuss different groups of nutritive fauna and flora in periphyton.

5. Discuss how different environmental parameters play an important role in periphyton development.

6. What are the substrates that are used for periphyton culture?

7. What types of nutrients are present in periphyton ? How nutrients greatly influence the growth of periphyton ?

8. With a suitable example discuss how periphytic organisms are responsible for fish culture in ponds.

9. What is microbial biofilm? state how microbial bioflim is beneficial to fish health.

10. Trace the problems that are related to periphyton-based fish culture.

11. What is micro-algae? Discuss the role of micro-algae in fish culture.

12. Discuss the factors responsible for the growth of micro-algae.

13. Discuss how agricultural practice affects the growth and activity of micro-algae.

14. Mention why micro-algae are superior than natural food organisms.
15. Trace the importance of micro-algae in fish production.
16. Discuss briefly why micro-algae and periphyton have enormous fish production potential.
17. State the problems of micro-algae generally involved in fish culture.